Treatise on Invertebrate Paleontology

Prepared under Sponsorship of
The Geological Society of America, Inc.

The Paleontological Society
The Palaeontographical Society

SEPM (The Sedimentological Society)
The Palaeontological Association

Raymond C. Moore
Founder

Roger L. Kaesler
Editor

Elizabeth Brosius, Jack Keim, Jane Priesner
Assistant Editors

Part R

Arthropoda 4

Volume 3: Superclass Hexapoda

By F. M. Carpenter

The Geological Society of America, Inc.
and
The University of Kansas
Boulder, Colorado, and Lawrence, Kansas
1992

© 1992 by The University of Kansas
and
The Geological Society of America, Inc.

All Rights Reserved

Library of Congress Catalogue Card Number 53-12913
ISBN 0-8137-3019-8

Distributed by the Geological Society of America, Inc., P.O. Box 9140, Boulder, Colorado 80301, from which current price lists of parts in print may be obtained and to which all orders and related correspondence should be directed. Editorial office of the *Treatise*: Paleontological Institute, 121 Lindley Hall, The University of Kansas, Lawrence, Kansas 66045.

The *Treatise on Invertebrate Paleontology* has been made possible by (1) funding principally from the National Science Foundation of the United States in its early stages, from the Geological Society of America through the bequest of Richard Alexander Fullerton Penrose, Jr., and from The Kansas University Endowment Association through the bequest of Raymond C. and Lillian B. Moore; (2) contribution of the knowledge and labor of specialists throughout the world, working in cooperation under sponsorship of the Geological Society of America, the Paleontological Society, the SEPM (The Sedimentological Society), the Palaeontographical Society, and the Palaeontological Association; and (3) acceptance by The University of Kansas of publication without any financial gain to The University.

TREATISE ON INVERTEBRATE PALEONTOLOGY

Parts of the *Treatise* are distinguished by assigned letters with a view to indicating their systematic sequence while allowing publication of units in whatever order each is made ready for the press. Copies are available on orders sent to the Publication Sales Department, The Geological Society of America, 3300 Penrose Place, P.O. Box 9140, Boulder, Colorado 80301.

VOLUMES ALREADY PUBLISHED

Part A. INTRODUCTION, xxiii + 569 p., 371 fig., 1979.

Part C. PROTISTA 2 (Sarcodina, chiefly "Thecamoebians" and Foraminiferida), xxxi + 900 p., 5,311 fig., 1964.

Part D. PROTISTA 3 (chiefly Radiolaria, Tintinnina), xii + 195 p., 1,050 fig., 1954.

Part E. ARCHAEOCYATHA, PORIFERA, xviii + 122 p., 728 fig., 1955.

Part E, Revised. ARCHAEOCYATHA, Volume 1, xxx + 158 p., 871 fig., 1972.

Part F. COELENTERATA, xvii + 498 p., 2,700 fig., 1956.

Part F. COELENTERATA, Supplement 1 (Rugosa and Tabulata), xl + 762 p., 3,317 fig., 1981.

Part G. BRYOZOA, xii + 253 p., 2,000 fig., 1953.

Part G, Revised. BRYOZOA, Volume 1 (Introduction, Order Cystoporata, Order Cryptostomata), xxvi + 626 p., 1,595 fig., 1983.

Part H. BRACHIOPODA, xxxii + 927 p., 5,198 fig., 1965.

Part I. MOLLUSCA 1 (Mollusca General Features, Scaphopoda, Amphineura, Monoplacophora, Gastropoda General Features, Archaeogastropoda, mainly Paleozoic Caenogastropoda and Opisthobranchia), xxiii + 351 p., 1,732 fig. 1960.

Part K. MOLLUSCA 3 (Cephalopoda General Features, Endoceratoidea, Actinoceratoidea, Nautiloidea, Bactritoidea), xxviii + 519 p., 2,382 fig., 1964.

Part L. MOLLUSCA 4 (Ammonoidea), xxii + 490 p., 3,800 fig., 1957.

Part N. MOLLUSCA 6 (Bivalvia), Volumes 1 and 2 (of 3), xxxvii + 952 p., 6,198 fig., 1969; Volume 3, iv + 272 p., 742 fig., 1971.

Part O. ARTHROPODA 1 (Arthropoda General Features, Protarthropoda, Euarthropoda General Features, Trilobitomorpha), xix + 560 p., 2,880 fig., 1959.

Part P. ARTHROPODA 2 (Chelicerata, Pycnogonida, Palaeoisopus), xvii + 181 p., 565 fig., 1955.

Part Q. ARTHROPODA 3 (Crustacea, Ostracoda), xxiii + 442 p., 3,476 fig., 1961.

Part R. ARTHROPODA 4, Volumes 1 and 2 (Crustacea exclusive of Ostracoda, Myriapoda, Hexapoda), xxxvi + 651 p., 1,762 fig., 1969.

Part S. ECHINODERMATA 1 (Echinodermata General Features, Homalozoa, Crinozoa, exclusive of Crinoidea), xxx + 650 p., 2,868 fig., 1967 [1968].

Part T. ECHINODERMATA 2 (Crinoidea), Volumes 1–3, xxxviii + 1027 p., 4,833 fig., 1978.

Part U. ECHINODERMATA 3 (Asterozoans, Echinozoans), xxx + 695 p., 3,485 fig., 1966.

Part V. GRAPTOLITHINA, xvii + 101 p., 358 fig., 1955.

Part V, Revised. GRAPTOLITHINA, xxxii + 163 p., 507 fig., 1970.

Part W. MISCELLANEA (Conodonts, Conoidal Shells of Uncertain Affinities, Worms, Trace Fossils, Problematica), xxv + 259 p., 1,058 fig., 1962.

Part W, Revised. MISCELLANEA, Supplement 1 (Trace fossils and Problematica), xxi + 269 p., 912 fig., 1975.

Part W, Revised. MISCELLANEA, Supplement 2 (Conodonta), xxviii + 202 p., frontis., 858 fig., 1981.

THIS VOLUME

Part R. ARTHROPODA 4, Volumes 3 and 4 (Hexapoda), xxii + 655 p., 1,489 fig., 1992.

VOLUMES IN PREPARATION

Part B. PROTISTA 1 (Chrysomonadida, Coccolithophorida, Charophyta, Diatomacea, etc.).
Part E, Revised. PORIFERA. Volume 2.
Part G, Revised. BRYOZOA (additional volumes).
Part H, Revised. BRACHIOPODA.
Part I. Introduction to MOLLUSCA (part).
Part J. MOLLUSCA 2 (Caenogastropoda, Streptoneura exclusive of Archaeogastropoda, Euthyneura).
Part L, Revised. MOLLUSCA 4 (Ammonoidea).
Part M. MOLLUSCA 5 (Coleoidea).
Part O, Revised. ARTHROPODA 1 (Trilobita).
Part Q, Revised. ARTHROPODA 3 (Ostracoda).

EDITORIAL PREFACE

FROM THE outset the aim of the *Treatise on Invertebrate Paleontology* has been to present a comprehensive and authoritative yet compact statement of knowledge concerning groups of invertebrate fossils. Typically, preparation of early *Treatise* volumes was undertaken by a single specialist with a synoptic view of the group being monographed. More rarely, two or perhaps three specialists worked together. Recently, however, both new *Treatise* volumes and revisions of existing ones have been undertaken increasingly by teams of specialists led by a coordinating author. Part R, Hexapoda, prepared by Professor Frank M. Carpenter, is certainly the last of the volumes that will be written by a single author rather than by a team of specialists. Few paleontologists have ever had such an all-encompassing command of a major group of fossils as Professor Carpenter's of the fossil insects. We are indeed privileged that he has found both the time and the energy over the years to compile this information and share it with the paleontological and entomological communities.

These volumes on the Hexapoda, the final section of Part R, are not a revision of previous work but are one of four remaining parts of the *Treatise* project that have not yet been covered for the first time. The others remaining to be done are Part B, Protista; Part J, Caenogastropoda; and Part M, Coleoidea, all of which are presently in preparation.

The fourth part of the arthropod *Treatise* has had a long history. Volumes 1 and 2, forming one unit, were published in 1969 and comprise an introduction to Hexapoda and an introduction and systematics sections on Onychophora, Crustacea other than Ostracoda, and Myriapoda. Volumes 3 and 4, originally planned for a single volume, cover the Hexapoda including, of course, the fossil insects, taxonomy of which fills most of the two volumes. The introduction to the insects is brief. The insects and their hexapod relatives are morphologically, physiologically, and ecologically quite complex organisms that abound in the modern world. Numerous excellent introductions are available. To reintroduce them here would require extensive duplication, and an adequate introductory section would be beyond the scope of the *Treatise on Invertebrate Paleontology*.

ZOOLOGICAL NAMES

Questions about the proper use of zoological names arise continually, especially questions regarding both the acceptability of names and alterations of names that are allowed or even required. Regulations prepared by the International Commission on Zoological Nomenclature (ICZN) and published in 1985 in the *International Code of Zoological Nomenclature*, hereinafter referred to as the *Code*, provide procedures for answering such questions. The prime objective of the *Code* is to promote stability and universality in the use of the scientific names of animals, ensuring also that each generic name is distinct and unique, while avoiding unwarranted restrictions on freedom of thought and action of systematists. Priority of names is a basic principle of the *Code*, but under specified conditions and by following prescribed procedures, priority may be set aside by the Commission. These procedures apply especially where slavish adherence to the principle of priority would hamper or even disrupt zoological nomenclature and the information it conveys.

The Commission, ever aware of the changing needs of systematists, is undertaking a revision of the *Code* that will enhance nomenclatorial stability. Nevertheless, the nomenclatorial tasks that confront zoological taxonomists are formidable and have often justified the complaint that the study of zoology and paleontology is too often merely the study of names rather than the study of animals. It is incumbent upon all systematists, therefore, to pay careful attention to the *Code* to enhance stability by minimizing the number of subsequent changes of names, too many of which are necessitated by insufficient attention to detail. To that end, several pages here are devoted to aspects of zoological nomenclature that are judged to have chief importance in relation to procedures adopted in the *Treatise*, especially in these two volumes. Terminology is explained, and examples are given of the style employed in the nomenclatorial parts of the systematic descriptions.

GROUPS OF TAXONOMIC CATEGORIES

Each taxon belongs to a category in the Linnean, hierarchical classification. The *Code* recognizes three groups of categories, a species-group, a genus-group, and a family-group. Taxa of lower rank than subspecies are excluded from the rules of zoological nomenclature, and those of higher rank than superfamily are not regulated by the *Code*. It is both natural and convenient to discuss nomenclatorial matters in general terms first and then to consider each of these three, recognized groups separately. Especially important is the provision that within each group the categories are coordinate, that is, equal in rank, whereas categories of different groups are not coordinate.

FORMS OF NAMES

All zoological names can be considered on the basis of their spelling. The first form of a name to be published is defined as the original spelling (*Code*, Article 32), and any form of the same name that is published later and is different from the original spelling is designated a subsequent spelling (Article 33). Not all original spellings are correct, just as is true of subsequent spellings.

Original Spellings

If the first form of a name to be published is consistent and unambiguous, the original is defined as correct unless it contravenes some stipulation of the *Code* (Articles 11, 27 to 31, and 34) or unless the original publication contains clear evidence of an inadvertent error in the sense of the *Code*, or, among names belonging to the family-group, unless correction of the termination or the stem of the type genus is required. An original spelling that fails to meet these requirements is defined as incorrect.

If a name is spelled in more than one way in the original publication, the form adopted by the first reviser is accepted as the correct original spelling, provided that it complies with mandatory stipulations of the *Code* (Articles 11 and 24 to 34).

Incorrect original spellings are any that fail to satisfy requirements of the *Code*, represent an inadvertent error, or are one of multiple original spellings not adopted by a first reviser. These have no separate status in zoological nomenclature and, therefore, cannot enter into homonymy or be used as replacement names; and they call for correction. For example, a name originally published with a diacritical mark, apostrophe, dieresis, or hyphen requires correction by deleting such features and uniting parts of the name originally separated by them, except that deletion of an umlaut from a vowel in a name derived from a German word or personal name requires the insertion of *e* after the vowel. Where original spelling is judged to be incorrect solely because of inadequacies of the Greek or Latin scholarship of the author, nomenclatorial changes conflict with the primary purpose of zoological nomenclature as an information retrieval system. One looks forward with hope to a revised *Code* wherein rules are emplaced that enhance stability rather than classical scholarship, thereby facilitating access to information.

Subsequent Spellings

If a subsequent spelling differs from an original spelling in any way, even by the omission, addition, or alteration of a single letter, the subsequent spelling must be defined as a different name. Exceptions include such changes as altered terminations of adjectival specific names to agree in gender with associated generic names; changes of family-group names to denote assigned taxonomic rank; and corrections that eliminate originally used diacritical marks, hyphens, and the like. Such changes are not regarded as spelling changes conceived to produce a different name. In some instances, however, species-group names having variable spellings are regarded as homonyms as specified in the *Code,* Article 58.

Altered subsequent spellings other than the exceptions noted may be either intentional or unintentional. If "demonstrably intentional" (*Code*, Article 33, p. 73), the change is designated as an emendation. Emendations may be either justifiable or unjustifiable. Justifiable emendations are corrections of incorrect original spellings, and these take the authorship and date of the original spellings. Unjustifiable emendations are names having their own status in nomenclature, with author and date of their publication. They are junior, objective synonyms of the name in its original form.

Subsequent spellings, if unintentional, are defined as incorrect subsequent spellings. They have no status in nomenclature, do not enter into homonymy, and cannot be used as replacement names.

AVAILABLE AND UNAVAILABLE NAMES

Editorial prefaces of previous volumes of the *Treatise* have discussed in appreciable detail the availability of the many kinds of zoological names that have been proposed under a variety of circumstances. Much of that information, while important, does not pertain to the present volumes in which the author has used only *nomen nudum* (plural *nomina nuda*, naked names). The reader is referred to *Part G Bryozoa* (*Revised*) of the *Treatise* and to the *Code* (Articles 10 to 20) for further details on availability of names. Here, besides the discussion of *nomina nuda* below, suffice it to say that an available zoological name is any that conforms to all mandatory provisions of the *Code*. All zoological names that fail to comply with mandatory provisions of the *Code* are unavailable and have no status in zoological nomenclature. Both available and unavailable names are classifiable into groups that have been recognized in previous volumes of the *Treatise*, although not explicitly differentiated in the *Code*. Among names that are available, these groups include inviolate names, perfect names, imperfect names, vain names, transferred names, improved or corrected names, substitute names, and conserved names. Kinds of unavailable names include naked names (see *nomina nuda* below), denied names, impermissible names, null names, and forgotten names.

Nomina nuda include all names that fail to satisfy provisions stipulated in Article 11 of the *Code*, which states general requirements of availability. In addition, they include names published before 1931 that were unaccompanied by a description, definition, or indication (Articles 12 and 16) and names published after 1930 that (1) lacked an accompanying statement of characters that differentiate the taxon, (2) were without a definite bibliographic reference to such a statement, (3) were not proposed expressly as a replacement (*nomen substitutum*) of a preexisting available name (Article 13a), or (4) for genus-group names, were unaccompanied by definite fixation of a type species by original designation or indication (Article 13b). *Nomina nuda* have no status in nomenclature and are not correctable to establish original authorship and date.

VALID AND INVALID NAMES

Important considerations distinguish valid from available names on the one hand and invalid from unavailable names on the other. Whereas determination of availability is based entirely on objective considerations guided by articles of the *Code*, conclusions as to validity of zoological names may be partly subjective. A valid name is the correct one for a given taxon, which may have two or more available names but only a single correct, hence valid, name, which is generally the oldest. Obviously, no valid name can also be an unavailable name, but invalid names may be either available or unavailable. It follows that any name for a given taxon other than the valid name, whether available or unavailable, is an invalid name.

One encounters a sort of nomenclatorial no-man's land in considering the status of such zoological names as *nomina dubia* (doubtful names), which may include both available and unavailable names. The unavailable ones can well be ignored, but names considered to be available contribute to uncertainty and instability in the systematic literature. These can ordinarily be removed only by appeal to the ICZN for special action. Because few systematists care to seek such remedy, invalid but available names persist in the literature.

NAME CHANGES IN RELATION TO GROUPS OF TAXONOMIC CATEGORIES

Species-Group Names

Detailed consideration of valid emendation of specific and subspecific names is unnecessary here, both because the topic is well understood and relatively inconsequential and because the *Treatise* deals with genus-group names and higher categories. When the form of an adjectival specific name is changed to agree with the gender of a generic name in transferring a species from one genus to another, one need never label the changed name as *nomen correctum*. Similarly, transliteration of a letter accompanied by a diacritical mark in the manner now called for by the *Code*, as in changing originally *bröggeri* to *broeggeri*, or eliminating a hyphen, as in changing originally published *cornu-oryx* to *cornuoryx*, does not require the designation *nomen correctum*.

Genus-Group Names

Conditions warranting change of the originally published, valid form of generic and subgeneric names are sufficiently rare that lengthy discussion is unnecessary. Only elimination of diacritical marks and hyphens in some names in this category and replacement of homonyms seem to furnish basis for valid emendation. Many names that formerly were regarded as homonyms are no longer so regarded, because two names that differ only by a single letter or in original publication by the presence of a diacritical mark in one are now construed to be entirely distinct.

As has been pointed out above, difficulty typically arises when one tries to decide whether a change of spelling of a name by a subsequent author was intentional or unintentional, and the decision has often to be made arbitrarily.

Family-Group Names: Authorship and Date

All family-group taxa having names based on the same type genus are attributed to the author who first published the name of any of these assemblages, whether tribe, subfamily, or family (superfamily being almost inevitably a later-conceived taxon). Accordingly, if a family is divided into subfamilies or a subfamily into tribes, the name of no such subfamily or tribe can antedate the family name. Also, every family containing differentiated subfamilies must have a nominotypical subfamily (*sensu stricto*), which is based on the same type genus as the family; and the author and date set down for the nominotypical subfamily invariably are identical with those of the family, irrespective of whether the author of the family or some subsequent author introduced subdivisions.

Corrections in the form of family-group names do not affect authorship and date of the taxon concerned, but in the *Treatise* recording the authorship and date of the correction is desirable because it provides a pathway to follow the thinking of the systematists involved.

Family-Group Names: Use of *nomen translatum*

The *Code* specifies the endings only for subfamily (-inae) and family (-idae) names, but all family-group taxa are defined as coordinate (Article 36, p. 77): "A name established for a taxon at any rank in the family group is deemed to be simultaneously established with the same author and date for taxa based upon the same name-bearing type (type genus) at other ranks in the family group, with appropriate mandatory change of suffix [Art. 34a]." Such changes of rank and concommitant changes of endings as elevation of a tribe to subfamily rank or of a subfamily to family rank, if introduced subsequent to designation of a subfamily or family based on the same nominotypical genus, are *nomina translata*. In the *Treatise* it is desirable to distinguish the valid alteration in the changed ending of each transferred family-group name by the term *nomen translatum*, abbreviated to *nom. transl.* Similarly for clarity, authors should record the author, date, and page of the alteration. This is especially important for superfamilies, for the information of interest is the author who initially introduced a taxon rather than the author of the superfamily as defined by the *Code*. The latter is merely the individual who first defined some lower-ranked, family-group taxon that contains the nominotypical genus of the superfamily. On the other hand, the publication that introduces the superfamily by *nomen translatum* is likely to furnish the information on taxonomic considerations that support definition of the taxon.

An example of the use of *nomen translatum* is the following.

Family HEXAGENITIDAE Lameere, 1917
[*nom. transl.* DEMOULIN, 1954, p. 566, *ex* Hexagenitinae LAMEERE, 1917, p. 74]

Family-Group Names: Use of *nomen correctum*

Valid name changes classed as *nomina correcta* do not depend on transfer from one category of family-group units to another but most commonly involve correction of the stem of the nominotypical genus. In addition, they include somewhat arbitrarily chosen modifications of endings for names of tribes or superfamilies. Examples of the use of *nomen correctum* are the following.

Family STREPTELASMATIDAE Nicholson, 1889
[*nom. correct.* WEDEKIND, 1927, p. 7, *pro* Streptelasmidae NICHOLSON in NICHOLSON & LYDEKKER, 1889, p. 297]

Family PALAEOSCORPIDAE Lehmann, 1944
[*nom. correct.* PETRUNKEVITCH, 1955, p. P73, *pro* Palaeoscorpionidae LEHMANN, 1944, p. 177]

Family-Group Names: Replacements

Family-group names are formed by adding combinations of letters, which are prescribed for family and subfamily, to the stem of the name belonging to the nominotypical genus first chosen as type of the assemblage. The type genus need not be the first genus in the family to have been named and defined, but among all those included it must be the first published as name giver to a family-group taxon. Once fixed, the family-group name remains tied to the nominotypical genus even if the generic name is changed by reason of status as a junior homonym or junior synonym, either objective or subjective. Seemingly, the *Code* requires replacement of a family-group name only if the nominotypical genus is found to have been a junior homonym when it was proposed (Article 39, p. 79), in which case " . . . it must be replaced either by the next oldest available name from among its synonyms, including those of its subordinate taxa, or, if there is no such name, by a new replacement name based on the valid name of the former type genus." Authorship and date attributed to the replacement family-group name are determined by first publication of the changed family-group name; but, for subsequent application of the rule of priority, the name takes the date of the replaced name (see Recommendation 40A). Many family-group names that have been in use for a long time are *nomina nuda,* since they fail to satisfy criteria of availability (Article 11f). These demand replacement by valid names.

The aim of family-group nomenclature is to yield the greatest possible stability and uniformity, just as in other zoological names. Both taxonomic experience and the *Code* (Article 40) indicate the wisdom of sustaining family-group names based on junior subjective synonyms if they have priority of publication, for opinions of the same worker may change from time to time. The retention of first-published, family-group names that are found to be based on junior objective synonyms, however, is less clearly desirable, especially if a replacement name derived from the senior objective synonym has been recognized very long and widely. To displace a widely used, family-group name based on the senior objective synonym by disinterring a forgotten and virtually unused family-group name based on a junior objective synonym because the latter happens to have priority of publication is unsettling.

A family-group name may need to be replaced if the nominotypical genus is transferred to another family-group. If so, the first-published of the generic names remaining in the family-group taxon is to be recognized in forming a replacement name.

Suprafamilial Taxa: Taxa above Family-Group

International rules of zoological nomenclature as given in the *Code* affect only lower-rank categories: subspecies to superfamily. Suprafamilial categories (suborder to phylum) are either unmentioned or explicitly placed outside of the application of zoological rules. The *Copenhagen Decisions on Zoological Nomenclature* (1953, Articles 59 to 69) proposed adopting rules for naming suborders and higher taxa up to and including phylum, with provision for designating a type genus for each, in such manner as not to interfere with the taxonomic freedom of workers. Procedures were outlined for applying the rule of priority and rule of homonymy to suprafamilial taxa and for dealing with the names of such taxa and their authorship, with assigned dates, if they should be transferred on taxonomic grounds from one rank to another. The adoption of terminations of names, different for each category but uniform within each, was recommended.

The Colloquium on Zoological Nomenclature, which met in London during the week just before the 15th International Congress of Zoology convened in 1958, thoroughly discussed the proposals for regulating suprafamilial nomenclature, as well as many others advocated for inclusion in the new *Code* or recommended for exclusion from it. A decision that was supported by a wide majority of the participants in the Colloquium was against the establishment of rules for naming taxa above family-group rank, mainly because it was judged that such regulation would unwisely tie the hands of taxonomists. For example, a class or order defined by an author at a given date, using chosen morphologic characters (e.g., gills of bivalves), should not be allowed to freeze nomenclature, taking precedence over another class or order that is proposed later and distinguished by different characters (e.g., hinge teeth of bivalves). Even the fixing of type genera for suprafamilial taxa would have little, if any, value, hindering taxonomic work rather than aiding it. No basis for establishing such types and for naming these taxa has yet been provided.

The considerations just stated do not prevent the editors of the *Treatise* from making rules for dealing with suprafamiliar groups of animals described and illustrated in this publication. Some uniformity is needed, especially for the guidance of *Treatise* authors. This policy should accord with recognized general practice among zoologists; but where general practice is indeterminate or nonexistent, our own procedure in suprafamilial nomenclature needs to be specified as clearly as possible. This pertains especially to decisions about names themselves, about citation of authors and dates, and about treatment of suprafamilial taxa that, on taxonomic grounds, are changed from their originally assigned rank. Accordingly, a few rules expressing *Treatise* policy are given here, some with examples of their application.

1. The name of any suprafamilial taxon must be a Latin or latinized, uninominal noun of plural form, or treated as such, with a capital initial letter and without diacritical mark, apostrophe, diaresis, or hyphen. If a component consists of a numeral, numerical adjective, or adverb, this must be written in full.

2. Names of suprafamilial taxa may be constructed in almost any manner. A name may indicate morphological attributes (e.g., Lamellibranchiata, Cyclostomata, Toxoglossa) or be based on the stem of an included genus (e.g., Bellerophontina, Nautilida, Fungiina) or on arbitrary combinations of letters (e.g., Yuania); none of these, however, can end in -idae or -inae, which terminations are reserved for family-group taxa. No suprafamilial name identical in form to that of a genus or to another published suprafamilial name should be employed (e.g., order Decapoda LATREILLE, 1803, crustaceans, and order Decapoda LEACH, 1818, cephalopods; suborder Chonetoidea MUIR-WOOD, 1955, and genus *Chonetoidea* JONES, 1928). Worthy of notice is the classificatory and nomenclatural distinction between suprafamilial and family-group taxa that, respectively, are named from the same type genus, since one is not considered to be transferable to the other (e.g., suborder Bellerophontina ULRICH & SCOFIELD, 1897; superfamily Bellerophontacea McCOY, 1851; family Bellerophontidae McCOY, 1851). Family-group names are not coordinate with suprafamilial names.

3. The rules of priority and homonymy lack any force of international agreement as applied to suprafamilial names, yet in the interest of nomenclatural stability and to avoid confusion these rules are widely applied by zoologists to taxa above the family-group level wherever they do not infringe on taxonomic freedom and long-established usage.

4. Authors who accept priority as a determinant in nomenclature of a suprafamilial taxon may change its assigned rank at will, with or without modifying the terminal letters of the name, but such changes cannot rationally be judged to alter the authorship and date of the taxon as published originally. A name revised from its previously published rank is a transferred name (*nomen translatum*), as illustrated in the following.

Order CORYNEXOCHIDA Kobayashi, 1935
[*nom. transl.* MOORE, 1959, p. O217, *ex* suborder Corynexochida KOBAYASHI, 1935, p. 81]

A name revised from its previously published form merely by adoption of a different termination without changing taxonomic rank is an altered name (*nomen correctum*).

Order DISPARIDA Moore & Laudon, 1943
[*nom. correct.* MOORE in MOORE, LALICKER, & FISCHER, 1952, p. 613, *pro* order Disparata MOORE & LAUDEN, 1943, p. 24]

A suprafamilial name revised from its previously published rank with accompanying change of termination, which signal the change of rank, is recorded as a *nomen translatum et correctum*.

Order HYBOCRINIDA Jackel, 1918
[*nom. transl. et correct.* MOORE in MOORE, LALICKER, & FISCHER, 1952, p. 613, *ex* suborder Hybocrinites JAEKEL, 1918, p. 90]

5. The authorship and date of nominotypical subordinate and supraordinate taxa among suprafamilial taxa are considered in the *Treatise* to be identical since each actually or potentially has the same type. Examples are given below.

Subclass ENDOCERATOIDEA Teichert, 1933
[*nom. transl.* TEICHERT in TEICHERT *et al.*, 1964, p. K128, *ex* order Endoceroidea TEICHERT, 1933, p. 214]

Order ENDOCERIDA Teichert, 1933
[*nom. correct.* TEICHERT in TEICHERT *et al.*, 1964, p. K165, *pro* order Endoceroidea TEICHERT, 1933, p. 214]

Suborder ENDOCERINA Teichert, 1933
[*nom. correct.*, TEICHERT in TEICHERT *et al.*, 1964, p. K165, *ex* Endoceratina SWEET, 1958, p. 33, suborder]

TAXONOMIC EMENDATION

Emendation has two distinct meanings as regards zoological nomenclature. These are (1) alteration of a name itself in various ways for various reasons, as has been reviewed, and (2) alteration of the taxonomic scope or concept for which a name is used. The *Code*

(Article 33a and Glossary, p. 148) concerns itself only with the first type of emendation, applying the term to either justified or unjustified changes, both intentional, of the original spelling of a name. The second type of emendation primarily concerns classification and inherently is not associated with change of name. Little attention generally has been paid to this distinction in spite of its significance.

Most zoologists, including paleontologists, who have emended zoological names, refer to what they consider a material change in application of the name such as may be expressed by an importantly altered diagnosis of the assemblage covered by the name. The abbreviation *emend.* then must accompany the name with statement of the author and date of the emendation. On the other hand, many systematists think that publication of *emend.* with a zoological name is valueless because alteration of a taxonomic concept is introduced whenever a subspecies, species, genus, or other assemblage of animals is incorporated into or removed from the coverage of a higher zoological taxon. Inevitably associated with such classificatory expansions and restrictions is some degree of emendation affecting diagnosis. Granting this, still it is true that now and then somewhat radical revisions are put forward, generally with published statement of reasons for changing the application of a name. To erect a signpost at such points of most significant change is worthwhile, both as aid to subsequent workers in taking account of the altered nomenclatural usage and to indicate where in the literature cogent discussion may be found. Authors of contributions to the *Treatise* are encouraged to include records of all especially noteworthy emendations of this nature, using the abbreviation *emend.* with the name to which it refers and citing the author, date, and page of the emendation.

Examples from *Treatise* volumes follow.

Order ORTHIDA Schuchert & Cooper, 1932
[*nom. transl. et correct.* MOORE in MOORE, LALICKER, & FISCHER, 1952, p. 220, *ex* suborder Orthoidea SCHUCHERT & COOPER, 1932, p. 43; *emend.*, WILLIAMS & WRIGHT, 1965, p. H299]

Subfamily ROVEACRININAE Peck, 1943
[Roveacrininae PECK, 1943, p. 465; *emend.*, PECK in MOORE & TEICHERT, eds. 1978, p. T921]

STYLE IN GENERIC DESCRIPTIONS

Citation of Type Species

The name of the type species of each genus and subgenus is given immediately following the generic name with its accompanying author, date, and page reference or after entries needed for definition of the name if it is involved in homonymy. The orginally published combination of generic and trivial names of this species is cited, accompanied by an asterisk (*), with notation of the author and date of original publication. An exception in this procedure is made, however, if the species was first published in the same paper and by the same author as that containing definition of the genus of which it is the type; in this instance, the initial letter of the generic name followed by the trivial name is given without repeating the name of the author and date. Examples of these two sorts of citations follow.

Orionastraea SMITH, 1917, p. 294 [*Sarcinula phillipsi* McCOY, 1849, p. 125; OD]
Schoenophyllum SIMPSON, 1900, p. 214 [*S. aggregatum*; OD]

If the cited type species is a junior synonym of some other species, the name of this latter also is given, as follows.

Actinocyathus D'ORBIGNY, 1849, p. 12 [*Cyathophyllum crenulate* PHILLIPS, 1836, p. 202; M; =*Lonsdalaeia floriformis* (MARTIN), 1809, pl. 43; validated by ICZN Opinion 419]

In the *Treatise* the name of the type species is always given in the exact form it had in the original publication except that diacritical marks have been removed. Where other mandatory changes are required, these are introduced later in the text, typically in a figure caption.

Fixation of Type Species Originally

It is desirable to record the manner of establishing the type species, whether by original designation (OD) or by subsequent designation (SD). The type species of a genus or subgenus, according to provisions of the *Code*, may be fixed in various ways in the original publication; or it may be fixed in ways specified by the *Code* (Article 68) and described in the next section. Type species fixed in the original publication include (1) *original designation* (in the *Treatise* indicated by "OD") when the type species is explicitly stated or (before 1931) indicated by "n. gen., n. sp." (or its equivalent) applied to a single species included in a new genus, (2) defined by use of *typus* or *typicus* for one of the species included in a new genus (adequately indicated in the *Treatise* by the specific name), (3) established by *monotypy* if a new genus or subgenus has only one originally included species (in the *Treatise* indicated as "M"), and (4) fixed by *tautonymy* if the genus-group name is identical to an included species name not indicated as the type.

Fixation of Type Species Subsequently

The type species of many genera are not determinable from the publication in which the generic name was introduced and therefore such genera can acquire a type species only by some manner of subsequent designation. Most commonly this is established by publishing a statement naming as type species one of the species originally included in the genus. In the *Treatise*, fixation of the type species in this manner is indicated by the letters "SD" accompanied by the name of the subsequent author (who may be the same person as the original author) and the date of publishing the subsequent designation. Some genera, as first described and named, included no mentioned species (for such genera established after 1930, see below); these necessarily lack a type species until a date subsequent to that of the original publication when one or more species are assigned to such a genus. If only a single species is thus assigned, it automatically becomes the type species. Of course, the first publication containing assignment of species to the genus that originally lacked any included species is the one concerned in fixation of the type species, and if this publication names two or more species as belonging to the genus but did not designate a type species, then a later "SD" designation is necessary. Examples of the use of "SD" as employed in the *Treatise* follow.

Hexagonaria GÜRICH, 1896, p. 171 [*Cyathophyllum hexagonum GOLDFUSS, 1826, p. 61; SD LANG, SMITH, & THOMAS, 1940, p. 69]

Mesephemera HANDLIRSCH, 1906, p. 600 [*Tineites lithophilus GERMAR, 1842, p. 88; SD CARPENTER, herein]

Another mode of fixing the type species of a genus is action of the International Commission of Zoological Nomenclature using its plenary powers. Definition in this way may set aside application of the *Code* so as to arrive at a decision considered to be in the best interest of continuity and stability of zoological nomenclature. When made, it is binding and commonly is cited in the *Treatise* by the letters "ICZN," accompanied by the date of announced decision and reference to the appropriate numbered opinion.

Subsequent designation of a type species is admissable only for genera established prior to 1931. A new genus-group name established after 1930 and not accompanied by fixation of a type species through original designation or original indication, is invalid (*Code*, Article 13b). Effort of a subsequent author to validate such a name by subsequent designation of a type species constitutes an original publication making the name available under authorship and date of the subsequent author.

Homonyms

Most generic names are distinct from all others and are indicated without ambiguity by citing their originally published spelling accompanied by name of the author and date of first publication. If the same generic name has been applied to two or more distinct taxonomic units, however, it is necessary to differentiate such homonyms. This calls for distinction between junior homonyms and senior homonyms. Because a junior homonym is invalid, it must be replaced by some other name. For example, *Callophora* HALL, 1852, introduced for Paleozoic trepostomate bryozoans, is invalid because GRAY in 1848 published the same name for Cretaceous-to-Holocene cheilostomate bryozoans. BASSLER in 1911 introduced the new name *Hallophora* to replace Hall's homonym. The *Treatise* style of entry is given below.

Hallophora BASSLER, 1911, p. 325, *nom. subst. pro Callophora* HALL, 1852, p. 144, *non* GRAY, 1848

In like manner, a needed replacement generic name may be introduced in the *Treatise* (even though first publication of generic names otherwise in this work is generally avoided). An exact bibliographic reference must be given for the replaced name as in the following example.

Mysterium DE LAUBENFELS, herein, *nom. subst. pro Mystrium* SCHRAMMEN, 1936, p. 183, *non* ROGER, 1862 [*Mystrium porosum* SCHRAMMEN, 1936, p. 183; OD]

Otherwise, no mention of the existence of a junior homonym generally is made.

Synonymous Homonyms

An author sometimes publishes a generic name in two or more papers of different date, each of which indicates that the name is new. This is a bothersome source of errors for later workers who are unaware that a supposed first publication that they have in hand is not actually the original one. Although the names were separately published, they are identical and therefore definable as homonyms; at the same time they are absolute synonyms. For the guidance of all concerned, it seems desirable to record such names as synonymous homonyms. In the *Treatise* the junior of one of these is indicated by the abbreviation "jr. syn. hom."

Not infrequently, identical family-group names are published as new names by different authors, the author of the later-introduced name being ignorant of previous publication(s) by one or more other workers. In spite of differences in taxonomic concepts as indicated by diagnoses and grouping of genera and possibly in assigned rank, these family-group taxa are nomenclatural homonyms, based on the same type genus; and they are also synonyms. Wherever encountered, such synonymous homonyms are distinguished in the *Treatise* as in dealing with generic names.

A rare but special case of homonymy exists when identical family names are formed from generic names having the same stem but differing in their endings. An example is the family name Scutellidae R. & E. RICHTER, 1925, based on *Scutellum* PUSCH, 1833, a trilobite. This name is a junior homonym of Scutellidae GRAY, 1825, based on the echinoid genus *Scutella* LAMARCK, 1816. The name of the trilobite family was later changed to Scutelluidae (ICZN, Opinion 1004, 1974).

Synonyms

In the *Treatise,* citation of synonyms is given immediately after the record of the type species. If two or more synonyms of differing date are recognized, these are arranged in chronological order. Objective synonyms are indicated by accompanying designation "obj.," others being understood to constitute subjective synonyms, of which the types are also indicated. Examples showing *Treatise* style in listing synonyms follow.

Mackenziephyllum PEDDER, 1971, p. 48 [*M. insolitum*; OD] [=*Zonastraea* TSYGANKO in SPASSKIY, KRAVTSOV, & TSYGANKO, 1971, p. 85, *nom. nud.*; *Zonastraea* TSYGANKO, 1972, p. 21 (type, *Z. graciosa*, OD)]

Kodonophyllum WEDEKIND, 1927, p. 34 [*Streptelasma Milne-Edwardsi* DYBOWSKI, 1873, p. 409; OD; =*Madrepora truncata* LINNÉ, 1758, p. 795, see SMITH & TREMBERTH, 1929, p. 368] [=*Patrophontes* LANG & SMITH, 1927, p. 456 (type, *Madrepora truncata* LINNÉ, 1758, p. 795, OD); *Codonophyllum* LANG, SMITH, & THOMAS, 1940, p. 39, obj.]

Some junior synonyms of either the objective or the subjective sort may take precedence desirably over senior synonyms whenever uniformity and continuity of nomenclature are served by retaining a widely used but technically rejectable name for a genus. This requires action of ICZN, which may use its plenary powers to set aside the unwanted name and validate the wanted one, with placement of the concerned names on appropriate official lists.

MATTERS OF STYLE SPECIFIC TO THESE VOLUMES

The Fossil Record of Hexapods

In spite of their being the most diverse group of organisms, the insects have a surprisingly poor fossil record. Their dominantly terrestrial mode of life and lack of mineralized skeletons have contributed to extensive taphonomic loss. Thus, whereas such *Treatise* volumes as Part Q, Ostracoda have sought to include all genera in the group whether or not they have a fossil record because of their potential for fossilization, to attempt to do so with the insects would be both beyond the scope of the *Treatise on Invertebrate Paleontology* and doomed to failure. Most of the recent genera of insects are not included herein. In fact, a recent genus with no fossil record is included only if it is the type genus of a family that contains fossil forms. Moreover, for recent genera that have a fossil record, we do not indicate type species or give diagnoses. Instead, we give only the last name of their author, the date of publication, and the page number. Although full citations of these author-date combinations are not in the bibliography, subsequent references to the literature are included.

Names of Taxa, Places, and Authors

Several matters relate specifically to the style of generic descriptions. Names of type species have been corrected only by having diacritical marks removed. For example, *Corydaloïdes* has been changed to *Corydaloides*. Throughout the text the author has used the solidus to indicate uncertainty with respect to age. "Oligo./Mio.," for example, indicates that the age of the genus is uncertain but is one of the two ages noted. The question mark is used when the age is still more uncertain.

Purists, *Treatise* editors among them, would like nothing better than a stable world with a stable geography that makes possible a stable biogeographical classification. Global events of the past two years have shown how rapidly geography can change, and in all likelihood we have not seen the last of such change. Throughout the text, the author has used the letters RSFSR to refer to the Russian Socialist Federated Soviet Republic with two parts, European and Asian, separated by the Ural Mountains. The RSFSR, of course, no longer exists as a political or geographical entity, but the strata containing fossil insects remain where they were. One expects confusion among readers in the future as they try to decipher such geographical terms as U.S.S.R. or Yugoslavia. Such confusion is unavoidable, as books must be completed and published at some time. Our libraries would be small indeed if publication were always delayed until the world had settled down.

Chinese scientists have become increasingly active in systematic paleontology in the past two decades. Chinese names cause English-language bibliographers headaches for two reasons. First, no scheme exists for one-to-one transliteration of Chinese characters into Roman letters. Thus, a Chinese author may change the Roman-letter spelling of his name from one publication to another. For example, the name Chang, which is the most common family name in the world, might also be spelled Zhang. The principal purpose of a bibliography is to provide the reader with entry into the literature. Quite arbitrarily, therefore, in the interest of information retrieval, the *Treatise* editorial staff has decided to retain the Roman spelling that the Chinese author used in each of his publications rather than attempting to adopt a common spelling of an author's name to be used in all citations to his work. It is entirely possible, therefore, that the publications of a Chinese author may be listed in more than one place in the bibliography.

Second, most but by no means all Chinese list their family name first followed by given names, but people with Chinese names who study in the West often reverse the order, putting the family name last. Thus, for example, Dr. Yi-Maw Chang, now on the staff of the Paleontological Institute, was Chang Yi-Maw when he lived in Taiwan. When he came to America, he became Yi-Maw Chang, and his subsequent bibliographic citations are listed as "Chang, Yi-Maw." The *Treatise* staff has adopted the convention of listing family names first, inserting a comma, and following this with given names or initials. We do this even for Chinese authors who have not reversed their names in the Western fashion.

Several specific systems exist for transliterating the Cyrillic alphabet into the Roman alphabet, so that this problem need not occur, for example, with names of Russian authors. We have adopted System II from J. Thomas Shaw's *Transliteration of Modern Russian for English-Language Publications*, which is the same as the Library of Congress system for transliteration of modern Russian with diacritical marks omitted.

Stratigraphical Range Charts

Readers may notice that stratigraphical range charts in this volume are somewhat different from those in previous volumes. Charts in this volume were prepared using RangeChart, an unpublished computer-software program developed by Kenneth C. Hood and David W. Foster, both now with Exxon, when they were graduate students at The University of Kansas. RangeChart sorts the taxa by their ranges and the degree of certainty of those ranges and uses different weights of lines for different categories. A revised version of the program, RangeChart 2.0, is in preparation.

Acknowledgments

The *Treatise* volumes on the Hexapoda have had a long history of development, having been the focus of Professor Carpenter's efforts with varying degrees of intensity since the early stages of the *Treatise* project. The staff of the Paleontological Institute has remained remarkably stable during the ensuing decades, but given the length of time quite a number of people have been involved with the volumes. They deserve special mention here, for without their efforts the *Treatise* project as a whole and these volumes specifically would not be what they are today. Not the least of these are the three previous Editors and Directors of the Paleontological Institute: the late Raymond C. Moore as well as Curt Teichert and Richard A. Robison. The two previous Assistant Editors for Text, Lavon McCormick and Virginia Ashlock, and the previous Assistant Editor for Illustrations, Roger B. Williams, worked closely with Professor Carpenter on the volumes. The present Assistant Editor for Text, Elizabeth Brosius, and the Assistant Editor for Illustrations, Jane Priesner, have faced admirably the formidable task of moving the volumes through the final stages of editing and into and beyond the production phase. In this they have been ably assisted by Jill Hardesty with word processing; Jill Krebs with editorial backup; and Jack Keim with photography, layout, and preparation of range charts. Yi-Maw Chang, the remaining member of the Paleontological Institute staff, is involved with preparation of PaleoBank, the paleontological data base for future *Treatise* volumes, and has not been closely involved with the hexapod *Treatise*. Margery Rowell edited the Russian titles in the bibliography, and Richard A. Leschen and George W. Byers, respectively, drew figures 173 and 204.

This Editorial Preface is an extensive revision of prefaces prepared for previous *Treatise* volumes by former editors, including the late Raymond C. Moore, Curt Teichert, and Richard A. Robison. I am indebted to them for preparing earlier prefaces and for the leadership they have provided in bringing the *Treatise* project to its present status.

Roger L. Kaesler
Lawrence, Kansas
May 1, 1992

AUTHOR'S PREFACE

Nearly thirty years ago Professor Raymond Moore, then editor of the *Treatise on Invertebrate Paleontology,* invited me to prepare the volume on the Hexapoda. Following considerable correspondence with him, I decided to undertake that assignment, although no definite date was set for its completion. My start on the project was slow, mainly because I was shortly asked to serve in several administrative positions at Harvard University, in addition to my regular teaching schedule. Not until 1974, when I became professor *emeritus,* was I able to devote full time to the preparation of the volume. At that time previously submitted manuscript was revised, and the first draft of the manuscript was sent to the editorial office of the *Treatise* in 1982. It was decided to set the end of 1983 as the terminal date for literature citations, since there had been an unusual amount of literature on fossil insects published during the preceding twenty years (1963 to 1983), and since a large part of that was in Russian and needed to be translated. In this connection I should mention that a bibliography of fossil insects, covering the years 1980 to 1990, is now in preparation by E. A. Jarzembowski and A. J. Ross (Booth Museum, Brighton, U. K.) and will be published in 1992 in *The Fossil Record* (eds., M. J. Benton & M. A. Whyte, Chapman & Hall, London).

I am deeply grateful to the editorial staff of the *Treatise,* especially to Elizabeth Brosius and Jane Priesner, for their indispensable assistance, particularly regarding the bibliography. I am equally indebted to Helen Vaitaitis, who has done all of the translating of the numerous Russian articles for me these many years. Dr. Laurie Burnham assisted me for several years with the preparation of the illustrations for the *Treatise,* and Dr. Curtis Sabrosky has provided helpful advice pertaining to special taxonomic problems. My wife, Ruth Carpenter, has been very supportive in many ways and especially with the preparation of an index to the genera in the early stages of the manuscript. I acknowledge with gratitude the cooperation of the following museums in the United States and Europe that have placed type material at my disposal at the institutions or have loaned such specimens when needed: the United States National Museum (Washington), the Field Museum (Chicago), the British Museum, Natural History (London), Museum d'Histoire Naturelle (Paris), and the Paleontological Institute (Moscow). Finally, I am indebted to our National Science Foundation for research grants that made these investigations possible.

Frank M. Carpenter
Cambridge, Massachusetts
January, 1992

SOURCES OF ILLUSTRATIONS

Some illustrations in this volume are new. Where previously published illustrations are used, the author and date of publication are given in parentheses in the figure explanation. Full citation of the publication is provided in the references.

In addition to the citation of the publication, additional credit was requested by those who supplied the following illustrations. Figure 10 is reproduced with permission, from the Annual Review of Entomology, Vol. 26, ©1981 by Annual Reviews Inc. Figure 193,2 is reproduced with permission, from the Museum of Comparative Zoology, Harvard University, ©President and Fellows of Harvard College. The artists responsible for the illustrations reproduced from *The Insects of Australia* are F. Nanninga (Figure 79,2), N. Key (Figure 82,2), B. Rankin (Figure 132), S. Curtis (Figure 134), and T. Binder (Figure 217).

STRATIGRAPHIC DIVISIONS

The major divisions of the geological time scale are reasonably well established throughout the world, but minor divisions (e.g., substages, stages, and subseries) are more likely to be provincial in application. The stratigraphical units listed here show the fairly coarse time resolution that is characteristic of the study of fossil hexapods.

CENOZOIC ERATHEM
  Quaternary System
    Holocene Series
    Pleistocene Series
  Tertiary System
    Pliocene Series
    Miocene Series
    Oligocene Series
    Eocene Series
    Paleocene Series
MESOZOIC ERATHEM
  Cretaceous System
  Jurassic System
  Triassic System
PALEOZOIC ERATHEM
  Permian System
  Carboniferous System
    Upper Carboniferous Subsystem
    Lower Carboniferous Subsystem
  Devonian System
  Silurian System
  Ordovician System
  Cambrian System
PRECAMBRIAN (undifferentiated herein)

PART R
ARTHROPODA 4
HEXAPODA
VOLUMES 3, 4

By Frank M. Carpenter

CONTENTS

Systematic Descriptions of the Superclass Hexapoda	1
Superclass Hexapoda	1
Class and Order Collembola	1
Class and Order Protura	3
Class and Order Diplura	4
Class Insecta	4
Introduction to the Insects	4
Subclass Apterygota	15
Order Archaeognatha	15
Order Zygentoma	17
Subclass Pterygota	18
Infraclass Palaeoptera	18
Order Ephemeroptera	19
Order Palaeodictyoptera	26
Order Megasecoptera	45
Order Diaphanopterodea	54
Order Protodonata	59
Order Odonata	62
Order Uncertain (Palaeoptera)	89
Infraclass Neoptera	92
Division Exopterygota	92
Order Perlaria	93
Order Protorthoptera	97
Order Blattaria	134
Order Isoptera	137
Order Manteodea	142
Order Protelytroptera	143
Order Dermaptera	150
Order Orthoptera	154
Order Grylloblattodea	181
Order Titanoptera	181
Order Phasmatodea	184
Order Embioptera	189
Order Psocoptera	191
Order Zoraptera	200
Order Mallophaga	200
Order Anoplura	200
Order Caloneurodea	200

Order Miomoptera	204
Order Thysanoptera	207
Order Hemiptera	212
Division Endopterygota	279
Order Coleoptera	279
Order Strepsiptera	337
Order Neuroptera	338
Order Glosselytrodea	356
Order Trichoptera	359
Order Lepidoptera	369
Order Mecoptera	380
Order Siphonaptera	395
Order Diptera	395
Order Hymenoptera	445
Order Uncertain (Neoptera)	499
Subclass Pterygota, Order Uncertain	502
Class Insecta, Subclass Uncertain	503
Superclass Hexapoda, Class Uncertain	503
OUTLINE OF CLASSIFICATION	504
RANGES OF TAXA	504
GLOSSARY OF ENTOMOLOGICAL TERMS	526
REFERENCES	529
INDEX	616

SYSTEMATIC DESCRIPTIONS OF THE SUPERCLASS HEXAPODA

Superclass HEXAPODA Latreille, 1825

[Hexapoda LATREILLE, 1825, p. 328]

Six-legged, tracheate arthropods, with thorax more or less demarcated from abdomen; head typically with 1 pair of antennae, and with mandibles, maxillae, and a labium; thorax usually strongly sclerotized, the coxa-body mechanisms diverse; abdomen with from 6 to 11 segments. Species mainly terrestrial, but some secondarily aquatic. Reproduction and life histories very diverse. *Dev.–Holo.*

The status of this group as a taxon is uncertain. MANTON (1969a, 1977, 1979), following her extensive investigations on functional morphology of the arthropods, was convinced that the classes she included in the Hexapoda were more akin to one another than to any other arthropod classes. At the same time, however, she was also convinced that there could not have been any one type of ancestral hexapod capable of giving rise to the existing hexapod classes. Although in recent years an unprecedented amount of literature has been published on arthropod evolution (see SCUDDER, 1973; BOUDREAUX, 1979; GUPTA, 1979; HENNIG, 1981), the relationships of the classes of Hexapoda seem as obscure as ever. In all probability this situation will not improve until we have a truly extensive record of the terrestrial arthropods in Lower Carboniferous (Mississippian) and Devonian strata. The four or five existing classes of six-legged arthropods have had a long history, apparently extending that far back; but as the present record stands only one species of hexapod is known earlier than the Late Carboniferous—*Rhyniella praecursor,* a collembolon from the Devonian of Scotland (see Fig. 2). Numerous fragments of other arthropods have been found in freshwater deposits of the Devonian, but for the most part they cannot be associated with any of the existing hexapod classes. It seems likely that the diversity of the wingless, noninsect hexapods during the Devonian was far greater than that represented by the few classes now in existence.

In the present treatment of the Hexapoda, I follow the classification proposed by MANTON (1969a) in the Introduction to the Arthropoda in this series of volumes, except that the Thysanura (*sensu lato*) are here included within the Insecta, as orders Archaeognatha and Zygentoma, instead of being separated into a distinct class.

Class and Order COLLEMBOLA Lubbock, 1871

[Collembola LUBBOCK, 1871, p. 295]

Mostly very small hexapods, body usually covered with hairs or, more rarely, with scales; head prognathous, with mandibulate, entognathous mouthparts; mandibles slender; maxillae and labium much reduced; antennae typically with 4 segments, the first 3 with intrinsic muscles; eyes consisting of a few ommatidia on each side of head, or entirely absent. Thorax diversely formed, pronotum usually much reduced; in some species, thorax fused with abdomen, the segmentation being obsolescent; legs lacking a distinct tarsal segment. Abdomen with only 6 segments, the first bearing a ventral, tubular, adhesive organ (collophore); fourth segment bearing a jumping organ (furcula), which at rest folds back under abdomen. Sperm transfer indirect, as in members of the Diplura. Adults and young occurring mostly in decaying vegetation; a few on foliage. *Dev.–Holo.*

This is a relatively small order of about 2,000 widely distributed species. Presumably because of their small size, Collembola are rarely preserved as fossils except in amber. All of the Baltic amber (Oligocene) species appear to belong to recent genera (HANDSCHIN, 1926a), but the only known

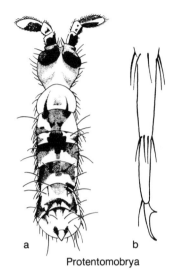

a   b
Protentomobrya

FIG. 1. Protentomobryidae (p. 2).

species in Cretaceous amber represents an extinct genus and family (Protentomobryidae), somewhat intermediate between certain existing families (DELAMARE DEBOUTTEVILLE & MASSOUD, 1968). The genus *Rhyniella* from the Rhynie chert of Scotland, apparently without question a collembolon (WHALLEY & JARZEMBOWSKI, 1981), is the only hexapod now known from the Devonian.

OLFERS in his 1907 article named several other genera in the Collembola, but, as HANDSCHIN (1926a) has shown, each of these was based on a mixed series of specimens belonging to several genera; no type species or specimens were designated, and HANDSCHIN felt compelled to reject the names. Also, MARI MUTT (1983) has reported the presence of several existing genera (*Cryptopygus, Isotomus, Lepidocyrtus, Pseudosinella, Seira, Salina, Paronella, Cyphoderus,* and *Sphyrotheca*) in the Miocene amber of the Dominican Republic but has not identified or named any species.

Family PROTENTOMOBRYIDAE
Folsom, 1937

[Protentomobryidae FOLSOM, 1937, p. 15]

Antennae short and stout, with 4 segments; body elongate, setose; pronotum weakly formed, concealed by mesonotum; first abdominal segment reduced; furcula consisting of a pair of long, simple stylets. *Cret.*

Protentomobrya FOLSOM, 1937, p. 15 [*P. walkeri*; OD]. Third abdominal segment almost as long as fourth; fifth abdominal segment not reduced. DELAMARE DEBOUTTEVILLE & MASSOUD, 1967, 1968. *Cret.*, Canada (Manitoba).——FIG. 1. *P. walkeri; a,* dorsal view, ×100; *b,* distal portion of left arm of furcula, ventral view, ×650 (both Delamare Deboutteville & Massoud, 1968).

Family ISOTOMIDAE
Schaeffer, 1896

[Isotomidae SCHAEFFER, 1896, p. 177]

Pronotum reduced and without setae; rest of body with scales or hairs; furcula usually present. *Oligo.–Holo.*

Isotoma BOURLET, 1839, p. 401. HANDSCHIN, 1926a; CHRISTIANSEN, 1971. *Oligo.,* Europe (Baltic); *Oligo./Mio.,* Mexico (Chiapas)–*Holo.*

Isotomurus BÖRNER, 1903, p. 171. CHRISTIANSEN, 1971. *Oligo./Mio.,* Mexico (Chiapas)–*Holo.*

Family HYPOGASTRURIDAE
Börner, 1913

[Hypogastruridae BÖRNER, 1913, p. 315]

Pronotum well developed, bearing setae; rest of body with scales; head prognathous. *Oligo.–Holo.*

Hypogastrura BOURLET, 1839, p. 404. HANDSCHIN, 1926a. *Oligo.,* Europe (Baltic)–*Holo.*

Family TOMOCERIDAE
Schaeffer, 1896

[Tomoceridae SCHAEFFER, 1896, p. 177]

Pronotum reduced; body with scales; antennae long. *Oligo.–Holo.*

Tomocerus NICOLET, 1841, p. 67. HANDSCHIN, 1926a. *Oligo.,* Europe (Baltic)–*Holo.*

Family ENTOMOBRYIDAE
Schaeffer, 1896

[Entomobryidae SCHAEFFER, 1896, p. 177]

Pronotum reduced; body usually with scales; antennae short; furca well developed. *Perm.–Holo.*

Entomobrya RONDANI, 1861, p. 40 [=*Stylonotus* OLFERS, 1907, p. 20 (type, *S. lanuginosus*); *Omo-*

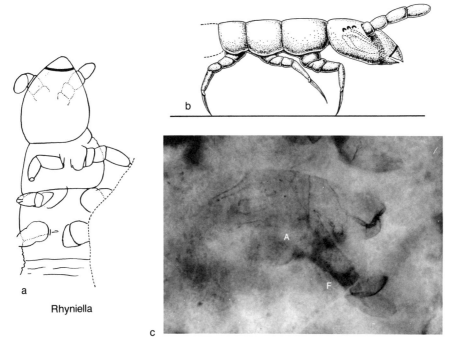

Fig. 2. Uncertain (p. 3).

phora OLFERS, 1907, p. 21 (type, *O. tricuspidata*); *Cuculliger* OLFERS, 1907, p. 24 (type, *C. longistylus*)]. CHRISTIANSEN, 1971. Oligo., Europe (Baltic); Oligo./Mio., Mexico (Chiapas)–*Holo.*
Lepidocyrtinus BÖRNER, 1903, p. 154. CHRISTIANSEN, 1971. Oligo./Mio., Mexico (Chiapas)–*Holo.*
Lepidocyrtus BOURLET, 1839, p. 391. KOCH & BERENDT, 1854; HANDSCHIN, 1926a. Oligo., Europe (Baltic)–*Holo.*
Orchesella TEMPLETON, 1835, p. 92. HANDSCHIN, 1926a. Oligo., Europe (Baltic)–*Holo.*
Permobrya RIEK, 1977, p. 141 [*P. mirabilis*; OD]. Similar in general appearance to *Lepidocyrtus*, but legs short and stout, and fourth antennal segment annulated. [Family assignment uncertain.] *Perm.*, South Africa.
Salina MACGILLIVRAY, 1894, p. 107. PIERCE, 1960; CHRISTIANSEN, 1971. Oligo./Mio., Mexico (Chiapas); Mio., USA (Colorado)–*Holo.*

Family SMINTHURIDAE
Lubbock, 1862

[Sminthuridae LUBBOCK, 1862, p. 430]

Antennae at least as long as head; thorax and the 4 basal abdominal segments fused into single unit. *Oligo.–Holo.*

Sminthurus LATREILLE, 1802, p. 71. STACH, 1923; HANDSCHIN, 1926a. Oligo., Europe (Baltic)–*Holo.*

Allacma BÖRNER, 1906, p. 183. HANDSCHIN, 1926a. Oligo., Europe (Baltic)–*Holo.*

Family UNCERTAIN

The genus described below, apparently belonging to the class and order Collembola, is too poorly known to permit family assignment.

Rhyniella HIRST & MAULIK, 1926, p. 69 [*R. praecursor*; OD]. Little-known genus; furcula well developed. TILLYARD, 1928c; SCOURFIELD, 1940a, 1940b; DELAMARE DEBOUTTEVILLE & MASSOUD, 1967; MASSOUD, 1967; WHALLEY & JARZEMBOWSKI, 1981. *Dev.*, Scotland.——FIG. 2. **R. praecursor*; *a*, head and thorax, ventral view, ×55 (Massoud, 1967); *b*, head and thorax, lateral view, reconstruction, ×55 (Delamare Deboutteville & Massoud, 1967); *c*, lateral view of abdomen (A) with furcula (F), ×100 (photograph courtesy of P. E. S. Whalley).

Class and Order PROTURA
Silvestri, 1907

[Protura SILVESTRI, 1907, p. 296]

Very small, slender hexapods; head prognathous, with entognathous mouthparts;

antennae and eyes absent; maxillary and labial palpi well developed; prothorax small; forelegs of moderate length, meso- and metathoracic legs short; abdomen with 12 segments in adult; small styli present on sterna; immature forms with anamorphic development, abdominal segments increasing from 9 to 12. Adults and young usually occurring in damp soil or leaf litter. TUXEN, 1964. *Holo.*

Class and Order DIPLURA
Börner, 1904

[Diplura BÖRNER, 1904, p. 524]

Small to large hexapods, with entognathous mouthparts. Antennae moniliform, with at least 20 segments, flagellar members with intrinsic muscles; mandibles elongate; maxillary and labial palpi much reduced; hypopharynx well developed; compound eyes and ocelli absent; thoracic segments slightly separated and free, prothorax smallest, meso- and metathorax nearly equal; tarsi consisting of a single segment; abdomen with 11 segments, the last bearing cerci; sterna of segments 2 through 7 with a pair of small, lateral, styliform appendages; cerci diversely formed, either multisegmented or modified to stout, heavily sclerotized forceps. So far as known, sperm transfer indirect, the male depositing stalked spermatophores on the substrate and the female taking up the sperm. Some species (Campodeidae) phytophagous, others (Japygidae) carnivorous. *Paleoc.–Holo.*

Family CAMPODEIDAE
Meinert, 1865

[Campodeidae MEINERT, 1865, p. 400]

Cerci long and multisegmented; thorax with 3 pairs of spiracles; abdominal styli soft. *Paleoc.–Holo.*

Campodea WESTWOOD, 1842, p. 71. SILVESTRI, 1913a; PACLT, 1957. *Oligo.,* Europe (Baltic)–*Holo.*
Onychocampodea PIERCE, 1951, p. 48 [**O. onychis;* OD]. Little-known genus; body length about 10 mm. [Family position doubtful.] PACLT, 1957. *Paleoc.–Plio.,* USA (Arizona).

Family UNCERTAIN

The following genera, apparently belonging to the class and order Diplura, are too poorly known to permit family assignment.

Onychojapyx PIERCE, 1950, p. 104 [**O. schmidti;* OD]. Little-known genus, with short, unsegmented cerci. PIERCE, 1951; PACLT, 1957; REDDELL, 1983. *Paleoc.–Plio.,* USA (Arizona).
Plioprojapyx PIERCE, 1951, p. 48 [**P. primitivus;* OD]. Little-known genus; cerci very short, apparently with only 3 segments. PACLT, 1957. *Paleoc.–Plio.,* USA (Arizona).

Class INSECTA Linné, 1758

[Insecta LINNÉ, 1758, p. 339]

Very small to large ectognathous hexapods; body composed of 20 embryonic segments, grouped into 3 main regions: head consisting of 6 segments, thorax of 3, and abdomen of 11; no abdominal segments added after embryonic stages; one pair of antennae, 2 compound eyes, and 3 ocelli usually present; mouthparts typically mandibulate but diversely modified in several orders; thoracic segments each bearing a pair of segmented legs; eleventh abdominal segment with a pair of segmented cerci, commonly much reduced or absent. Immature stages of primitive insects similar to adults, but those of most existing species greatly modified. *U. Carb.–Holo.*

INTRODUCTION TO THE INSECTS

GENERAL MORPHOLOGY

The class Insecta is not only the largest of all the existing classes of animals but is larger than all other classes combined. As a consequence, there is very great morphological diversity within the class. The following survey is concerned with those structures that are generally used in the higher classification of the insects. Detailed accounts of insect morphology are available in such basic works as *The Principles of Insect Morphology* by R. E. SNODGRASS (1935), *The Insects of Australia* by the Commonwealth Scientific and

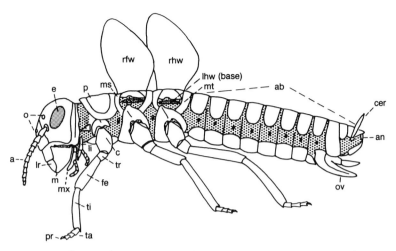

Fig. 3. Lateral view of typical female insect. Head: antenna (a); compound eye (e); labium (li); labrum (lr); mandible (m); maxilla (mx); ocelli (o). Thorax: coxa (c); femur (fe); left hind wing, base (lhw); mesonotum (ms); metanotum (mt); pronotum (p); pretarsus (pr); right fore wing (rfw); right hind wing (rhw); tarsus (ta); tibia (ti). Abdomen: anus (an); cercus (cer); ovipositor (ov) (Carpenter, new).

Industrial Research Organization (CSIRO), Canberra (1970), and *Imms' General Textbook of Entomology,* tenth edition, edited by O. W. RICHARDS and R. G. DAVIES (1977), as well as in many more specialized works on insects. Basic elements in the external morphology of a typical insect are diagrammed in Figure 3.

HEAD

The antennae are usually the most conspicuous structures on the head. In generalized insects, the antennae are usually long and filamentous, the numerous segments showing little differentiation. In most insects, however, the basal segment (scape) is at least a little longer than the others, and in many species the distal segments are much enlarged, forming club-shaped or comb-shaped antennae. In a few orders, such as the Odonata, the antennae are very small to minute.

The mouthparts of the primitive insects were obviously used for chewing, consisting in part of two pairs of plates articulated to the head capsule and controlled by muscles. In several orders, such as the Lepidoptera, however, they have been modified for sucking liquid food (haustellate), the mandibles and maxillae forming stylets though which the food is drawn. In others, such as the Hemiptera and Diptera, the mouthparts are adapted for both piercing and sucking.

Three types of heads are usually recognized, based on the position of the mouthparts relative to the longitudinal axis of the head. A head is termed hypognathous when its longitudinal axis is vertical and the mouthparts ventral. This is the commonest and probably the most generalized type, occurring among foliage feeders, such as nearly all Orthoptera and Dermaptera, and many Coleoptera. A prognathous head has the longitudinal axis of the head horizontal and the mouthparts anterior. This usually occurs in predaceous species. An opisthognathous head has its axis nearly horizontal but the mouthparts are posterior, the mouthparts arising near the base of the prothoracic legs. This occurs chiefly among some of the Hemiptera.

The two compound eyes are the main visual organs of insects. Each eye is usually divided into numerous visual units (ommatidia), ranging in number from a few to over 20,000. The eyes are usually located dorsally on the sides of the head. The 3 ocelli, each of which comprises a single visual unit, are located on the front area of the head; the median ocellus lies near the center of the frons, and the other two are positioned slightly more dorsally.

THORAX

The three thoracic segments, termed the prothorax, mesothorax, and metathorax, are very similar in the primitively wingless insects (subclass Apterygota), but in most of the winged species (subclass Pterygota) there is a marked differentiation. The prothorax, which bears no functional wings, is much smaller than the mesothorax and the metathorax. The two latter segments may be different from each other, depending on the relative sizes of their wings. Such insects as the Diptera, in which hind wings are greatly reduced, have a small metathorax. On the other hand, the metathorax of the Coleoptera and Dermaptera, in which the hind wings are the main organs of flight, is much larger than the mesothorax.

The legs typically consist of 5 segments: coxa, the basal segment, followed by the trochanter, femur, tibia, and tarsus. The coxa and trochanter are usually very short, but the other segments are diversely modified. The basic function of the legs was presumably walking (gressorial) or running (cursorial). One or more pairs of the legs are often modified for special functions, such as jumping, swimming, burrowing, or seizing prey. The tarsus is typically further subdivided into 5 segments, the last of which is the pretarsus, usually consisting of a pair of claws.

The wings are the most notable structures of the insects. They develop laterally on the meso- and metathoracic segments in the immature stages (nymphs or larvae) as expansions of the integument and resemble flat pouches with an upper and lower layer (HOLDSWORTH, 1940, 1941, 1942). Spaces (lacunae) containing blood are formed in the wing pads, and the integument near the lacunae produces the veins. In the final stages of development, as the adult insect emerges, the wings are inflated by increased blood pressure in the veins, the cuticle hardens, and the wings become functional in a surprisingly short time.

The venational patterns of the wings are of much importance in the systematics of most orders of insects, especially in the study of fossil insects, since the cuticle of the wings is usually much better preserved than the soft parts of the insects' bodies. Early attempts to use the venation in systematics were unsuccessful, mainly because there was no generally accepted concept of the evolution and homology of the veins in the several orders. (See, for example, the *Principles of Zoology*, by LOUIS AGASSIZ and A. A. GOULD, 1871, second edition, p. 237–239.) HAGEN (1870) tried in a preliminary way to homologize the wing veins of insects, but REDTENBACHER (1886) followed with the most significant contribution to the subject. He recognized six main veins, termed the costa, subcosta, radius, media, cubitus, and anal vein, a terminology that is still used. He based his homologies in part on the topographic positions of the veins, having noted that some of the veins were on ridges (convex) and others in depressions (concave). In 1895, COMSTOCK and NEEDHAM began their studies of wing venation, using REDTENBACHER's terminology for the main veins (COMSTOCK, 1918). Their homology of the veins, however, was based on the assumption that the venational pattern was determined by the tracheal pattern in the developing wing pads, and this led to some erroneous conclusions (COMSTOCK & NEEDHAM, 1898–1899). Actually, as later shown by HOLDSWORTH (1940, 1941), HENKE (1951), and LESTON (1962), the tracheae do not enter the wing pads until the lacunae have already determined the positions of the veins.

In 1922, LAMEERE, while studying the Carboniferous insects from Commentry, France, was impressed by the alternate convexity and concavity of the main wing veins, and he was convinced that COMSTOCK and NEEDHAM had included two distinct veins in their media and two in their cubitus, one of each being convex and the other concave. He accordingly termed the convex media the *anterior*

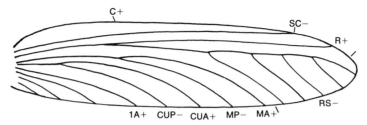

FIG. 4. Fore wing of *Stenodictya* sp., Palaeodictyoptera, Upper Carboniferous of France (Carpenter, new).

media (MA) and the concave media the *posterior media* (MP). Similarly, he termed the convex cubitus the *anterior cubitus* (CUA) and the concave cubitus the *posterior cubitus* (CUP). His studies led a large number of entomologists interested in insect evolution to their own investigations of venation, which ultimately fully supported LAMEERE's conclusions (TILLYARD, 1923d; MARTYNOV, 1924a; SPIETH, 1932; HOLDSWORTH, 1940, 1941). Among such primitive pterygotes as the Ephemeroptera, the convex veins are formed on the dorsal membrane of the wing pouch, and the concave veins on the ventral membrane. Among more specialized insects, at least most of the cuticular material forming the convex veins is produced on the dorsal layer, and most of that of the concave veins on the ventral layer. This results in the alternation of the convex and concave veins when the two layers are fused together.

The venational interpretation and terminology advocated by WOOTTON (1979) are followed here: costa (C, convex), subcosta (SC, concave), radius (R, convex), radial sector (RS, concave), anterior media (MA, convex), posterior media (MP, concave), anterior cubitus (CUA, convex), posterior cubitus (CUP, concave), anal vein (1A, convex) (Figs. 4 and 5). Thickened wings, such as tegmina and elytra, tend to lose the convexity or concavity of the media veins. If both veins of the median system are flat, they are simply designated as the media (M). In addition to these main longitudinal veins, there are often many small veins, such as crossveins, that occur in various parts of the wings, especially the anterior areas; but these are not part of the system of main longitudinal veins discussed above (Fig. 6).

In many insects the hind wings have been secondarily lost, as in the Diptera, or much reduced, as in many Hymenoptera. In some others, the fore wings have been lost, the hind wings being much enlarged, as in the Strepsiptera. In two orders, Siphonaptera and Grylloblattodea, all existing species have lost their wings, and it is noteworthy that at least some secondarily wingless species occur in all existing orders of insects except the Ephemeroptera and the Odonata, both of which are members of the Palaeoptera.

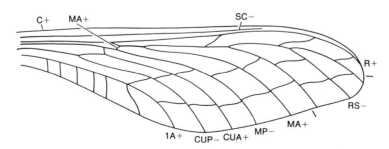

FIG. 5. Fore wing of *Psilothorax* sp., Megasecoptera, Upper Carboniferous of France (Carpenter, 1951).

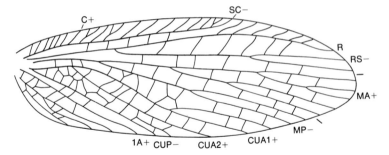

Fig. 6. Fore wing of *Liometopum* sp., Protorthoptera, Permian of Kansas (Carpenter, 1950).

ABDOMEN

Evidence from embryos indicates that the primitive insects had 11 abdominal segments, but in most existing species the 3 terminal segments are commonly much reduced or modified. In some insects, the eleventh segment is represented by a pair of segmented appendages, the cerci, which are very prominent in some orders, as in Ephemeroptera, but much reduced in most others. In a very few species (some Apterygota and Ephemeroptera) a median process or style also arises from the eleventh segment. The female abdomen typically has three pairs of unsegmented processes arising from the eighth and ninth segments and forming an ovipositor. The male abdomen has a pair of claspers, apparently arising from the ninth segment and used for holding the female during mating.

REPRODUCTION AND DEVELOPMENT

Among the most primitive of the living insects, the order Archaeognatha of the subclass Apterygota, the transfer of sperm to the female is indirect, the sperm being deposited in droplets, usually on the ground. These are picked up by the females and inserted into their genital tracts. In all other existing insects the sperm is transferred directly into the female tract, usually after a specific pattern of courtship behavior. The eggs are deposited in environments appropriate for the species concerned, as in soil, on foliage, in water, or, in the case of parasitic species, on the bodies of host species. Parthenogenesis does occur in several orders. In some of these the unfertilized eggs produce males, as in certain Hymenoptera and Hemiptera (Homoptera), the cycle of parthenogenesis and normal mating being involved with their social behavior (ALEXANDER, 1964; ENGLEMANN, 1970).

The postembryonic development of insects is characterized by a series of cuticular molts. The newly hatched young of the Apterygota closely resemble the adults, except in size, but they molt many times, even after the adult stage has been attained (DELANY, 1961). The immature stages of the Pterygota differ, at least in form, from the adults, and in most species they are strikingly different (Fig. 7).

The great majority of insects are terrestrial in their immature stages, but aquatic species occur in several orders, such as Diptera, Coleoptera, Hemiptera, and all species are aquatic in a few orders, such as Ephemeroptera, Odonata, Trichoptera, and Perlaria. The food of immature forms is very diverse; in some it is similar to that of the adults, but in most species it is very different.

ORIGIN OF INSECTS

Although more than two hundred research papers have been published on this subject, there is still no convincing evidence regarding the ancestral stock that produced the insects. TIEGS and MANTON (1958) have provided a very useful discussion of the subject, and MANTON (1969a, 1969b, 1977, 1979) has summarized her conclusions, after many years of research, on the evolution of the Arthropoda, including the insects. The present account is a brief synopsis of the diverse views of zoologists and entomologists on the subject.

The most unlikely theories are those of WALTON (1927) and HANDLIRSCH (1908a). WALTON was of the opinion that the insects had evolved from the polychete annelids, and HANDLIRSCH proposed that the pterygotes were directly evolved from the trilobites, the apterygotes having subsequently developed from the pterygotes. MÜLLER (1864), HANSEN (1893), and CARPENTER (1903, 1905) believed that the insects arose from the larvae of decapod crustaceans; and CRAMPTON (1920, 1938) was convinced that they were descended from adult Crustacea allied to the Syncarida. TILLYARD (1930) was of the opinion, from his own research, that they were derived from the Collembola, through the Protura. PACKARD (1873), IMMS (1936), SNODGRASS (1952, 1958), WILLE (1960), and SHAROV (1966b) favored the Symphyla as the ancestors of all the hexapods, including the insects, whereas MANTON (1979) concluded that the Hexapoda and Symphyla could not have shared an immediate, common ancestor, and that the present myriopod and insect faunas represent the isolated descendents of a once widespread, early radiation of terrestrial arthropods. Unfortunately, the present geological record of the insects is no help in this connection, since the earliest insects now known (Late Carboniferous) are true insects, belonging to the subclasses Apterygota and Pterygota.

EVOLUTION OF INSECTA

The present concept of the evolution of insects after the appearance of the Apterygota recognizes two major events: the development of wings and the acquisition of a complicated metamorphosis during the immature stages.

The literature on the origin of wings is nearly as extensive as that on the origin of the insects. The several theories have been proposed and discussed by WIGGLESWORTH and others (1963), WIGGLESWORTH (1963, 1973, 1976), WOOTTON (1976), KUKALOVÁ-PECK, (1978, 1983) and RASNITSYN (1981). Although there are obvious differences in opinions, the theory generally accepted assumes that the wings were derived from small meso- and metathoracic, paranotal lobes, which may have originally functioned as sex attractants (ALEXANDER & BROWN, 1963), as thermoregulators (DOUGLAS, 1980), or as stationary aids in aerial migrations of small insects (RASNITSYN, 1981). There is some experimental evidence that such lobes, even without muscular movements, could have had selective survival value. Once formed, the lobes could have been modified to wings. Unfortunately, the geological record of the insects provides no actual record of the evolution of wings, although some species of Paleozoic orders, such as the Palaeodictyoptera, Protorthoptera, and Ephemeroptera, had small prothoracic lobes similar to those postulated above on the meso- and metathoracic segments.

Whatever their origin, the development of wings, which obviously occurred before the beginning of the Late Carboniferous, must be regarded as the most significant event in the evolution of the insects, which so far as we know, were the first animals to develop organs of flight. They provided a unique means of dispersal and of escape from predators. It is not surprising that the winged insects, comprising the subclass Pterygota, have been the predominate insects since the beginning of the Late Carboniferous, at least.

From their first appearance in the Carboniferous, the pterygotes have included two groups of orders, which MARTYNOV (1924) designated the infraclasses Palaeoptera and Neoptera. The first of these includes species that have a somewhat limited articulation of the wings with the thorax, with the result that they are unable to fold their wings back over the abdomen at rest. The evolutionary significance of this was first noted by WOODWORTH (1907) and was much later extensively discussed by MARTYNOV (1924, 1925e, 1938b), CRAMPTON (1924), and

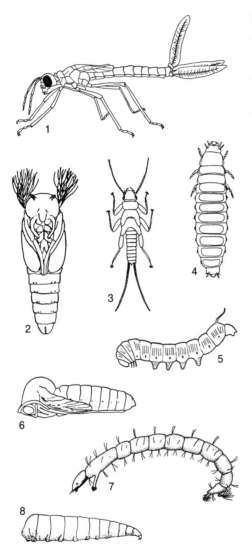

Fig. 7. Immature stages of diverse insects, all Holocene.——*1.* Nymph of *Lestes* sp., Odonata, lateral view (Essig, 1942).——*2.* Pupa of *Simulium* sp., Diptera, ventral view (Brues, Melander, & Carpenter, 1954).——*3.* Nymph of *Perla* sp., Perlaria, dorsal view (Brues, Melander, & Carpenter, 1954).——*4.* Larva of *Calosoma* sp., Coleoptera, dorsal view (Brues, Melander, & Carpenter, 1954).——*5.* Larva of *Acherontia* sp., Lepidoptera, lateral view (Brues, Melander, & Carpenter, 1954).——*6.* Pupa of *Leia* sp., Diptera, lateral view (Brues, Melander, & Carpenter, 1954).——*7.* Larva of *Tanypus* sp., Diptera, lateral view (Brues, Melander, & Carpenter, 1954).——*8.* Larva of *Musca* sp., Diptera, lateral view (Brues, Melander, & Carpenter, 1954).

Snodgrass (1935). Two existing orders, Ephemeroptera and Odonata, belong in the infraclass Palaeoptera along with several extinct orders, including the Palaeodictyoptera, Protodonata, and Megasecoptera. The remaining orders of the Pterygota, which constitute the infraclass Neoptera, have a more complicated wing articulation that allows the wings to be placed back over the abdomen when the insect is at rest. Since these insects are not hindered in their activities by outstretched wings, they are able to crawl among dense foliage, under stones, and even in tunnels in the soil. This was apparently a significant development in the evolution of the insects, since 99 percent of all living species of the Pterygota belong to the infraclass Neoptera. In this connection it is interesting to note that the extinct order Diaphanopterodea, known only from the Upper Carboniferous and Permian, apparently developed wing folding independently of the Neoptera, as shown by the position of the wings in the fossils. The Diaphanopterodea, however, have the venational features and the long haustellate mouthparts characteristic of the Palaeodictyoptera and Megasecoptera, both members of the Palaeoptera. The articular plates of their wings are very different from those of the Neoptera (Kukalová-Peck, 1974a, 1974b).

The more primitive members of the Neoptera are characterized by having the wings develop externally in the immature stages. With few exceptions, the nymphs resemble the adults closely, live in the same environments, and feed on similar foods (Fig. 7,*1,3*). Most of the existing orders of insects belong in this category, termed the division Exopterygota (Carpenter & Burnham, 1985). However, the interrelationships of some of these orders are uncertain, and the exopterygotes may not constitute a monophyletic group.

The development of a complicated metamorphosis in the postembryonic stages apparently occurred within the Neoptera. This involved major changes. The wings, instead of developing externally, are invag-

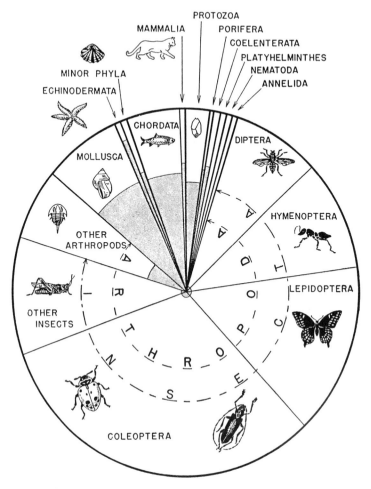

Fig. 8. Relative numbers of living and extinct species of animals. The percentage of extinct species is shown by the relative size of the gray area for each taxon (adapted from Muller & Campbell, 1954).

inated beneath the thoracic cuticle. After a series of molts and ecdyses, the insects pass into the pupal stage, in which the wing pads are evaginated and become external (Fig. 7,2,6). This is a quiescent stage, during which there is no feeding, although extensive internal changes occur. With the final ecdysis, the adults emerge and the wings expand as in the exopterygotes. The significance of this metamorphosis is that the immature stages (larvae) are very different in appearance from the adults, occupy very different environments, and feed on different foods (Fig. 7,4,5,7,8). Although only nine existing orders belong in this division, termed the Endopterygota, they comprise about 85 percent of all living Neoptera, including such large orders as the Diptera, Hymenoptera, and Coleoptera. No endopterygotes are known from the Upper Carboniferous, but four existing orders are well represented in the Permian. The endopterygotes are generally considered to comprise a monophyletic group.

THE GEOLOGICAL RECORD AND PHYLOGENY OF THE INSECTA

Although the fossil record of the insects includes about six thousand genera, our

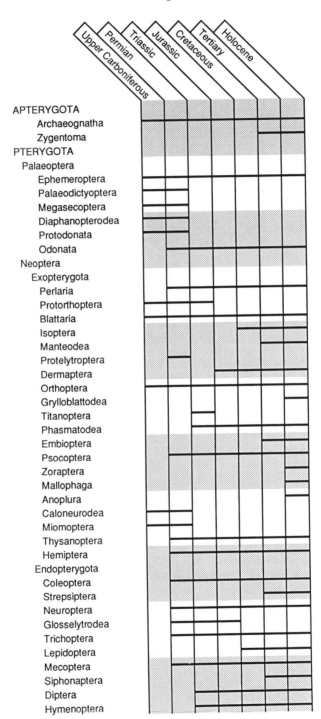

Fig. 9. Geological ranges of orders of insects (Carpenter, new).

knowledge of the geological history of the class is actually very limited. This is apparent from the analysis by MULLER & CAMPBELL (1954) of the relative numbers of known species, both existing and extinct, in all animal phyla, and the percentage of those known as fossils (Fig. 8). For most existing phyla, at least 35 percent of the total species known are extinct. For the insects, however, the number of extinct species in the record is only one percent. Since most of insects are terrestrial, they are ordinarily preserved as fossils only under special environmental conditions.

At this time, insects are unknown in deposits older than the Upper Carboniferous, but the presence of eleven orders in those rocks, including representatives of the Apterygota, Palaeoptera, and Neoptera as well as of four existing orders, indicates that the class existed at least in the Lower Carboniferous and possibly in the Devonian (Fig. 9). Apart from the Carboniferous, the least known of the extinct insect faunas is that of the Cretaceous, a very long and unusually important period in the history of the existing insect families. The Tertiary fossils are the most numerous, but their generic and even family identifications, as recorded, are not always reliable. Many of them were named a century or more ago and placed into existing genera, long before the current concepts of those genera were reached. Restudy of the early type collections by specialists in the families concerned is probably the most urgent need in paleoentomology. There are also many differences of opinion about the systematic positions of some of the extinct genera, especially those based on fragmented specimens. Restudy of additional material is essential, and, until more is known about them, such genera are best assigned to the category of family Uncertain.

In citing the geological ranges of the Cenozoic genera, I have recorded the names of the series, but for the Mesozoic and Paleozoic genera, I have intentionally omitted the series. In many cases the precise ages of the insect deposits within those systems are not definitely known. The one exception to this policy is my use of the series term Upper Carboniferous; this is done because there is at present no record of insects in the Lower Carboniferous. The precise ages of some of the insect-bearing ambers, mostly Tertiary, are not certain. In general I have followed the ages cited by BURLEIGH and WHALLEY (1983). In referring to the insects in the Baltic amber, I have used the term "Baltic," as is usually done, without specifying the several countries in western Europe in which the resin occurs.

The number of existing orders of insects currently recognized by entomologists varies considerably, although the range is usually between twenty-five and thirty. In the present account I recognize twenty-eight, all but four (Grylloblattodea, Zoraptera, Mallophaga, Anoplura) being represented in the fossil record. In contrast, fifty-five extinct orders have been named, most of them from the Carboniferous and Permian. The majority of these extinct orders, however, were based on small fragments or otherwise poorly preserved specimens that have subsequently been placed in other orders or in the category of order unknown. In this treatise I recognize ten extinct orders as valid (Table 1). Additional extinct orders will almost certainly become known as new collections of fossils are studied.

The relationships of the existing orders have been extensively discussed in the literature. In the past there have been many differences of opinion but in recent years the main lines of insect evolution, discussed above, have been generally accepted; and in most respects the more detailed concept of the phylogeny of existing orders proposed by KRISTENSEN (1981) has been widely adopted (Fig. 10). Although the phylogeny of the endopterygote orders is apparently clear, that of the more primitive exopterygotes remains uncertain. The relationships of most of the ten extinct orders seem obvious. Four of these are palaeopterous, five exopterygote, and one endopterygote. Their relationships are discussed below in detail.

TABLE 1. *Extinct Orders of Insects.* Chronological list of extinct orders of insects recorded in the literature. The ordinal names printed in boldface are accepted as valid in this publication; the rest are included in other orders or as indicated.

1. **Palaeodictyoptera** GOLDENBERG, 1877
2. **Megasecoptera** BRONGNIART, 1885a
3. **Protodonata** BRONGNIART, 1893
4. Palaeohemiptera HANDLIRSCH, 1904b (Hemiptera)
5. Protoblattoidea HANDLIRSCH, 1906a (Protorthoptera)
6. Hadentomoidea HANDLIRSCH, 1906a (Protorthoptera)
7. Mixotermitoidea HANDLIRSCH, 1906a (Neoptera uncertain)
8. Reculoidea HANDLIRSCH, 1906b (Protorthoptera)
9. Hapalopteroidea HANDLIRSCH, 1906a (Protorthoptera)
10. Protephemeroidea HANDLIRSCH, 1906b (Ephemeroptera)
11. Protohemiptera HANDLIRSCH, 1906b (Palaeodictyoptera)
12. **Protorthoptera** HANDLIRSCH, 1906a
13. Sypharopteroidea HANDLIRSCH, 1911 (Palaeoptera uncertain)
14. Protomecoptera TILLYARD, 1917a (Neoptera uncertain)
15. Paratrichoptera TILLYARD, 1919a (Mecoptera)
16. Paramecoptera TILLYARD, 1919b (Mecoptera)
17. Synarmogoidea HANDLIRSCH, 1919b (Palaeodictyoptera)
18. **Diaphanopterodea** HANDLIRSCH, 1919b
19. Aeroplanoptera TILLYARD, 1923b (Phasmatodea)
20. Protohymenoptera TILLYARD, 1924a (Megasecoptera)
21. Protocoleoptera TILLYARD, 1924b (Protelytroptera)
22. **Miomoptera** MARTYNOV, 1927d
23. Protoperlaria TILLYARD, 1928b (Protorthoptera)
24. Pruvostitoptera M. D. ZALESSKY, 1928b (Orthoptera)
25. Permodonata G. M. ZALESSKY, 1931 (Odonata)
26. **Protelytroptera** TILLYARD, 1931
27. Archodonata MARTYNOV, 1932 (Palaeodictyoptera)
28. Meganisoptera MARTYNOV, 1932 (Protodonata)
29. Hemipsocoptera ZALESSKY, 1937e (Hemiptera)
30. **Caloneurodea** HANDLIRSCH, 1937
31. Cnemidolestoidea HANDLIRSCH, 1937 (Protorthoptera)
32. Strephocladodea MARTYNOV, 1938b (Protorthoptera)
33. Paraplecoptera MARTYNOV, 1938b (Protorthoptera)
34. **Glosselytrodea** MARTYNOV, 1938c
35. Protocicadida HAUPT, 1941 (Palaeodictyoptera, Protorthoptera)
36. Protofulgorida HAUPT, 1941 (Protorthoptera, Blattaria)
37. Archaehymenoptera HAUPT, 1941 (Palaeodictyoptera)
38. Palaeohymenoptera HAUPT, 1941 (Diaphanopterodea)
39. Perielytrodea ZALESSKY, 1943 (Neoptera uncertain)
40. Anisaxia FORBES, 1943 (Palaeodictyoptera)
41. Permodictyoptera ZALESSKY, 1944a (Palaeoptera uncertain)
42. Aphelophlebia PIERCE, 1945 (Ephemeroptera)
43. Hemiodonata ZALESSKY, 1946a (Palaeodictyoptera)
44. Breyerida HAUPT, 1949 (Palaeodictyoptera)
45. Eopalaeodictyoptera LAURENTIAUX, 1952a (Palaeodictyoptera)
46. Syntonopterodea LAURENTIAUX, 1953 (Palaeodictyoptera)
47. Permoneurodea LAURENTIAUX, 1953 (Palaeoptera uncertain)
48. Paracoleoptera LAURENTIAUX, 1953 (Neoptera uncertain)
49. Eubleptidodea LAURENTIAUX, 1953 (Palaeodictyoptera)
50. Campylopterodea ROHDENDORF, 1962a (Palaeoptera uncertain)
51. **Titanoptera** SHAROV, 1968
52. Dictyoneurida ROHDENDORF, 1977 (Palaeodictyoptera)
53. Permothemistida SINITSHENKOVA, 1980a (Palaeodictyoptera)
54. Hypoperlida RASNITSYN, 1980f (Neoptera uncertain)
55. Blattinopseida RASNITSYN, 1980f (Neoptera uncertain)

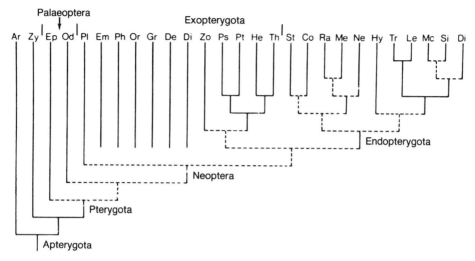

Fig. 10. Phylogeny of existing orders of insects. Higher categories: Apterygota; Pterygota; Palaeoptera; Neoptera; Exopterygota; Endopterygota. Orders: Archaeognatha (Ar); Zygentoma (Zy); Ephemeroptera (Ep); Odonata (Od); Perlaria (Plecoptera) (Pl); Embioptera (Em); Phasmatodea (Ph); Orthoptera (Or); Grylloblattodea (Gr); Dermaptera (De); Dictyoptera (Manteodea, Blattaria, Isoptera) (Di); Zoraptera (Zo); Psocoptera (Ps); Phthiaptera (Mallophaga, Anoplura) (Pt); Hemiptera (He); Thysanoptera (Th); Strepsiptera (St); Coleoptera (Co); Raphidioptera (Ra); Megaloptera (Me); Neuroptera (Ne); Hymenoptera (Hy); Trichoptera (Tr); Lepidoptera (Le); Mecoptera (Mc); Siphonaptera (Si); Diptera (Di) (adapted from Kristensen, 1981).

Subclass APTERYGOTA Brauer, 1885

[Apterygota BRAUER, 1885, p. 290]

Primitively wingless insects. Antennae usually well developed; mouthparts mandibulate; thoracic segments not united, similar in size and form; ventral styli commonly present on abdominal segments 2 through 9; cerci prominent, typically very long, rarely reduced or absent; median caudal process usually well developed. Reproduction indirect. Ecdysis and molting occurring throughout life. *U. Carb.–Holo.*

Order ARCHAEOGNATHA Börner, 1904

[Archaeognatha BÖRNER, 1904, p. 523]

Body cylindrical, with a covering of hairs, scales, or both; head usually hypognathous; compound eyes large; mandibles with a single articulation; antennae filiform, usually long and multisegmented, rarely short; maxillary palpi long to very long, with 7 segments in the Machiloidea (and probably in the Monura); thorax strongly arched dorsally; tarsi with from 1 to 3 segments; abdomen with 11 segments, the last bearing a median caudal process (appendix dorsalis) and in the Machiloidea bearing a pair of cerci, usually somewhat shorter than median caudal process; eighth and ninth abdominal segments of females each with a pair of prominent gonopophyses, combining to form an ovipositor; abdominal segments 2 through 9 with ventral styli. Reproduction indirect, sperm deposited on substrate (often in a spermatophore) by male and then gathered by female, with transference to her spermatheca. Postembryonic development involving only minor external changes; sexual maturity reached after about 10 molts, but ecdyses continuing throughout life, the number of molts often reaching 60.

Nocturnal insects, feeding mainly on algae and vegetable debris, they are fast runners and they can also jump by a downward flexing of the abdomen. *U. Carb.–Holo.*

Hexapoda

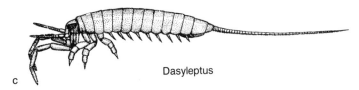

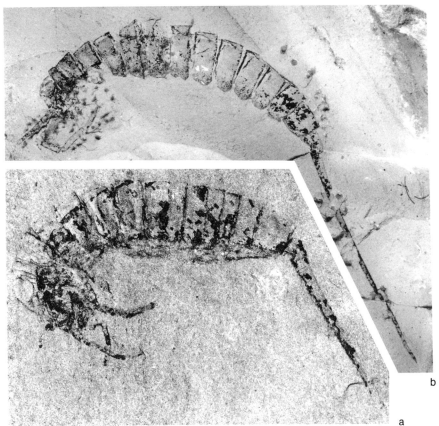

Fig. 11. Dasyleptidae (p. 16–17).

Suborder MONURA
Sharov, 1957

[Monura SHAROV, 1957c, p. 796]

Antennae well developed but may be relatively short; maxillary palpi with at least 5 segments (incompletely known); thoracic segments showing little differentiation dorsally from abdominal segments; cerci absent; median caudal process stout and about as long as body; legs relatively short, tarsi unsegmented, bearing only a single claw. *U. Carb.–Perm.*

Family DASYLEPTIDAE
Sharov, 1957

[Dasyleptidae SHAROV, 1957c, p. 797]

Antennae short, with only a few segments; ovipositor short, extending only to level of hind margin of eleventh segment. *U. Carb.–Perm.*

Dasyleptus BRONGNIART, 1885b, p. cii [*D. lucasi*; OD]. Compound eyes with upper margin nearly straight and lower margin convex; prothorax about half as long as mesothoracic segment; body with an extensive covering of short hairs. SHAROV, 1957c; ROHDENDORF, 1962a. *U. Carb.,* Europe

(France); *Perm.*, USSR (European RSFSR).——FIG. 11,*a*. **D. lucasi*, U. Carb.; lateral view, ×10 (Carpenter, new).——FIG. 11,*b,c*. *D. brongniarti* SHAROV, Perm.; *b*, lateral view, ×10 (Carpenter, new); *c*, reconstruction, lateral view, ×7 (Sharov, 1957c).

Lepidodasypus DURDEN, 1978, p. 1 [**L. sharovi*; OD]. Similar to *Dasyleptus* but with scales apparently present, in addition to hairs, on body and appendages. *Perm.*, USA (Kansas).

Suborder MACHILOIDEA Handlirsch, 1904

[Machiloidea HANDLIRSCH, 1904c, p. 758]

Antennae long, multisegmented; thoracic segments clearly differentiated dorsally from abdominal segments; tarsi usually with 3 segments (rarely only 2) and bearing 2 claws; cerci present but usually shorter than median caudal process. *Trias.–Holo.*

Family MACHILIDAE Grassi, 1888

[Machilidae GRASSI, 1888, p. 582] [=Triassomachilidae SHAROV, 1948, p. 517]

Abdominal sterna large. *Trias.–Holo.*

Machilis LATREILLE, 1802, p. 70 [=*Lepismodion* OLFERS, 1907, p. 16 (type, *L. machilops*); *Machilodes* OLFERS, 1907, p. 11 (type, *M. diastatica*)]. GADEAU DE KERVILLE, 1893; SILVESTRI, 1913a; PACLT, 1972. *Oligo.*, Europe (Baltic)–*Holo.*

Praemachilis SILVESTRI, 1905, p. 8. SILVESTRI, 1913a. *Oligo.*, Europe (Baltic)–*Holo.*

Triassomachilis SHAROV, 1948, p. 517 [**T. uralensis*; OD]. Little-known genus, similar to *Machilis*. PACLT, 1972. *Trias.*, USSR (European RSFSR).——FIG. 12. **T. uralensis*; dorsal view, ×7 (Sharov, 1948).

Order ZYGENTOMA Börner, 1904

[Zygentoma BÖRNER, 1904, p. 524]

Body distinctly flattened dorsoventrally, with or without a covering of scales; compound eyes small or absent; mandibles with both anterior and posterior articulations; maxillary palpi with 5 segments; cerci and median caudal process present, but cerci usually longer than median process; styli usually present on abdominal segments 2 through 9, rarely absent from segments 8 and 9. Reproduction and postembryonic development much as in Archaeognatha. Nocturnal,

Triassomachilis

FIG. 12. Machilidae (p. 17).

omnivorous apterygotes (silverfish), capable of running with extreme rapidity but without jumping ability of Archaeognatha. *Oligo.–Holo.*

Family LEPIDOTRICHIDAE Silvestri, 1913

[*nom. transl. et correct.* ANDER, 1942, p. 57, *ex* Lepidothricinae SILVESTRI, 1913a, p. 51]

Body lacking scales; tarsi with 5 segments. *Oligo.–Holo.*

Lepidotrix MENGE, 1854, p. 117, footnote [**L. pilifera*; OD] [=*Lepidothrix* SILVESTRI, 1913a, p. 49, unjustified emend.; *Lepidion* MENGE, 1854, p. 117 (type, *L. pisciculus*); *Klebsia* OLFERS, 1907, p. 8 (type, *K. horrens*); *Micropa* OLFERS, 1907, p. 8 (type, *M. stylifera*)]. Similar to *Tricholepidion* (recent) but lacking ocelli. KOCH & BERENDT, 1854; SILVESTRI, 1913a; WYGODZINSKY, 1961a; PACLT, 1967. *Oligo.*, Europe (Baltic).——FIG. 13. **L. pilifera*; dorsal view of whole insect as preserved, ×10 (Silvestri, 1913a).

Family LEPISMATIDAE Latreille, 1802

[Lepismatidae LATREILLE, 1802a, p. 70]

Compound eyes present; body covered with scales; tarsi with 3 or 4 segments. *Oligo.–Holo.*

Lepisma LINNÉ, 1758, p. 608. *Holo.*
Allocrotelsa SILVESTRI, 1935, p. 307, *nom. subst. pro Lampropholis* MENGE, 1854, p. 117, *non* FITZINGER, 1843. SILVESTRI, 1913a; COCKERELL, 1917g; WYGODZINSKY, 1961b; PACLT, 1967. *Oligo.*, Europe (Baltic); *Mio.*, Burma–*Holo.*

Subclass PTERYGOTA
Brauer, 1885

[Pterygota BRAUER, 1885, p. 290]

Wings present or secondarily absent. Antennae usually long, rarely short; mouthparts mandibulate or haustellate; prothorax much smaller than the meso- or metathorax, the latter two united to form the pterothorax; abdominal segments 2 through 9 without styli; cerci prominent in more primitive orders but much reduced or absent in others; median caudal process well developed in Ephemeroptera but absent in other orders. Reproduction direct. Ecdysis and molting occurring only in nymphal or larval stages. *U. Carb.–Holo.*

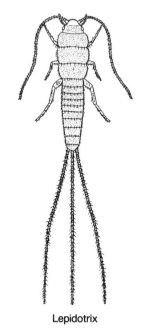

Lepidotrix

FIG. 13. Lepidotrichidae (p. 17).

Infraclass PALAEOPTERA
Martynov, 1923

[Palaeoptera MARTYNOV, 1923, p. 89]

Wings articulated to thorax by sclerotized plates (axillaries) fused to bases of main veins; flexor muscles absent; wings with main veins forming a complete alternation of convexities and concavities; vein MA distinctly convex; cerci commonly well developed and long; median caudal process present only in Ephemeroptera. Nymphs diversely formed, some aquatic; wings developing in cuticular sheaths, much as in exopterygote Neoptera. Subimaginal stage or stages present in some orders. *U. Carb.–Holo.*

This infraclass, now represented by only two orders (Ephemeroptera and Odonata), was apparently much more diverse in the Paleozoic. The fusion of the axillary sclerites to the bases of the main veins and the absence of wing-flexing muscles prevent the placing of the wings back along the abdomen at rest. Consequently, complete palaeopterous fossils, with wings and body, are nearly always preserved with the wings outstretched (see Figs. 26,3c and 39). The exceptions to this are of unusual interest. Members of the extinct order Diaphanopterodea are consistently preserved with the wings resting back along the abdomen. The structural mechanisms that make this possible are not known, but the wing axillaries were apparently not arranged as they are in the Neoptera (KUKALOVÁ-PECK, 1975), indicating that the wing flexing in the Diaphanopterodea was developed independently of that in the Neoptera. The known members of the Triassic odonate suborder Triadophlebiomorpha are likewise preserved with their wings placed along the abdomen, but the wing axillaries are unknown (PRITYKINA, 1981). In this connection it is noteworthy that the members of the existing odonate suborder Zygoptera also hold their wings along the abdomen at rest. However, this posture has been achieved by tilting of the pterothorax at an obtuse angle with reference to the abdomen.

Six orders are considered to belong to the Palaeoptera: Palaeodictyoptera, Megasecoptera, Diaphanopterodea, Ephemeroptera, Protodonata, and Odonata. With the single exception of the Odonata, all of these are known as far back as the Late Carboniferous.

Order EPHEMEROPTERA
Hyatt & Arms, 1890

[Ephemeroptera HYATT & ARMS, 1890, p. 69] [=Protephemeroidea HANDLIRSCH, 1906b, p. 311; Aphelophlebia PIERCE, 1945, p. 4] [Several names have been proposed for this order. Ephemeroptera is the term consistently used now by specialists in the order.]

Delicate insects with short, filiform antennae; mouthparts vestigial in existing families, mandibulate and functional in Paleozoic families; compound eyes large, 3 ocelli present; abdomen slender, terminating in a pair of long, segmented cerci and usually with a long, median caudal process; legs usually weak in recent species, the mesothoracic and metathoracic pairs often much reduced, but all legs well developed in Paleozoic families; wings very delicate, with a complete set of all main veins in addition to intercalary veins (indicated by an I prefix) and crossveins; base of costal area supported in some families by a stout crossvein or a series of crossveins (costal brace; see Figs. 14,4a and 15,a); in all recent and Tertiary species, as well as those from the Mesozoic, hind wings much smaller than fore pair and in some genera completely absent; in known Paleozoic species, pairs of wings similar in size and venation; digestive tract modified to form aerostatic organ; reproductive ducts paired in both sexes. Nymphs aquatic, occurring in ponds and streams, usually with at least 7 pairs of abdominal tracheal gills; cerci and median caudal process present; mostly herbivorous. Postembryonic development slow, with 20 or more ecdyses and a single molt from winged subimago to imago. *U. Carb.–Holo.*

The Ephemeroptera is a relatively small order of about 2,000 species. Although basically primitive, the recent members are highly adapted to living in aquatic environments in the nymphal stages and to a very brief imaginal life. The nymphal gills are unusually large compared with those of other aquatic insects and are capable of rapid movements, aiding the circulation of water. Nymphal development is slow, taking at least a few months and commonly as long as three years. In contrast, most imagoes live for only a few hours to a few days. The process of mating is hastened by swarming.

The earliest record of the Ephemeroptera is a single imago of *Triplosoba pulchella* (BRONGNIART) from the Upper Carboniferous of Commentry, France, but representatives of five extinct families, including nymphs as well as adults, are known from the Permian. The peak of diversity appears to have been reached in the Jurassic, from which nine families have been obtained, including the existing families Siphlonuridae, Leptophlebiidae, Palingeniidae, Behningiidae, and possibly the Ephemerellidae. So far as known, the imagoes of all the Permian species had fully developed mouthparts with functional, dentate mandibles, normally developed legs, and similar fore and hind wings. These imagoes appear to be the most primitive of the known pterygote insects. The nymphs, however, were apparently as well adapted to an aquatic life as those now existing.

The classification of the Ephemeroptera has been discussed by several specialists in the order in recent years (TSHERNOVA, 1970; LANDA, 1979; MCCAFFERTY & EDMUNDS, 1979; RIEK, 1979), mainly with reference to the existing families. There seems to be general agreement that division of the order into the suborders Permoplectoptera and Euplectoptera, separating the Permian families from the later ones, as proposed by TILLYARD (1932b), is unsatisfactory. In the following account the sequence of families follows that of MCCAFFERTY and EDMUNDS (1979) and TSHERNOVA (1970).

Family TRIPLOSOBIDAE
Handlirsch, 1906

[Triplosobidae HANDLIRSCH, 1906b, p. 312]

Fore and hind wings apparently similar in form and venation; crossveins numerous; costal brace apparently absent; vein SC extending to wing apex; RS arising directly from

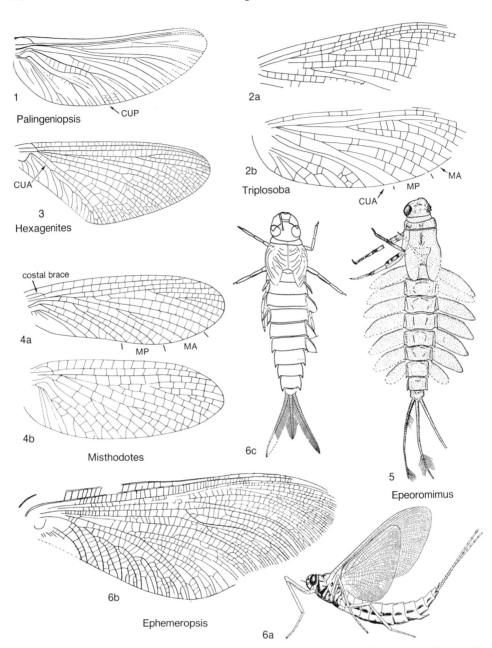

Fig. 14. Triplosobidae, Misthodotidae, Palingeniopsidae, Hexagenitidae, and Epeoromimidae (p. 20–24).

R, free from MA and including 2 intercalary veins; abdomen slender, with prominent cerci and a median caudal process. *U. Carb.*

Triplosoba HANDLIRSCH, 1906b, p. 312, *nom. subst. pro Blanchardia* BRONGNIART, 1893, p. 325, *non* CASTELNAU, 1875 [*Blanchardia pulchella* BRONGNIART, 1893, p. 325; OD]. MA and CUA unbranched in both wings; MP branched. CARPENTER, 1963c. *U. Carb.,* Europe (France).—— FIG. 14,2. *T. pulchella* (BRONGNIART); *a,* fore and *b,* hind wings, ×2.5 (Carpenter, new).

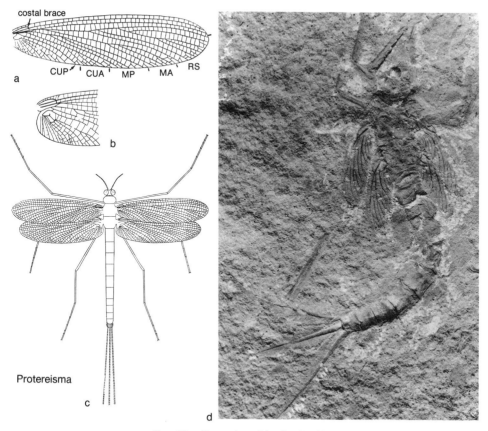

FIG. 15. Protereismatidae (p. 21–22).

Family PROTEREISMATIDAE
Lameere, 1917

[*nom. correct.* TILLYARD, 1932b, *ex* Protereismidae LAMEERE, 1917a, p. 45] [=Kukalovidae DEMOULIN, 1970b, p. 6]

Adults moderate to large in size. Wings elongate-oval; fore and hind wings similar in form and venation, hind pair only slightly shorter; crossveins numerous; costal brace strongly developed in both wings (Fig. 15,*a*); vein SC extending almost to wing apex; RS coalesced with MA immediately after its origin, and including 3 intercalary veins; MP and CUA each with a single triad; antennae short but longer than in existing mayflies; mandibles sclerotized and dentate; compound eyes large; legs very long and slender, with 5 tarsal segments; cerci and median caudal process elongate; males with prominent claspers. Nymphs with well-developed mandibles; legs subequal, with 5 tarsal segments; cerci and median caudal process well developed; abdomen with 9 pairs of tracheal gills; wing pads independent of each other, attached to thorax only along equivalent of the articular area of adult wings, projecting obliquely. *Perm.*

Protereisma SELLARDS, 1907, p. 347 [*P. permianum*; OD] [=*Protechma* SELLARDS, 1907, p. 349 (type, *P. acuminatum*); *Prodromus* SELLARDS, 1907, p. 349 (type, *P. rectus*); *Bantiska* SELLARDS, 1907, p. 349 (type, *B. elongata*); *Pinctodia* SELLARDS, 1907, p. 352 (type, *P. curta*); *Recter* SELLARDS, 1909, p. 151, *pro Rekter* SELLARDS, 1907, p. 349 (type, *R. arcuatus*); *Esca* SELLARDS, 1909, p. 151 (type, *Therates planus* SELLARDS, 1907, p. 354); *Mecus* SELLARDS, 1909, p. 151 (type, *Scopus gracilis* SELLARDS, 1907, p. 352); *Loxophlebia* MARTYNOV, 1928b, p. 8, *non* BUTLER, 1876 (type, *L. apicalis*)]. MP forked more deeply than RS. TILLYARD, 1932b; CAR-

PENTER, 1933a, 1979; ROHDENDORF, 1962a; GUTHÖRL, 1965; DEMOULIN, 1970b. *Perm.*, USA (Kansas, Oklahoma), ?Europe (Germany).——FIG. 15,*a–c*. **P. permianum*, Kansas; *a*, fore wing, *b*, base of hind wing, both ×3.5; *c*, reconstruction, dorsal view, ×1.5 (all Carpenter, 1933a).——FIG. 15,*d*. *P. americanum* DEMOULIN, Oklahoma; photograph of nymph, dorsal view, ×5 (Carpenter, new).

Family MISTHODOTIDAE Tillyard, 1932

[Misthodotidae TILLYARD, 1932b, p. 260] [=Eudoteridae DEMOULIN, 1954c, p. 561]

Adults small to moderate in size. Wings broadly oval; fore and hind wings similar in venation but hind wings distinctly broader; vein CUA unbranched, lacking a triad; crossveins less numerous than in Protereismatidae; legs of moderate length; tarsi with 4 segments; cerci and median caudal process very long. Nymphs with 9 pairs of tracheal gills. *Perm.*

Misthodotes SELLARDS, 1909, p. 151, *nom. subst. pro Dromeus* SELLARDS, 1907, p. 351, *non* REICHE, 1854 [**Dromeus obtusus* SELLARDS, 1907, p. 351; OD] [=*Eudoter* TILLYARD, 1936, p. 443 (type, *E. delicatulus*)]. Posterior margin of hind wing strongly convex. LAMEERE, 1917a; TILLYARD, 1932b; CARPENTER, 1933a, 1979; DEMOULIN, 1954c; TSHERNOVA, 1965. *Perm.*, USA (Kansas, Oklahoma), USSR (Asian RSFSR).——FIG. 14,*4a*. **M. obtusus* (SELLARDS), Kansas; fore wing, ×5.5 (Carpenter, 1933a).——FIG. 14,*4b*. *M. edmundsi* CARPENTER, Oklahoma; hind wing, ×5.5 (Carpenter, 1979).

Family JARMILIDAE Demoulin, 1970

[Jarmilidae DEMOULIN, 1970b, p. 7]

Little-known family (nymph only); mesothorax and metathorax nearly twice as broad as long; mesonotum larger than metanotum; tracheal gills narrow and elongate. *Perm.*

Jarmila DEMOULIN, 1970b, p. 7 [**J. elongata*; OD]. Diagnostic characters same as for family. *Perm.*, Europe (Czechoslovakia).

Family OBORIPHLEBIIDAE Hubbard & Kukalová-Peck, 1980

[Oboriphlebiidae HUBBARD & KUKALOVÁ-PECK, 1980, p. 29]

Little-known family (nymph only); mesothorax slightly larger than metathorax; wing pads divergent. *Perm.*

Oboriphlebia HUBBARD & KUKALOVÁ-PECK, 1980, p. 30 [**O. tertia*; OD]. Diagnostic characters same as for family. *Perm.*, Europe (Czechoslovakia).

Family PALINGENIOPSIDAE Martynov, 1938

[Palingeniopsidae MARTYNOV, 1938b, p. 35]

Little-known family, based on hind wing only; vein CUP strongly sigmoidal. *Perm.*

Palingeniopsis MARTYNOV, 1932, p. 10 [**P. praecox*; OD]. Little-known genus; intercalary veins incompletely known. MARTYNOV, 1938b; ROHDENDORF, 1962a. *Perm.*, USSR (European RSFSR).——FIG. 14,*1*. **P. praecox*; hind wing, ×1.7 (Martynov, 1932).

Family MESEPHEMERIDAE Lameere, 1917

[Mesephemeridae LAMEERE, 1917a, p. 47]

Little-known family. Fore and hind wings apparently similar in size and venation; inner and outer margins of fore wing forming a smooth curve; costal brace apparently absent. *Jur.*

Mesephemera HANDLIRSCH, 1906b, p. 600 [**Tineites lithophilus* GERMAR, 1842, p. 88; SD CARPENTER, herein]. Little-known genus; hind wings apparently at least as broad as fore wings. CARPENTER, 1932a; DEMOULIN, 1955b; TSHERNOVA, 1970. *Jur.*, Europe (Germany).

Family HEXAGENITIDAE Lameere, 1917

[*nom. transl.* DEMOULIN, 1954c, p. 566, *ex* Hexagenitinae LAMEERE, 1917a, p. 74] [=Paedephemeridae LAMEERE, 1917a, p. 49; Ephemeropsidae COCKERELL, 1927a, p. 1; Stenodicranidae DEMOULIN, 1954c, p. 567]

Mayflies of moderate to very large size. Fore wing triangular owing to well-developed tornus of hind margin; vein CUA of fore wing forked, one of its branches with a series of loop-shaped veinlets leading to wing margin. Nymphs with 7 pairs of gills along sides of abdomen. *Jur.–Cret.*

Hexagenites SCUDDER, 1880, p. 6 [**H. weyenberghi*; OD; =*Ephemera cellulosa* HAGEN, 1862, p. 115] [=*Paedephemera* HANDLIRSCH, 1906b, p. 601 (type, *Ephemera multivenosa* OPPENHEIM, 1888, p. 225)]. Adults of moderate size. Fore wing about twice as long as wide; MA1 and MA2 forming a symmetrical fork; few crossveins. CARPENTER, 1932a; TSHERNOVA, 1961; DEMOULIN, 1970c; TSHERNOVA & SINITSHENKOVA, 1974. *Jur.,* Europe (Germany).——FIG. 14,3. **H. weyenberghi;* fore wing, ×3.5 (Carpenter, 1932a).

Ephemeropsis EICHWALD, 1864, p. 21 [**E. tristalis;* OD] [=*Phacelobranchus* HANDLIRSCH, 1906b, p. 604 (type, *P. braueri*)]. Adults very large. Fore wing more than 2.5 times as long as its width; hind wing more than half as long as fore wing. PING, 1928; UENO, 1935; DEMOULIN, 1954a, 1956d; MESHKOVA, 1961; TSHERNOVA, 1961; SINITSHENKOVA, 1975. *Cret.,* USSR (Asian RSFSR).——FIG. 14,6. **E. tristalis; a,* restoration, ×0.8 (Tshernova, 1961); *b,* fore wing, ×2.0 (Tshernova & Sinitshenkova, 1974); *c,* nymph, restoration, dorsal view, ×2.0 (Meshkova, 1961).

Hexameropsis TSHERNOVA & SINITSHENKOVA, 1974, p. 131 [**H. selini;* OD]. Similar to *Hexagenites,* but MA1 and MA2 forming asymmetrical fork; hind wing less than half length of fore wing. SINITSHENKOVA, 1975. *Cret.,* USSR (Ukraine), Africa (Algeria).

Family SIPHLONURIDAE
Banks, 1900

[Siphlonuridae BANKS, 1900, p. 246]

Fore wings narrow and triangular; hind wings relatively large; crossveins numerous in both wings; vein CUA of fore wing connected to hind margin by several veinlets; forks of MP and CUA almost symmetrical. *Jur.–Holo.*

Siphlonurus EATON, 1868, p. 89. DEMOULIN, 1968c. *Oligo.,* Europe (Baltic)–*Holo.*

Baltameletus DEMOULIN, 1968c, p. 238 [**B. oligocaenicus;* OD]. Little-known genus, based on subimago; apparently related to *Ameletus* (recent). *Oligo.,* Europe (Baltic).

Balticophlebia DEMOULIN, 1968c, p. 237 [**B. hennigi;* OD]. Based on female imago; similar to *Chaquihua* (recent) but with hind wings more elongate. *Oligo.,* Europe (Baltic).

Cronicus EATON, 1871, p. 133 [**Baetis anomala* PICTET in PICTET & HAGEN, 1856, p. 75; OD]. Gonostyle of male subimago with 5 segments, the third about twice as long as the second and as long as segments 4 and 5 combined. DEMOULIN, 1955a, 1968c, 1974. *Oligo.,* Europe (Baltic). ——FIG. 16,2. **C. anomalus* (PICTET); fore and hind wings and part of body, dorsal view, ×3.5 (Demoulin, 1968c).

Isonychia EATON, 1871, p. 33. LEWIS, 1977b. *Oligo.,* USA (Montana)–*Holo.*

Oligisca DEMOULIN, 1970c, p. 6 [**Paedephemera schwertschlageri* HANDLIRSCH, 1906b, p. 602; OD]. Little-known genus, based on poorly preserved wing; similar to *Stackelbergisca,* but branches of CUA simple; MP with long branches. *Jur.,* Europe (Germany).

Proameletus SINITSHENKOVA, 1976, p. 86 [**P. caudata;* OD]. Imago: fore wing similar to that of *Oligisca* but with an intercalary vein between RS1 and RS2; median caudal process long, with 10 segments. Nymph: legs long and slender; 7 pairs of oval gills along abdomen. *Cret.,* USSR (Asian RSFSR).

Siphlurites COCKERELL, 1923d, p. 170 [**S. explanatus;* OD]. Little-known genus, apparently related to *Murphyella* (recent). DEMOULIN, 1970d, 1974. *Oligo.,* USA (Colorado).

Stackelbergisca TSHERNOVA, 1967, p. 323 [**S. sibirica;* OD]. Imago: fore wing triangular; anal margin long; CUA straight and connected to wing margin by a series of veinlets; CUP slightly curved. Nymph: with 7 pairs of foliate gills along sides of abdomen. DEMOULIN, 1968a. *Jur.,* USSR (Asian RSFSR).——FIG. 16,3. **S. sibirica; a,* fore wing, *b,* nymph, dorsal view, both ×3.5 (Tshernova, 1967).

Family AMETROPODIDAE
Bengtsson, 1913

[Ametropodidae BENGTSSON, 1913, p. 305]

Fore tarsi of male very long; hind tarsi with 4 segments; basal tarsal segment fused to tibia; fore wing with only 1 or 2 unattached cubital intercalaries; vein 1A of fore wing connected to hind margin by several veinlets. *Oligo.–Holo.*

Ametropus ALBARDA, 1878, p. 129. *Holo.*

Brevitibia DEMOULIN, 1968c, p. 245 [**B. intricans;* OD]. Similar to *Ametropus* (recent), with shorter median caudal process. *Oligo.,* Europe (Baltic).

Metretopus EATON, 1901, p. 253. DEMOULIN, 1968c. *Oligo.,* Europe (Baltic)–*Holo.*

Siphloplecton CLEMENS, 1915, p. 245. DEMOULIN, 1968c, 1970a. *Oligo.,* Europe (Baltic)–*Holo.*

Family BAETIDAE Leach, 1815

[Baetidae LEACH, 1815, p. 137]

Eyes of males divided; fore wing with veins IMA, MA2, IMP, and MP2 detached basally; hind wing reduced or absent; median caudal process absent. *Plio.–Holo.*

Baetis LEACH, 1815, p. 137. *Holo.*

Cleon LEACH, 1815, p. 137. [Generic assignment of fossil (nymph) doubtful.] RIEK, 1954b. *Plio.,* Australia (New South Wales)–*Holo.*

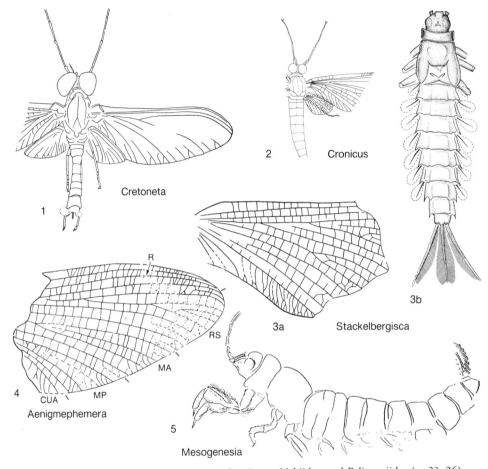

Fig. 16. Siphlonuridae, Aenigmephemeridae, Leptophlebiidae, and Palingeniidae (p. 23–26).

Family EPEOROMIMIDAE Tshernova, 1969

[Epeoromimidae Tshernova, 1969, p. 154]

Known only from nymphs, apparently related to the Heptageniidae. Head and thorax short; abdomen elongate; legs thin and short; abdomen 3 times as long as thorax; fifth abdominal segment 2 or 3 times as wide as long; 7 pairs of gill plates along sides of abdomen. *Jur.–Cret.*

Epeoromimus Tshernova, 1969, p. 155 [*E. kuzlauskasi*; OD]. Anterior margin of pronotum strongly concave; gill plates long and foliaceous. Sinitshenkova, 1976. *Jur.–Cret.*, USSR (Asian RSFSR).——Fig. 14,5. *E. kuzlauskasi*; dorsal view of nymph, ×5.5 (Tshernova, 1969).

Family HEPTAGENIIDAE Needham, 1901

[Heptageniidae Needham in Needham & Betten, 1901, p. 419]

Cubitus of fore wing with 2 pairs of intercalary veins; MP1 and MP2 forming symmetrical fork; hind tarsi with 5 segments; median caudal process absent. *Oligo.–Holo.*

Heptagenia Walsh, 1863, p. 197. Demoulin, 1968c, 1970a. *Oligo.*, Europe (Baltic)–*Holo.*
Cinygma Eaton, 1885, p. 236. [Generic assignment of fossil doubtful.] Demoulin, 1968c. *Oligo.*, Europe (Baltic)–*Holo.*
Electrogenia Demoulin, 1956a, p. 95 [*E. dewalschei*; OD]. MA of hind wing unbranched; crossveins dense over fore wing; third tarsi with first segment longer than second; gonostyle with 4 segments. *Oligo.*, Europe (Baltic).

Miocoenogenia Tshernova, 1962, p. 943 [*M. gorbunovi*; OD]. Little-known genus, nymph only; similar to *Heptagenia* but with relatively small head; pronotum broad, with anterior angles projecting forward. *Mio.*, USSR (Asian RSFSR).

Rhithrogena Eaton, 1881, p. 23. Demoulin, 1968c. *Oligo.*, Europe (Baltic)–*Holo.*

Succinogenia Demoulin, 1965, p. 151 [*S. larssoni*; OD]. Little-known genus, based on young nymph. *Oligo.*, Europe (Baltic).

Family AENIGMEPHEMERIDAE Tshernova, 1968

[Aenigmephemeridae Tshernova, 1968, p. 23]

Apparently related to the Heptageniidae. Fore wing narrow, inner and outer margins forming a smooth curve; longitudinal veins straight, almost equidistant from each other; fork of MA very deep. *Jur.*

Aenigmephemera Tshernova, 1968, p. 23 [*A. demoulini*; OD]. Fore wing with forking of MA at level of origin of RS; 5 longitudinal veins between MA2 and CUA; crossveins numerous. Demoulin, 1969a. *Jur.*, USSR (Kazakh).——Fig. 16,4. *A. demoulini*; fore wing, ×3 (Tshernova, 1968).

Family LEPTOPHLEBIIDAE Banks, 1900

[Leptophlebiidae Banks, 1900, p. 246]

Eyes of male divided; 2 to 4 long intercalary veins between veins CUA and CUP; CUP usually strongly curved; median caudal process present. *Jur.–Holo.*

Leptophlebia Westwood, 1840, p. 31. *Holo.*
Atalophlebia Eaton, 1881, p. 193. Etheridge & Olliff, 1890; Riek, 1954b. *Plio.*, Australia (New South Wales)–*Holo.*
Blasturophlebia Demoulin, 1968c, p. 268 [*B. hirsuta*; OD]. Little-known genus, based on a subimaginal exuvium of a male. [Family assignment doubtful.] Hubbard & Savage, 1981. *Oligo.*, Europe (Baltic).
Cretoneta Tshernova, 1971, p. 614 [*C. zherichini*; OD]. Fore wing with MA about half length of stem M; base of MP2 connected to MP1; cubital area very narrow; eyes of male not divided. Hubbard & Savage, 1981. *Cret.*, USSR (Asian RSFSR).——Fig. 16,1. *C. zherichini*; dorsal view, male, ×10 (Tshernova, 1971).
Lepismophlebia Demoulin, 1968b, p. 7 [*Lepisma platymera* Scudder, 1890, p. 102; OD]. Little-known nymph. [Family assignment doubtful.] Demoulin, 1956b. *Oligo.*, USA (Colorado).
Mesoneta Brauer, Redtenbacher, & Ganglbauer, 1889, p. 4 [*M. antiqua*; OD]. Little-known genus, nymph only. Head small, thorax very short; femur longer than tibia; 7 pairs of tracheal gills along sides of abdomen. Tshernova, 1971; Sinitshenkova, 1976. *Jur.–Cret.*, USSR (Asian RSFSR).
Paraleptophlebia Lestage, 1917, p. 340 [=*Oligophlebia* Demoulin, 1965, p. 146 (type, *O. calliarcys*)]. Demoulin, 1968c, 1970a; Hubbard & Savage, 1981. *Oligo.*, Europe (Baltic)–*Holo.*
Xenophlebia Demoulin, 1968c, p. 267 [*X. aenigmatica*; OD]. Only male adult known. Forking of MA and MP in fore wing symmetrical; median caudal process absent. Demoulin, 1970a; Hubbard & Savage, 1981. *Oligo.*, Europe (Baltic).

Family EPHEMERELLIDAE Klapálek, 1909

[Ephemerellidae Klapálek, 1909, p. 13]

Fore wing with 1 or 2 long intercalary veins between veins MP and CUA and usually with detached marginal intercalary veins; crossveins usually absent or very weak. *Jur.–Holo.*

Ephemerella Walsh, 1862, p. 377. *Holo.*
Philolimnias Hong, 1979, p. 336 [*P. sinica*; OD]. Similar to Ephemerella (recent), but costal area narrower and CUA1 with 5 branches. *Eoc.*, China (Liaoning).
Turfanerella Demoulin, 1954a, p. 324 [*Ephemeropsis tingi* Ping, 1935, p. 107; OD]. Little-known genus, based on nymph. *Jur.*, China (Sinkiang).

Family BEHNINGIIDAE Motas & Bocasco, 1938

[Behningiidae Motas & Bocasco, 1938, p. 25]

Legs of adults much reduced; forelegs of nymphs resembling palpi; middle and hind legs modified to protect the tracheal gills; gills ventral. *Jur.–Holo.*

Behningia Lestage, 1930, p. 436. *Holo.*
Archaeobehningia Tshernova, 1977, p. 94 [*A. edmundsi*; OD]. Little-known genus, based on nymph. Similar to *Protobehningia* (recent) but with claws present on all tarsi, and forelegs not functionally part of trophi. *Jur.*, USSR (Asian RSFSR).

Family NEOEPHEMERIDAE Needham, 1935

[Neoephemeridae NEEDHAM in NEEDHAM, TRAVER, & HSU, 1935, p. 288]

Adults resembling those of ephemerids, but crossveins in basal half of fore wing weak or atrophied; costal angle of hind wing acute. Nymphs as in caenids but gills operculate, fused medially. *Oligo.–Holo.*

Neoephemera McDUNNOUGH, 1925, p. 168. *Holo.*
Potamanthellus LESTAGE, 1930, p. 120. LEWIS, 1977b. *Oligo.*, USA (Montana)–*Holo.*

Family EPHEMERIDAE Leach, 1815

[Ephemeridae LEACH, 1815, p. 137]

Legs well developed; veins MP2 and CUA abruptly diverging from MP1 basally; 1A unbranched but connected to hind margin of wing by at least 3 veinlets. *Oligo.–Holo.*

Ephemera LINNÉ, 1758, p. 546. COCKERELL, 1908e. *Oligo.*, USA (Colorado)–*Holo.*

Family POLYMITARCIDAE Banks, 1900

[Polymitarcidae BANKS, 1900, p. 246]

Adults as in Euthyplociidae (recent) but with veins MP2 and CUA strongly divergent from MP1 basally; middle and hind legs weakly developed. Nymphs with fossorial legs; gills dorsal. *Mio.–Holo.*

Ephoron WILLIAMSON, 1802, p. 71. *Holo.*
Asthenopodichnium THENIUS, 1979, p. 185 [*A. xylobiontum*; OD]. Trace fossils; burrows in fossil wood, resembling those now made by polymitarcid nymphs. *Mio.*, Europe (Austria).

Family PALINGENIIDAE Selys-Longchamps, 1888

[Palingeniidae SELYS-LONGCHAMPS, 1888, p. 147]

Main veins of fore wings arranged in pairs, converging at wing margin; crossveins numerous; forelegs of nymphs flattened and fossorial in nature; tibiae toothed. *Jur.–Holo.*

Palingenia BURMEISTER, 1839, p. 802. *Holo.*
Mesogenesia TSHERNOVA, 1977, p. 92 [*M. petersae*; OD]. Little-known genus, nymph only; similar to *Heterogenesia* (recent), with very short mandibles and lacking a distinct frontal process. *Jur.*, USSR (Asian RSFSR).——FIG. 16,5. *M. petersae*; lateral view of nymph, ×7.5 (Tshernova, 1977).

Family UNCERTAIN

The following genera, apparently belonging to the order Ephemeroptera, are too poorly known to permit assignment to families.

Aphelophlebodes PIERCE, 1945, p. 3 [*A. stocki*; OD]. Little-known genus, based on small fragment of wing. [Type of family Aphelophlebodidae and order Aphelophlebia PIERCE, 1945.] CARPENTER, 1960b; DEMOULIN, 1962. *Mio.*, USA (California).
Mesobaetis BRAUER, REDTENBACHER, & GANGLBAUER, 1889, p. 5 [*M. sibirica*; OD]. Little-known nymph. DEMOULIN, 1954a, 1968b; ROHDENDORF, 1962a; TSHERNOVA, 1970; HUBBARD & SAVAGE, 1981. *Jur.*, USSR (Asian RSFSR).
Mesoplectopteron HANDLIRSCH, 1918, p. 112 [*M. longipes*; OD]. Little-known genus, based on nymph. *Trias.*, Europe (Germany).
Parabaetis HAUPT, 1956, p. 32 [*P. eocaenicus*; OD]. Little-known genus, based on small fragment of wing. DEMOULIN, 1957. *Eoc.*, Europe (Germany).
Phthartus HANDLIRSCH, 1904b, p. 6 [*P. rossicus* HANDLIRSCH, 1904b, p. 6; SD CARPENTER, herein]. Little-known genus, based on nymph. HANDLIRSCH, 1906b. *Perm.*, USSR (Asian RSFSR).
Protoligoneuria DEMOULIN, 1955d, p. 270 [*P. limai*; OD]. Little-known genus, based on nymph only, possibly related to Oligoneuridae. COSTA LIMA, 1950. *Paleoc.–Plio.*, Brazil.

Order PALAEODICTYOPTERA Goldenberg, 1877

[Palaeodictyoptera GOLDENBERG, 1877, p. 8] [=Protohemiptera HANDLIRSCH, 1906b, p. 387; Synarmogoidea HANDLIRSCH, 1919b, p. 28; Protocicadida HAUPT (in part), 1941, p. 75; Archaehymenoptera HAUPT, 1941, p. 102; Archodonata MARTYNOV, 1932, p. 12; Anisaxia FORBES, 1943, p. 403; Hemiodonata ZALESSKY, 1946a, p. 63; Breyerida HAUPT, 1949, p. 23; Eopalaeodictyoptera LAURENTIAUX, 1952a, p. 234; Eubleptidodea LAURENTIAUX, 1953, p. 423; Syntonopterodea LAURENTIAUX, 1953, p. 425; Dictyoneurida ROHDENDORF, 1977, p. 20; Permothemistida SINITSHENKOVA, 1980a, p. 49]

Palaeoptera of moderate to very large size. Wings containing all main veins, including MA, MP, CUA, and CUP, with alternation of convexities and concavities; main veins usually without coalescence and always arising independently; area between veins with a delicate, irregular network (archedictyon) or with true crossveins, or with a combination of both; intercalary veins present in a very few families (e.g., Syntonopteridae); fore and hind wings similar in form and venation in some families (e.g., Dictyoneuridae); in others (e.g., Spilapteridae) hind wings much broader than fore pair with basic venational pattern remaining the same; in some others (e.g., Eugereonidae and Megaptilidae) hind wings only about half as long as fore wings; in one family (Diathemidae) hind wings minute, in a related family (Permothemistidae) hind wings completely absent; front margin of wing commonly serrate, costa with or without setae; wings in some families with prominent pigment markings. Antennae setaceous, usually of moderate length but may be long and threadlike; head typically small, hypognathous (in some slightly prognathous), with prominent eyes, and with well-developed haustellate beak, enclosing 5 stylets derived from mandibles, maxillae, and presumably hypopharynx; maxillary palpi usually well developed, labial palpi apparently absent. Thoracic segments typically subequal, but prothorax in most species with a pair of lateral winglike lobes, usually membranous and commonly with veinlike supports; legs (known in very few genera) short, with 5 tarsal segments; abdomen of moderate length, slender, segments showing little differentiation; in some species pleurites apparently separated from tergites by longitudinal ridges; in others tergites strongly sclerotized and bearing lateral extensions; cerci long and multisegmented in both sexes, densely covered with hairs; ovipositor broad and short, strongly curved. Nymphs apparently terrestrial, without indications of aquatic modifications; mouthparts haustellate like those of adults; wing pads of nymphs held in an oblique-lateral position, independent of each other in all stages (so far as known), and articulated to thorax like wings of adult.

The food of nymphs and adults of the Palaeodictyoptera has been the subject of much speculation (SHAROV, 1973). It seems virtually certain that their mouthparts were adapted for obtaining liquid food from plants. Those with short beaks could have fed on the juices of foliage; those with longer beaks may have fed on contents of the developing inflorescences of the Cordaitales, which were abundant in Late Carboniferous and Permian forests. *U. Carb.–Perm.*

The Palaeodictyoptera comprise one of the major orders of the Upper Carboniferous (beginning with the Namurian) and to a lesser extent of the Permian. During the past 20 years our knowledge of the order has been greatly extended, and our present concept of the group is far different from that given by HANDLIRSCH in 1920. However, the classification of the Palaeodictyoptera is necessarily an arbitrary one to a large extent. Eighty-one genera, placed in twenty families, are recognized here, along with another forty-odd genera that are too poorly known for assignment to family. Most of these genera are based almost entirely on wings, details of the body structures being only rarely preserved. The chief difficulty in developing a satisfactory phylogenetic classification is the lack of enough material (specimens and species) to permit evaluation of the characteristics of the several levels of taxa within the order. A few groups of related families can readily be recognized (e.g., the Eugereonidae, Archaemegaptilidae, and Calvertiellidae, for one group; and the Permothemistidae and Diathemidae, for another), but there is not enough evidence to support the designation of a series of suborders or superfamilies. For the same reason it is difficult to determine with confidence the evolutionary level of the families within the order. The Dictyoneuridae, which have homonomous wings with a dense archedictyon, and which are known almost exclusively from the Upper Carboniferous (including the Namurian), appear to be the least specialized of the families now known. The most obvious specialization among the Palaeodictyoptera is the reduction of the hind wings, which occurs in the Eugereonidae and the Megaptilidae, as well as in the Diathemidae and Permothemistidae.

The Palaeodictyoptera are obviously closely related to the Megasecoptera, which have similar haustellate beaks and many other morphological features of the Palaeodictyoptera. Indeed, as more genera of these orders become known, distinctions between them are increasingly difficult to find, except for the nature of the articular plates (pteralia). Ultimately, these two orders will probably be merged into one, without even subordinal separation. The Diaphanopterodea also share the haustellate mouthparts and several other structural characters with the Palaeodictyoptera and Megasecoptera but are isolated from them by their unique ability, as Palaeoptera, to fold their wings over the abdomen at rest (CARPENTER & RICHARDSON, 1971; SHAROV, 1973; KUKALOVÁ-PECK, 1974b).

Family DICTYONEURIDAE Handlirsch, 1906

[Dictyoneuridae HANDLIRSCH, 1906a, p. 670]

Fore wing moderately slender, apex slightly pointed; costal area often broad up to midwing; main veins without coalescence; vein SC terminating well beyond midwing; R ending near apex; RS with several branches; MA unbranched, usually strongly curved; MP with or without branches; CUA unbranched; CUP with or without branches; archedictyon well developed over most of wing, usually dense but rarely coarse. Hind wing usually similar to fore wing but costal area narrower. Head small; antennae multisegmented; prothoracic lobes relatively large; legs short, with 5 tarsal segments; cerci long and multisegmented; ovipositor short and curved; males with claspers. *U. Carb.–Perm.*

Dictyoneura GOLDENBERG, 1854, p. 33 [*D. libelluloides*; OD]. Hind wing broad, with strongly curved hind margin; RS dichotomously branched; MP with 4 branches. HANDLIRSCH, 1906b; GUTHÖRL, 1934. *U. Carb.,* Europe (Germany).——FIG. 17,*9.* *D. libelluloides*; hind wing, ×0.8 (Guthörl, 1934).

Cleffia GUTHÖRL, 1931, p. 91 [*C. sarana*; OD] [=*Pseudocleffia* GUTHÖRL, 1940, p. 48 (type, *P. palatina*)]. Little-known genus. Wings slender; CUP with 2 branches. GUTHÖRL, 1934. *U. Carb.,* Europe (Germany).——FIG. 17,*12.* *C. sarana*; wing, ×1.5 (Guthörl, 1934).

Dictyoneurula HANDLIRSCH, 1906b, p. 75 [*Dictyoneura gracilis* KLIVER, 1886, p. 107; SD HANDLIRSCH, 1922, p. 30]. Little-known genus, apparently similar to *Microdictya* but anal area narrower. GUTHÖRL, 1934. *U. Carb.,* Europe (Germany).——FIG. 17,*8.* *D. gracilis* (KLIVER); wing, ×1 (Guthörl, 1934).

Goldenbergia SCUDDER, 1885a, p. 170 [*Dictyoneura elongata* GOLDENBERG, 1877, p. 50; SD HANDLIRSCH, 1906b, p. 71]. Fore wing elongate; costal margin moderately curved; SC extending nearly to wing apex; R and RS contiguous at wing base but apparently not fused; MA separating from MP at basal one-fourth of wing; MP deeply forked; CUA and CUP unbranched; several anal veins; dense archedictyon over wing surface. GUTHÖRL, 1934; SHAROV & SINITSHENKOVA, 1977. *U. Carb.,* Europe (Germany); *Perm.,* USSR (Kazakh).

Kallenbergia GUTHÖRL, 1930, p. 147 [*K. handlirschi*; OD]. Little-known genus, based on wing fragment. Wing broad, but narrowed basally; posterior margin strongly curved; RS pectinately branched. GUTHÖRL, 1934. *U. Carb.,* Europe (Germany).——FIG. 17,*4.* *K. handlirschi*; wing, ×1.5 (Guthörl, 1934).

Macrodictya GUTHÖRL, 1940, p. 46 [*M. stenomediali*; OD]. Little-known genus, based on wing fragment. RS with 5 terminal branches; MA very close to MP. *U. Carb.,* Europe (Germany).

Microdictya BRONGNIART, 1893 (atlas), p. 28, *nom. subst. pro Heeria* BRONGNIART, 1893, p. 388, *non* SCUDDER, 1890 [*Heeria vaillanti* BRONGNIART, 1893, p. 399; SD HANDLIRSCH, 1922, p. 25]. Fore and hind wings similar, broadest at about middle; posterior margin of hind wing smoothly curved; RS dichotomously forked; MP forked at least once; CUP usually branched; archedictyon nearly uniform over wings. LAURENTIAUX & TEIXEIRA, 1958a; KUKALOVÁ, 1970. *U. Carb.,* Europe (France).——FIG. 17,*1.* *M. hamyi* BRONGNIART; fore wing, ×0.8 (Kukalová, 1970).

Polioptenus SCUDDER, 1885a, p. 170 [*Dictyoneura elegans* GOLDENBERG, 1877, p. 9; OD] [=*Acanthodictyon* HANDLIRSCH, 1906b, p. 72 (type, *Termes decheni* GOLDENBERG)]. Similar to *Stenodictyoneura*, but RS with 4 terminal branches. GUTHÖRL, 1934. *U. Carb.,* Europe (Germany).——FIG. 17,*10.* *P. elegans* (GOLDENBERG); wing, ×1 (Guthörl, 1934).

Rotundopteris GUTHÖRL, 1940, p. 44 [*R. multimediali*; OD]. Little-known genus. RS with only 3 branches; MA and branches of MP strongly curved. *U. Carb.,* Europe (Germany).——FIG. 17,*11.* *R. multimediali*; wing, ×1 (Guthörl, 1940).

Sagenoptera HANDLIRSCH, 1906b, p. 72 [*Termes formosus* GOLDENBERG, 1854, p. 30; OD] [=*Arltia* GUTHÖRL, 1934, p. 56 (type, *Dictyoneura schmitzi* GOLDENBERG)]. Little-known genus. Posterior margin of wing smoothly curved; RS dichotomously branched in distal third of wing; MP dichotomously branched. BRAUCKMANN & HAHN, 1983. *U. Carb.,* Europe (Germany).——

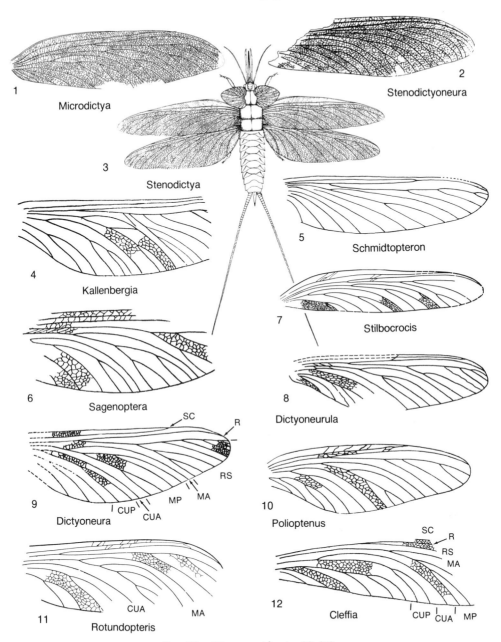

FIG. 17. Dictyoneuridae (p. 28–30).

FIG. 17,6. *S. formosa* (GOLDENBERG); wing, ×1 (Guthörl, 1934).

Schmidtopteron BRAUCKMANN & HAHN, 1978, p. 14 [**S. adictyon*; OD]. Hind(?) wing with costal area very narrow; SC terminating about two-thirds wing length from base, just beyond level of first fork of RS; RS arising in basal quarter of wing, with 3 dichotomous forks in distal area; MA and MP separating at about level of origin of RS; MA, CUA, and CUP unbranched; MP forked at about half of its length; CUA and CUP nearly straight, not strongly curved toward hind margin; 3 to 4 anal veins. *U. Carb.*, Europe (Germany).——Fig. 17,5. **S. adictyon*; hind(?) wing, ×1.8 (Brauckmann & Hahn, 1978).

Siberiodictya Sharov & Sinitshenkova, 1977, p. 51 [**S. gigas*; OD]. Little-known genus, apparently similar to *Stenodictya*. Origin of RS, separation of MA and MP, and division of CUA and CUP occurring near wing base and at almost same level. *U. Carb.*, USSR (Asian RSFSR).

Stenodictya Brongniart, 1893, p. 383, *nom. subst. pro Scudderia* Brongniart, 1885a, p. 61, *non* Grote, 1873 [**Scudderia lobata* Brongniart, 1885a, p. 61; SD Handlirsch, 1922, p. 24]. Wings: RS arising at about midwing, its branches pectinate; MA, CUA, and CUP unbranched; MP usually unbranched; anal area broad; archedictyon irregular; costal area with thin but dense, regular crossveins. Laurentiaux, 1952a; Kukalová, 1970. *U. Carb.*, Europe (France). ——Fig. 17,3. *Stenodictya*; reconstruction, ×0.8 (Kukalová, 1970).

Stenodictyoneura Leriche, 1911, p. 195 [**S. belgica*; OD]. Little-known genus. RS with 8 terminal branches; MP and CUP with 4 branches. *U. Carb.*, Europe (Belgium).——Fig. 17,2. **S. belgica*; wing, ×0.7 (Leriche, 1911).

Stilbocrocis Handlirsch, 1906b, p. 74 [**Termes heeri* Goldenberg, 1854, p. 29; OD] [=*Longivenapteris* Guthörl, 1940, p. 52 (type, *L. pulchra*)]. Little-known genus. Wings very slender; RS dichotomously branched; CUP with 3 branches. *U. Carb.*, Europe (Germany).——Fig. 17,7. *S. lanceolata* Guthörl; fore wing, ×1 (Guthörl, 1940).

Family LITHOMANTEIDAE
Handlirsch, 1906

[*nom. correct.* Carpenter in Brues, Melander, & Carpenter, 1954, p. 790, *pro* Lithomantidae Handlirsch, 1906a, p. 673] [=Macropteridae Laurentiaux, 1949b, p. 217; Lusiellidae Laurentiaux & Teixeira, 1958a, p. 6]

Fore wing: anterior margin and stems of main veins strongly curved basally; veins MA and CUA unbranched; crossveins slightly irregular, but mostly unbranched. Hind wing much broader than fore wing basally. *U. Carb.*

Lithomantis Woodward, 1876, p. 63 [**L. carbonaria*; OD] [=*Hadroneuria* Handlirsch, 1906b, p. 84 (type, *Gryllacris bohemica* Novák, 1880, p. 69); *Lithosialis* Scudder, 1881a, p. 167 (type, *Corydalis brongniarti* Mantell, 1839, p. 680)]. Fore wing with CUP more extensive than MP. Kukalová, 1969c. *U. Carb.*, England, Europe (Holland, Czechoslovakia).——Fig. 18,1. **L. carbonaria*, England; dorsal view, ×0.6 (Woodward, 1876).

Lusiella Laurentiaux & Teixeira, 1958a, p. 6, *nom. subst. pro Macroptera* Laurentiaux, 1949b, p. 217, *non* Lioy, 1863 [**Lusiella fariai*; OD]. Similar to *Lithomantis*, but hind wing much broader than fore wing and crossveins in both wings fewer and more regular. Kukalová, 1969c. *U. Carb.*, Europe (France, Portugal).——Fig. 18,2. **L. fariai*, Portugal; hind wing, ×0.6 (Laurentiaux & Teixeira, 1958a).

Synarmoge Handlirsch, 1910d, p. 250 [**S. ferrarii*; OD] [=*Climacoptera* Laurentiaux, 1949b, p. 214, *non* Redtenbacher, 1895 (type, *C. antiqua*)]. Fore wing with posterior margin of anal area concave; crossveins diversely formed but mostly unbranched. Handlirsch, 1919b; Laurentiaux, 1953; Kukalová, 1969c. *U. Carb.*, Europe (Germany, France).——Fig. 18,3. **S. ferrarii*; wing, ×0.6 (Laurentiaux, 1953).

Family MEGAPTILIDAE
Handlirsch, 1906

[Megaptilidae Handlirsch, 1906b, p. 80] [=Lithoptilidae Handlirsch, 1922, p. 44]

Fore wing large and broad; vein SC apparently extending nearly to wing apex; RS with at least 5 terminal branches; branches of M and CU strongly curved toward posterior wing margin; MA unbranched; CUP with at least one fork; crossveins very dense and reticulate. Hind wing apparently much shorter than fore wing, with reduced branching of RS and MP. *U. Carb.*

Megaptilus Brongniart, 1885a, p. 61 [**M. blanchardi*; SD Handlirsch, 1906b, p. 80]. Fore wing with RS3+4 arising before level of midwing; M more extensively branched than CU. Brongniart, 1893; Lameere, 1917b; Kukalová, 1969c. *U. Carb.*, Europe (France).

Lithoptilus Lameere, 1917b, p. 157 [**Archaeoptilus boulei* Meunier, 1909a, p. 131; OD] [=*Anaxion* Handlirsch, 1919b, p. 529, obj.]. Little-known genus, based on distal part of hind wing. Wing apparently short and broad; costal area broad; RS short, with only 6 terminal branches; MP with 2 forks; crossveins numerous, coarsely reticulate. Kukalová, 1969c. *U. Carb.*, Europe (France).——Fig. 18,4. **L. boulei*; hind wing, ×0.8 (Kukalová, 1969c).

Family ARCHAEMEGAPTILIDAE
Handlirsch, 1919

[Archaemegaptilidae Handlirsch, 1919b, p. 523]

Little-known family, based on a hind(?) wing fragment. Vein SC long, extending almost to wing apex; R very close to SC except at wing base; RS arising about one-quarter wing length from base; M dividing a short distance from level of origin of RS. U. Carb.

Archaemegaptilus MEUNIER, 1908g, p. 174 [*A. kiefferi; OD]. Hind(?) wing: RS with 5 terminal branches; MA unbranched; MP with 6 terminal branches; CUP with at least 3 terminal branches; crossveins coarsely reticulate. LAMEERE, 1917b; KUKALOVÁ, 1969c. U. Carb., Europe (France).

Family EUGEREONIDAE Handlirsch, 1906

[Eugereonidae HANDLIRSCH, 1906b, p. 389] [=Peromapteridae HANDLIRSCH, 1906b, p. 79; Dictyoptilidae LAMEERE, 1917b, p. 191]

Fore wing long and narrow; precostal area present; furrow extending from anal area to vein R; SC ending at or nearly at wing apex; stems of R and M independent at base of wing but coalesced for short distance beyond that; MA unbranched; MP usually with 4 branches; CUA typically unbranched; CUP with one fork; anal area extending about one-third wing length from base; crossveins forming dense pattern with much reticulation. Hind wing distinctly shorter than fore wing, with stem of M strongly curved toward posterior margin; CUA recurved toward anal area. Beak long; pronotal lobes small but distinct. U. Carb.–Perm.

Eugereon DOHRN, 1866, p. 333 [*E. boeckingi; OD]. Little-known genus, represented by bases of wings and details of head and thorax. Fore wing with 1A not strongly curved toward posterior wing margin. Hind wing with area between stem RS and MA relatively broad. HANDLIRSCH, 1906b; GUTHÖRL, 1934; LAMEERE, 1935; CARPENTER, 1964a; KUKALOVÁ, 1969c; MÜLLER, 1978a. Perm., Europe (Germany).——FIG. 19. *E. boeckingi; a, dorsal view of head, thorax, and wing bases, ×0.5 (Laurentiaux, 1953); b, fore wing base, ×1; c, hind wing base, ×1 (both Carpenter, 1964a).

Dictyoptilus BRONGNIART, 1893, p. 390 [*D. renaulti; OD] [=Cockerelliella MEUNIER, 1909a, p. 132, nom. subst. pro Cockerellia MEUNIER, 1908b, p. 154, non ASHMEAD, 1898 (type, C. peromapteroides)]. Fore wing similar to that of Eugereon, but 1A strongly curved toward posterior wing margin. Hind wing with area between stems of RS and M narrow. LAMEERE, 1917b; HANDLIRSCH, 1919b; CARPENTER, 1964a; KUKALOVÁ, 1969c. U. Carb., Europe (France). ——FIG. 20,2a. D. sepultus (MEUNIER); fore wing, ×1 (Kukalová, 1969c).——FIG. 20,2b. D. per-

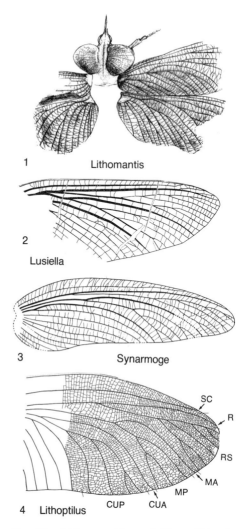

FIG. 18. Lithomanteidae and Megaptilidae (p. 30).

omapteroides (MEUNIER); fore and hind wings, ×0.8 (Kukalová, 1969c).

Peromaptera BRONGNIART, 1893, p. 391 [*P. filholi; OD]. Fore wing with CUA forked; MP with 3 terminal branches. Hind wing only about half as long as fore wing and much broader. MARTYNOV, 1931a; KUKALOVÁ, 1969c. U. Carb., Europe (France).

Sandiella CARPENTER, 1970, p. 405 [*S. readi; OD]. Similar to Dictyoptilus but with a coarser reticulation of crossveins and without rows of crossveins between R and RS, and R and SC; SC ending well before apex of wing. U. Carb., USA (New Mexico).——FIG. 20,1. *S. readi; fore wing, ×2 (Carpenter, 1970).

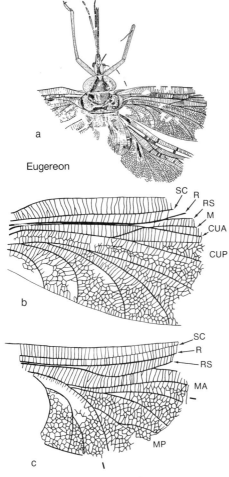

Fig. 19. Eugereonidae (p. 31).

stem R and M separate at wing base but contiguous near origin of RS; M forking near level of RS; MA unbranched, MP branched; stem CU independent of M; CUA diverging toward M near origin of RS and then running close to posterior branch of MP; 6 anal veins; crossveins numerous, with some reticulation. Hind wing very broad basally; anterior margin concave; venation essentially as in fore wing except for modifications associated with wing shape. Body unknown. *U. Carb.–Perm.*

Calvertiella TILLYARD, 1925b, p. 43 [**C. permiana*; OD]. Fore wing moderately slender; CUA coalesced with M for short distance just before origin of MP; MP forked twice, with 3 branches; crossveins in distal part of wing coarsely reticulate. KUKALOVÁ, 1964a. *Perm.*, USA (Kansas).——FIG. 21,2. **C. permiana*; fore wing, ×2 (Kukalová, 1964a).

Carrizopteryx KUKALOVÁ-PECK in KUKALOVÁ-PECK & PECK, 1976, p. 87 [**C. arroyo*; OD]. Hind wing apparently similar to that of *Moravia*, but stems of R, M, and CU fused from base to point of divergence of the three veins, just before origin of RS. *U. Carb.*, USA (New Mexico).

Moravia KUKALOVÁ, 1964a, p. 159 [**M. convergens*; OD]. Fore wing broad, nearly oval, apex broadly rounded, anterior margin convex; MP forked 3 or 4 times, with at least 4 branches; crossveins in distal area finely reticulate. Hind wing much broader basally than fore wing; crossveins reticulate in basal region also. Nymphal wings and presumed subimaginal wings with venation like that of imaginal forms. KUKALOVÁ-PECK & PECK, 1976; CARPENTER, 1979. *Perm.*, Europe (Czechoslovakia), USA (Oklahoma).——FIG. 21,7. **M. convergens*; *a*, fore and *b*, hind wings, ×1.5 (Kukalová, 1964a).

Moraviptera KUKALOVÁ, 1955a, p. 547 [**M. reticulata*; OD]. Little-known genus, based on apical wing fragment. Similar to *Moravia*, but wing much narrower and with pointed apex. KUKALOVÁ, 1964a. *Perm.*, Europe (Czechoslovakia).

Sharovia SINITSHENKOVA in SHAROV & SINITSHENKOVA, 1977, p. 61 [**S. sojanica*; OD]. Little-known genus, based on apical portions of fore and hind wings. Apex of wing more pointed than in *Moravia*; anterior border of wing more nearly straight; MP with 4 branches; CUA only slightly curved distally. *Perm.*, USSR (European RSFSR).——FIG. 21,9. **S. sojanica*; fore wing, ×1.2 (Sharov & Sinitshenkova, 1977).

Valdeania TEIXEIRA, 1941, p. 15 [**V. medeirosi*; OD]. Little-known genus. Fore wing elongate, front and hind margins nearly parallel for most of their lengths; R terminating at apex of wing; RS with 6 terminal branches; MA unbranched; MP with 3 terminal branches. Hind wing only half as long as fore wing but much broader; all veins, including R, strongly curved toward hind margin. LAURENTIAUX, 1953. *U. Carb.*, Europe (Portugal).

Family CALVERTIELLIDAE
Martynov, 1931

[Calvertiellidae MARTYNOV, 1931b, p. 146]

Fore wing with vein SC terminating on R just beyond midwing; RS originating in basal third of wing, with 3 or 4 main branches;

Family LYCOCERCIDAE
Handlirsch, 1906

[Lycocercidae HANDLIRSCH, 1906a, p. 675] [=Polycreagridae HANDLIRSCH, 1906b, p. 110; Apopappidae LAMEERE, 1917a, p. 42; Patteiskyidae LAURENTIAUX, 1958, p. 302]

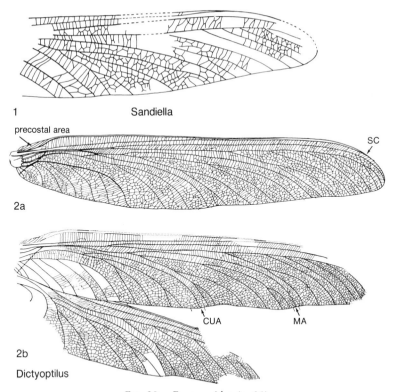

Fig. 20. Eugereonidae (p. 31).

Fore wing with anterior margin nearly straight for most of its length; vein SC extending nearly to wing apex; RS with many branches; MA unbranched, apparently arising from anterior branch of MP; MP with many branches, mostly arising distally; CUA unbranched; CUP with at least 2 branches; anal area extensive; crossveins very numerous, commonly reticulate. Hind wing apparently similar to fore wing. Head small, with prominent eyes; beak broad basally. *U. Carb.*

Lycocercus HANDLIRSCH, 1906b, p. 89 [*Dictyoneura goldenbergi* BRONGNIART, 1883, p. 268; SD HANDLIRSCH, 1922, p. 39] [=*Patteiskya* LAURENTIAUX, 1958, p. 302 (type, *P. bouckaerti*)]. Fore wing: RS with at least 10 terminal branches; MP with 10 to 20 forks; usually 8 anal veins. DEMOULIN, 1958a; KUKALOVÁ, 1969c. *U. Carb.,* Europe (France, Germany).——FIG. 21,6. *L. goldenbergi* (BRONGNIART); fore wing, ×0.75 (Kukalová, 1969c).

Apopappus HANDLIRSCH, 1906b, p. 100 [*Spilaptera guernei* BRONGNIART, 1893, p. 463; OD]. Little-known genus, based on apical fragment of wing. Similar to *Lycocercus* but with CUP more extensively branched and with crossveins more regularly arranged. LAMEERE, 1917b; KUKALOVÁ, 1969c. *U. Carb.,* Europe (France).

Lycodemas CARPENTER & RICHARDSON, 1971, p. 268 [*L. adolescens*; OD]. Little-known genus, based on nymphs; venation similar to that of *Lycocercus,* but MP with fewer branches. *U. Carb.,* USA (Illinois).

Madera CARPENTER, 1970, p. 402 [*M. mamayi*; OD]. Similar to *Lycocercus,* but fore and hind wings relatively broad; MA arising in both wings at about level of origin of RS; CUP consisting of 2 large branches, without marginal forks. *U. Carb.,* USA (New Mexico).——FIG. 21,5. *M. mamayi*; fore wing, ×3.5 (Carpenter, 1970).

Notorachis CARPENTER & RICHARDSON, 1971, p. 272 [*N. wolfforum*; OD]. Pronotal lobes heavily sclerotized and bearing long spines. Venation similar to that of *Lycocercus*; MA arising before origin of RS; MP with 5 terminal branches. *U. Carb.,* USA (Illinois).——FIG. 21,3. *N. wolfforum*; dorsal view, ×1.6 (Carpenter & Richardson, 1971).

Polycreagra HANDLIRSCH, 1906a, p. 678 [*P. elegans*; OD]. Hind(?) wing: RS and MP with numerous, fine branches; about 15 terminal anal veins. KUKALOVÁ, 1969c. *U. Carb.,* USA (Rhode

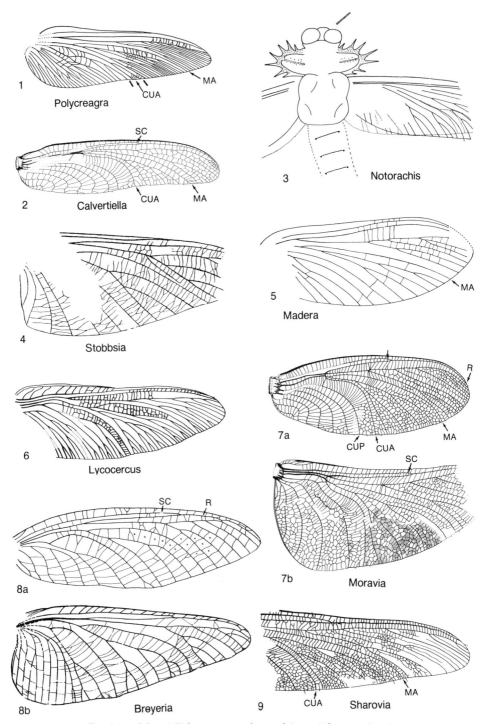

FIG. 21. Calvertiellidae, Lycocercidae, and Breyeriidae (p. 32–35).

Island).——FIG. 21,*1*. **P. elegans*; hind(?) wing, ×0.7 (Handlirsch, 1906a).

Family GRAPHIPTILIDAE
Handlirsch, 1906

[Graphiptilidae HANDLIRSCH, 1906b, p. 99] [=Rhabdoptilidae HANDLIRSCH, 1919b, p. 525]

Little-known family. Hind(?) wing with vein SC extending nearly to apex of wing; RS with several pectinate branches; M forking just before level of midwing; MA usually unbranched; MP with 3 branches; CUA unbranched; CUP with 3 branches; crossveins numerous and fine. *U. Carb.*

Graphiptilus BRONGNIART, 1893, p. 348 [**G. heeri*; OD] [=*Graphiptiloides* HANDLIRSCH, 1906b, p. 92 (type, *Graphiptilus williamsoni* BRONGNIART, 1893, p. 349)]. Little-known genus. RS with 5 terminal branches, the first ending at wing apex; MP forking shortly after its separation from MA. KUKALOVÁ, 1969c. *U. Carb.*, Europe (France).
Rhabdoptilus BRONGNIART, 1893, p. 364 [**R. edwardsi*; OD]. Little-known genus. Hind(?) wing with RS arising in basal third of wing, with at least 5 terminal branches; MP with 6 terminal branches; crossveins more irregular than in *Graphiptilus*. HANDLIRSCH, 1906b, 1919b; LAMEERE, 1917b; KUKALOVÁ, 1969c. *U. Carb.*, Europe (France).

Family BREYERIIDAE
Handlirsch, 1906

[Breyeriidae HANDLIRSCH, 1906b, p. 95]

Fore wing broad; costal margin strongly curved; vein SC terminating at midwing or well before wing apex; stems of R and M very close or in contact basally; R extending almost to apex; RS with 5 or 6 terminal branches, widely separated; MA unbranched; CUA usually unbranched, at most with a marginal fork; CUP usually forked at least once; crossveins very fine, irregular, and numerous, commonly anastomosed. Hind wing much broader than fore wing, posterior margin strongly curved. *U. Carb.*

Breyeria BORRE, 1875b, p. lxvi [**Pachytylopsis borinensis* BORRE, 1875a, p. xli; OD]. Fore wing with SC terminating on R; branches of M and CU strongly curved toward hind margin of wing. BRONGNIART, 1893; MEUNIER, 1910a; HANDLIRSCH, 1919b; STRAND, 1929; KELLER, 1935; LAURENTIAUX, 1949a; KUKALOVÁ, 1959a, 1969c; LAURENTIAUX-VIEIRA & LAURENTIAUX, 1963; CARPENTER, 1967a. *U. Carb.*, Europe (France, Belgium, Germany, Czechoslovakia, Holland), England, USA (Tennessee).——FIG. 21,*8a. B. rappi* CARPENTER, Tennessee; fore wing, ×0.8 (Carpenter, 1967a).——FIG. 21,*8b. B. barborae* KUKALOVÁ, Czechoslovakia; hind wing, ×0.7 (Kukalová, 1959a).
Megaptiloides HANDLIRSCH, 1906b, p. 97 [**Megaptilus brodiei* BRONGNIART, 1893, p. 375; OD]. Little-known genus, based on wing fragment; similar to *Breyeria*, but crossveins numerous and more irregular. KUKALOVÁ, 1969c. [Family assignment doubtful.] *U. Carb.*, Europe (France).
Stobbsia HANDLIRSCH, 1908a, p. 1347 [**S. woodwardiana*; OD]. Similar to *Breyeria*, but SC terminating on costa; branches of M and CU not strongly curved toward posterior margin of wing. [Family assignment doubtful.] LAMEERE, 1917b; LAURENTIAUX & LAURENTIAUX-VIEIRA, 1951; KUKALOVÁ, 1969c. *U. Carb.*, England.——FIG. 21,*4*. **S. woodwardiana*; wing, ×1 (Handlirsch, 1908a).

Family TCHIRKOVAEIDAE
Sinitshenkova, 1979

[Tchirkovaeidae SINITSHENKOVA, 1979, p. 74]

Similar to Breyeriidae, but vein MP unbranched or with only a short fork; branches of M and CU only slightly curved distally. *U. Carb.*

Tchirkovaea M. D. ZALESSKY, 1931, p. 403 [**T. guttata*; OD]. Fore wing with anterior margin strongly convex; SC extending well beyond two-thirds of wing length; MP forked to about half its length; crossveins forming coarse reticulation in several areas. Hind wing with anterior margin almost straight. ZALESSKY, 1932a; SINITSHENKOVA, 1979, 1981a. *U. Carb.*, USSR (Asian RSFSR).
Paimbia SINITSHENKOVA, 1979, p. 82 [**P. fenestrata*; OD]. Similar to *Tchirkovaea*, but fore wing with anterior margin straight and costal area narrow. Hind wing with concave anterior margin. SINITSHENKOVA, 1981a. *U. Carb.*, USSR (Asian RSFSR).

Family HOMOIOPTERIDAE
Handlirsch, 1906

[Homoiopteridae HANDLIRSCH, 1906b, p. 91] [=Rocklingiidae GUTHÖRL, 1934, p. 188; Thesoneuridae CARPENTER, 1944, p. 10]

Large insects. Fore wing with stems of veins SC, R, and M distinctly curved basally; SC long, terminating near wing apex; RS with relatively few branches; MA unbranched or

Hexapoda

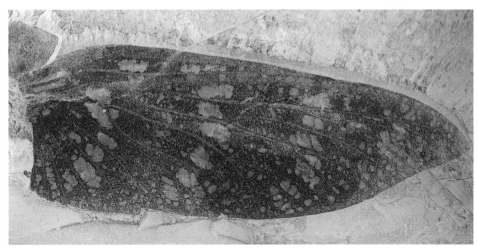

Homoioptera

FIG. 22. Homoiopteridae (p. 36).

with very few short branches; MP with several branches; CUA with several short branches; stems of CUA and CUP more or less parallel for most of their lengths. Hind wing somewhat broader than fore wing, with narrow costal area. Head apparently small, eyes bulging; prothoracic lobes well developed. *U. Carb.*

Homoioptera BRONGNIART, 1893, p. 353 [*H. woodwardi*; OD] [=*Homoeophlebia* HANDLIRSCH, 1906b, p. 92 (type, *Homoioptera gigantea* AGNUS, 1902, p. 259); *Anthracentomon* HANDLIRSCH, 1904a, p. 6 (type, *A. latipenne*)]. Fore wing: RS with only 3 main branches; M dividing near level of midwing; MA usually unbranched or with distal twig; MP with 3 or 4 branches; 6 to 8 anal veins. Crossveins numerous, weak, and with much reticulation. HANDLIRSCH, 1906a; MEUNIER, 1908f, 1910a; KUKALOVÁ, 1969c; BRAUCKMANN & KOCH, 1982. *U. Carb.*, Europe (France, Belgium, Germany).——FIG. 22. *H. gigantea* AGNUS, France; fore wing and pronotal lobe, ×0.7 (Carpenter, new).

Adolarryia KUKALOVÁ-PECK & RICHARDSON, 1983, p. 1677 [*A. bairdi*; OD]. Little-known genus, based on nymph consisting of thoracic segments and partially developed wings. *U. Carb.*, USA (Illinois).

Boltopruvostia STRAND, 1929, p. 20, nom. subst. pro *Boltonia* PRUVOST, 1919, p. 284, non KOENIG, 1820 [*Boltonia robusta* PRUVOST; OD] [=*Ostrava* KUKALOVÁ, 1960, p. 241 (type, *O. nigra*)]. Little-known genus. Similar to *Homoioptera* but wings longer; RS more extensively branched and with much more extensive anal area. GUTHÖRL, 1934; HAUPT, 1949; KUKALOVÁ, 1969c. *U. Carb.*, Europe (Germany, France, Czechoslovakia).

Larryia KUKALOVÁ-PECK & RICHARDSON, 1983, p. 1678 [*L. osterbergi*; OD]. Little-known genus, based on the metathorax and a hind wing with distinct bend in costal margin near midwing. *U. Carb.*, USA (Illinois).

Mazonopterum KUKALOVÁ-PECK & RICHARDSON, 1983, p. 1674 [*M. wolfforum*; OD]. Hind wing similar to that of *Boltopruvostia*, but costal margin straight and costal area narrow; space between CUA and CUP conspicuously wide. *U. Carb.*, USA (Illinois).

Mazothairos KUKALOVÁ-PECK & RICHARDSON, 1983, p. 1672 [*M. enormis*; OD]. Little-known genus, based on single thoracic segment. *U. Carb.*, USA (Illinois).

Parathesoneura SINITSHENKOVA in SHAROV & SINITSHENKOVA, 1977, p. 48 [*P. carpenteri*; OD]. Hind wing similar to *Thesoneura*, but M dividing nearer wing base; CUA unbranched; archedictyon coarse. *U. Carb.*, USSR (Kazakh).

Thesoneura CARPENTER, 1944, p. 10 [*T. americana*; OD]. Hind wing with MA unbranched; MP with only 3 short branches; CUA with several long branches, arising pectinately and directed anteriorly; CUP sinuously curved; several anal veins, mostly unbranched. KUKALOVÁ, 1969c. *U. Carb.*, USA (Illinois).——FIG. 23. *T. americana*; hind wing, ×0.6 (Carpenter, 1944).

Turneropterum KUKALOVÁ-PECK & RICHARDSON, 1983, p. 1680 [*T. turneri*; OD]. Little-known genus, based on thorax and basal parts of fore and hind wings. Fore wing with costal margin strongly concave basally; stems of R, M, and CU separating close to base of wing and dividing early. *U. Carb.*, USA (Illinois).

Family MECYNOSTOMATIDAE Kukalová, 1969

[Mecynostomatidae KUKALOVÁ, 1969b, p. 208]

Fore wing with costal area very broad basally; vein SC short, terminating on R at about midwing; branches of RS curving posteriorly; MA, MP, CUA, and CUP branched; crossveins numerous, many irregular. Hind wing similar to fore wing, but costal area much narrower. Head small, with prominent eyes; beak elongate. *U. Carb.*

Mecynostomata METCALF, 1952, p. 230, *nom. subst. pro Mecynostoma* BRONGNIART, 1893, p. 451, *non* GRAFF, 1882 [*Mecynostoma dohrni* BRONGNIART, 1893, p. 452; OD]. Fore wing: MA with 2 terminal branches, MP with 3; CUA more deeply forked than CUP. HANDLIRSCH, 1906b; KUKALOVÁ, 1969b. *U. Carb.*, Europe (France).——FIG. 24,2. *M. dohrni* (BRONGNIART); fore wing, head, and foreleg, ×1 (Kukalová, 1969b).

Family FOUQUEIDAE Handlirsch, 1906

[Fouqueidae HANDLIRSCH, 1906b, p. 98]

Wing venation similar to that of Spilapteridae, but crossveins very numerous and reticulate over entire wing. *U. Carb.*

Fouquea BRONGNIART, 1893, p. 372, *nom. subst. pro Oustaletia* BRONGNIART, 1885a, p. 66, *non* TROUESSART, 1885 [*Oustaletia lacroixi*; OD] [=*Archaecompsoneura* MEUNIER, 1909b, p. 139 (type, *A. superba*)]. Fore wing: RS with 4 to 7 branches; MA with a long fork; MP with several branches; CUP with 4 terminal branches; at least 3 branched anal veins. Hind wing much broader than fore wing, with similar venation. KUKALOVÁ, 1969b. *U. Carb.*, Europe (France).——FIG. 24,3. *F. lacroixi* (BRONGNIART); fore wing, ×1 (Kukalová, 1969b).
Neofouquea CARPENTER, 1967a, p. 62 [*N. suzanneae*; OD]. Little-known genus. Hind(?) wing similar to *Fouquea*, but CUP with single long fork. *U. Carb.*, USA (Illinois).

Family EUBLEPTIDAE Handlirsch, 1906

[Eubleptidae HANDLIRSCH, 1906a, p. 679]

Small Palaeodictyoptera with slender, pointed wings. Fore wing with vein SC

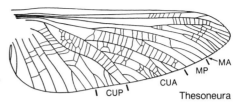

FIG. 23. Homoiopteridae (p. 36).

extending nearly to wing apex; RS dichotomously forked, usually with 4 branches; M forking just before origin of RS; MA with long fork; MP with 3 terminal branches; CUA with short fork; CUP usually with 3 terminal branches; relatively few crossveins, forming distinct pattern. Hind wing similar to fore wing in venation, but slightly broader. Body slender, pronotal lobes small. *U. Carb.*

Eubleptus HANDLIRSCH, 1906a, p. 681 [*E. danielsi*; OD]. Fork of MA nearly at level of first fork of RS; first fork of MP well before midwing, its posterior branch forked near wing margin. CARPENTER, 1983. *U. Carb.*, USA (Illinois).——FIG. 24,1. *E. danielsi*; dorsal view, ×2.5 (Carpenter, 1983).

Family SPILAPTERIDAE Handlirsch, 1906

[Spilapteridae HANDLIRSCH, 1906b, p. 101] [=Lamproptiliidae HANDLIRSCH, 1906b, p. 109; Dunbariidae HANDLIRSCH, 1937, p. 81; Doropteridae ZALESSKY, 1946a, p. 64; Neuburgiidae ROHDENDORF, 1961a, p. 72]

Fore wing: anterior margin more or less concave; vein SC long, usually extending to wing apex; RS with 3 to 6 terminal branches, usually pectinate; MA and MP with at least 2 branches; CUA with several branches, CUP with few or, rarely, unbranched; several anal veins. Hind wing: broader than fore wing, with larger anal area; venation similar to that of fore wing. Both wings commonly marked with spots or bands. Body structure: head broad, with bulging eyes; beak long; antennae long, multisegmented; pronotal lobes usually well developed and with radiating support veins; metathorax usually slightly longer than mesothorax; legs short, cursorial; abdomen usually slender, female with 10 visible segments and a short, curved ovipositor, male apparently with 11 abdominal segments; cerci well developed. *U. Carb.–Perm.*

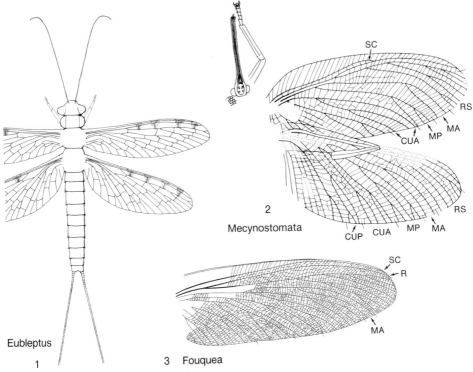

Fig. 24. Mecynostomatidae, Fouqueidae, and Eubleptidae (p. 37).

Spilaptera BRONGNIART, 1885a, p. 63 [**S. packardi*; SD HANDLIRSCH, 1922, p. 45]. Fore wing: R without terminal branches; RS with 4 to 6 terminal branches; area between R and RS with several strong, oblique crossveins; M free from RS; MP with several forks; CUA pectinately branched; CUP usually forked; relatively few crossveins, commonly forming distinct pattern. BRONGNIART, 1893; HANDLIRSCH, 1906b; KUKALOVÁ, 1969b; CARPENTER & RICHARDSON, 1971; SHAROV & SINITSHENKOVA, 1977. *U. Carb.*, Europe (France), USSR (Ukraina), USA (Illinois).——FIG. 25,*4*. **S. packardi*; fore wing, ×0.9 (Kukalová, 1969b).

Abaptilon ZALESSKY, 1946c, p. 58 [**A. sibiricum*; OD]. Little-known genus, based on apical fragment of hind(?) wing. MA and MP each with 3 terminal branches; RS apparently originating near midwing. [Family assignment doubtful.] ROHDENDORF, 1962a. *U. Carb.*, USSR (Asian RSFSR).

Baeoneura SINITSHENKOVA in SHAROV & SINITSHENKOVA, 1977, p. 58 [**B. obscura*; OD]. RS of hind wing with pectinate branches as in *Dunbaria* but with several branches forked. *U. Carb.*, USSR (Asian RSFSR).

Becquerelia BRONGNIART, 1893, p. 356 [**B. superba*; SD HANDLIRSCH, 1922, p. 46] [=*Pseudobecquerelia* HANDLIRSCH, 1919b, p. 534 (type, *Becquerelia elegans* BRONGNIART)]. Similar to *Homaloneura*, but MA apparently coalesced with RS for considerable interval; CUA pectinate; R with short terminal branches. HANDLIRSCH, 1919b; KUKALOVÁ, 1969b. *U. Carb.*, Europe (France).——FIG. 25,*3*. **B. superba*; hind wing, ×0.6 (Kukalová, 1969b).

Dunbaria TILLYARD in DUNBAR & TILLYARD, 1924, p. 203 [**D. fasciipennis*; OD] [=*Doropteron* ZALESSKY, 1946a, p. 64 (type, *D. mirum*)]. Fore wing with anterior margin serrate and distinctly concave; branches of RS without forks; MA and MP arising close to wing base, MA with single fork, MP with 2 or 3 branches; CUA with several branches; CUP unbranched; cuticular thickenings between CUP and 1A. Hind wing venation as in fore wing; anterior margin more deeply concave than that of fore wing; anal area very broad. ROHDENDORF, 1962a; KUKALOVÁ-PECK, 1971; SHAROV & SINITSHENKOVA, 1977. *Perm.*, USA (Kansas), USSR (European RSFSR).——FIG. 26,*3*. **D. fasciipennis*; *a*, fore wing, ×3.6; *b*, hind wing, ×3.7; *c*, dorsal view, ×1.7 (all Kukalová-Peck, 1971).

Epitethe HANDLIRSCH, 1906b, p. 103 [**Spilaptera*

meunieri BRONGNIART, 1893, p. 343; OD]. Similar to *Palaeoptilus,* but R without terminal branches; area between R and RS with straight (not sigmoidal) crossveins. KUKALOVÁ, 1969b. *U. Carb.,* Europe (France).

Homaloneura BRONGNIART, 1885a, p. 66 [**H. elegans*; OD] [=*Homaloneurina* HANDLIRSCH, 1906b, p. 106 (type, *Homaloneura bonnieri* BRONGNIART); *Homaloneurites* HANDLIRSCH, 1906b, p. 107 (type, *Homaloneura joannae* BRONGNIART)]. Anterior margins of wings usually with only slight concavity; venation very similar to that of *Spilaptera*; a cuticular ridge extending from near base to 1A to R. BRONGNIART, 1893; HANDLIRSCH, 1922; CARPENTER, 1964b; KUKALOVÁ, 1969b; CARPENTER & RICHARDSON, 1971. *U. Carb.,* Europe (France), USA (Illinois).——FIG. 27,*a*. **H. elegans,* France; dorsal view of wings, head, and part of thorax, ×1.3 (Kukalová, 1969b). ——FIG. 27,*b*. *H. dabasinskasi* CARPENTER, Illinois; fore and hind wings, ×1.5 (Carpenter, new).

Lamproptilia BRONGNIART, 1885a, p. 63 [**L. grandeuryi*; OD]. Fore wing unusually broad; hind wing broader than fore; cubital-anal area forming distinct lobe; R without terminal branches; cuticular thickenings near wing base apparently absent. BRONGNIART, 1893; HANDLIRSCH, 1906b; KUKALOVÁ, 1969b. *U. Carb.,* Europe (France). ——FIG. 25,*5*. **L. grandeuryi; a,* fore and *b,* hind wings, ×0.8 (Kukalová, 1969b).

Mcluckiepteron RICHARDSON, 1956, p. 20 [**M. luciae*; OD]. Little-known genus, based on isolated hind wing. Costal margin serrate and strongly concave; SC and R very close together distally; RS with pectinate branching; MA with only small fork; MP more extensively branched; CUA with many branches; CUP unbranched. [Family assignment doubtful.] *U. Carb.,* USA (Illinois).——FIG. 26,*2*. **M. luciae*; hind wing, ×0.8 (Richardson, 1956).

Neuburgia MARTYNOV, 1931a, p. 74 [**N. altaica*; OD]. Fore wing unusually slender; RS arising near wing base and M forked at level of origin of RS; CUP unbranched. [Family assignment uncertain.] ROHDENDORF & others, 1961; ROHDENDORF, 1962a; KUKALOVÁ, 1969b; SHAROV & SINITSHENKOVA, 1977. *U. Carb.,* USSR (Asian RSFSR).——FIG. 25,*1*. **N. altaica*; fore wing, ×1.3 (Martynov, 1931a).

Palaeoptilus BRONGNIART, 1893, p. 352 [**P. brullei*; OD]. Similar to *Becquerelia,* but MA apparently not coalesced with RS. [Probably a synonym of *Becquerelia*.] HANDLIRSCH, 1906b; KUKALOVÁ, 1969b. *U. Carb.,* Europe (France).

Paradunbaria SHAROV & SINITSHENKOVA, 1977, p. 54 [**P. pectinata*; OD]. Similar to *Dunbaria,* but RS and CUA with more extensive branching. *Perm.,* USSR (Asian RSFSR).——FIG. 26,*1*. **P. pectinata*; ventral view, ×1.5 (Sharov & Sinitshenkova, 1977).

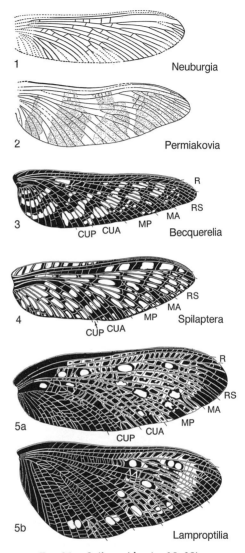

FIG. 25. Spilapteridae (p. 38–39).

Permiakovia MARTYNOV, 1940, p. 7 [**P. quinquefasciata*; OD]. Similar to *Dunbaria* but with several branches of RS deeply forked. ROHDENDORF, 1962a; SHAROV & SINITSHENKOVA, 1977. *Perm.,* USSR (European and Asian RSFSR). ——FIG. 25,*2*. **P. quinquefasciata*; fore wing, ×1.3 (Martynov, 1940).

Spiloptilus HANDLIRSCH, 1906b, p. 100 [**Graphiptilus ramondi* BRONGNIART, 1893, p. 351; OD]. Little-known genus. RS originating almost at level of midwing; M dividing much nearer to wing base. [Family assignment doubtful.] KUKALOVÁ, 1969b. *U. Carb.,* Europe (France).

Tectoptilus KUKALOVÁ, 1969b, p. 193 [**Becquer-*

Hexapoda

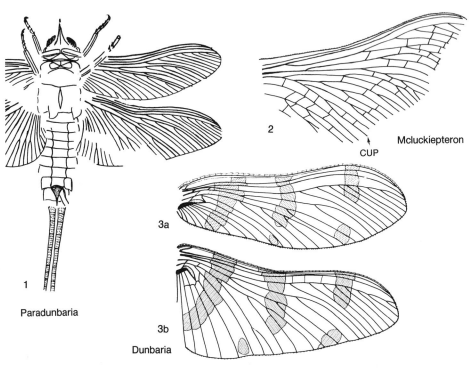

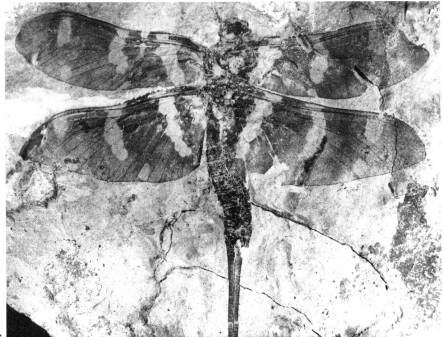

FIG. 26. Spilapteridae (p. 38–39).

elia grehanti BRONGNIART, 1893, p. 359; OD].
Wings without cuticular thickenings between 1A
and CUP, with fewer branches than in *Epitethe*
and *Palaeoptilus*. HANDLIRSCH, 1906b. *U. Carb.,*
Europe (France).

Vorkutoneura SINITSHENKOVA in SHAROV & SINITSHENKOVA, 1977, p. 60 [*V. variabilis*; OD].
Hind wing very broad basally; RS, MA, MP, and
CUA multibranched. *Perm.,* USSR (European
RSFSR).

Family ELMOBORIIDAE
Carpenter, 1976

[Elmoboriidae CARPENTER, 1976, p. 349]

Fore wing slender, at least 4 times as long as wide, broadest distally; vein R close to and parallel to SC, except distally; RS with several long branches; M forking nearly at same level as origin of RS; MP forking almost immediately after its origin from M, with 2 or 3 branches; CU forking near base of wing; CUA and CUP deeply forked; crossveins weak, apparently generally distributed over wing area. Hind wing and body unknown. *Perm.*

Elmoboria CARPENTER, 1976, p. 350 [*E. piperi*; OD]. Fore wing with R extending almost to wing apex; RS dichotomously branched, with 4 terminal branches; MA unbranched; MP3+4 deeply forked. *Perm.,* USA (Kansas).——FIG. 28,2. *E. piperi;* fore wing, (Carpenter, 1976).

Oboria KUKALOVÁ, 1960, p. 245 [*O. longa*; OD]. Similar to *Elmoboria,* but RS apparently with 7 or 8 terminal branches and MA deeply forked. CARPENTER, 1976. *Perm.,* Europe (Czechoslovakia).

Family SYNTONOPTERIDAE
Handlirsch, 1911

[Syntonopteridae HANDLIRSCH, 1911, p. 299]

Fore wing broadest near midwing; anterior margin with slight curvature basally; vein RS arising near wing base and forking just before midwing; stem of M independent of R basally; MA and MP separating a short interval from wing base, with MA diverging at about 45° angle toward RS and coalescing with it for short interval; MA and MP forked; CUA and CUP diverging near wing base; CUA forked; CUP unbranched; 3 anal veins; crossveins numerous, with some reticulation basally; intercalary veins present between some

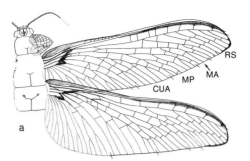

FIG. 27. Spilapteridae (p. 39).

branches of RS, MA, MP, and CUA. Hind wing much broader basally than fore wing, hind margin strongly curved; vein SC terminating just before wing apex; stem of M coalesced with basal part of CUA near wing base; immediately after its origin, MA coalescing with RS for short interval, as in fore wing; CUP with prominent bend directed toward hind margin of wing; crossveins and intercalary veins present as in fore wing. Body little known; antennae very thin and pronotal lobes apparently small. *U. Carb.*

[The Syntonopteridae are considered by some investigators (EDMUNDS & TRAVER, 1954; EDMUNDS, 1972; WOOTTON, 1981) to

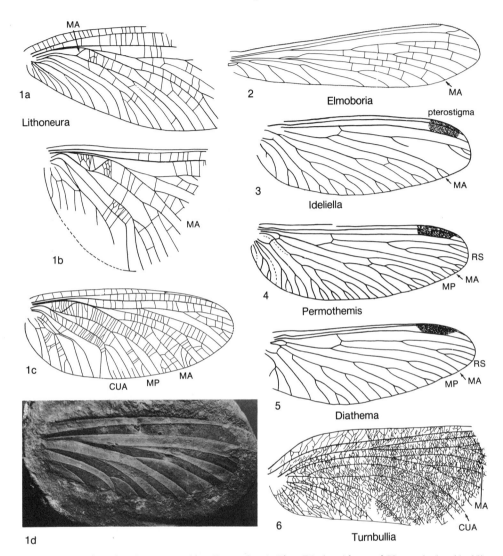

FIG. 28. Elmoboriidae, Syntonopteridae, Permothemistidae, Diathemidae, and Uncertain (p. 41–45).

belong to the order Ephemeroptera, chiefly because of the presence of intercalary veins. However, KUKALOVÁ-PECK (HUBBARD & KUKALOVÁ-PECK, 1980) has reported the presence of a haustellate beak in the type specimen of the syntonopterid genus *Lithoneura*. All the Paleozoic Ephemeroptera known had normal, dentate mandibles.]

Syntonoptera HANDLIRSCH, 1911, p. 299 [*S. schucherti*; OD]. Little-known genus, based on fragment of fore wing. Costal area broad basally, narrowed near midwing; RS coalesced with MA near wing base. CARPENTER, 1938, 1944; LAURENTIAUX, 1953. U. Carb., USA (Illinois).

Lithoneura CARPENTER, 1938, p. 446 [*L. lameerei*; OD]. Fore wing similar to that of *Syntonoptera*, but costal area much narrower basally; coalescence of MA and RS more remote from wing base. Hind wing broadly oval; RS with 1 long and 3 short intercalary sectors; MA, MP, and CUA each with 1 intercalary vein. CARPENTER, 1944; RICHARDSON, 1956. U. Carb., USA (Illinois).——FIG. 28,1a,b. *L. lameerei; a,* fore and *b,* hind wings, ×1.6 (Carpenter, 1938).——FIG. 28,1c. *L. mirifica* CARPENTER; hind wing, ×0.7 (Carpenter, 1944).——FIG. 28,1d. *L. carpenteri* RICHARDSON; fore wing, ×1.5 (Carpenter, new).

Family PERMOTHEMISTIDAE
Martynov, 1938

[nom. correct. ROHDENDORF, 1962a, p. 55, ex Permothemidae MARTYNOV, 1938b, p. 37]

Fore wing with vein SC long, extending to wing apex; pterostigma present; RS arising in basal third of wing; CUA and CUP separating at base of wing; RS, MP, CUA, and CUP branched and usually with marginal forks. Hind wings absent. Antennae long and slender; eyes large; prothoracic lobes apparently absent; cerci long; ovipositor short, curved. *Perm.*

Permothemis MARTYNOV, 1934, p. 995, *nom. subst.* pro *Palaeothemis* MARTYNOV, 1932, p. 12, *non* FRASER, 1923 [*Palaeothemis libelluloides*; OD]. Fore wing: pterostigma at least 3 times as long as wide; MP with several dichotomous branches; CUA with 3 or 4 branches. ROHDENDORF, 1962a; SINITSHENKOVA, 1980b. *Perm.*, USSR (European RSFSR).——FIG. 28,*4*. *P. libelluloides*; fore wing, ×3 (Sinitshenkova, 1980b).
Ideliella ZALESSKY, 1937c, p. 107 [*I. decora*; OD]. Fore wing: pterostigma broader than in *Permothemis*; MP and CUA each with only 1 long fork. ROHDENDORF, 1962a; SINITSHENKOVA, 1980b. *Perm.*, USSR (Asian RSFSR).——FIG. 28,*3*. *I. decora*; fore wing, ×4.5 (Sinitshenkova, 1980b).
Pauciramus SINITSHENKOVA, 1980b, p. 99 [*P. demoulini*; OD]. Similar to *Permothemis*, but pterostigma longer; MA unbranched, MP forked once. *Perm.*, USSR (Asian RSFSR).
Permothemidia ROHDENDORF, 1940b, p. 109 [*P. caudata*; OD] [=*Uralothemis* ZALESSKY, 1951c, p. 270 (type, *U. bellus*)]. Fore wing with pterostigma short, about twice as long as wide; RS with 5 to 6 branches directed posteriorly. SINITSHENKOVA, 1980b. *Perm.*, USSR (Asian RSFSR).

Family DIATHEMIDAE
Sinitshenkova, 1980

[Diathemidae SINITSHENKOVA, 1980b, p. 101]

Fore wing with pterostigma about 4 times as long as wide; veins MP and CUA not anastomosed; 3 anal veins. Hind wing present but greatly reduced, with little venation. *Perm.*

Diathema SINITSHENKOVA, 1980b, p. 102 [*D. tenerum*; OD]. MP of fore wing with 2 branches. *Perm.*, USSR (Asian RSFSR).——FIG. 28,*5*. *D. tenerum*; fore wing, ×4 (Sinitshenkova, 1980b).
Diathemidia SINITSHENKOVA, 1980b, p. 105 [*D. monstruosa*; OD]. Fore wing similar to *Diathema*, but MP with 3 branches. *Perm.*, USSR (Asian RSFSR).

Family PSYCHROPTILIDAE
Riek, 1976

[Psychroptilidae RIEK, 1976c, p. 230]

Palaeodictyoptera of moderate size; hind wing slightly broader than fore wing. Vein RS with 3 branches; MA, CUA, CUP unbranched; MP branched. Body structure little known; pronotal lobes apparently absent. *U. Carb.*

Psychroptilus RIEK, 1976c, p. 230 [*P. burrettae*; OD]. SC ending on costa about two-thirds wing length from base. [Ordinal assignment uncertain; the family was originally placed in the Megasecoptera.] *U. Carb.*, Tasmania.

Family UNCERTAIN

The following genera, apparently belonging to the order Palaeodictyoptera, are too poorly known to permit family assignments.

Althansia GUTHÖRL, 1934, p. 61 [*A. sahneri*; OD]. Fragment of wing with archedictyon. *U. Carb.*, Europe (Germany).
Ametretus HANDLIRSCH, 1911, p. 303 [*A. laevis*; OD]. Little-known genus, based on basal fragment of large wing. KUKALOVÁ, 1969c. *U. Carb.*, USA (Illinois).
Amousus HANDLIRSCH, 1911, p. 301 [*A. mazonus*; OD]. Little-known genus, based on basal fragment of large wing. KUKALOVÁ, 1969c. *U. Carb.*, USA (Illinois).
Anagesthes HANDLIRSCH, 1906b, p. 70 [*Termes affinis* GOLDENBERG, 1854, p. 31; OD]. Small fragment of wing with archedictyon. GOLDENBERG, 1873. *U. Carb.*, Europe (Germany).
Anthracosta PRUVOST, 1930, p. 147 [*A. dubois*; OD]. Small fragment of large wing. *U. Carb.*, Europe (Germany).
Asiodictya ROHDENDORF, 1961a, p. 70 [*A. rossica*; OD]. Apical fragment of wing with archedictyon. ROHDENDORF, 1962a. *U. Carb.*, USSR (Asian RSFSR).
Bathytaptus HANDLIRSCH, 1906a, p. 686 [*B. falcipennis*; OD]. Apical wing fragment. HANDLIRSCH, 1906b. *U. Carb.*, USA (Alabama).
Bojoptera KUKALOVÁ, 1958c, p. 235 [*B. colorata*; OD]. Fore wing with costal area very broad basally; SC terminating well before apex of wing; RS arising before midwing, with many branches; MA with only 3 branches; MP extensively branched; CUA and its branches strongly curved, forming prominent loop toward CU; CUP also curved but with only 2 terminal branches. Hind wing little known; costal area narrower than in

fore wing; CUA and CUP more nearly normal in form. *U. Carb.*, Europe (Czechoslovakia).

Boltonocosta CARPENTER, 1986, p. 575, *nom. subst. pro Orthocosta* BOLTON, 1912, p. 310, *non* FRITSCH, 1879 [**Orthocosta splendens*; OD]. Little-known genus, based on wing fragment. RS with first fork well beyond midwing; M dividing just beyond point of origin of RS. *U. Carb.*, England.

Catadyesthus HANDLIRSCH, 1906b, p. 87 [**Acridites priscus* ANDREE, 1864, p. 163; OD]. Basal fragment of wing. AGNUS, 1902. *U. Carb.*, Europe (Czechoslovakia).

Compsoneura BRONGNIART, 1893, p. 334, *nom. subst. pro Zeilleria* BRONGNIART, 1885a, p. 63, *non* BAYLE, 1878 [**C. fusca*; OD]. Little-known genus, based on wing fragment. RS, MP, and CUA with several branches; MA with 1 fork; crossveins numerous, curved and anastomosed, especially dense distally. KUKALOVÁ, 1969b. *U. Carb.*, Europe (France).

Diexodus HANDLIRSCH, 1911, p. 302 [**D. debilis*; OD]. Basal fragment of wing. *U. Carb.*, USA (Illinois).

Eumecoptera HANDLIRSCH, 1906b, p. 73 [**Termes laxus* GOLDENBERG, 1877, p. 50; OD]. Fragment of wing with archedictyon. *U. Carb.*, Europe (Germany).

Eurydictyella CARPENTER, 1986, p. 575, *nom. subst. pro Eurydictya* GUTHÖRL, 1934, p. 49, *non* ULRICH, 1889 [**Eurydictya richteri*; OD]. Little-known genus, based on wing fragment. *U. Carb.*, Europe (Germany).

Eurythmopteryx HANDLIRSCH, 1906a, p. 675 [**E. antiqua*; OD]. Little-known genus, based on poorly preserved wing; SC ending on costal margin before apex; crossveins numerous over entire wing, without reticulation. *U. Carb.*, USA (Alabama).

Gegenemene HANDLIRSCH, 1906b, p. 76 [**Dictyoneura sinuosa* KLIVER, 1883, p. 259; OD]. Little-known genus, based on poorly preserved wing. BRONGNIART, 1893; GUTHÖRL, 1934. *U. Carb.*, Europe (Germany).

Haplophlebium SCUDDER, 1867, p. 151 [**H. barnesii*; OD]. Wing fragment with archedictyon. HANDLIRSCH, 1906b. *U. Carb.*, Canada (Nova Scotia).

Heolus HANDLIRSCH, 1906b, p. 94 [**H. providentiae*; OD]. Little-known genus, based on small wing fragment. [Type of family Heolidae HANDLIRSCH.] *U. Carb.*, USA (Rhode Island).

Idoptilus WOOTTON, 1972, p. 666 [**I. onisciformis*; OD]. Little-known genus, based on nymph. Similar to *Rochdalia* but with different venational pattern in wing pads. [Ordinal assignment uncertain.] *U. Carb.*, England.

Jongmansia LAURENTIAUX, 1950, p. 18 [**Mecynoptera tuberculata* BOLTON, 1921, p. 37; OD]. Little-known genus, based on wing fragment. RS much reduced, apparently with series of short branches extending anteriorly toward wing apex and R; MA unbranched; MP with long branches, extending apically only; CUP apparently branched distally. [Type of family Jongmansiidae LAURENTIAUX.] ROHDENDORF, 1962a. *U. Carb.*, England, Europe (Holland).

Kansasia TILLYARD, 1937a, p. 85 [**K. pulchra*; OD]. Little-known genus, based on apical fragment; probably related to Diathemidae. [Type of family Kansasiidae DEMOULIN, 1954b, p. 334.] SINITSHENKOVA, 1980b. *Perm.*, USA (Kansas).

Mammia HANDLIRSCH, 1906a, p. 671 [**M. alutacea*; OD]. Little-known genus, based on wing fragment. HANDLIRSCH, 1906b. *U. Carb.*, USA (Illinois).

Mecynoptera HANDLIRSCH, 1904a, p. 7 [**M. splendida*; OD]. Little-known genus, based on poorly preserved wing. [Type of family Mecynopteridae HANDLIRSCH (1904a); placed in Dictyoneuridae by LAMEERE (1917b), and in new order, Archaehymenoptera, by HAUPT (1941).] *U. Carb.*, Europe (Belgium).

Mecynostomites HANDLIRSCH, 1919b, p. 535 [**M. brongniarti*; OD]. Little-known genus, based on wing fragment. KUKALOVÁ, 1969b. *U. Carb.*, Europe (France).

Monsteropterum KUKALOVÁ-PECK, 1972, p. 259 [**M. moravicum*; OD]. Little-known genus, based mainly on body structures and wing bases, including details of beak and legs. *Perm.*, Europe (Czechoslovakia).

Palaiotaptus HANDLIRSCH, 1906a, p. 687 [**P. mazonus*; OD]. Apical wing fragment with archedictyon. HANDLIRSCH, 1906b. *U. Carb.*, USA (Illinois).

Palapteris GUTHÖRL, 1940, p. 41 [**P. stenodictyus*; OD]. Little-known genus, based on wing fragment. MA, MP, CUA, and CUP apparently unbranched. [Ordinal assignment doubtful.] *U. Carb.*, Europe (Germany).

Paramecynostoma HANDLIRSCH, 1919b, p. 535 [**P. dohrnianum*; OD]. Little-known genus, based on small fragment of wing. KUKALOVÁ, 1969b. *U. Carb.*, Europe (France).

Paramegaptilus HANDLIRSCH, 1906b, p. 118 [**Megaptilus scudderi* BRONGNIART, 1893, p. 325; OD]. Small fragment of wing. KUKALOVÁ, 1969c. *U. Carb.*, Europe (France).

Platephemera SCUDDER, 1867, p. 151 [**P. antiqua*; OD]. Little-known genus, based on small fragment of wing. [Ordinal assignment doubtful.] SCUDDER, 1868c, 1880; HANDLIRSCH, 1906a. *U. Carb.*, Canada (New Brunswick).

Propalingenia HANDLIRSCH, 1906b, p. 86 [**Palingenia feistmanteli* FRITSCH, 1880, p. 241; OD]. Small fragment of wing. FRITSCH, 1889. *U. Carb.*, Europe (Czechoslovakia).

Pseudomecynostoma HANDLIRSCH, 1919b, p. 535 [**P. dubium*; OD]. Little-known genus, based on small fragment of wing. KUKALOVÁ, 1969b. *U. Carb.*, Europe (France).

Pteronidia BOLTON, 1912, p. 314 [**P. plicatula*;

OD]. Little-known genus, based on wing fragment. [Type of family Pteronidiidae BOLTON, 1912.] HANDLIRSCH, 1919b. *U. Carb.,* England.

Rochdalia WOODWARD, 1913, p. 352 [**R. parkeri*; OD]. Little-known genus, based on nymph. Pronotum extended laterally; fore wing with very broad triangular costal area; cerci prominent. [Ordinal assignment uncertain.] ROLFE, 1967; WOOTTON, 1972. *U. Carb.,* England.

Saarlandia GUTHÖRL, 1930, p. 154 [**S. flexsubcostata*; OD]. Little-known genus, based on poorly preserved wing fragment. [Type of family Saarlandiidae GUTHÖRL, 1930.] *U. Carb.,* Europe (Germany).

Sabitaptus PRUVOST, 1930, p. 149 [*ptsS. lagagei*; OD]. Little-known genus, based on poorly preserved hind wing. *U. Carb.,* Europe (Belgium).

Scepasma HANDLIRSCH, 1911, p. 302 [*ptsS. gigas*; OD]. Little-known genus, based on small fragment of large wing. KUKALOVÁ, 1969c. *U. Carb.,* USA (Illinois).

Schedoneura CARPENTER, 1963b, p. 62 [*Brodioptera amii* COPELAND, 1957, p. 54; OD]. Little-known genus, based on incomplete hind wing. *U. Carb.,* Canada (Nova Scotia).

Severinopsis KUKALOVÁ, 1958c, p. 232 [*ptsS. vetusta*; OD]. Little-known genus, based on wing fragment with strongly concave anterior margin. [Possibly a spilapterid.] *U. Carb.,* Europe (Czechoslovakia).

Titanodictya HANDLIRSCH, 1906a, p. 671 [*Titanophasma jucunda* SCUDDER, 1885a, p. 169; OD]. Distal fragment of wing, with archedictyon. HANDLIRSCH, 1906b. *U. Carb.,* USA (Pennsylvania).

Turnbullia RICHARDSON, 1956, p. 27 [*T. luciae*; OD]. Little-known genus, based on basal portion of large wing, probably a fore wing. Costal margin strongly arched basally, costal area relatively wide; RS arising before separation of MA and MP; MP with at least 4 branches; CUA arising near wing base, unbranched and strongly arched; crossveins forming fine network over entire wing, except for basal part of costal area. [Possibly related to the Megaptilidae.] *U. Carb.,* USA (Illinois).——FIG. 28,6. *T. luciae*; fore(?) wing, ×0.9 (Carpenter, new).

Order MEGASECOPTERA Brongniart, 1885

[*nom. correct.* HANDLIRSCH, 1906a, p. 691, *pro* Megasecopterida BRONGNIART, 1885a, p. 63] [=Protohymenoptera TILLYARD, 1924a, p. 111]

Small to large palaeopterous insects; wings homonomous or nearly so; all main longitudinal veins present, forming an alternation of convexities and concavities; archedictyon only rarely present; crossveins usually well developed, numerous, and evenly distributed in some families (e.g., Aspidothoracidae) but reduced in number and arranged in rows in others (e.g., Protohymenidae); wing membrane hyaline, macrotrichia rarely well developed (e.g., Bardohymenidae); maculations may be present; veins SC and R very close together and usually close to costal margin. Body structure known in very few genera; head small; antennae setaceous, of moderate length; mouthparts haustellate, as in the Palaeodictyoptera; legs and abdomen usually slender, cerci very long, usually longer than body proper; prothoracic lobes and median caudal process absent. Nymphs, best known in the Mischopteridae, with haustellate mouthparts like those of adults; tracheal gills and other aquatic modifications absent. *U. Carb.–Perm.*

The order Megasecoptera, as treated here, comprises only the families formerly included in the suborder Eumegasecoptera (CARPENTER, 1947); the others, previously contained in the suborder Paramegasecoptera (BRUES, MELANDER, & CARPENTER, 1954), are placed in the order Diaphanopterodea. The separation of these two taxa into orders is based mainly on the palaeopterous condition of the wings in the Megasecoptera (i.e., Eumegasecoptera) and the flexed or folded condition in the Diaphanopterodea (i.e., Paramegasecoptera). The palaeopterous condition of the wings in the Megasecoptera is conclusively shown in whole insects by the consistent preservation of the wings in outstretched position, as in the Palaeodictyoptera.

Wings of the Megasecoptera are diverse in both form and venation. In the evolution of the order there have apparently been several lines of change: (1) veins SC and R have become closer together and have finally merged with the costa along the anterior margin of the wings (e.g., Aspidothoracidae, Protohymenidae, Bardohymenidae); (2) veins MA and MP have coalesced for varying intervals with their neighboring veins (e.g., Corydaloididae, Mischopteridae, Protohymenidae); (3) crossveins became fewer and developed in definite rows (e.g., Mischopteridae, Protohymenidae); (4) the wings became slender and petiolate (e.g., Brodiidae and Sphecopteridae). These changes obviously took place several times quite independently. The Corydaloididae appear to have had the most generalized wing form and venation, although the coalescence of RS with MA and of MP with CUA had already started.

The body structure, except for general features, is known in only a few genera, chiefly *Mischoptera* and *Protohymen*. The presence of a haustellate beak, in the nymphs as well as the adults, has now been definitely established, although details are not so well known as in the Palaeodictyoptera. None of the Megasecoptera seem to have had pronotal lobes comparable to those of the Palaeodictyoptera.

The nymphs are known in the Mischopteridae and Brodiidae, as well as in *Lameereites* (family uncertain). Their most striking features are found in the wing pads: the divergent position, the nature of their articulation to the thorax, and the advanced state of the venation. Unlike the developing wings in nymphs of existing insects, those of the Megasecoptera are joined to the thorax only at the articular areas of the adult wing, and they extend obliquely to the sides. The wings appear to have had some freedom of movement and the early development of the venation enhances that view. None of the nymphs appears to have had tracheal gills, swimming legs, or other modifications for an aquatic existence.

The Megasecoptera are obviously close relatives of the Palaeodictyoptera. In fact, during the past thirty years, as the Megasecoptera have become better known, separation of the Palaeodictyoptera and Megasecoptera, on wing venation alone, has become increasingly difficult (CARPENTER, 1962; SINITSHENKOVA, 1980a). Eventually, we may come to recognize these two taxa as representing one order, although KUKALOVÁ-PECK (1974b) has indicated that the articular sclerites at the bases of the wings are different in the two groups. In any case, it seems advisable to continue to recognize the two taxa as separate orders until we know more about the body structure and its diversity in both groups.

Family ASPIDOTHORACIDAE Handlirsch, 1919

[Aspidothoracidae HANDLIRSCH, 1919b, p. 579]

Venation of fore and hind wings similar; vein SC terminating well before wing apex; SC and R very close together and submarginal; stem of M very close to R basally, but separate from it; MA free from RS and not diverging toward it; stem of CU very close to that of M but not fused with it; 1 anal vein. Crossveins numerous and nearly uniformly distributed over wings. Prothorax with a conspicuous, thickened notum armed with stout spines. U. Carb.

Aspidothorax BRONGNIART, 1893, p. 304 [**A. triangularis*; SD HANDLIRSCH, 1922, p. 202] [=*Protocapnia* BRONGNIART, 1885a, p. 63, *nom. nud.*]. RS arising slightly basad of midwing, with 3 to 5 terminal branches; MA and CUA unbranched, MP forked. HANDLIRSCH, 1906b, 1919b; CARPENTER, 1951. *U. Carb.*, Europe (France).——FIG. 29,8. **A. triangularis*; *a,* fore and *b,* hind wings, ×1.7 (Carpenter, 1951).

Family ANCHINEURIDAE Carpenter, 1963

[Anchineuridae CARPENTER, 1963a, p. 44]

Wing elongate-oval (base unknown), anterior margin smoothly curved; vein SC very close to costal margin and terminating near wing apex; R parallel and close to SC; RS with numerous branches; MA free from RS and CUA free from MP; crossveins numerous, irregular, in some areas forming a reticulation; costal margin serrate, with setae; hind margin and some veins with small setae. U. Carb.

Anchineura CARPENTER, 1963a, p. 46 [*A. hispanica*; OD]. RS with 6 main branches; MA and CUP unbranched; MP and CUA branched. *U. Carb.*, Europe (Spain).——FIG. 29,10. **A. hispanica*; wing, ×1.5 (Carpenter, 1963a).

Family ASPIDOHYMENIDAE Martynov, 1930

[Aspidohymenidae MARTYNOV, 1930a, p. 80]

Veins SC and R close together; costal space very narrow; MA not anastomosed with RS; anterior branch of RS apparently coalesced

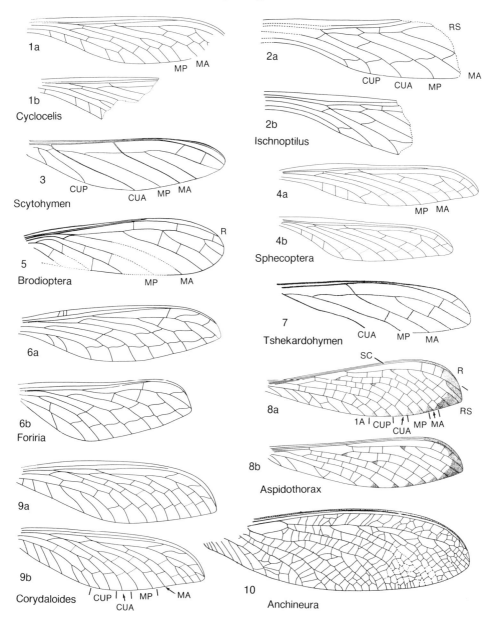

Fig. 29. Aspidothoracidae, Anchineuridae, Corydaloididae, Brodiopteridae, Foririidae, Sphecopteridae, Ischnoptilidae, and Scytohymenidae (p. 46–51).

with R for a short interval; few crossveins, arranged in 2 rows. *Perm.*

Aspidohymen MARTYNOV, 1930a, p. 80 [**A. extensus*; OD]. MA and MP unbranched. CARPENTER, 1930d; MARTYNOV, 1937b; ZALESSKY, 1937c. *Perm.,* USSR (European RSFSR).——FIG. 30,2. **A. extensus*; distal half of wing, ×2 (Martynov, 1930a).

Family CORYDALOIDIDAE Handlirsch, 1906

[Corydaloididae HANDLIRSCH, 1906b, p. 314]

Wings broad, as in Mischopteridae. Vein MA anastomosed for very short distance with RS, and CUA anastomosed with M for longer distance. *U. Carb.*

Corydaloides BRONGNIART, 1885a, p. 64 [*C. scudderi; SD HANDLIRSCH, 1922, p. 201] [=Corydaliodes HANDLIRSCH, 1919b, p. 579 (type, Corydaloides simplex BRONGNIART, 1893, p. 476, pl. 32, fig. 8,9)]. MA unbranched, MP deeply forked; crossveins in main part of wing arranged in 3 rows. HANDLIRSCH, 1906b; CARPENTER, 1951. U. Carb., Europe (France).——FIG. 29,9. *C. scudderi; a, fore and b, hind wings, ×1 (Carpenter, 1951).

Family BRODIOPTERIDAE
Carpenter, 1963

[Brodiopteridae CARPENTER, 1963b, p. 59]

Little-known family, probably related to the Corydaloididae. Wing broad basally, but vein SC clearly terminating on costa and MA not fused with RS. U. Carb.

Brodioptera COPELAND, 1957, p. 53 [*B. cumberlandensis; OD]. SC terminating a little beyond midwing; RS with several branches; MA, MP, CUA, and CUP unbranched. CARPENTER, 1963b. U. Carb., Canada (Nova Scotia).——FIG. 29,5. *B. cumberlandensis; wing, ×3.5 (Carpenter, 1963b).

Family FORIRIIDAE
Handlirsch, 1919

[Foririidae HANDLIRSCH, 1919b, p. 577]

Vein SC clearly terminating on R; MA free from RS, and CUA free from MP. U. Carb.

Foriria MEUNIER, 1908g, p. 172 [*F. maculata; OD]. RS with 3 terminal branches; other main veins unbranched; crossveins mainly sigmoidal, arranged in 2 rows. HANDLIRSCH, 1919b; CARPENTER, 1951. U. Carb., Europe (France).——FIG. 29,6. *F. maculata; a, fore and b, hind wings, ×1.5 (Carpenter, 1951).

Family SPHECOPTERIDAE
Carpenter, 1951

[Sphecopteridae CARPENTER, 1951, p. 345]

Wings slender, petiolate; vein SC clearly terminating on R; MA anastomosing with RS for very short distance in fore wing and usually in hind wing; crossveins fewer than in Mischopteridae and forming only 1 complete row in main part of wing. U. Carb.

Sphecoptera BRONGNIART, 1893, p. 294 [*S. gracilis; SD HANDLIRSCH, 1922, p. 205]. Crossveins slightly sigmoidal; MP unbranched. HANDLIRSCH, 1906b; CARPENTER, 1951; ROHDENDORF, 1962a. U. Carb., Europe (France).——FIG. 29,4. *S. gracilis; a, fore and b, hind wings, ×1 (Carpenter, 1951).

Cyclocelis BRONGNIART, 1893, p. 290 [*C. chatini; SD HANDLIRSCH, 1922, p. 204]. Crossveins straight; MP deeply forked. HANDLIRSCH, 1906b; CARPENTER, 1951. U. Carb., Europe (France). ——FIG. 29,1. *C. chatini; a, fore and b, hind wings, ×1 (Carpenter, 1951).

Family ISCHNOPTILIDAE
Carpenter, 1951

[Ischnoptilidae CARPENTER, 1951, p. 349]

Wings slender, petiolate; vein MA anastomosed with RS for much greater interval than in Sphecopteridae; CUA anastomosed with MP for short interval. U. Carb.

Ischnoptilus BRONGNIART, 1893, p. 296 [*I. elegans; OD]. Crossveins sigmoidal, forming single row. HANDLIRSCH, 1906b; CARPENTER, 1951; ROHDENDORF, 1962a. U. Carb., Europe (France). ——FIG. 29,2. *I. elegans; a, fore and b, hind wings, ×2.6 (Carpenter, 1951).

Family MISCHOPTERIDAE
Handlirsch, 1906

[Mischopteridae HANDLIRSCH, 1906b, p. 316]

Vein SC more remote from margin of wing than in Aspidothoracidae; MA anastomosed for very short interval with RS; crossveins regularly arranged, forming 2 or 3 rows over most of wing. Prothorax very short, with or without lateral spines. U. Carb.

Mischoptera BRONGNIART, 1893, p. 283, nom. subst. pro Woodwardia BRONGNIART, 1885a, p. 64, non CROSSE & FISCHER, 1861 [*Woodwardia nigra BRONGNIART, 1885a, p. 64; SD HANDLIRSCH, 1922, p. 203]. Both fore and hind wings falcate; most crossveins strongly sigmoidal; circular, cuticular thickenings regularly distributed over each wing. Prothorax with lateral spines. Nymph with similar lateral spines and similar venation in wing buds. CARPENTER, 1951. U. Carb., Europe (France), USA (Illinois).——FIG. 30,3a–c. *M. nigra (BRONGNIART); a, whole insect, ×0.4, b, fore wing, ×0.7, c, hind wing, ×0.7 (Carpenter, 1951).——FIG. 30,3d,e. M. douglassi CARPENTER & RICHARDSON, Illinois; nymph, d, ×1.5, e, ×1.9 (Carpenter & Richardson, 1968).

Psilothorax BRONGNIART, 1893, p. 288 [*Woodwardia longicauda BRONGNIART, 1885a, p. 64; OD]. Fore wing with an evenly rounded posterior margin, not falcate; crossveins only slightly sigmoidal; cuticular thickenings absent. HANDLIRSCH, 1906b; CARPENTER, 1951; ROHDENDORF, 1962a. U. Carb., Europe (France).

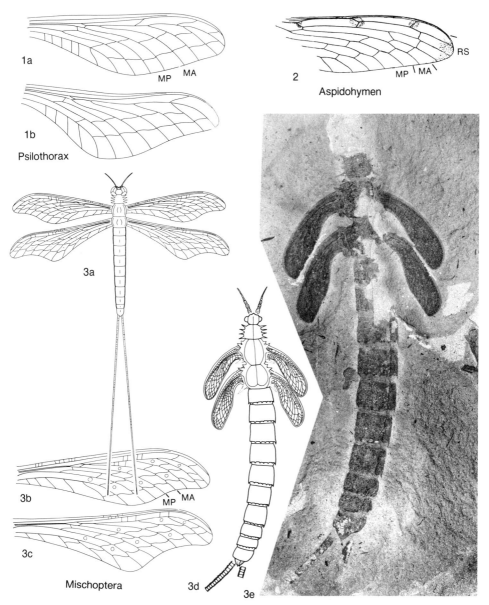

FIG. 30. Aspidohymenidae and Mischopteridae (p. 47–49).

——FIG. 30,*1*. **P. longicauda* (BRONGNIART); *a*, fore and *b*, hind wings, ×0.8 (Carpenter, 1951).

Family PROTOHYMENIDAE Tillyard, 1924

[Protohymenidae TILLYARD, 1924a, p. 112]

Veins SC and R close to one another and to costal margin; several crossveins (usually 10 to 12) present; MA and RS anastomosed. *Perm.*

Protohymen TILLYARD, 1924a, p. 113 [**P. permianus*; OD] [=*Pseudohymen* MARTYNOV, 1932, p. 5 (type, *P. angustipennis*); *Pseudohymenopsis* ZALESSKY, 1956b, p. 1089 (type, *P. concinna*)]. Wings petiolate or subpetiolate; crossvein between 1A and hind margin remote from wing

Hexapoda

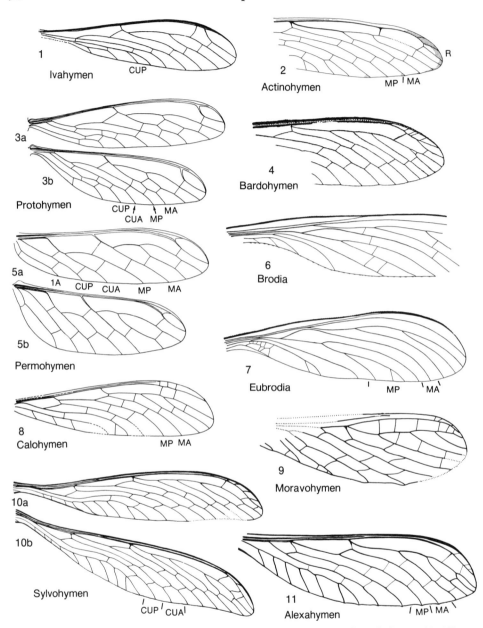

FIG. 31. Protohymenidae, Bardohymenidae, Moravohymenidae, and Brodiidae (p. 49–52).

base; CUP unbranched. CARPENTER, 1947; ROHDENDORF, 1962a. *Perm.,* USA (Kansas, Oklahoma), USSR (Asian RSFSR).——FIG. 31,*3*. **P. permianus; a,* fore and *b,* hind wings, ×4 (Carpenter, 1947).

Ivahymen MARTYNOV, 1932, p. 9 [**I. constrictus;* OD]. Wings petiolate; CUP forked distally.

ROHDENDORF, 1962a. *Perm.,* USSR (European RSFSR).——FIG. 31,*1*. **I. constrictus;* wing, ×3 (Martynov, 1932).

Permohymen TILLYARD, 1924a, p. 115 [**P. schucherti;* OD]. Wings broad basally; crossveins between 1A and margin near base of wing; CUP unbranched. CARPENTER, 1930d, 1947. *Perm.,*

USA (Kansas, Oklahoma).——FIG. 31,5. *P. schucherti; a, fore and b, hind wings, ×4 (Carpenter, 1930d).

Family SCYTOHYMENIDAE Martynov, 1937

[Scytohymenidae MARTYNOV, 1937b, p. 58]

Veins SC and R close to one another and to costal margin; crossveins very few (usually less than 6); MA and RS anastomosed. *Perm.*

Scytohymen MARTYNOV, 1937b, p. 58 [*S. extremus; OD]. Stem of RS beyond MA about one-third as long as branches of RS; no crossveins between branches of RS or between MA and MP. ROHDENDORF, 1962a. *Perm.,* USSR (Asian RSFSR).——FIG. 29,3. *S. extremus; wing, ×1.3 (Martynov, 1937b).
Tshekardohymen ROHDENDORF, 1940a, p. 106 [*T. martynovi; OD]. Stem of RS beyond MA about half as long as branches of RS; a single crossvein between branches of RS and between MA and MP. ROHDENDORF, 1962a. *Perm.,* USSR (Asian RSFSR).——FIG. 29,7. *T. martynovi; wing, ×1.7 (Rohdendorf, 1962a).

Family BARDOHYMENIDAE Zalessky, 1937

[Bardohymenidae ZALESSKY, 1937a, p. 601]

Fore and hind wings similar in shape and venation. Vein SC weak and very close to costa, usually obsolescent at least by midwing; R close to SC and costa, except distally; R with short terminal branches; RS arising somewhat before midwing, with from 2 to 5 branches; stem of M very close to R basally, but diverging well before origin of RS; a strong crossvein connecting MA to R or RS; CU coalesced at base with stem of M; CUA connected to M by strong crossvein; 1A well developed, with series of veinlets leading to hind margin of wing; veins and wing margin bearing rows of setae or setal sockets. Head apparently short and broad; antennae long, setaceous; meso- and metathorax large; female with prominent ovipositor. *Perm.*

Bardohymen ZALESSKY, 1937a, p. 602 [*B. magnipennifer; OD]. RS with 5 terminal branches. ROHDENDORF, 1962a; KUKALOVÁ-PECK, 1972. *Perm.,* USSR (Asian RSFSR).——FIG. 31,4. *B. magnipennifer; wing, ×1 (Zalessky, 1937a).
Actinohymen CARPENTER, 1962, p. 37 [*A. russelli; OD]. RS with 3 terminal branches; wing broadest near midwing. *Perm.,* USA (Texas).——FIG. 31,2. *A. russelli; apical half of wing, ×1.7 (Carpenter, 1962).
Alexahymen KUKALOVÁ-PECK, 1972, p. 254 [*A. maruska; OD]. Similar to *Sylvohymen* but wings short and broad; 1A not sigmoidal; apex broadly rounded. *Perm.,* Europe (Czechoslovakia).——FIG. 31,11. *A. maruska; hind wing, ×2 (Kukalová-Peck, 1972).
Calohymen CARPENTER, 1947, p. 30 [*C. permianus; OD]. RS with 3 terminal branches; wing broadest apically. *Perm.,* USA (Oklahoma).——FIG. 31,8. *C. permianus; wing, ×2 (Carpenter, 1947).
Sylvohymen MARTYNOV, 1940, p. 10 [*S. robustus; OD]. Wings long and slender, tapered markedly in basal third; hind margin of fore wing nearly straight, that of hind wing smoothly curved; RS with 4 terminal branches; 1A sigmoidal in both wings; thorax and abdomen with prominent spines. CARPENTER, 1947, 1962; KUKALOVÁ-PECK, 1972; ROHDENDORF, 1962a. *Perm.,* USSR (Asian RSFSR), USA (Oklahoma).——FIG. 31,10. *S. sibiricus* KUKALOVÁ-PECK, USSR; a, fore and b, hind wings, ×1.3 (Kukalová-Peck, 1972).

Family MORAVOHYMENIDAE Kukalová-Peck, 1972

[Moravohymenidae KUKALOVÁ-PECK, 1972, p. 256]

Little-known family, apparently related to Bardohymenidae. Hind wing broadest beyond midwing; vein SC more remote from costa than in Bardohymenidae and R remote from RS distally; MA, MP, and branches of RS curving slightly anteriorly in distal portions. *Perm.*

Moravohymen KUKALOVÁ-PECK, 1972, p. 256 [*M. vitreus; OD]. Little-known genus; MA, MP, CUA, and CUP unbranched; 1A not parallel to hind margin of wing. *Perm.,* Europe (Czechoslovakia).——FIG. 31,9. *M. vitreus; hind wing, ×4 (Kukalová-Peck, 1972).

Family BRODIIDAE Handlirsch, 1906

[Brodiidae HANDLIRSCH, 1906b, p. 113]

Wings petiolate; entire wing margin serrate, costal margin more distinctly so; vein SC close to R basally and parallel to it for most of its length, apparently merging with costa beyond midwing, but retaining its identity until near wing apex; R nearly straight to level of midwing, unbranched, and curving slightly away from costal margin distally; RS arising before midwing, with 3 or 4 branches; M independent of R basally; MA unbranched, diverging anteriorly toward RS but not coalescing with it; MP dividing shortly after level of origin of RS, with at least 2 terminal branches; CUA unbranched; 1A present. Only a few distinct crossveins on wing, but a fine archedictyon present in some genera over much of wing. Differences between fore and hind wings unknown, body structure unknown. *U. Carb.*

Brodia SCUDDER, 1881b, p. 293 [*B. priscocincta*; OD]. Wings broadest at midwing; several distinct, transverse crossveins present, but most of wing surface with uniform pattern of weak crossveins, very close together, not forming an archedictyon. CARPENTER, 1967a. *U. Carb.*, England.——FIG. 31,6. *B. priscocincta*; wing, ×1.2 (Carpenter, 1967a).

Eubrodia CARPENTER, 1967a, p. 73 [*E. dabasinskasi*; OD]. Similar to *Brodia*, but wing broadest beyond middle; no distinct transverse crossveins present, but archedictyon covering most of wing surface. CARPENTER & RICHARDSON, 1971. *U. Carb.*, USA (Illinois).——FIG. 31,7. *E. dabasinskasi*; wing, ×1.2 (Carpenter & Richardson, 1971).

Family ANCOPTERIDAE
Kukalová-Peck, 1975

[Ancopteridae KUKALOVÁ-PECK, 1975, p. 10]

Wings slender, but apparently not petiolate; hind margin slightly undulated; vein SC extending well beyond midwing; bases of R and M not coalesced; RS arising well before midwing; MA and CUA unbranched; MP and CUP branched; crossveins numerous, sometimes reticulate and forming intercalary veins. *Perm.*

Ancoptera KUKALOVÁ-PECK, 1975, p. 12 [*A. permiana*; OD]. Apex of wing broadly rounded; SC and R close to costal margin of wing distally. *Perm.*, Europe (Czechoslovakia).——FIG. 32,1. *A. permiana*; wing, ×1.8 (Kukalová-Peck, 1975).

Family VORKUTIIDAE
Rohdendorf, 1947

[Vorkutiidae ROHDENDORF, 1947, p. 391]

Little-known family, based on wing fragments. Vein SC coalesced with R beyond level of origin of RS; 2 crossveins between RS and MA. *U. Carb.–Perm.*

Vorkutia ROHDENDORF, 1947, p. 391 [*V. tshernovi*; OD]. Little-known genus. No crossvein between stem of RS and MA. ROHDENDORF, 1962a. *Perm.*, USSR (Asian RSFSR).

Sibiriohymen ROHDENDORF, 1961a, p. 76 [*S. asiaticus*; OD]. One crossvein between stem of RS and MA. *U. Carb.*, USSR (Asian RSFSR).

Family ALECTONEURIDAE
Kukalová-Peck, 1975

[Alectoneuridae KUKALOVÁ-PECK, 1975, p. 15]

Little-known family, based on wing fragment. Wings narrow basally but not petiolate; vein SC extending beyond midwing; RS arising before midwing; MA coalescing briefly with RS just after its origin; 1A with long branches. *Perm.*

Alectoneura KUKALOVÁ-PECK, 1975, p. 15 [*A. europaea*; OD]. Subcostal area relatively broad in basal half of wing, with oblique veinlets to wing margin; MA and MP separating at about level of origin of RS. *Perm.*, Europe (Czechoslovakia).——FIG. 32,3. *A. europaea*; wing, ×5 (Kukalová-Peck, 1975).

Family HANIDAE
Kukalová-Peck, 1991

[Hanidae KUKALOVÁ-PECK, 1991, p. 193]

Little-known family, based on wing fragments. Wings apparently very slender and petiolate, broadest near midwing; hind margin undulated beyond midwing; vein RS arising at about midwing; CUA apparently unbranched; crossveins very numerous, forming a network in posterior area of wing. *Perm.*

Hana KUKALOVÁ-PECK, 1991, p. 193 [*H. filia*; OD]. RS with 3 or 4 branches distally; 1A closely following posterior margin of wing; wing membrane with a dense covering of tubercles. *Perm.*, Europe (Czechoslovakia).——FIG. 32,7a. *H. filia*; distal part of fore wing, ×1.4 (Kukalová-Peck, 1975).——FIG. 32,7b. *H. lineata* KUKALOVÁ-PECK; basal part of wing, ×1.4 (Kukalová-Peck, 1975).

Family ARCIONEURIDAE
Kukalová-Peck, 1975

[Arcioneuridae KUKALOVÁ-PECK, 1975, p. 8]

Little-known family, based on nymphal wing pad and fragments of adult wing. Wing

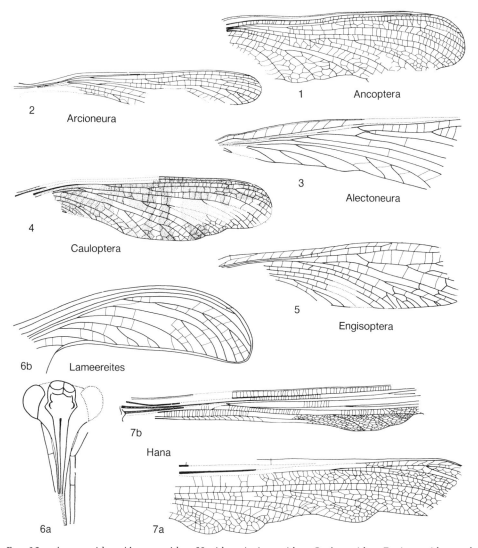

FIG. 32. Ancopteridae, Alectoneuridae, Hanidae, Arcioneuridae, Caulopteridae, Engisopteridae, and Uncertain (p. 52–54).

slender, petiolate; vein SC short; RS arising before midwing; MA and MP with several branches; CUA and CUP with few branches; crossveins numerous, forming intercalary veins. *Perm.*

Arcioneura Kukalová-Peck, 1975, p. 9 [**A. juveniles*; OD]. Little-known genus, based on nymphal wing pad. R close to costal margin of wing distally; MA with short branches; MP forked at about level of origin of RS. *Perm.,* Europe (Czechoslovakia).——Fig. 32,2. **A. juveniles*; nymphal wing pad, ×3.6 (Kukalová-Peck, 1975).

Anconeura Kukalová-Peck, 1975, p. 10 [**A. havlatai*; OD]. Little-known genus, based on isolated adult wing. Similar to *Arcioneura* but with more branches on RS and CUA. *Perm.,* Europe (Czechoslovakia).

Family CAULOPTERIDAE
Kukalová-Peck, 1975

[Caulopteridae Kukalová-Peck, 1975, p. 4]

Little-known family, based on nymphal wing pad; wing slender, probably petiolate; hind margin slightly undulated; vein SC long, extending nearly to apex of wing; bases of R and M apparently not coalesced; RS arising before midwing; both MA and MP branched; MP coalesced with CUA for short distance, just before level of origin of RS; CUA unbranched; CUP branched; crossveins numerous, irregular, forming in some areas short intercalary veins. *Perm.*

Cauloptera Kukalová-Peck, 1975, p. 4 [*C. colorata*; OD]. RS with 2 long branches; MA with several short branches. *Perm.*, Europe (Czechoslovakia).——Fig. 32,4. *C. colorata*; nymphal wing pad, ×1.8 (Kukalová-Peck, 1975).

Family ENGISOPTERIDAE Kukalová-Peck, 1975

[Engisopteridae Kukalová-Peck, 1975, p. 13]

Little-known family, based on wing fragment. Wing elongate-oval; vein SC extending well beyond midwing; RS arising just beyond midwing; MA with numerous branches; CUA unbranched, CUP with long branches; crossveins numerous. *Perm.*

Engisoptera Kukalová-Peck, 1975, p. 13 [*E. simplices*; OD]. RS remote from R for most of its length, with 2 main branches. *Perm.*, Europe (Czechoslovakia).——Fig. 32,5. *E. simplices*; wing, ×4 (Kukalová-Peck, 1975).

Family UNCERTAIN

The genus described below, apparently belonging to the order Megasecoptera, is too poorly known to permit family assignment.

Lameereites Handlirsch, 1911a, p. 375 [*L. curvipennis*; OD]. Little-known genus, based on nymphal wing pads and parts of body; crossveins not arranged in rows. Carpenter & Richardson, 1968. *U. Carb.*, USA (Illinois).——Fig. 32,6. *L. curvipennis*; a, head, front view, ×6; b, wing pad, ×3.8 (Carpenter & Richardson, 1968).

Order DIAPHANOPTERODEA Handlirsch, 1919

[*nom. correct.* Rohdendorf, 1962a, p. 69, *pro* Diaphanopteroidea Handlirsch, 1919b, p. 575] [=Palaeohymenoptera Haupt, 1941, p. 99]

Palaeoptera resembling Megasecoptera but with wings held backward along abdomen at rest; fore and hind wings similar in general form and venation; all main longitudinal veins present, with alternation of convexities and concavities; archedictyon absent, crossveins distinct; in most families, stems of veins R and M very close together, even contiguous, forming a distinct curve; at distal end of this curve MA and MP separating from R, MA nearly bisecting the angle formed by R and MP; head with haustellate beak; cerci very long, as in Megasecoptera; ovipositor present and well developed. Immature stages unknown. *U. Carb.–Perm.*

The ordinal relationships of this series of families have been problematical. Most of the species included have previously been assigned to the Megasecoptera because of their very similar wing venation and body structure (e.g., the haustellate beak and long cerci). However, specimens of all families of Diaphanopterodea in which both wings and body are known (i.e., Diaphanopteridae, Prochoropteridae, Martynoviidae, and Asthenohymenidae) are preserved with the wings placed backward along the abdomen, much as in the neopterous insects. On the other hand, the similarity of their wing venation and haustellate mouthparts to those of the Megasecoptera shows that they are actually members of the Palaeoptera that developed a mechanism for flexing the wings back along the abdomen when at rest. That this mechanism was developed independently of the Neoptera seems virtually certain; their haustellate mouthparts eliminate them as ancestral stock of the primitive Neoptera (i.e., Perlaria and Orthopteroidea). Also, the articular plates (pteralia) of the wing bases of the Diaphanopterodea lack the third axillary characteristic of the wing-flexing mechanism of the Neoptera (Kukalová-Peck, 1974b).

The wing venation seems to have evolved along similar lines in the Diaphanopterodea and Megasecoptera, as shown in the Protohymenidae of the Megasecoptera and the Asthenohymenidae of the Diaphanopterodea. Consequently, specimens consisting of isolated wings or especially of fragments of wings cannot be assigned to either order with confidence. The most characteristic feature of the diaphanopterodean venation is the curvature of R+M basally and the separation of MA and MP just beyond that point.

Family DIAPHANOPTERIDAE Handlirsch, 1906

[Diaphanopteridae HANDLIRSCH, 1906b, p. 313] [=Diaphanopteritidae HANDLIRSCH, 1919b, p. 575]

Fore and hind wings similar. Vein SC terminating on R slightly beyond midwing; MA diverging from MP directly after its origin and just touching RS before continuing as an independent vein; CUA coalesced with base of M. Several large, thickened, circular spots on membrane of both wings. *U. Carb.*

Diaphanoptera BRONGNIART, 1893, p. 308 [*D. munieri*; SD HANDLIRSCH, 1922, p. 200] [=*Diaphanopterites* HANDLIRSCH, 1919b, p. 576 (type, *Diaphanoptera superba* MEUNIER, 1908b); *Pseudanthracothremma* HANDLIRSCH, 1906b, p. 324 (type, *Anthracothremma scudderi* BRONGNIART, 1893, p. 329)]. RS3+4 and distal part of MA nearly parallel. CARPENTER, 1963d; CARPENTER & RICHARDSON, 1978. *U. Carb.*, Europe (France).——FIG. 33,*5a,b. D. superba*; *a*, fore and *b*, hind wings, ×1.5 (Carpenter, 1963d).——FIG. 33,*5c.* *D. munieri*; fore wing, ×1.5 (Carpenter, 1963d).——FIG. 34. *D. superba*; dorsal view, ×1.8 (Carpenter, new).
Philiasptilon ZALESSKY, 1932a, p. 217 [*P. maculosum*; OD]. Apparently similar to *Diaphanoptera*, but RS3+4 and distal part of MA convergent. [Family assignment uncertain.] PINTO & ORNELLAS, 1978a. *U. Carb.*, USSR (Asian RSFSR), Argentina (Province San Luis).——FIG. 33,*8.* *P. maculosum*; wing, ×1 (Zalessky, 1932a).

Family PROCHOROPTERIDAE Handlirsch, 1911

[Prochoropteridae HANDLIRSCH, 1911, p. 375]

Little-known family. Wings slender; fore wing with vein SC terminating on R well beyond origin of RS; MA anastomosed with RS for short interval; MP with 3 short, terminal branches; crossveins numerous. Hind wing slightly broader than fore wing, but with similar venation basally. Ovipositor long; cerci about twice as long as entire body. *U. Carb.*

Prochoroptera HANDLIRSCH, 1911, p. 376 [*P. calopteryx*; OD]. Fore wing broadest distally. CARPENTER & RICHARDSON, 1978. *U. Carb.*, USA (Illinois).——FIG. 33,*2.* *P. calopteryx*; *a*, fore and *b*, hind wings, ×2.5 (Carpenter & Richardson, 1978).
Euchoroptera CARPENTER, 1940b, p. 638 [*E. longipennis*; OD]. Fore wing broadest near midwing. *U. Carb.*, USA (Kansas).——FIG. 33,*6.* *E. longipennis*; wing, ×2.5 (Carpenter, 1940b).

Family ELMOIDAE Tillyard, 1937

[Elmoidae TILLYARD, 1937a, p. 82]

Fore wing narrow; costal margin slightly arched basally; vein SC terminating on R; RS with 3 terminal branches; MA not coalesced with RS; MP with deep fork; CUA and CUP unbranched; 2 anal veins. Hind wing oval, shorter than fore wing but with similar venation. *Perm.*

Elmoa TILLYARD, 1937a, p. 82 [*E. trisecta*; OD]. SC terminating only slightly beyond level of origin of RS; MP forked to about half its length. ZALESSKY, 1937b; CARPENTER, 1943a, 1947. *Perm.*, USA (Kansas, Oklahoma).——FIG. 33,*1.* *E. trisecta*; *a*, fore and *b*, hind wings, ×2.2 (Carpenter, 1943a).

Family PARELMOIDAE Rohdendorf, 1962

[Parelmoidae ROHDENDORF, 1962a, p. 71]

Vein SC terminating on costal margin a short distance beyond midwing; costal margin strongly curved basally; MA not coalesced with RS; RS with 3 long branches; MP deeply forked; CUA unbranched; 3 anal veins; hind margin of wing angular basally. *Perm.*

Parelmoa CARPENTER, 1947, p. 28 [*P. revelata*; OD]. MA and MP diverging just beyond level of origin of RS. ROHDENDORF, 1962a. *Perm.*, USA (Oklahoma).——FIG. 33,*3.* *P. revelata*; fore wing, ×3.5 (Carpenter, 1947).
Pseudelmoa CARPENTER, 1947, p. 29 [*P. ampla*; OD]. MA and MP diverging far beyond level of origin of RS and nearly at level of first fork of RS. ROHDENDORF, 1962a. *Perm.*, USA (Oklahoma).——FIG. 33,*7.* *P. ampla*; fore wing, ×2.5 (Carpenter, 1947).

Family MARTYNOVIIDAE Tillyard, 1932

[Martynoviidae TILLYARD, 1932a, p. 13]

Fore wing moderately slender; costal area broad as far as midwing, much narrowed distally; vein SC terminating on R at about

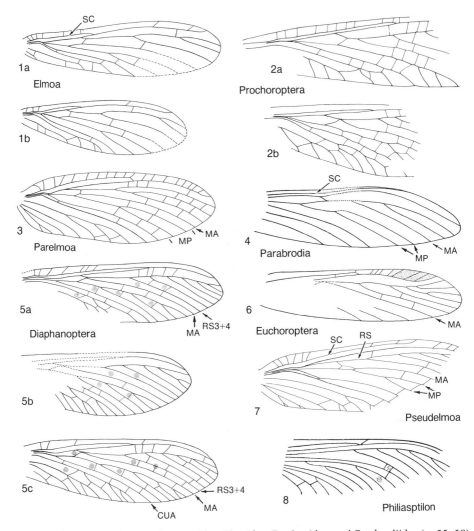

FIG. 33. Diaphanopteridae, Prochoropteridae, Elmoidae, Parelmoidae, and Parabrodiidae (p. 55–58).

midwing; RS with 3 to 5 terminal branches; MA coalesced basally with R or RS or both; MP, CUA, and CUP unbranched; 2 or 3 anal veins. Hind wing similar to fore wing in venation but broader, with strongly curved posterior margin. *Perm.*

Martynovia TILLYARD, 1932a, p. 14 [*M. insignis*; OD] [=*Martynoviella* TILLYARD, 1932a, p. 17 (type, *M. protohymenoides*)]. MA coalesced with RS only, not with R. CARPENTER, 1931b, 1943a, 1947. *Perm.,* USA (Kansas, Oklahoma).──FIG. 35,5. *M. insignis*; *a,* fore and *b,* hind wings, ×3.5 (Carpenter, 1943a).

Eumartynovia CARPENTER, 1947, p. 33 [*E. raaschi*; OD]. MA coalesced with stem of R and with RS for similar lengths. *Perm.,* USA (Oklahoma).──FIG. 35,1. *E. raaschi*; fore wing, ×2.5 (Carpenter, 1947).

Phaneroneura CARPENTER, 1947, p. 33 [*P. martynovae*; OD]. MA coalesced with R for a much greater interval than with RS. *Perm.,* USA (Oklahoma).── FIG. 35,3. *P. martynovae*; fore wing, ×4.3 (Carpenter, 1947).

Family BIARMOHYMENIDAE Zalessky, 1937

[Biarmohymenidae ZALESSKY, 1937b, p. 609]

Little-known family. Vein SC remote from R; costal space very broad; MA coalesced with RS from origin of RS nearly to its mid-

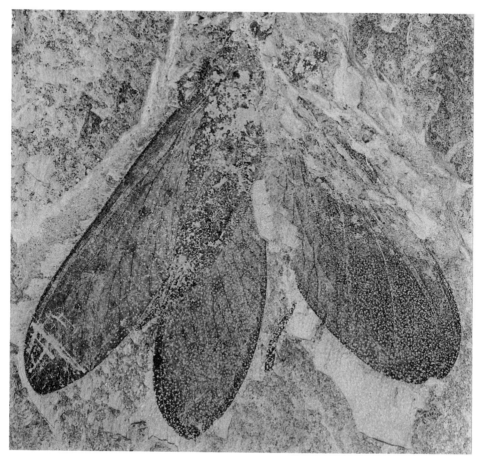

Diaphanoptera

FIG. 34. Diaphanopteridae (p. 55).

point; M coalesced with CUA proximally; RS with 2 dichotomous forks. *Perm.*

Biarmohymen ZALESSKY, 1937b, p. 609 [*B. bardense*; OD]. MA and MP unbranched; pterostigma very long. ROHDENDORF, 1962a. *Perm., USSR (Asian RSFSR).*——FIG. 35,4. *B. bardense*; wing, ×2.2 (Zalessky, 1937b).

Family ASTHENOHYMENIDAE Tillyard, 1924

[Asthenohymenidae TILLYARD, 1924a, p. 117]

Small species, with similar fore and hind wings. Stems of all main veins crowded together toward anterior margin of wing base; MA coalesced with R and part of RS; MP coalesced with CUA. Antennae long, with about 24 segments; ovipositor short; cerci very long, about twice as long as body proper, consisting of about 85 segments. *Perm.*

Asthenohymen TILLYARD, 1924, p. 117 [*A. dunbari*; OD] [=*Karoohymen* RIEK, 1976a, p. 757 (type, *K. delicatulus*)]. RS with 2 branches; MA, MP, CUA, and CUP unbranched. CARPENTER, 1930d, 1931b, 1933a, 1939, 1943a, 1947. *Perm., USA (Kansas, Oklahoma), South Africa (Natal).*——FIG. 35,2. *A. apicalis* CARPENTER, Oklahoma; fore wing, ×12.0 (Carpenter, 1947). ——FIG. 36. *A. dunbari*, Kansas; dorsal view of complete insect, ×4.5 (Carpenter, 1939).

Family RHAPHIDIOPSIDAE Handlirsch, 1906

[Rhaphidiopsidae HANDLIRSCH, 1906b, p. 319]

Vein SC close to costal margin, apparently terminating at wing apex; MA coalesced with

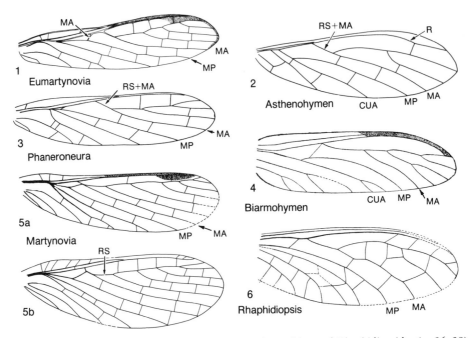

FIG. 35. Martynoviidae, Biarmohymenidae, Asthenohymenidae, and Rhaphidiopsidae (p. 56–58).

RS for short interval; crossveins few, irregular. *U. Carb.*

Rhaphidiopsis SCUDDER, 1893b, p. 11 [**R. diversipenna*; OD]. MA and MP unbranched; RS with deep fork. CARPENTER, 1933a. *U. Carb.*, USA (Rhode Island).——FIG. 35,6. **R. diversipenna*; wing, ×1.6 (Carpenter, 1933a).

Family PARABRODIIDAE Carpenter, 1933

[Parabrodiidae CARPENTER, 1933b, p. 365]

Vein SC terminating well before apex of wing; MA anastomosed with RS. RS with a single long fork. [Ordinal position uncertain.] *U. Carb.*

Parabrodia CARPENTER, 1933b, p. 366 [**P. carbonaria*; OD]. MA unbranched; MP deeply forked. *U. Carb.*, USA (Kansas).——FIG. 33,4. **P. carbonaria*; wing, ×2 (Carpenter, 1933b).

Family UNCERTAIN

The following genera, which were originally placed in the family Elmoidae, show so much diversity in wing venation and wing form that their separation into a distinct family (or families) seems advisable. Most are based on isolated wings, virtually nothing being known of their body structure.

Diapha KUKALOVÁ-PECK, 1974a, p. 323 [**P. candida*; OD]. Fore wing slender and long; SC terminating on R just beyond midwing; MA coalesced for a short distance with RS; RS with 6 or 7 terminal branches. *Perm.*, Europe (Czechoslovakia).——FIG. 37,2. **P. candida*; fore wing, ×3 (Kukalová-Peck, 1974a).

Elmodiapha KUKALOVÁ-PECK, 1974a, p. 320 [**E. ovata*; OD]. Fore wing broadly rounded; SC terminating on R beyond midwing; RS with 6 terminal branches; RS and MA not coalesced. *Perm.*, Europe (Czechoslovakia).——FIG. 37,1. **E. ovata*; fore wing, ×3 (Kukalová-Peck, 1974a).

Paradiapha KUKALOVÁ-PECK, 1974a, p. 329 [**P. delicatula*; OD]. Fore wing little known. Hind wing moderately slender, broadest beyond midwing; SC terminating on R beyond midwing; RS with 3 terminal branches; MA very slightly coalesced with RS. *Perm.*, Europe (Czechoslovakia).——FIG. 37,4. **P. delicatula*; hind wing, ×4 (Kukalová-Peck, 1974a).

Permodiapha KUKALOVÁ-PECK, 1974a, p. 323 [**P. carpenteri*; OD]. Fore wing broad; SC terminating on R well beyond midwing; MA coalesced with RS for short interval; RS apparently with 4 terminal branches. Hind wing much broader than fore wing. *Perm.*, Europe (Czechoslovakia). ——FIG. 37,6. **P. carpenteri*; hind wing, ×3.75 (Kukalová-Peck, 1974a).

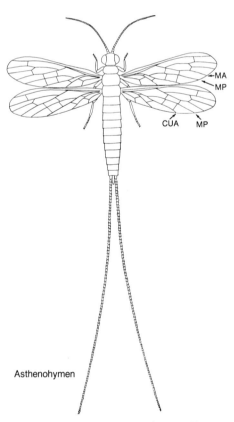

Fig. 36. Asthenohymenidae (p. 57).

Protodiapha KUKALOVÁ-PECK, 1974a, p. 321 [*P. maculifera*; OD]. Hind wing very broad; SC terminating on R beyond midwing; MA and RS not coalesced but connected by very short crossvein; RS with 5 terminal branches. *Perm.*, Europe (Czechoslovakia).——FIG. 37,5. *P. maculifera*; hind wing, ×3.7 (Kukalová-Peck, 1974a).

Stenodiapha KUKALOVÁ-PECK, 1974a, p. 327 [*S. angusta*; OD]. Fore wing elongate and slender; SC terminating on R well beyond midwing; RS and MA coalesced for short interval; RS with 4 terminal branches. *Perm.*, Europe (Czechoslovakia).——FIG. 37,3. *S. angusta*; fore wing, ×3 (Kukalová-Peck, 1974a).

Order PROTODONATA Brongniart, 1893

[Protodonata BRONGNIART, 1893, p. 394] [=Meganisoptera MARTYNOV, 1932, p. 17] [FRASER's erroneous comment (1957, p. 21) on the name of this order has caused much confusion. For a full account of this subject, see CARPENTER, 1960b.]

Large to very large insects. Wings subequal, with similar venation; fore wing usually more slender and slightly longer than hind wing; setal bases rarely present on wing membrane of some Meganeuridae; anterior margin of wings usually serrate; nodus and pterostigma absent; precostal area present, usually well developed; vein SC extending at least to midwing, usually nearly to wing apex; R long, extending to apex, unbranched; RS arising near wing base and forking before or near midwing, with many branches, including intercalary veins; MA coalesced with R basally, separating from R along with RS (Fig. 38,4a) or fused with RS for a short distance, forming an incipient arculus (Fig. 38,3); MA with numerous branches and intercalary veins; MP apparently absent; CUA also absent or reduced to a short vestigial vein at wing base; CUP strong, sinuously curved, unbranched; 1A long, extending to about midwing, with numerous branches extending to hind margin of wing. Body structure known only in a few Meganeuridae; head globose, with large, dentate mandibles; thorax large, legs stout and spinous; abdomen long and slender, apparently similar to that of the Odonata in general form. Immature stages unknown. *U. Carb.–Perm.*

The Protodonata are obviously closely related to the Odonata and are considered by some entomologists to comprise a suborder of that order. However, the absence from the protodonate wing of a pterostigma, nodus, and a well-developed arculus justifies the separation of the two orders. Detailed structure of the abdomen, which is highly modified in the Odonata, is not known in the Protodonata, and, since the immature stages of the Protodonata are also not known, there is no evidence that the immature stages were similar in the two groups.

The large, dentate mandibles and the spinous legs, with the fore pair extending anteriorly, strongly indicate that the protodonate adults were predaceous, like those of the Odonata. As such, they were probably important predators on other insects. All known Protodonata were large and some species of the family Meganeuridae, from both the Upper Carboniferous and Permian, had a wing span of about 700 mm.

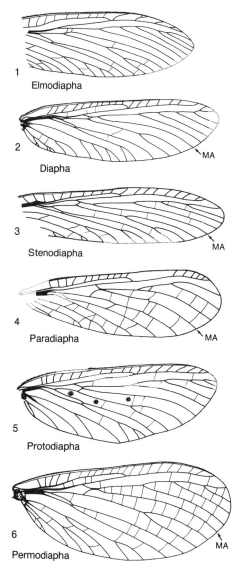

Fig. 37. Uncertain (p. 58–59).

The Protodonata and Odonata share two venational peculiarities. One of these is the presence of intercalary veins, which are also found in some other insects, notably the Ephemeroptera. The other is the absence of two main veins, which have consistently been identified as MP and CUA. The regular alternation of convexities and concavities of the main veins, charactistic of other palaeopterous insects, has been retained. In some protodonates and Palaeozoic odonates, one or two very short veins can be seen near the wing bases and these have been interpreted as vestiges of the missing veins. It seems probable that the venational patterns of the Protodonata and Odonata were derived from common ancestral stock, presumably during the Early Carboniferous, since the Protodonata are known from the Namurian of the Upper Carboniferous.

As defined above, the Protodonata are unknown after the Permian. Several Triassic odonates, recently described by PRITYKINA (1981) from the Soviet Union, had certain features suggestive of the Protodonata, but all had a well-developed nodus.

Family MEGANEURIDAE Handlirsch, 1906

[Meganeuridae HANDLIRSCH, 1906b, p. 306] [=Typidae HANDLIRSCH, 1919b, p. 572; Kohlwaldiidae GUTHÖRL, 1962b, p. 51]

Crossveins very numerous; vein SC long, extending nearly to wing apex; RS1+2 and RS3+4 gradually divergent after their origins. *U. Carb.–Perm.*

Meganeura BRONGNIART, 1885a, p. 60 [*Dictyoneura monyi* BRONGNIART, 1884, p. 833; OD] [=*Meganeurella* HANDLIRSCH, 1919b, p. 569 (type, *M. rapax*)]. Precostal area long, extending nearly to wing apex; very large species. BRONGNIART, 1893; HANDLIRSCH, 1906a; CARPENTER, 1943a; GUTHÖRL, 1962b. *U. Carb.,* Europe (France).——FIG. 38,5. **M. monyi* (BRONGNIART); base of fore wing, ×1 (Carpenter, new).

Boltonites HANDLIRSCH, 1919b, p. 571 [*Meganeura radstockensis* BOLTON, 1914, p. 125; OD]. Little-known genus, based on wing fragment. Precostal area short; anal crossing a short, heavy, oblique vein; 2A arising independently, not as a branch of 1A. *U. Carb.,* England.

Kohlwaldia GUTHÖRL, 1962b, p. 52 [*K. kuhni*; OD]. Little-known genus, based on wing fragment, similar to *Tupus*. *U. Carb.,* Europe (Germany).

Meganeuropsis CARPENTER, 1939, p. 39 [*M. permiana*; OD]. Precostal area much narrower than in *Meganeura*. CARPENTER, 1947. *Perm.,* USA (Kansas, Oklahoma).——FIG. 38,6. *M. americana* CARPENTER, Oklahoma; fore wing, ×0.4 (Carpenter, 1947).

Megatypus TILLYARD, 1925b, p. 52 [*M. schucherti*; OD]. Precostal area as in *Tupus*; anal crossing, at least in hind wing, strongly developed. CAR-

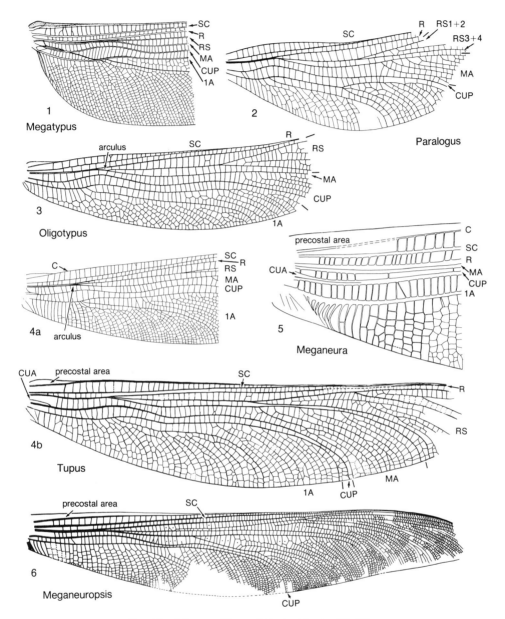

FIG. 38. Meganeuridae and Paralogidae (p. 60–62).

PENTER, 1931a, 1939, 1940b. *Perm.*, USA (Kansas).——FIG. 38,*1*. **M. schucherti*; base of hind wing, ×0.9 (Tillyard, 1925b).

Tupus SELLARDS, 1906, p. 249 [=*Typus* SELLARDS, 1909, p. 151, unjustified emendation, Opinion 1317, ICZN, 1984] [**Tupus permianus*; OD] [=*Meganeurula* HANDLIRSCH, 1906b, p. 309 (type, *Meganura selysii* BRONGNIART, 1893, p. 395); *Gilsonia* MEUNIER, 1908c, p. 243 (type, *G. titana*); *Meganeurina* HANDLIRSCH, 1919b, p. 570 (type, *M. confusa*); *Meganeurites* HANDLIRSCH, 1919b, p. 570 (type, *M. gracilipes*); *Arctotypus* MARTYNOV, 1932, p. 18 (type, *A. sinuatus*)]. Precostal area short, extending at most to one-fourth wing length from base; anal crossing not present in either fore or hind wing; 2A a distinct branch of 1A. COCKERELL, 1913b; CARPENTER, 1931a, 1933a, 1939, 1947, 1960b; WHALLEY, 1980b. *U. Carb.*, Europe (France), England, USA (Georgia); *Perm.*, USSR (European and Asian RSFSR), USA (Kansas, Oklahoma, Arizona).——FIG. 38,*4a*. *T. gracilis* CARPENTER, Perm., Oklahoma; base of fore wing, ×0.6 (Carpenter, 1947).——FIG. 38,*4b*. **T. permianus* (SELLARDS), Perm., Kansas; fore wing, ×1.3 (Carpenter, 1931a).

Family PARALOGIDAE Handlirsch, 1906

[Paralogidae HANDLIRSCH, 1906b, p. 310]

Crossveins numerous; vein SC short, extending about to midwing; RS1+2 and RS3+4 widely divergent after origins. *U. Carb.–Perm.*

Paralogus SCUDDER, 1893b, p. 20 [**P. aeschnoides*; OD]. Wing nearly oval, with a strongly curved posterior margin. ROHDENDORF, 1940b; CARPENTER, 1960b. *U. Carb.*, USA (Rhode Island).——FIG. 38,*2*. **P. aeschnoides*; fore wing, ×1.3 (Carpenter, 1960b).

Oligotypus CARPENTER, 1931a, p. 106 [**O. tillyardi*; OD]. Wing slender, hind margin only moderately curved. CARPENTER, 1947, 1960b; CARPENTER & RICHARDSON, 1971. *U. Carb.*, USA (Illinois); *Perm.*, USA (Kansas, Oklahoma).——FIG. 38,*3*. **O. tillyardi*, Perm., Oklahoma; fore wing, ×2 (Carpenter, 1947).

Family UNCERTAIN

The following genera, apparently belonging to the order Protodonata, are too poorly known to permit assignment to families.

Palaeotherates HANDLIRSCH, 1906a, p. 690 [**P. pennsylvanicus*; OD]. Little-known genus, based on wing fragment. HANDLIRSCH, 1906b; CARPENTER, 1980. *U. Carb.*, USA (Pennsylvania).

Paralogopsis HANDLIRSCH, 1911, p. 374 [**P. longipes*; OD]. Little-known genus, based on fragments of fore and hind wings. CARPENTER, 1960b. *U. Carb.*, USA (Pennsylvania, Illinois).

Petrotypus ZALESSKY, 1950, p. 100 [**P. multivenosus*; OD]. Little-known genus, based on basal part of wing. ROHDENDORF, 1962a. *Perm.*, USSR (Asian RSFSR).

Truemania BOLTON, 1934, p. 183, *nom. subst. pro Tillyardia* BOLTON, 1922, p. 145, *non* CARTER, 1913 [**Tillyardia multiplicata* BOLTON; OD]. Little-known genus, based on distal fragment of wing. *U. Carb.*, England.

Typoides ZALESSKY, 1948a, p. 49 [**T. uralicus*; OD]. Little-known genus, based on wing fragment. *Perm.*, USSR (Asian RSFSR).

Order ODONATA Fabricius, 1793

[Odonata FABRICIUS, 1793, p. 373] [=Permodonata G. M. ZALESSKY, 1931, p. 855]

Predaceous Palaeoptera, mostly large to very large; head unusually large, on a flexible cervix; compound eyes large, bulging; 3 ocelli present; antennae filiform, very short, with at most 7 segments; mandibles large, conspicuously dentate; maxillae spinose, palpi reduced; labial palpi forming a pair of large lobes, each with a prominent spine. Prothorax small, not fused to mesothorax; meso- and metathorax fused into a rigid, oblique pterothorax; legs homonomous, attached far forward on their thoracic segments and conspicuously spinose; tarsi with 3 segments. Wings homonomous or nearly so, with a very distinctive venation, a nodus at end of vein SC, and a conspicuous pterostigma on the anterior margin of the wings distally. Abdomen slender, elongate; second and third sterna of male with accessory reproductive structures; cerci short, consisting of a single segment; females with a short (rarely long) ovipositor. Eggs deposited in or near fresh water. Nymphs aquatic, with tracheal gills or filaments; labium of nymphs greatly modified, forming a long, dentate, grasping appendage. *Perm.–Holo.*

Other than the Ephemeroptera, this is the only order of the Palaeoptera that still exists. The internal structure of the odonates is that of a primitive pterygote, but the reduced cerci and antennae, the oblique position of the thoracic segments, and the complexity of their wing venation indicate a long evolutionary history of the order, especially since these specializations had been acquired before the end of the Jurassic Period. The geological record of the Odonata is, in fact, extensive and long. The aquatic nymphal stages and the tendency for the adults to remain near water have undoubtedly favored the preservation of specimens as fossils.

Until a few years ago, however, our knowledge of the Mesozoic odonates was based on a very few, poorly preserved specimens. The

Fig. 39. Odonata; *Protolindenia wittei* (GIEBEL), Anisoptera, Gomphidae; ×2 (Carpenter, new). This specimen shows the typical odonate body form and the convexity and concavity of the main wing veins as well as intercalary veins and triads.

recent investigations by Dr. L. N. PRITYKINA of the Paleontological Institute in Moscow of extensive collections from Mesozoic deposits in the Soviet Union have greatly improved our knowledge of the order during that important period in the history of these insects. It is now apparent, as we could previously only assume, that the Odonata were at the peak of their diversity during the Jurassic and perhaps even the Triassic.

The existing Odonata and the extinct species from Tertiary, Cretaceous, and Jurassic deposits show little diversity in general body structure (Fig. 39, *Protolindenia*). Unfortunately, we know virtually nothing of their body structure earlier than that.

The wing venation, on the other hand, shows great diversity from the beginning of the Jurassic, providing many diagnostic characters for genera and higher taxonomic categories. The venational pattern is complicated, and in the past there has been considerable controversy about the homologies of the veins with those of other insects. Subsequent studies, chiefly by LAMEERE (1922), MARTYNOV (1924a), CARPENTER (1931a), and TILLYARD and FRASER (1938–1940) have provided the interpretation of the venation now in general use. As previously noted, an important feature of the venational pattern that is shared with the Protodonata is the apparent absence of two main veins, the posterior media (MP, concave) and the anterior cubitus (CUA, convex). A presumed vestige of CUA is present at the base of the wings of at least some members of three extinct suborders, Protanisoptera (Fig. 40,*1*, *Ditaxineura*; see Fig. 42,*1a*), Archizygoptera (see Fig. 43,*1,5*, *Kennedya*, *Permolestes*), and Anisozygoptera (Fig. 41, *Tarsophlebiopsis*). The longitudinal veins in the odonate wings are therefore the costa (C), the subcosta (SC), the radius (R), the radial sector (RS), the anterior media (MA), the posterior cubitus (CUP), and the anal vein (A). In addition, intercalary veins (indicated by an I prefix) are commonly present both between RS1 and RS2 and between RS2 and RS3 (or RS3+4), forming groups of three veins or triads (Fig. 41; see Fig. 42,*1a*, *Ditaxineura*). A precostal area, comparable to that of the Protodonata, occurs in several of the odonate suborders (Fig. 40,*1*).

In all odonates the junction of vein SC with the costal margin of both pairs of wings is marked by the presence of the nodus, a slight cuticular thickening associated with a bend in the wing margin and commonly with a definite break in the sclerotization of the margin. In some of the extinct suborders, such as the Archizygoptera (see Fig. 43,*6*, *Progoneura*), the nodus is incipient, with little sclerotization; in others the nodus is very distinct (Fig. 41, *Tarsophlebiopsis*; see Fig. 43,*3*, *Selenothemis*). Its position on the wings of course varies, depending on the length of SC. Two crossveins are commonly associated with the nodus: the nodal crossvein connects the nodus to R, and the subnodal crossvein joins R to RS1 (Fig. 41). In most odonates these crossveins are aligned (see Fig. 47,*3*, *Heterophlebia*), but in species with an incipient nodus they are not aligned or even near the nodus.

The odonate wing almost always has a series of crossveins along its front margin. Those in the costal area basal of the nodus are termed antenodals, and those distal of the nodus, between veins C and R, are termed the postnodals (Fig. 41). In some families two of the antenodals are consistently thicker than the others; these are the primary antenodals and the others are the secondaries.

In all odonate wings the basal parts of veins RS and MA are fused and are also coalesced with the stem of R, forming for a short distance a thick compound vein. The fused RS+MA then diverges from R, and after an even shorter distance RS and MA separate. The short segment of RS+MA, termed the arculus, is the center of the most diversified part of the wing (Fig. 40,*1,2,4–6*). In the Protanisoptera (see Fig. 42,*1a*, *Ditaxineura*) the arculus is almost parallel to the longitudinal axis of the wing, but in most odonate wings it is more oblique in position and directed toward the hind margin of the wing (see Fig. 43,*6*, *Progoneura*). In the fore wings of the more generalized members of the Anisozygoptera, the arculus is connected to CUP by a discoidal vein (see Figs. 40,*2* and 46,*1*, *Tarsophlebia eximia* and Fig. 41, *Tarsophlebiopsis*), which is often aligned with the base of MA and the arculus. This produces a small space, the open discoidal cell, just basal of the discoidal vein (Fig. 40,*2*). In most Anisozygoptera the base of MA is also joined to CUP by another crossvein at the apex of the curve in CUP, forming a "closed" discoidal cell (Fig.

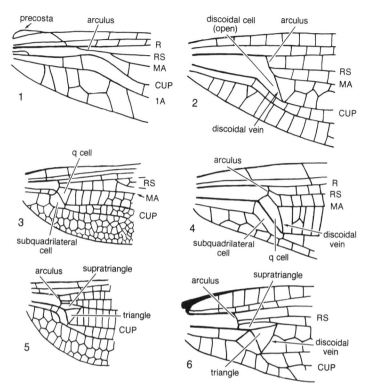

Fig. 40. Odonata; homology of wing structures in the region of the arculus in three suborders. Discoidal cell generally termed the q cell when closed.——*1*. Fore wing of *Ditaxineura anomalostigma*, Protanisoptera (Carpenter, new).——*2*. Fore wing of *Tarsophlebia eximia*, Anisozygoptera (Carpenter, new).——*3*. Fore wing of *Turanothemis nodalis*, Anisozygoptera (after Pritykina, 1968).——*4,5*. *Heterophlebia buckmani*, Anisozygoptera; *4*, fore and *5*, hind wings (after Tillyard, 1925a).——*6*. Hind wing of *Gomphus exilis*, recent, Anisoptera (Carpenter, new).

40,*3,4*); since this cell is a quadrilateral, it is usually termed the q cell to distinguish it from the subquadrilateral (sq) cell below it. These cells occur in most fore wings and nearly all hind wings of the Anisozygoptera as well as in all wings of the Zygoptera. In the hind wings of some families of the Anisozygoptera the q cell is divided by a crossvein that joins CUP to MA (Fig. 40,*5,6*), forming a triangle and a supratriangle. Homologous triangles occur in the fore and hind wings of all Anisoptera.

The positions, shapes, and sizes of these various structures in the wings provide the greater part of the basis for the classification of the fossil Odonata.

The order is here divided into six suborders: Protanisoptera, Archizygoptera, Triadophlebiomorpha, Anisozygoptera, Anisoptera, and Zygoptera. Only the last two are extant. The existing Odonata, estimated to be somewhat more than 5,000 species, are generally grouped into 16 families. The present geological record of the order comprises 42 families, of which 31 are extinct, mostly known only from the Mesozoic. The phylogenetic position of some of these families is uncertain. There has obviously been a great deal of convergence in the evolution of the wing venation. The closing of the discoidal cell and its division into two triangles, for example, have clearly occurred several times independently. The Triadophlebiomorpha are the most unusual of the known odonates. They possess a well-developed nodus, with nodal and subnodal crossveins, but com-

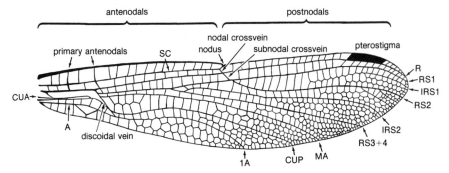

Fig. 41. Odonata; fore wing of *Tarsophlebiopsis mayi,* Anisozygoptera (after Fraser, 1955b).

pletely lack a pterostigma. In addition, when at rest, these odonates placed their wings over the abdomen with the dorsal surface inward like the Zygoptera (Pritykina, 1981). The general aspect of the wings of these triadophlebiomorphs resembles that of the Protodonata. However, the protodonate wing had no nodus, had a long subcosta, and could not be flexed over the abdomen.

The existing Odonata spend by far the greater part of their lives as nymphs. Only a few species reach the adult stage in one year; most species take two or three years and some require four or five. Fossil remains of nymphs are not uncommon in Tertiary deposits, but except for a few poorly preserved nymphs from the Jurassic, they appear to be absent from all pre-Cretaceous deposits.

Suborder PROTANISOPTERA
Carpenter, 1931

[Protanisoptera Carpenter, 1931a, p. 122] [=Permanisoptera Martynov, 1931b, p. 146]

Wings moderately broad, nonpetiolate; hind wings much broader basally than fore wings; precostal area well developed; nodus weakly formed but wing margin at nodus with a distinct bend; at least 4 antenodals; arculus incipient, more longitudinal than oblique; pterostigma traversed by vein R; intercalary veins IRS1, IRS2, and IMA present; vestige of CUA at wing base; CUP with only a slightly sinuous curve; 1A long, extending to about midwing. Body and immature stages unknown. Fraser, 1957; Rohdendorf, 1962a; Prit ykina, 1980b. *Perm.*

Family DITAXINEURIDAE
Tillyard, 1926

[Ditaxineuridae Tillyard, 1926b, p. 69]

Wings with 4 to 6 antenodals; crossveins few, regularly arranged, forming 2 graduate series in distal part of wing. *Perm.*

Ditaxineura Tillyard, 1926b, p. 69 [*$D.$ anomalostigma*; OD]. Nodal crossvein slightly distal of end of SC; postnodals absent. *Perm.,* USA (Kansas).——Fig. 42,*1.* *$D.$ anomalostigma; a,* fore wing, ×2.5 (Carpenter, 1931a); *b,* hind wing, ×2.5 (Carpenter, 1939).

Family PERMAESCHNIDAE
Martynov, 1931

[Permaeschnidae Martynov, 1931b, p. 141] [=Pholidoptilidae G. M. Zalessky, 1931, p. 855; Polytaxineuridae Tillyard, 1935b, p. 375; Callimokaltaniidae Zalessky, 1955a, p. 630; Hemizygopteridae Zalessky, 1955a, p. 632]

Wings with numerous antenodals; crossveins irregularly arranged, forming an irregular network in some parts of wings. *Perm.*

Permaeschna Martynov, 1931b, p. 141 [*$P.$ dolloi*; OD] [=*Pholidoptilon* G. M. Zalessky, 1931, p. 855 (type, *P. camense*)]. Postnodals apparently absent; pterostigma remote from apex of wing; indentation of wing margin near end of RS3+4. *Perm.,* USSR (European RSFSR).——Fig. 42,*2a.* *$P.$ dolloi;* wing as preserved, ×1.2 (Martynov, 1931b).——Fig. 42,*2b. P. camense* (G. M. Zalessky); wing as preserved, ×1 (G. M. Zalessky, 1931).

Callimokaltania Zalessky, 1955a, p. 630 [*$C.$ martynovi*; OD]. Pterostigma very close to wing apex;

posterior margin of wing smoothly curved. ROHDENDORF, 1962a. *Perm.*, USSR (Asian RSFSR).——FIG. 42,3. **C. martynovi*; wing as preserved, ×1.8 (Zalessky, 1955a).

Ditaxineurella MARTYNOV, 1940, p. 11 [**D. stigmalis*; OD] [=*Hemizygopteron* ZALESSKY, 1955a, p. 632 (type, *H. uralensis*)]. Little-known genus, based on apical wing fragments. Several postnodals present; pterostigma nearer wing apex than in *Permaeschna*; no indentation of wing margin at end of RS3+4. ROHDENDORF, 1961a, 1962a. *Perm.*, USSR (Asian RSFSR).——FIG. 42,4. **D. stigmalis*; wing as preserved, ×1.4 (Martynov, 1940).

Polytaxineura TILLYARD, 1935b, p. 375 [**P. stanleyi*; OD]. Similar to *Permaeschna*, but hind margin of wing smoothly curved. *Perm.*, Australia (New South Wales).——FIG. 42,5. **P. stanleyi*; wing as preserved, ×1.4 (Tillyard, 1935b).

Suborder ARCHIZYGOPTERA Handlirsch, 1906

[Archizygoptera HANDLIRSCH, 1906b, p. 471] [=Protozygoptera TILLYARD, 1925b, p. 62]

Small species, with petiolate wings; petiole usually very slender; hind wings either similar to fore wings in form or somewhat broader; precostal area absent; nodus commonly incipient, much nearer to arculus than to pterostigma; arculus incipient or more nearly oblique; pterostigma between vein R and wing margin, well developed but slender; intercalary veins IRS1 and IRS2 usually present; MA without a concave, intercalary branch; vestige of CUA commonly present at wing base; CUP frequently abruptly curved near arculus; 1A commonly long but rarely very short or absent. Body and immature stages unknown. PRITYKINA, 1980a. *Perm.–Jur.*

Family KENNEDYIDAE Tillyard, 1925

[Kennedyidae TILLYARD, 1925b, p. 63]

Fore and hind wings long and slender; costal margin with or without a distinct bend at end of vein SC, but with no definite nodal crossvein; only 4 postnodals; a single row of cells between main veins; 1A short, extending at most only to slightly beyond level of nodus. *Perm.–Trias.*

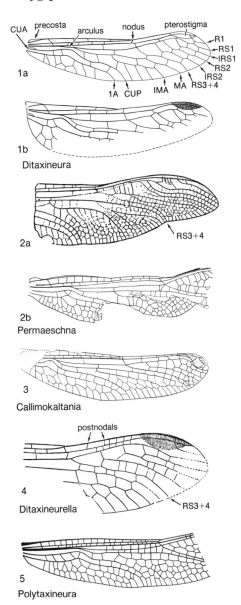

FIG. 42. Ditaxineuridae and Permaeschnidae (p. 66–67).

Kennedya TILLYARD, 1925b, p. 63 [**K. mirabilis*; OD]. Costal margin of wing with a distinct break at end of SC; 1A terminating slightly beyond level of nodus; numerous crossveins between CUP and hind margin of wing. CARPENTER, 1931a, 1947; FRASER, 1957; PRITYKINA, 1980a, 1981. *Perm.*, USA (Kansas, Oklahoma); *Trias.*, USSR (Kirghiz).——FIG. 43,*1. K. fraseri* CARPENTER, Perm., Oklahoma; wing, ×3 (Carpenter, 1947).

Progoneura CARPENTER, 1931a, p. 119 [*P. minuta*; OD]. Anterior margin of fore wing without nodal break; few crossveins between CUP and wing margin; 1A much shorter than in *Kennedya*. CARPENTER, 1947. *Perm.*, USA (Oklahoma).——FIG. 43,6. *P. nobilis* CARPENTER; wing, ×4 (Carpenter, 1947).

Family PERMOLESTIDAE
Martynov, 1932

[Permolestidae MARTYNOV, 1932, p. 33] [=Solikamptilonidae ZALESSKY, 1948a, p. 49]

Wings similar to those of Kennedyidae but with definite nodal and subnodal crossveins, usually aligned or nearly so; crossveins more numerous over entire wing; vein 1A much longer than in Kennedyidae. *Perm.*

Permolestes MARTYNOV, 1932, p. 33 [*P. gracilis*; OD]. Arculus more nearly longitudinal than transverse; numerous cellules in distal and posterior parts of wings, including area between 1A and hind margin. ROHDENDORF, 1962a. *Perm.*, USSR (European RSFSR).——FIG. 43,5. *P. gracilis*; wing, ×1 (Martynov, 1932).
Epilestes MARTYNOV, 1937b, p. 16 [*E. kargalensis*; OD]. Wings with only a few small cellules, almost none between 1A and hind margin; pterostigma very long. ROHDENDORF, 1962a. *Perm.*, USSR (European RSFSR).——FIG. 43,8. *E. kargalensis*; wing, ×1.7 (Martynov, 1937b).
Scytolestes MARTYNOV, 1937b, p. 18 [*S. stigmalis*; OD]. Similar to *Permolestes*, but wings with arculus more nearly transverse than longitudinal; cellules numerous only in area between 1A and wing margin; pterostigma short. ROHDENDORF, 1962a. *Perm.*, USSR (European RSFSR).——FIG. 43,7. *S. stigmalis*; wing, ×2 (Martynov, 1937b).
Solikamptilon ZALESSKY, 1948a, p. 49 [*S. remuliforme*; OD]. Little-known genus, based on wing fragment; 1A very long, parallel to hind margin for most of its length. Pterostigma not preserved. *Perm.*, USSR (Asian RSFSR).——FIG. 43,9. *S. remuliforme*; wing, ×2 (Zalessky, 1948a).
Sushkinia MARTYNOV, 1930a, p. 71 [*S. parvula*; OD]. Little-known genus, based on wing fragment; pterostigma unusually long. [Family assignment doubtful.] *Perm.*, USSR (European RSFSR).

Family PERMAGRIONIDAE
Tillyard, 1928

[Permagrionidae TILLYARD, 1928a, p. 56]

Similar to Kennedyidae, but nodus more pronounced; nodal and subnodal crossveins aligned; arculus more transverse than in Kennedyidae, and vein 1A longer. [The relationships of this family are uncertain. TILLYARD (1928a) and FRASER (1957) considered it to belong to the Zygoptera, FRASER placing it in the recent superfamily Coenagrionidae. PRITYKINA (1980a) has placed it in the suborder Archizygoptera, close to Kennedyidae and Permolestidae.] *Perm.*

Permagrion TILLYARD, 1928a, p. 56 [*P. falklandicum*; OD]. Wings with 8 postnodals, all aligned with crossveins below; pterostigma rhomboidal. PRITYKINA, 1980a. *Perm.*, South America (Falkland Islands).——FIG. 43,4. *P. falklandicum*; wing, ×1.5 (Tillyard, 1928a).

Family PERMEPALLAGIDAE
Martynov, 1938

[Permepallagidae MARTYNOV, 1938b, p. 58]

Similar to Kennedyidae, but wings extremely slender; antenodals and postnodals numerous; several intercalary veins between branches of RS. *Perm.*

Permepallage MARTYNOV, 1938b, p. 50 [*P. angustissima*; OD]. Crossveins between 1A and hind margin of wing numerous and unbranched. ROHDENDORF, 1962a. *Perm.*, USSR (European RSFSR).

Family PROTOMYRMELEONTIDAE
Handlirsch, 1906

[*nom. correct.* TILLYARD, 1925a, p. 36, *ex* Protomyrmeleonidae HANDLIRSCH, 1906b, p. 471, *nom. imperf.*] [=Triassagrionidae TILLYARD, 1922b, p. 454]

Fore and hind wings long and slender; costal margin without a distinct bend at end of vein SC and without a definite nodal crossvein; postnodals numerous; many crossveins between R and RS1; IRS2 weakly developed or absent; 1A extending well beyond level of nodus. *Trias.–Jur.*

Protomyrmeleon GEINITZ, 1887, p. 204 [*P. brunonis*; OD]. Wings with a single row of cells between RS1 and IRS1; IRS2 weakly developed. HANDLIRSCH, 1906b; MARTYNOV, 1927b; PRITYKINA, 1980b. *Jur.*, Europe (Germany), USSR (Kazakh).——FIG. 43,*12a*. *P. brunonis*, Germany; wing, ×2.5 (Handlirsch, 1906b).——FIG. 43,*12b*. *P. handlirschi* MARTYNOV, USSR; wing, ×2 (Martynov, 1927b).
Tillyardagrion MARTYNOV, 1927b, p. 762 [*Protomyrmeleon anglicanus* TILLYARD, 1925a, p. 37; OD]. Little-known genus, similar to *Triassagrion* but lacking small cellules between RS1 and

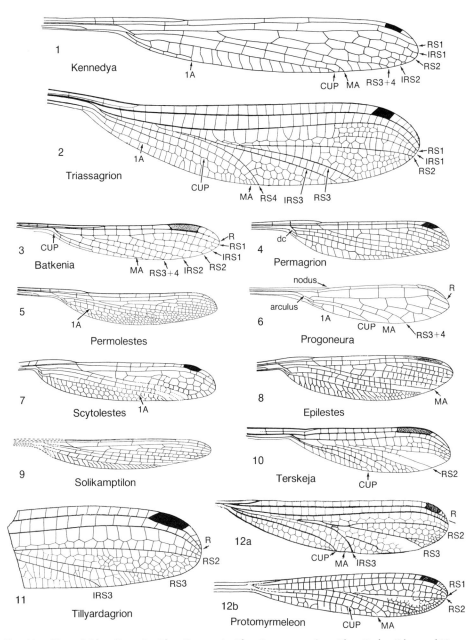

FIG. 43. Kennedyidae, Permolestidae, Permagrionidae, Protomyrmeleontidae, Batkeniidae, and Uncertain (p. 67–70).

IRS1 distally; RS2 and RS3 only slightly divergent; IRS3 and RS3 slightly divergent. *Jur.*, England.——FIG. 43,*11*. **T. anglicanus* (TILLYARD); wing, ×4 (Tillyard, 1925a).

Triassagrion TILLYARD, 1922b, p. 455 [**T. australiense*; OD]. Wings with several rows of cells between RS1 and IRS1 distally; IRS3 strongly developed, close to RS3 and nearly parallel to it; RS2 and RS3 widely divergent. MARTYNOV, 1927b. *Trias.*, Australia (Queensland).——FIG. 43,*2*. **T. australiense*; wing, ×4.5 (Tillyard, 1922b).

Family BATKENIIDAE
Pritykina, 1981

[Batkeniidae Pritykina, 1981, p. 38]

Small species, with petiolate wings, hind pair much broader than fore pair; nodus incipient; pterostigma relatively broad; costal margin without a bend at end of vein SC; SC much longer than in Protomyrmeleontidae, extending well beyond level of origin of RS3+4; CUP very short, not as long as SC; 1A absent. *Trias.*

Batkenia Pritykina, 1981, p. 38 [*B. pusilla*; OD]. Fore wing with 3 antenodals and 4 postnodals; 2 cells below pterostigma. *Trias.*, USSR (Kirghiz).——Fig. 43,3. *B. pusilla*; fore wing, ×3.5 (Pritykina, 1981).

Family UNCERTAIN

The following genera, apparently belonging to the order Odonata, suborder Archizygoptera, are too poorly known to permit assignment to families.

Terskeja Pritykina, 1981, p. 35 [*T. paula*; OD]. Wings as in Protomyrmeleontidae, but vein SC longer and distal branches of RS more evenly spaced and curving posteriorly more strongly; pterostigma slender and strongly developed; SC and R with a distinct bend before nodus, costal margin of wing with a slight bend at same level; IRS forming a triad complex of several branches, all curving posteriorly; RS2 ending on posterior margin of wing, remote from apex; RS3+4 unbranched. [This genus was placed in the Protomyrmeleontidae by its author. However, if *Terskeja* were included, the definition of that family would require drastic changes. It therefore seems advisable to separate *Terskeja* from the Protomyrmeleontidae, at least until additional genera connecting *Terskeja* to the Protomyrmeleontidae have been found.] *Trias.*, USSR (Kirghiz).——Fig. 43,10. *T. paula*; fore wing, ×4 (Pritykina, 1981).

Suborder TRIADOPHLEBIOMORPHA
Pritykina, 1981

[Triadophlebiomorpha Pritykina, 1981, p. 11]

Insects of moderate to large size. Wings petiolate; pterostigma absent; nodus and arculus well developed; bases of longitudinal veins very close together in petiole, almost fused; triads present between veins RS1 and RS2, and between RS3 and RS4; crossveins forming a fine network in posterior areas of wings; vestige of CUA apparently absent. Hind wings apparently similar to fore wings. Wings held back over abdomen at rest. Immature stages unknown. *Trias.*

Family TRIADOPHLEBIIDAE
Pritykina, 1981

[Triadophlebiidae Pritykina, 1981, p. 11]

Postnodal margin of wing straight; hind margin smoothly curved; vein IRS1 arising very close to origin of RS2. *Trias.*

Triadophlebia Pritykina, 1981, p. 12 [*T. madygenica*; OD]. Antenodals and postnodals very numerous; large species. *Trias.*, USSR (Kirghiz).——Fig. 44,4. *T. madygenica*; wing, ×1.4 (Pritykina, 1981).

Cladophlebia Pritykina, 1981, p. 20 [*C. parvula*; OD]. Similar to *Triadophlebia* but much smaller and with relatively fewer crossveins in anterior part of wings. *Trias.*, USSR (Kirghiz).——Fig. 44,5. *C. parvula*; wing, ×2.5 (Pritykina, 1981).

Neritophlebia Pritykina, 1981, p. 16 [*N. elegans*; OD]. Wings much more slender than those of *Triadophlebia*; crossveins in anterior part of wing more widely spaced. *Trias.*, USSR (Kirghiz).——Fig. 44,1. *N. longa* Pritykina; wing, ×1.3 (Pritykina, 1981).

Nonymophlebia Pritykina, 1981, p. 24 [*N. venosa*; OD]. Similar to *Triadophlebia*, but venation even more dense, with double rows of cells between veins forming triads. *Trias.*, USSR (Kirghiz).

Paurophlebia Pritykina, 1981, p. 21 [*P. lepida*; OD]. Wings similar to those of *Cladophlebia* but more slender; crossveins more dense; R curving anteriorly about one-fourth wing length from apex and touching or nearly touching costal margin. *Trias.*, USSR (Kirghiz).——Fig. 44,3. *P. lepida*; wing, ×2.5 (Pritykina, 1981).

Family TRIADOTYPIDAE
Grauvogel & Laurentiaux, 1952

[Triadotypidae Grauvogel & Laurentiaux, 1952, p. 124]

Large species; nodus distinct, with nodal and subnodal crossveins; antenodals and postnodals numerous; nodus about one-third wing length from base; vein RS3+4 forking near midwing, with 2 sets of triads; MA curving posteriorly toward wing margin, nearly touching end of CUP; anal area extensive. *Trias.*

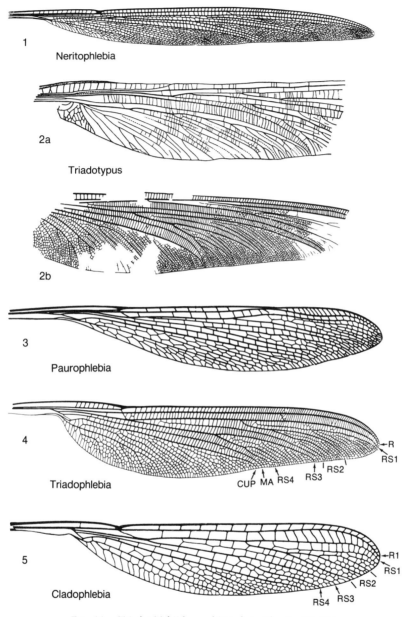

FIG. 44. Triadophlebiidae and Triadotypidae (p. 70–71).

Triadotypus GRAUVOGEL & LAURENTIAUX, 1952, p. 124 [**T. guillaumei*; OD]. RS2 arising at level of subnodal crossvein. PRITYKINA, 1981. Trias., Europe (France), USSR (Kirghiz).——FIG. 44,*2a*. **T. guillaumei,* France; fore wing, ×0.75 (Grauvogel & Laurentiaux, 1952).——FIG. 44,*2b*. *T. sogdianus* PRITYKINA, USSR; hind wing, ×1 (Pritykina, 1981).

Family MITOPHLEBIIDAE Pritykina, 1981

[Mitophlebiidae PRITYKINA, 1981, p. 24]

Wings with very thick veins; front margin of wings strongly convex in distal third, with a broad costal area; hind margin strongly lobed, widest at about midwing. *Trias.*

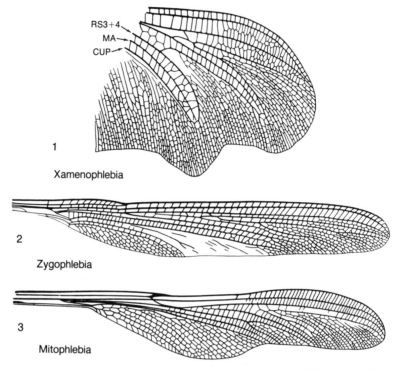

Fig. 45. Mitophlebiidae, Zygophlebiidae, and Xamenophlebiidae (p. 72).

Mitophlebia PRITYKINA, 1981, p. 25 [*P. enormis; OD]. Wings with about 60 postnodals, none aligned with crossveins below. *Trias.*, USSR (Kirghiz).——FIG. 45,3. *M. enormis*; wing, ×2.3 (Pritykina, 1981).

Family ZYGOPHLEBIIDAE Pritykina, 1981

[Zygophlebiidae PRITYKINA, 1981, p. 27]

Insects of moderate size; wings very long and slender; venation dense; distal two-thirds of wing of uniform width; nodus small. *Trias.*

Zygophlebia PRITYKINA, 1981, p. 27 [*Z. ramosa; OD]. Nodal and subnodal crossveins aligned; RS3+4 arising from RS at level of nodus and continuing unbranched until near hind margin of wing. *Trias.*, USSR (Kirghiz).——FIG. 45,2. *Z. ramosa*; wing, ×2.5 (Pritykina, 1981).
Cyrtophlebia PRITYKINA, 1981, p. 30 [*C. sinuosa; OD]. Little-known genus. Wing much broader than in *Zygophlebia*; crossveins more dense, area between R and RS reticulate; IRS2 strongly curved. *Trias.*, USSR (Kirghiz).
Mixophlebia PRITYKINA, 1981, p. 29 [*M. mixta; OD]. Wing similar to that of *Zygophlebia* but relatively broader; RS4+5 much longer. *Trias.*, USSR (Kirghiz).

Zygophlebiella PRITYKINA, 1981, p. 29 [*Z. curta; OD]. Wings similar to those of *Zygophlebia*, but RS3 arising beyond nodus and 1A extending much farther distally. *Trias.*, USSR (Kirghiz).

Family XAMENOPHLEBIIDAE Pritykina, 1981

[Xamenophlebiidae PRITYKINA, 1981, p. 32]

Little-known family. Wings very broad, apex blunt, hind margin with broad undulations; base of wing unknown. *Trias.*

Xamenophlebia PRITYKINA, 1981, p. 32 [*X. ornata; OD]. Crossveins very dense along hind margin and apical region of wing but relatively open in more anterior areas. *Trias.*, USSR (Kirghiz).——FIG. 45,1. *X. ornata*; distal part of wing, ×1.5 (Pritykina, 1981).

Suborder ANISOZYGOPTERA Handlirsch, 1906

[Anisozygoptera HANDLIRSCH, 1906b, p. 463]

Wings moderately broad, narrowed basally but rarely petiolate; hind wings commonly much broader basally; nodus well formed, usually distal of midwing; nodal and subnodal crossveins commonly aligned; arculus transverse or nearly so in most; discoidal cell usually closed in fore and hind wings, forming quadrilateral cell, but rarely open in fore wing and more rarely open in hind wing; quadrilateral cell of fore wing not divided; that of hind wing rarely divided by 1 or more crossveins; vestige of vein CUA rarely present; intercalary veins of RS well developed. Immature stages unknown. *Trias.–Cret.*

This is the most diverse of the suborders of the Odonata and was the dominant one during the Jurassic. Many years ago HANDLIRSCH (1906b) placed here the extant genus *Epiophlebia* of the family Epiophlebiidae (=*Palaeophlebia* SELYS), thus extending the range of the Anisozygoptera to the present. Since then the two species known in *Epiophlebia*, from Japan and India, have been studied in detail by ASAHINA (1954, 1958, 1963). As our knowledge of the Jurassic Anisozygoptera advanced, it became increasingly clear that *Epiophlebia* is not a member of that suborder (CARPENTER, 1931a). More recently, PRITYKINA (1980a) concluded that it is a derivative of an ancient aeshnoid line (Anisoptera) and placed it in a new superfamily, Epiophlebioidea.

Family TARSOPHLEBIIDAE Handlirsch, 1906

[Tarsophlebiidae HANDLIRSCH, 1906b, p. 467]

Discoidal cell open in fore and hind wings; arculus more oblique in fore wing than in hind wing. Body and legs slender. *Jur.*

Tarsophlebia HAGEN, 1866a, p. 58 [**Heterophlebia eximia* HAGEN, 1862, p. 106; OD]. Basal bend of CUP (near arculus) very abrupt and angular; 1A nearly parallel to CUP basally; fore wing very narrow at base. *Jur.*, Europe (Germany), USSR (Kazakh).——FIG. 46,*1a,b*. **T. eximia* (HAGEN), Germany; *a*, hind wing, ×1.8 (Hagen, 1866a); *b*, base of fore wing, ×5.3 (Carpenter, 1932a). ——FIG. 46,*1c*. *T. neckini* MARTYNOV, USSR; fore wing base, ×2 (Martynov, 1927b).
Sphenophlebia BODE, 1953, p. 41 [**S. interrupta*; OD]. Similar to *Tarsophlebiopsis*, but 1A shorter and CUP with definite branches. *Jur.*, Europe (Germany).——FIG. 46,*2*. **S. interrupta*; fore wing, ×1.5 (Bode, 1953).
Tarsophlebiopsis TILLYARD, 1923d, p. 149 [**T. mayi*; OD]. Basal bend of CUP (near arculus) rounded; 1A only slightly curved basally. *Jur.*, England. —— FIG. 41. **T. mayi*; fore wing, ×3.5 (Fraser, 1955b).
Turanophlebia PRITYKINA, 1968, p. 42 [**T. martynovi*; OD]. Similar to *Tarsophlebia*, but pterostigma larger and wing more slender. PRITYKINA, 1977. *Jur.*, USSR (Kazakh).——FIG. 46,*4*. **T. martynovi*; hind wing, ×1.6 (Pritykina, 1968).

Family ISOPHLEBIIDAE Handlirsch, 1906

[Isophlebiidae HANDLIRSCH, 1906b, p. 582]

Discoidal cell closed in fore and hind wings, rectangular, without crossveins. Legs not so long as in Tarsophlebiidae. *Jur.*

Isophlebia HAGEN, 1866a, p. 68 [**I. aspasia* HAGEN, 1866a, p. 70; SD CARPENTER, herein]. Crossveins below distal side of discoidal cell aligned to form apparent continuation of that side, in both fore and hind wings; proximal and distal sides of cell nearly parallel in fore wing; 1A very short. *Jur.*, Europe (Germany).——FIG. 46,*3*. **I. aspasia*; bases of *a*, fore and *b*, hind wings, ×1 (Deichmüller, 1886).

Family LIASSOPHLEBIIDAE Tillyard, 1925

[Liassophlebiidae TILLYARD, 1925a, p. 11]

Discoidal cell open in fore wing, closed in hind wing; anterodistal angle of discoidal cell of hind wing slightly acute. *Jur.*

Liassophlebia TILLYARD, 1925a, p. 13 [**L. magnifica*; OD]. Two primary antenodals; MA separated from CUP by 4 to 7 rows of cells; CUA abruptly bent near arculus. ZEUNER, 1962a; PRITYKINA, 1970. *Jur.*, Europe (Germany), England. —— FIG. 47,*2*. **L. magnifica*; *a*, basal half of fore wing, ×0.7; *b*, hind wing, ×0.7 (both Tillyard, 1925a).
Bathmophlebia PRITYKINA, 1970, p. 111 [**B. unica*; OD]. Little-known genus, based on fragment of hind wing; CUP with an abrupt bend anteriorly, at about level of nodus; RS3+4 arising at level of distal antenodal. *Jur.*, USSR (Kirghiz).
Caraphlebia CARPENTER, 1969, p. 419 [**C. antarctica*; OD]. Hind wing similar to that of *Liassophlebia* but with several weak antenodals in addition to primary ones; IRS1 weakly developed; space between MA and CUP narrow. *Jur.*, Antarctica (South Victoria Land).
Ferganophlebia PRITYKINA, 1970, p. 116 [**F. insignis*; OD]. Little-known genus, based on fragment of fore wing; CUP smoothly curved; cells between CUP and M uncommonly large. *Jur.*, USSR (Kirghiz).

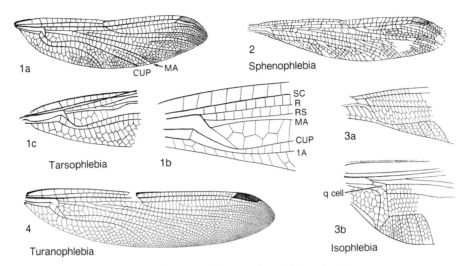

Fig. 46. Tarsophlebiidae and Isophlebiidae (p. 73).

Hypsophlebia PRITYKINA, 1970, p. 114 [*H. scalaris*; OD]. Little-known genus, similar to *Xanthohypsa*, but hind wing narrow basally and subquadrilateral cell extending close to hind margin of wing. *Jur.*, USSR (Kirghiz).

Oreophlebia PRITYKINA, 1970, p. 110 [*O. lata*; OD]. Venational pattern as in *Xanthohypsa*, but wing much broader; CUP and 1A extensively branched. *Jur.*, USSR (Kirghiz).——FIG. 47,7. *S. lata*; hind wing, ×1.2 (Pritykina, 1970).

Petrophlebia TILLYARD, 1925a, p. 11 [*P. anglicana*; OD]. Similar to *Liassophlebia* but with 2 regular rows of cells between veins MA and CUP proximally; CUP only slightly bent near arculus. ZEUNER, 1962a; PRITYKINA, 1970. *Jur.*, England.——FIG. 47,1. *P. anglicana*; base of hind wing, ×1.5 (Tillyard, 1925a).

Pternopteron PRITYKINA, 1970, p. 112 [*P. mirabile*; OD]. Wings long and narrow; 2 strong antenodals, aligned with subcostal crossveins; subquadrilateral cell abruptly geniculate; pterostigma long, narrow; posterior margin of hind wing in anal area with a prominent, recurved spur. *Jur.*, USSR (Kirghiz).

Sagulia PRITYKINA, 1970, p. 113 [*S. ansinervis*; OD]. Little-known genus, apparently related to *Xanthohypsa*, but with a semicircular loop formed by branches of 1A. *Jur.*, USSR (Kirghiz).

Sarytashia PRITYKINA, 1970, p. 115 [*S. gracilis*; OD]. Little-known genus, based on fore wing fragment; similar to *Xanthohypsa* but with shorter pterostigma and few costal veinlets distal of pterostigma. *Jur.*, USSR (Kirghiz).

Sogdophlebia PRITYKINA, 1970, p. 108 [*S. singularis*; OD]. Hind wing similar to that of *Xanthohypsa*, but subquadrilateral cell with a longitudinal vein; 1A with distinct branches. *Jur.*, USSR (Kirghiz).

Xanthohypsa PRITYKINA, 1970, p. 107 [*X. tillyardi*; OD]. Hind wing broad, with strongly curved hind margin; nodus slightly proximal of midwing; 2 antenodals, aligned with subcostal crossveins; 5 or 6 postnodals; pterostigma narrow and long; q cell narrow; subquadrilateral cell long, containing several crossveins, without a longitudinal vein; 1A with weakly defined branches. *Jur.*, USSR (Kirghiz).——FIG. 47,5. *X. tillyardi*; hind wing, ×1.2 (Pritykina, 1970).

Family HETEROPHLEBIIDAE Handlirsch, 1906

[Heterophlebiidae HANDLIRSCH, 1906b, p. 466]

Discoidal cell of fore wing closed but not divided; discoidal cell of hind wing closed and divided, forming supratriangle and triangle. Two strong, primary antenodals present, and usually a few weak secondary ones. *Jur.*

Heterophlebia BRODIE, 1849, p. 35 [*Agrion buckmani* BRODIE, 1845, p. 102; SD TILLYARD, 1925a, p. 27]. MA in fore wing ending at level of middle of pterostigma; IRS1 about equidistant from RS1 and RS2. ZESSIN, 1982. *Jur.*, England, Europe (Germany).——FIG. 47,3. *H. buckmani* (BRODIE), England; *a*, fore wing, ×1.6; *b*, base of hind wing, ×4 (both Tillyard, 1925a).

Clydonophlebia COWLEY, 1942, p. 70 [*Heterophlebia megapolitana* HANDLIRSCH, 1939, p. 26; OD]. Anterior side of triangle of hind wing ending at distal angle of supratriangle. *Jur.*, Europe (Germany).——FIG. 47,6. *C. megapolitana* (HANDLIRSCH); hind wing, ×1.4 (Handlirsch, 1939).

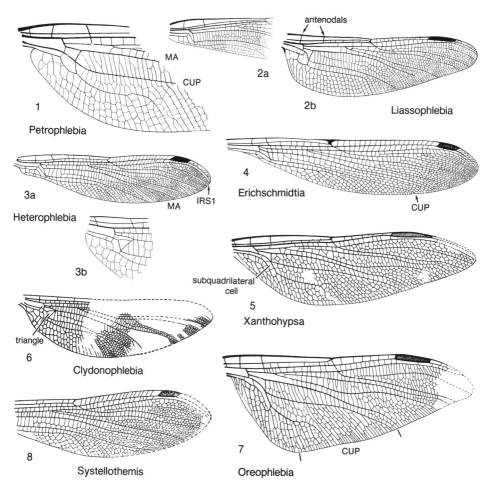

FIG. 47. Liassophlebiidae and Heterophlebiidae (p. 73–75).

Erichschmidtia PRITYKINA, 1968, p. 37 [*E. nigrimontana; OD]. Fore wing as in *Heterophlebia*, but nodus only about one-third wing length from base; 6 to 8 subcostal crossveins in antenodal area; CUP with only a slight bend near discoidal cell. *Jur.*, USSR (Kazakh).——FIG. 47,4. *E. nigrimontana; fore wing, ×1.7 (Pritykina, 1968).

Systellothemis HANDLIRSCH, 1939, p. 27 [*S. reticulata; OD]. Little-known wing, probably synonymous with *Heterophlebia*. COWLEY, 1942. *Jur.*, Europe (Germany).——FIG. 47,8. *S. reticulata; wing, ×2 (Handlirsch, 1939).

Family PROGONOPHLEBIIDAE Tillyard, 1925

[Progonophlebiidae TILLYARD, 1925a, p. 8]

Discoidal cell of hind wing closed and thus a q cell, undivided; 2 strong antenodals; at most only a few weak secondary antenodals; subnodal crossvein not aligned with nodal crossvein. *Jur.*

Progonophlebia TILLYARD, 1925a, p. 9 [*T. woodwardi; OD]. Hind wing: nodus at about level of midwing; q cell small, almost square. ZEUNER, 1959a. *Jur.*, England.——FIG. 48,1. *P. woodwardi; hind wing, ×1.6 (Tillyard, 1925a).

Cyclothemis PRITYKINA, 1980a, p. 126 [*C. sagulica; OD]. Based on incomplete wing; wings apparently similar to those of *Shurabiola*, but CUP less curved and subquadrilateral cell shorter. [Originally placed in Archithemistidae.] *Jur.*, USSR (Kirghiz).——FIG. 48,6. *C. sogjutensis* PRITYKINA; hind wing, ×4 (Pritykina, 1980a).

Shurabiola PRITYKINA, 1980a, p. 123 [*S. nana; OD]. Small species, with broad wings and fewer crossveins than in *Progonophlebia*; q cell large, distal side twice as long as proximal side;

subquadrilateral cell long; CUP smoothly curved. [Originally placed in Archithemistidae.] *Jur.,* USSR (Kirghiz).——Fig. 48,7. **S. nana*; wing, ×6 (Pritykina, 1980a).

Family ARCHITHEMISTIDAE
Tillyard, 1917

[nom. transl. HANDLIRSCH, 1920, p. 177, et correct. COWLEY, 1942, p. 65, ex Architheminae TILLYARD, 1917c, p. 307] [=Campterophlebiidae HANDLIRSCH, 1920, p. 178; Selenothemistidae HANDLIRSCH, 1939, p. 20]

Discoidal cell closed and undivided in fore and hind wings; primary antenodals absent; numerous weak secondary antenodals. *Jur.*

Archithemis HANDLIRSCH, 1906b, p. 466 [**Libellula brodiei* GEINITZ, 1884, p. 581; OD] [=*Diastatommites* TILLYARD, 1925a, p. 21 (type, *Aeschna liassina* STRICKLAND, 1840, p. 301)]. Numerous oblique, parallel veins between RS3+4 and hind margin distally; RS3+4 and CUP smoothly curved. *Jur.,* Europe (Germany), England.——Fig. 48,*4a*. **A. brodiei* (GEINITZ), Germany; fore wing, ×1.8 (Handlirsch, 1906b).——Fig. 48,*4b*. *A. liassina* (STRICKLAND), England; hind wing, ×0.8 (Tillyard, 1925a).
Campterophlebia BODE, 1905, p. 226 [**C. elegans*; OD]. RS3+4 and CUP strongly undulated distally. *Jur.,* Europe (Germany).
Selenothemis HANDLIRSCH, 1920, p. 178 [**S. liadis*; OD]. Area between RS3+4 and hind margin without series of long, oblique, parallel veinlets; RS3+4 and CUP smoothly curved. *Jur.,* Europe (Germany).——Fig. 48,*3*. **S. liadis*; hind wing, ×2 (Handlirsch, 1920).

Family KARATAWIIDAE
Martynov, 1925

[Karatawiidae MARTYNOV, 1925b, p. 589]

Similar to Turanothemistidae, but wings commonly with 9 to 12 subcostal crossveins in antenodal area; discoidal cell incomplete (open) in fore wing but closed in hind wing, forming the q cell. *Jur.–Cret.*

Karatawia MARTYNOV, 1925b, p. 587 [**K. turanica*; OD]. Fore wing: 1A short and very close to hind margin; arculus about midway between wing base and origin of RS3+4. PRITYKINA, 1968, 1980a. *Jur.,* USSR (Kazakh, Kirghiz).
Adelophlebia PRITYKINA, 1980a, p. 130 [**A. obsoleta*; OD]. Little-known genus; area between M and CUP with 2 rows of cells. *Jur.,* USSR (Kirghiz).
Gampsophlebia PRITYKINA, 1980a, p. 131 [**G. modica*; OD]. Little-known genus, based on hind wing fragment; area between MA and CUP with only a single row of cells; posterior side of q cell strongly curved; 1A sigmoidally curved; at least 3 rows of cells between vein A and wing margin. *Jur.,* USSR (Kirghiz).
Hypsomelana PRITYKINA, 1968, p. 40 [**H. sepulta*; OD]. Hind wing similar to that of *Melanohypsa*, but distal angle of q cell nearly a right angle, posterior-distal angle slightly acute; only 1 row of cells between 1A and hind margin; hind margin of wing with a gently curved incision basally. *Jur.,* USSR (Kazakh).
Hypsothemis PRITYKINA, 1968, p. 41 [**H. jurassica*; OD]. Similar to *Hypsomelana* but hind margin of wing without an incision basally. *Jur.,* USSR (Kazakh).
Melanohypsa PRITYKINA, 1968, p. 39 [**M. angulata*; OD]. Hind wing with distal angle of q cell acute, posterior-distal angle slightly obtuse; 2 rows of cells between vein A and wing margin, 3 rows of cells between 1A and wing margin; hind margin of hind wing with a deep, abrupt incision basally. *Jur.,* USSR (Kazakh).——Fig. 48,*2*. **M. angulata; a,* fore and *b,* hind wings, ×3.3 (Pritykina, 1968).
Nacholonda PRITYKINA, 1977, p. 83 [**N. crassicosta*; OD]. Hind wing: only 2 antenodals, aligned with subcostal crossveins; at least 6 postnodals; RS3+4 arising slightly beyond level of second antenodal; CUP sigmoidal; subquadrilateral cell long and wide, extending almost to hind margin of wing. [Family assignment doubtful.] *Cret.,* USSR (Asian RSFSR).——Fig. 48,*5*. **N. crassicosta*; hind wing as preserved, ×1 (Pritykina, 1977).

Family OREOPTERIDAE
Pritykina, 1968

[Oreopteridae PRITYKINA, 1968, p. 29]

Fore and hind wings of similar width, hind wing more petiolate; 2 or 3 thickened antenodals, aligned with subcostal crossveins, basal one before level of arculus; commonly several subcostal crossveins, 9 to 12 postnodals, not aligned with subcostal crossveins; discoidal cell open in fore wing, closed in hind wing; vein RS2 arising distally of nodus; pterostigma short. PRITYKINA, 1980a. *Jur.*

Oreopteron PRITYKINA, 1968, p. 29 [**O. asiaticum*; OD]. Four subcostal crossveins in antenodal area; subquadrilateral cell of hind wing about same width as q cell. *Jur.,* USSR (Kazakh).——Fig. 49,*2a*. **O. asiaticum*; base of fore wing, ×4 (Pritykina, 1968).——Fig. 49,*2b*. *O. simile* PRITYKINA; hind wing, ×4 (Pritykina, 1980a).
Amblyopteron PRITYKINA, 1980a, p. 123 [**A. breve*; OD]. Apex of wing bluntly rounded; 3 crossveins below pterostigma; only 1 row of cells between RS1 and IRS1; cubitoanal area of wing narrow. *Jur.,* USSR (Kirghiz).
Oreopterella PRITYKINA, 1968, p. 33 [**O. paula*;

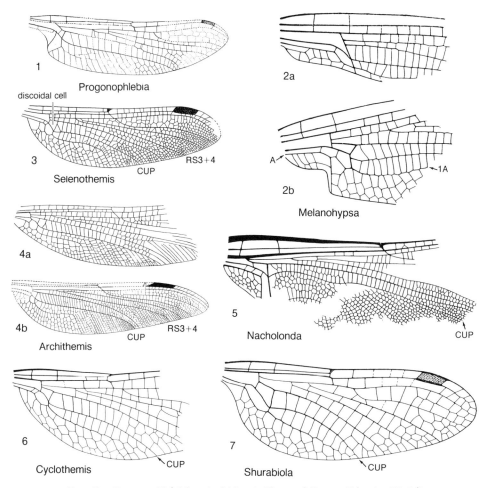

FIG. 48. Progonophlebiidae, Archithemistidae, and Karatawiidae (p. 75–76).

OD]. Little-known genus, based on fragments of fore and hind wings. Similar to *Oreopteron*, but pterostigma much longer; RS2 arising near nodal crossvein. *Jur.*, USSR (Kazakh).

Pauropteron PRITYKINA, 1980a, p. 124 [**P. miserum*; OD]. Similar to *Sogdopteron* but with only 1 row of cells between RS1 and IRS1; only 1 complete cell below pterostigma; vein A submarginal. *Jur.*, USSR (Kirghiz).

Sogdopteron PRITYKINA, 1980a, p. 121 [*S. leve*; OD]. Similar to *Oreopteron*, but petiole of wings more narrow; vein A marginal; 2 rows of cells between RS1 and IRS1. *Jur.*, USSR (Kirghiz).
——FIG. 49,6. *S. leve*; wing, ×2.3 (Pritykina, 1980a).

Sogjutella PRITYKINA, 1980a, p. 122 [*S. mollis*; OD]. Antenodal area of wings with only 2 subcostal crossveins; MA and CUP parallel but diverging distally; RS2 arising far distal of level of nodus. *Jur.*, USSR (Kirghiz).

Turanopteron PRITYKINA, 1968, p. 31 [*T. minor*;

OD]. Similar to *Oreopteron* but with only 2 antenodals and with a longer pterostigma. *Jur.*, USSR (Kazakh).

Family ASIOPTERIDAE Pritykina, 1968

[Asiopteridae PRITYKINA, 1968, p. 34]

Hind wing slender, but petiole very short; 2 well-developed antenodals, aligned with crossveins below; 4 additional crossveins in antenodal area; 9 postnodals; q cell small; subquadrilateral cell with 6 sides; a single row of cells between veins M and CUP; anal area broad, with several rows of cells between 1A and wing margin. *Jur.*

Asiopteron PRITYKINA, 1968, p. 34 [*A. antiquum*; OD]. Hind wing: distal and proximal sides of q cell parallel, former a little longer than latter; RS2 arising at second cell distal to nodus. *Jur.,* USSR (Kazakh).——FIG. 49,*1*. **A. antiquum*; hind wing, ×2 (Pritykina, 1968).

Family EUTHEMISTIDAE Pritykina, 1968

[Euthemistidae PRITYKINA, 1968, p. 44]

Fore wing narrow but not petiolate; 2 primary antenodals and numerous other crossveins in antenodal area; discoidal cell open. *Jur.*

Euthemis PRITYKINA, 1968, p. 44 [**E. multivenosa*; OD]. Fore wing with about 20 crossveins in postnodal area; M and RS almost contiguous at arculus; arculus, 1A, and a hind marginal crossvein forming a straight line. *Jur.,* USSR (Kazakh).—— FIG. 49,*4*. *E. cellulata* PRITYKINA; fore wing, ×1.6 (Pritykina, 1968).

Family TURANOTHEMISTIDAE Pritykina, 1968

[Turanothemistidae PRITYKINA, 1968, p. 38]

Hind wing with 2 thick antenodals; subcostal area in antenodal region with 2 crossveins aligned with antenodals but no other crossveins; vein RS2 arising directly from subnodal crossvein; q cell without crossveins. *Jur.*

Turanothemis PRITYKINA, 1968, p. 38 [**T. nodalis*; OD]. RS3+4 arising slightly nearer to nodus than to arculus. *Jur.,* USSR (Kazakh).——FIG. 49,*3*. **T. nodalis*; hind wing, ×1.8 (Pritykina, 1968).

Family TRIASSOLESTIDAE Tillyard, 1918

[*nom. transl.* PRITYKINA, 1981, p. 39, *ex* Triassolestinae TILLYARD, 1918c, p. 418]

Species of moderate size. Wings: fore wing slender, hind wing much broader; nodus slightly nearer to wing base than to pterostigma; discoidal cell open in fore wing, apparently closed in hind wing; anal veins much reduced. PRITYKINA, 1981. *Trias.*

Triassolestes TILLYARD, 1918c, p. 418 [**T. epiophlebioides*; OD]. Little-known genus, based on small wing fragment. PRITYKINA, 1981. *Trias.,* Australia (Queensland).
Triassolestodes PRITYKINA, 1981, p. 40 [**T. asiaticus*; OD]. Fore wing: discoidal crossvein aligned with arculus; pterostigma much longer than that of hind wing. RS3+4 in both wings arising at level of nodus. Hind wing more than twice as broad as fore wing. *Trias.,* USSR (Asian RSFSR).——FIG. 49,*5*. **T. asiaticus; a,* fore and *b,* hind wings, ×2.2 (Pritykina, 1981).
Triassothemis CARPENTER, 1960c, p. 71 [**T. mendozensis*; OD]. Little-known genus, based on wing fragment; nodus incipient, remote from wing base. [Family assignment doubtful.] PRITYKINA, 1981. *Trias.,* South America (Argentina).

Family UNCERTAIN

The following genera, apparently belonging to the order Odonata, suborder Anisozygoptera, are too poorly known to permit assignment to families.

Acrophlebia COWLEY, 1942, p. 71 [**Heterophlebia geinitzi* HANDLIRSCH, 1906b, p. 467]. Little-known genus, based on wing fragment. *Jur.,* Europe (Germany).
Anisophlebia HANDLIRSCH, 1906b, p. 584 [**Heterophlebia helle* HAGEN, 1862, p. 105; OD]. Little-known genus, based on poorly preserved fore wing; nodus weakly formed, nodal break absent; discoidal cell closed, containing a few crossveins; costal wing margin thick and spinous. *Jur.,* Europe (Germany).
Anomothemis HANDLIRSCH, 1906b, p. 470 [**A. brevistigma*; OD]. Little-known genus, based on apical wing fragment. *Jur.,* Europe (Germany).
Dialothemis COWLEY, 1942, p. 68 [**Liadothemis dubis* HANDLIRSCH, 1939, p. 22; OD]. Little-known genus, based on wing fragment. *Jur.,* Europe (Germany).
Ensphingophlebia BODE, 1953, p. 45 [**E. undulata*; OD]. Little-known genus, based on wing fragments; probably related to *Liassophlebia. Jur.,* Europe (Germany).
Hemerobioides BUCKLAND, 1838, p. 688 [**H. giganteus*; OD]. Little-known genus, based on wing fragment. *Jur.,* England.
Heterothemis HANDLIRSCH, 1906b, p. 468 [**H. germanica*; OD]. Little-known genus, based on wing fragment. *Jur.,* Europe (Germany).
Isophlebiodes PRITYKINA, 1968, p. 46 [**I. obscurus*; OD]. Little-known genus, based on basal fragment of hind wing; q cell as in *Kazachophlebia* but less irregular. *Jur.,* USSR (Kazakh).
Kazachophlebia PRITYKINA, 1968, p. 47 [**K. curvata*; OD]. Little-known genus, based on basal fragment of fore(?) wing; q cell long and irregular, its anterodistal corner forming a right angle. *Jur.,* USSR (Kazakh).
Liadothemis HANDLIRSCH, 1906b, p. 469 [**L. hydrodictyon*; OD]. Little-known genus, based on wing fragment. ZESSIN, 1982. *Jur.,* Europe (Germany).
Oryctothemis HANDLIRSCH, 1906b, p. 469 [**O.

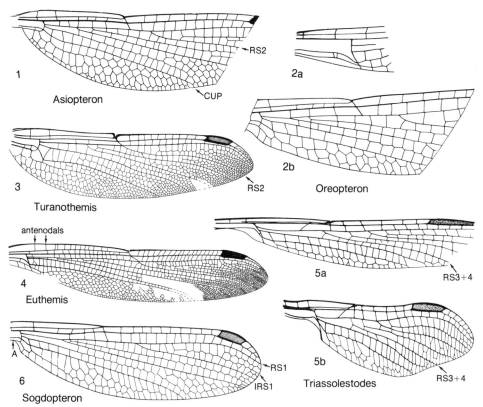

FIG. 49. Oreopteridae, Asiopteridae, Euthemistidae, Turanothemistidae, and Triassolestidae (p. 76–78).

hageni; OD]. Little-known genus, based on wing fragment. *Jur.,* Europe (Germany).

Parelthothemis HANDLIRSCH, 1906b, p. 470 [**P. dobbertinensis*; OD]. Little-known genus, based on wing fragment. *Jur.,* Europe (Germany).

Petrothemis HANDLIRSCH, 1906b, p. 469 [**P. singularis*; OD]. Little-known genus, based on wing fragment. *Jur.,* Europe (Germany).

Plagiophlebia BODE, 1953, p. 52 [**P. praecostarea*; OD]. Little-known genus, based on wing fragments. *Jur.,* Europe (Germany).

Pycnothemis HANDLIRSCH, 1939, p. 28 [**P. densa*; OD]. Little-known genus, based on wing fragment. *Jur.,* Europe (Germany).

Rhabdothemis HANDLIRSCH, 1939, p. 28 [**R. strigivena*; OD]. Little-known genus, based on apical wing fragment. *Jur.,* Europe (Germany).

Sogdothemis MARTYNOV, 1937a, p. 116 [**S. moderata*; OD]. Little-known genus, based on wing fragment. *Jur.,* USSR (Tadzhik).

Temnostigma HANDLIRSCH, 1939, p. 28 [**T. singulare*; OD]. Little-known genus, based on wing fragments. *Jur.,* Europe (Germany).

Triassoneura RIEK, 1976b, p. 794 [**T. andersoni*; OD]. Little-known genus, based on basal fragment of wing; discoidal cell open; vein A apparently coalesced with CUP basally. PRITYKINA, 1981. *Trias.,* South Africa, USSR (Asian RSFSR).

Triassophlebia TILLYARD, 1922b, p. 454 [**T. stigmatica*; OD]. Little-known genus, based on small wing fragment showing pterostigmal area. [Possibly related to *Triassolestes.*] PRITYKINA, 1981. *Trias.,* Australia (New South Wales).

Suborder ANISOPTERA
Selys-Longchamps 1854

[Anisoptera SELYS-LONGCHAMPS in SELYS-LONGCHAMPS & HAGEN, 1854, p. 1]

Wings broad basally, never petiolate; hind wings commonly markedly broadened; nodus very well developed, usually situated nearer to apex than to base of wing, occasionally near midwing; pterostigma well developed, commonly elongate; arculus specialized; discoidal cells of both wings closed and usually divided into a supratriangle (anterior) and a

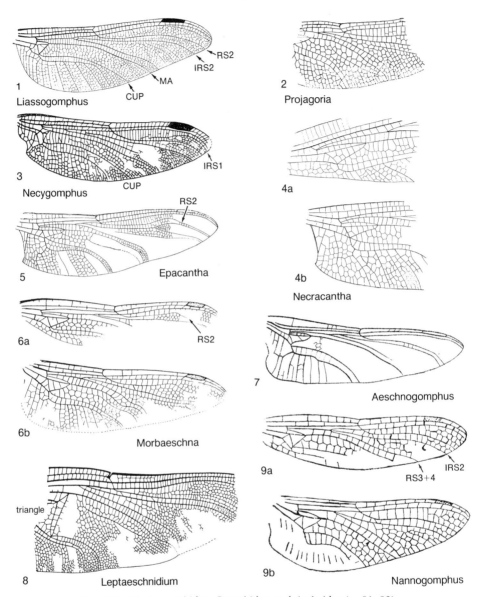

FIG. 50. Liassogomphidae, Gomphidae, and Aeshnidae (p. 81–82).

triangle (posterior), which is commonly divided by additional veins that may even form a reticulation; a third triangle, the subtriangle, more rarely present just posterior to triangle; vein RS with 3 main branches; vestige of CUA absent; cells in hind wing commonly clustered in a loop resembling a boot. Nymphs with rectal gills. *Jur.–Holo.*

Family LIASSOGOMPHIDAE Tillyard, 1935

[Liassogomphidae Tillyard, 1935b, p. 381] [=Gomphitidae Tillyard, 1925a, p. 33; Myopophlebiidae Bode, 1953, p. 63]

Wings apparently similar to those of Gomphidae, but hind wing with an extensive area behind vein 1A reaching to level of nodus; pterostigma slender. *Jur.*

Liassogomphus COWLEY, 1934c, p. 275, *nom. subst.* pro *Gomphites* TILLYARD, 1925a, p. 35, *non* CARTER, 1871 [**Aeschna brodiei* BUCKMAN, 1844, p. 82; OD]. Two rows of cells between RS2 and IRS2, beginning about halfway between nodus and pterostigma; 3 rows of cells between MA and CUP to level of origin of RS3+4. COWLEY, 1942. *Jur.*, England.——FIG. 50,*1*. **L. brodiei* (BUCKMAN); hind wing, ×1.5 (Tillyard, 1925a).

Necygomphus COWLEY, 1942, p. 75, *nom. subst.* pro *Necrogomphus* HANDLIRSCH, 1939, p. 31, *non* CAMPION, 1923 [**Necrogomphus brunswigiae* HANDLIRSCH, 1939, p. 31; OD] [=*Myopophlebia* BODE, 1953, p. 63 (type, *M. libera*)]. Hind wing: intercalcary vein IRS1 long, arising halfway between nodus and pterostigma. *Jur.*, Europe (Germany).——FIG. 50,*3*. *N. libera* (BODE); hind wing, ×1.5 (Bode, 1953).

Palaeogomphus HANDLIRSCH, 1939, p. 31 [**Heterophlebia propingua* BODE, 1905, p. 233; OD]. Fore wing: MA and CUP subparallel for most of their lengths but converging abruptly distally; upper side of triangle ending below distal angle of supratriangle. *Jur.*, Europe (Germany).

Phthitogomphus COWLEY, 1942, p. 71, *nom. subst.* pro *Paragomphus* HANDLIRSCH, 1939, p. 31, *non* COWLEY, 1934 [**Paragomphus angulatus* HANDLIRSCH, 1939, p. 31; OD]. Two rows of cells between RS2 and IRS2 beginning just proximad of pterostigma; 4 rows of cells to about level of nodus between MA and CUP. *Jur.*, Europe (Germany).

Proinogomphus HANDLIRSCH, 1939, p. 31 [**P. bodei*; OD]. Hind wing: 2 rows of cells between MA and CUP from triangle to about 8 cells distal to it, then increasing to 3 or more rows. BODE, 1905; COWLEY, 1942. *Jur.*, Europe (Germany).

Family GOMPHIDAE Rambur, 1842

[Gomphidae RAMBUR, 1842, p. 152]

Eyes widely separated; antenodal system similar to that of Aeshnidae; triangles usually short, not elongate along longitudinal axis of wing and about equidistant from arculus in both pairs of wings; pterostigma with a distinct brace vein; anal loop, if present, very small. *Jur.–Holo.*

Gomphus LEACH, 1815, p. 37. [Generic assignment of fossil (nymph) doubtful.] HAGEN, 1848. *Oligo.*, Europe (Baltic)–*Holo.*

Aeschnogomphus HANDLIRSCH, 1906b, p. 590 [**Aeschna charpentieri* HAGEN, 1848, p. 11; SD COWLEY, 1934b, p. 249]. Little-known genus, based on fragments of fore and hind wings. Triangle of hind wing more slender than that of fore wing. TILLYARD, 1932b. *Jur.*, Europe (Germany).——FIG. 50,*7*. **A. charpentieri* (HAGEN); hind wing, ×0.6 (Hagen, 1861–1863).

Aeschnopsis HANDLIRSCH, 1939, p. 153 [**Aeschna perampla* BRODIE, 1845, p. 33; OD]. Little-known genus, based on hind wing. Apparently similar to *Protolindenia*; triangle greatly extended longitudinally and containing a crossvein. COWLEY, 1942. *Jur.*, England.

Gomphoides SELYS-LONGCHAMPS in SELYS-LONGCHAMPS & HAGEN, 1850, p. 360. [Generic assignment of fossil doubtful.] PICTET & HAGEN, 1856. *Oligo.*, Europe (Baltic)–*Holo.*

Nannogomphus HANDLIRSCH, 1906b, p. 586 [**N. bavaricus*; SD COWLEY, 1934b, p. 252]. Fore and hind wings with relatively few cells in most areas; only 2 rows of cells between MA and CUP proximally; in fore wing IRS2 and RS3+4 only slightly divergent distally, with only 3 or 4 rows of cells between them; RS2 smoothly curved; 1A close to posterior margin of wing. *Jur.*, Europe (Germany).——FIG. 50,*9*. **N. bavaricus; a*, fore and *b*, hind wings, ×2.5 (Handlirsch, 1906b).

Protolindenia DEICHMÜLLER, 1886, p. 37 [**Aeschna wittei* GIEBEL, 1860, p. 127; OD]. Similar to *Nannogomphus*, but fore and hind wings with numerous small cells; 3 or more cells between MA and CUP proximally; in fore wing, IRS2 and RS2 strongly divergent, with many rows of cells between them distally; RS2 abruptly curved distally; 1A curving away from hind margin. FRASER, 1957; PRITYKINA, 1968. *Jur.*, Europe (Germany), USSR (Kazakh).——FIG. 39. **P. wittei* (GIEBEL), Germany; fore and hind wings and body, ×2 (Carpenter, new).

Family AESHNIDAE Leach, 1815

[Aeshnidae LEACH, 1815, p. 126]

Head large, nearly hemispherical; eyes very large, meeting at middorsal region. Wings with 2 distinct primary antenodals; other antenodals usually not aligned with crossveins below; triangles of both pairs of wings similar in shape, elongate along longitudinal axis of wing; triangle of fore wing slightly longer than that of hind wing; triangles and supratriangles of both wings with several crossveins; vein RS2 arched forward near level of pterostigma; brace vein well developed at proximal end of pterostigma. *Jur.–Holo.*

Aeshna FABRICIUS, 1775, p. 424. COCKERELL, 1908q; PITON, 1935a; THÉOBALD, 1937a; TIMON-DAVID, 1946. *Oligo.*, USA (Colorado), Europe (France); *Mio.*, Europe (France)–*Holo.*

Baissaeschna PRITYKINA, 1977, p. 85 [**B. prisca*; OD]. Similar to *Oligoaeschna* (recent), but network of crossveins more dense; anal vein more extensive. *Cret.*, USSR (Asian RSFSR).

Basiaeschna SELYS-LONGCHAMPS, 1883, p. 735. MARTYNOV, 1929; FRASER, 1957. *Oligo.*, USSR (Kazakh)–*Holo.*

Epacantha MARTYNOV, 1929, p. 190 [*E. magnifica*; OD]. Hind wing unusually broad basally; apex subacute; RS2 with a prominent anterior curve just before level of pterostigma; IRS2 deeply forked; triangle with 5 cells. FRASER, 1957. *Oligo.*, USSR (Kazakh).——FIG. 50,5. **E. magnifica*; hind wing, ×1 (Martynov, 1929).

Epiaeschna HAGEN, 1873, p. 271. MARTYNOV, 1927c. *Mio.*, USSR (European RSFSR)–*Holo.*

Gobiaeschna PRITYKINA, 1977, p. 85 [*G. occulta*; OD]. Fore wing as in *Hoplonaeschna* (recent) but with larger pterostigma; RS2 and IRS2 nearly parallel, with only 1 row of cells between them. *Cret.*, Asia (Mongolia).

Gomphaeschna SELYS-LONGCHAMPS, 1871, p. 413. PRITYKINA, 1977. *Cret.*, USSR (Asian RSFSR)–*Holo.*

Heliaeschna SELYS-LONGCHAMPS, 1882, p. 667. MARTYNOV, 1927c. [Generic assignment of fossil doubtful.] *Mio.*, USSR (European RSFSR)–*Holo.*

Leptaeschnidium PRITYKINA, 1977, p. 88 [*L. latum*; OD]. Hind wing: venation less dense than in other genera of family; only 1 row of cells in costal area; triangle very narrow. *Cret.*, USSR (Asian RSFSR).——FIG. 50,8. **L. latum*; hind wing as preserved, ×2 (Pritykina, 1977).

Lithaeschna COCKERELL, 1907b, p. 133 [*L. needhami*; OD]. Similar to *Gomphaeschna* (recent), but anal loop with 5 cells; triangle with 3 cells. *Oligo.*, USA (Colorado).

Morbaeschna NEEDHAM, 1907, p. 141 [*Aeshna muensteri* GERMAR, 1839, p. 215; OD]. In fore and hind wings, RS2 strongly arched just proximad of pterostigma; triangle with a few cells. FRASER, 1957. *Jur.*, Europe (Germany).——FIG. 50,6. **M. muensteri* (Germar); *a*, fore and *b*, hind wings, ×0.8 (Needham, 1907).

Necracantha MARTYNOV, 1929, p. 193 [*N. composita*; SD COWLEY, 1934d, p. 243]. Little-known genus, based on fragments of fore and hind wings; apparently similar to *Epacantha*, but triangle with 7 or 8 cells. FRASER, 1957. *Oligo.*, USSR (Kazakh). ——FIG. 50,4. **N. composita*; *a*, fore and *b*, hind wings, ×1.5 (Martynov, 1929).

Oligaeschna PITON & THÉOBALD, 1939, p. 6 [*O. jungi*; OD] [=*Neoligaeschna* CARPENTER, 1986, p. 576, obj.] Little-known genus, based on wing fragment. *Oligo.*, Europe (France).

Oligoaeschna SELYS-LONGCHAMPS, 1889, p. 470. ESAKI & ASAHINA, 1957; NEL & PAPAZIAN, 1983. *Oligo.*, Europe (France); *Plio.*, Japan–*Holo.*

Oplonaeschna SELYS-LONGCHAMPS, 1883, p. 375. COCKERELL, 1913c; HENRIKSEN, 1922a; PONGRÁCZ, 1928. *Oligo.*, USA (Colorado); *Mio.*, Europe (Denmark, Yugoslavia)–*Holo.*

Projagoria MARTYNOV, 1929, p. 186 [*P. conjuncta*; OD]. Little-known genus, apparently similar to *Oligoaeschna* (recent) but with reticulation of crossveins more dense. *Oligo.*, USSR (Kazakh).——FIG. 50,2. **P. conjuncta*; hind wing, ×1.5 (Martynov, 1929).

Triaeschna CAMPION, 1916, p. 230 [*T. gossi*; OD]. Fore wing: nodus almost exactly at midwing; RS2 arching anteriorly just before level of pterostigma, then curving posteriorly; IRS2 forking before proximal end of pterostigma, not arched anteriorly; triangle very long, curved. *Eoc.*, England.

Family AESCHNIDIIDAE Handlirsch, 1906

[Aeschnidiidae HANDLIRSCH, 1906b, p. 593]

Apparently related to the Cordulegastridae (recent). Wings very broad; nodus at about middle of wing; crossveins very numerous, forming a dense reticulation; triangles very narrow and elongate, with numerous cells; ovipositor very long. *Jur.–Cret.*

Aeschnidium WESTWOOD, 1854, p. 394 [*A. bubas*; OD] [=*Estemoa* GIEBEL, 1856, p. 286, obj.]. Triangles of both wings with a nearly straight distal side; arculus connected to CuP by a short crossvein. DEICHMÜLLER, 1886; TILLYARD, 1918a; RIEK, 1954b; FRASER, 1957. *Jur.*, England, Europe (Germany).——FIG. 51,7. *A. densum* HAGEN, Germany; fore and hind wings, ×1.5 (Hagen, 1862).

Aeschnidiella ZALESSKY, 1953a, p. 165 [*A. abanovi*; OD]. Little-known genus, based on hind wing fragments. Similar to *Aeschnidiopsis*, but distal side of triangles only very slightly concave. *Cret.*, USSR (Kazakh).

Aeschnidiopsis TILLYARD, 1918a, p. 690 [*A. flindersiensis* WOODWARD, 1884, p. 337; OD]. Triangles with a strongly curved distal side; arculus not connected to CuP. FRASER, 1957. *Cret.*, Australia (Queensland).——FIG. 51,8. **A. flindersiensis*; base of hind wing, ×1.3 (Tillyard, 1918a).

Family AKTASSIIDAE Pritykina, 1968

[Aktassiidae PRITYKINA, 1968, p. 48]

Large species; hind wing with nearly straight posterior margin except near wing base; pterostigma elongate, slender; intercalary vein IRS1 well developed, parallel to RS1; triangle almost equilateral; supratriangle without crossveins. *Jur.*

Aktassia PRITYKINA, 1968, p. 48 [*A. magna*; OD]. Hind wing: distal and posterior areas with fine meshwork of crossveins; M terminating on hind margin slightly beyond level of midwing. *Jur.*,

USSR (Kazakh).——FIG. 51,5. *A. magna*; hind wing as preserved, ×0.9 (Pritykina, 1968).

Family PETALURIDAE Needham, 1903

[Petaluridae NEEDHAM, 1903, p. 739]

Large to very large species, with long, reticulate, and slightly falcate wings; pterostigma long to very long, with a strong brace vein; 2 primary antenodals, with numerous secondary antenodals, as in Gomphidae; triangles of both wings usually similar; anal loop not well developed. Eyes separated, as in Gomphidae. *Jur.–Holo.*

Petalura LEACH, 1815, p. 95. *Holo.*
Cymatophlebiella PRITYKINA, 1968, p. 51 [*C. euryptera*; OD]. Similar to *Libellulium,* but triangle of hind wing equilateral, with 3 cells; 1A with 12 branches; 12 antenodals and 12 postnodals. *Jur.,* USSR (Kazakh).
Cymatophlebiopsis HANDLIRSCH, 1939, p. 153 [*C. pseudobubas*; OD]. Little-known genus, based on fragment of hind wing. Similar to *Libellulium,* but triangle more elongate along longitudinal axis of wing and divided by oblique crossvein; 1A with about 6 descending branches. [Family assignment doubtful; placed in Gomphidae by HANDLIRSCH (1939) and COWLEY (1942).] NEEDHAM, 1907; COWLEY, 1942. *Jur.,* England.
Libellulium WESTWOOD, 1854, p. 394 [*L. agrias*; OD] [=*Cymatophlebia* DEICHMÜLLER, 1886, p. 49 (type, *Libellula longialata* GERMAR, 1839, p. 216, *vide* FRASER, 1957)]. RS2 only slightly bent just basad of pterostigma; triangle with many cells. [Family assignment uncertain; placed in Gomphidae by HANDLIRSCH (1906b), in Aeschnidae by NEEDHAM (1907), and in Petaluridae by FRASER (1957).] *Jur.,* England, Europe (Germany).——FIG. 51,*1. L. longialata* (GERMAR); *a,* fore and *b,* hind wings, ×0.8 (Needham, 1907).
Mesuropetala HANDLIRSCH, 1906b, p. 588 [*Gomphus koehleri* HAGEN, 1848, p. 8; SD COWLEY, 1934b, p. 252]. Little-known genus, apparently similar to *Petalura* (recent); RS2 smoothly curved, close and parallel to IRS1. DEICHMÜLLER, 1886; FRASER, 1957; PRITYKINA, 1968. *Jur.,* Europe (Germany), USSR (Kazakh).

Family CORDULIIDAE Selys-Longchamps, 1850

[Corduliidae SELYS-LONGCHAMPS in SELYS-LONGCHAMPS & HAGEN, 1850, p. 66]

Antenodals aligned with subcostal crossveins below, not differentiated into primaries and secondaries; triangle of fore wing elongate anteroposteriorly; that of hind wing slightly elongate along longitudinal axis of wing; anal loop of hind wing reduced or absent. *Eoc.–Holo.*

Cordulia LEACH, 1815, p. 136. *Holo.*
Croatcordulia KIAUTA, 1969, p. 86 [*Libellula platyptera* CHARPENTIER, 1843, p. 408; OD]. Similar to *Cordulia* (recent); 7 postnodals in both fore and hind wings; 5 antenodals in hind wing; pterostigma 4 times as long as wide; 2 cells in triangle of hind wing. *Mio.,* Europe (Yugoslavia).
Miocordulia KENNEDY, 1931, p. 314 [*M. latipennis*; OD]. Apparently related to *Somatochlora* and *Epicordulia* (both recent), but hind wing with 3 rows of cells extending outward from triangle. FRASER, 1957. *Mio.,* USA (Washington).——FIG. 51,*3. *M. latipennis*; hind wing, ×1.5 (Kennedy, 1931).
Stenogomphus SCUDDER, 1892, p. 13 [*S. carletoni*; OD]. Fore wing: triangle relatively remote from arculus; nodus slightly nearer to pterostigma than to arculus; MA sigmoidally curved. [Family position uncertain; considered by HAGEN and SELYS-LONGCHAMPS (SCUDDER, 1892) to be a gomphid but by RIS and MUTTKOWSKI (RIS, 1910) to be a corduliid.] COCKERELL, 1921e. *Eoc.,* USA (Colorado, Wyoming).——FIG. 51,*2. *S. carletoni*; fore wing, ×1.4 (Scudder, 1892).

Family LIBELLULIDAE Leach, 1815

[Libellulidae LEACH, 1815, p. 136]

Adults similar to those of the Corduliidae, but with anal loop of hind wing well developed and boot-shaped. *Oligo.–Holo.*

Libellula LINNÉ, 1758, p. 543. [Generic assignment of fossils doubtful.] HANDLIRSCH, 1906b. *Oligo.,* Europe (Germany, France); *Mio.,* Europe (Germany)–*Holo.*
Celithemis HAGEN, 1861, p. 147. STATZ, 1937. *Mio.,* Europe (Germany)–*Holo.*
Lithemis FRASER, 1951, p. 51 [*L. lejeunecarpentieri*; OD]. Related to *Neurothemis* (recent) but with only a single row of cells between RS2 and IRS2; arculus at level of first antenodal. *Mio.,* Europe (Germany).
Oligocaemia FRASER, 1951, p. 52 [*O. imperfecta*; OD]. Little-known genus, based on fragment of hind wing; apparently related to *Rhyothemis* (recent). Basal side of triangle at level of arculus; arculus between first and second antenodals. *Mio.,* Europe (Germany).——FIG. 51,*4. *O. imperfecta*; hind wing, ×1 (Fraser, 1951).
Trameobasileus ZEUNER, 1938, p. 109 [*T. moguntiacus*; OD]. Similar to *Hydrobasileus* (recent), but triangle of hind wing with 5 cells. *Mio.,* Europe (Germany).

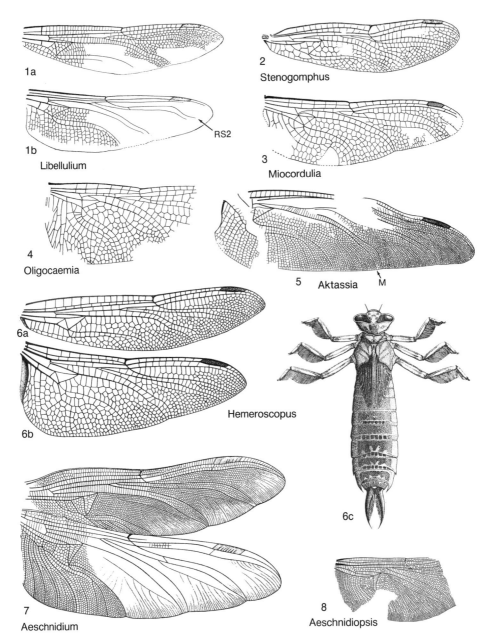

FIG. 51. Aeschnidiidae, Aktassiidae, Petaluridae, Corduliidae, Libellulidae, and Hemeroscopidae (p. (p. 82–85).

Family HEMEROSCOPIDAE Pritykina, 1977

[Hemeroscopidae Pritykina, 1977, p. 89]

Related to Cordulegastridae (recent). Adults with triangles of fore and hind wings similar in form and size in both sexes; hind wing much broader than fore wing; branches of vein 1A forming a large loop. Nymphs with streamlined body; eyes unusually large; tibiae with large brushes of long hairs. *Cret.*

Hemeroscopus Pritykina, 1977, p. 88 [*H. baissicus*; OD]. Wings: RS2 smoothly curved, slightly divergent from IRS2 distally; RS3+4 more sigmoidal in fore wing than in hind. *Cret.,* USSR (Asian RSFSR).——Fig. 51,6. *H. baissicus; a,* fore wing, *b,* hind wing, *c,* adult nymph, all ×1 (Pritykina, 1977).

Family UNCERTAIN

The following genera, apparently belonging to the order Odonata, suborder Anisoptera, are too poorly known to permit assignment to families.

Elattogomphus Bode, 1953, p. 58 [*E. latus*; OD]. Little-known genus, based on incomplete, broad wing. *Jur.,* Europe (Germany).
Lithogomphus Beier, 1952, p. 129 [*L. munzenbergianus*; OD]. Little-known genus, based on incomplete wing. *Mio.,* Europe (Germany).
Necrogomphus Campion, 1923, p. 669, *nom. subst.* pro Mesogomphus Handlirsch, 1906b, p. 592, non Förster, 1906 [*Gomphus petrifactus* Hagen in Selys-Longchamps & Hagen, 1850, p. 359; SD Cowley, 1934a, p. 202]. Little-known genus, based on wing fragment. *Jur.,* England.
Paleoaeschna Meunier, 1914c, p. 180 [*P. vidal*; OD]. Little-known genus, based on nymph. *Jur.,* Europe (Spain).
Protopaltothemis Pongrácz, 1928, p. 122 [*P. hageni*; OD]. Little-known genus, based on wing fragments. *Mio.,* Europe (Yugoslavia).
Sinaeschnidia Hong, 1965, p. 171 [*S. heishankowensis*; OD]. Little-known genus, based on wing fragment. Zhou & Wei, 1980. *Jur.,* China (Zhejiang).
Strongylogomphus Bode, 1953, p. 62 [*S. grasselianus*; OD]. Little-known genus, based on wing fragment. *Jur.,* Europe (Germany).
Urogomphus Handlirsch, 1906b, p. 594 [*Aeschna gigantea* Germar, 1839, p. 216; SD Cowley, 1934b, p. 253]. Little-known genus, based on incomplete wings. *Jur.,* Europe (Germany).

Suborder ZYGOPTERA Selys-Longchamps, 1854

[Zygoptera Selys-Longchamps in Selys-Longchamps & Hagen, 1854, p. 2]

Wings petiolate, subpetiolate, or nonpetiolate, but not greatly widened near base; fore and hind wings closely similar in size, shape, and venation; nodus well developed and situated at or slightly basad of middle of costal margin; arculus and pterostigma well developed; discoidal cell situated below arculus, either open or closed, formed by space between oblique basal part of vein MA above and curving part of CUP below; discoidal cell not divided into triangle and supratriangle; RS with 3 main branches and at least 2 intercalaries; no vestige of CUA present at wing base. Nymphs with 3 caudal gill plates. *Jur.–Holo.*

Family COENAGRIONIDAE Kirby, 1890

[Coenagrionidae Kirby, 1890, p. 119]

Wings petiolate and narrow; primary antenodals well developed and extending to vein R; rarely a few accessory antenodals present; discoidal cell complete but short; MA zigzagging for at least a considerable part of its length; veins IRS2 and RS3 arising closer to subnodal crossvein than to arculus. *Oligo.–Holo.*

Coenagrion Kirby, 1890, p. 148. *Holo.*
Argia Rambur, 1842, p. 254. [Generic assignment of fossil doubtful.] Scudder, 1892. *Oligo.,* USA (Colorado)–*Holo.*
Enallagma Charpentier, 1840, p. 21 [=*Sobobapteron* Pierce, 1965, p. 160 (type, *S. kirkbyae*)]. Cockerell, 1925b; Carpenter, 1968. *Oligo.,* USA (Colorado); *Pleist.,* USA (California)–*Holo.*
Hesperagrion Calvert, 1902, p. 103. Scudder, 1892; Cockerell, 1907b, 1908j; Fraser, 1957. *Oligo.,* USA (Colorado)–*Holo.*

Family LESTIDAE Calvert, 1901

[Lestidae Calvert, 1901, p. 32]

Wings petiolate and slender; primary antenodals well developed; accessory nodals rarely present; postnodals aligned with crossveins below; pterostigma much longer than wide, not usually pointed distally; discoidal cell usually closed. *Oligo.–Holo.*

Lestes Leach, 1815, p. 137. Heer, 1847, 1849, 1853a; Hagen, 1858; Théobald, 1937a; Schmidt, 1958. *Oligo.,* Europe (Germany, France)–*Holo.*
Oligolestes Schmidt, 1958, p. 3 [*Lestes grandis* Statz, 1930, p. 11; OD]. Similar to *Lestes* but with more intercalary veins. *Oligo.,* Europe (Germany).——Fig. 52,4. *O. grandis* (Statz); wings, ×1.4 (Schmidt, 1958).

Family MEGAPODAGRIONIDAE Tillyard, 1917

[Megapodagrionidae Tillyard, 1917c, p. 278]

Wings petiolate; venation similar to that of *Lestes* (recent), but veins IRS2 and RS3

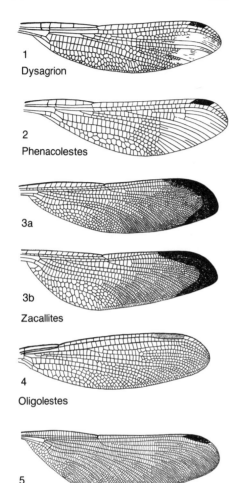

Fig. 52. Lestidae, Pseudolestidae, and Euphaeidae (p. 85–88).

arising nearer to nodus than to arculus; pterostigma usually pointed distally. *Eoc.–Holo.*

Megapodagrion SELYS-LONGCHAMPS, 1885, p. 29. *Holo.*
Eopodagrion COCKERELL, 1920c, p. 237 [**E. scudderi*; OD]. Little-known genus, based on wing fragment; related to *Megapodagrion* (recent) but with an oblique brace at proximal end of pterostigma. KENNEDY, 1925. *Eoc.*, USA (Wyoming).
Lithagrion SCUDDER, 1890, p. 134 [**L. hyalinum*; SD COCKERELL, 1907b, p. 138]. Similar to *Melanagrion*, but pterostigma bounded by 3 or 4 cells below. COCKERELL, 1908j; MARTYNOV, 1929; FRASER, 1957. *Oligo.*, USA (Colorado), USSR (Kazakh).

Melanagrion COCKERELL, 1907b, p. 138 [**Lithagrion umbratum* SCUDDER, 1890, p. 136; OD]. Wings dark; pterostigma bounded by 5 cells below; cells of first 2 rows between nodus and pterostigma higher than long. FRASER, 1957. *Oligo.*, USA (Colorado).
Miopodagrion KENNEDY, 1925, p. 297 [**Lithagrion optimum* COCKERELL, 1916c, p. 101; OD]. Little-known genus, based on wing fragment; possibly close to *Argiolestes* (recent). FRASER, 1957. *Oligo.*, USA (Colorado).
Oligoargiolestes KENNEDY, 1925, p. 296 [**O. oligocenum*; OD]. Little-known genus, based on wing fragment; possibly close to *Megapodagrion*. FRASER, 1957. *Oligo.*, England.
Stenolestes SCUDDER, 1895a, p. 119 [**Agrion iris* HEER, 1865, p. 395; OD]. Little-known genus, based on wing fragment. *Mio.*, Europe (Germany).

Family PSEUDOLESTIDAE
Fraser, 1957

[Pseudolestidae FRASER, 1957, p. 62]

Wings petiolate; primary antenodals well developed and commonly aligned with crossveins below; accessory antenodals few or absent; veins IRS2 and RS3 ordinarily arising nearer to arculus than to subnodal crossvein. *Eoc.–Holo.*

Pseudolestes KIRBY, 1900, p. 537. *Holo.*
Dysagrion SCUDDER, 1878a, p. 534 [**D. frederici*; OD]. Apparently related to *Thaumatoneura* (recent), but family assignment uncertain. Wings with 2 accessory antenodals; postnodals not aligned with crossveins below. CALVERT, 1913; FRASER, 1957. *Eoc.*, USA (Wyoming).——FIG. 52,1. **D. frederici*; wing, ×1.4 (Fraser, 1957).
Phenacolestes COCKERELL, 1908p, p. 61 [**P. mirandus*; OD]. Similar to *Dysagrion* but with 3 accessory antenodals. CALVERT, 1913. *Oligo.*, USA (Colorado).——FIG. 52,2. **P. mirandus*; wing, ×2 (Fraser, 1957).

Family AMPHIPTERYGIDAE
Tillyard, 1926

[Amphipterygidae TILLYARD, 1926d, p. 79]

Wings petiolate; primaries distinct, extending to vein R; only a few accessory antenodals; postnodals not aligned with crossveins below; RS1+2 not arched toward R basally. *Eoc.–Holo.*

Amphipteryx SELYS-LONGCHAMPS, 1853, p. 66. *Holo.*
Petrolestes COCKERELL, 1927c, p. 81 [**P. hendersoni*; OD]. Little-known genus, based on wing

Odonata—Zygoptera

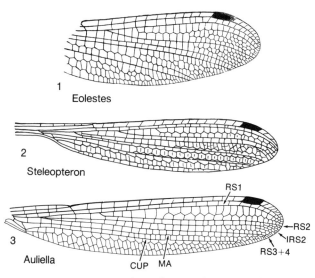

Fig. 53. Steleopteridae and Uncertain (p. 87–88).

fragment. [Family assignment doubtful.] Fraser, 1957. *Eoc.,* USA (Colorado).

Protamphipteryx Cockerell, 1920c, p. 236 [*P. basalis*; OD]. Little-known genus, based on wing fragment; arculus midway between first and second antenodals. [Family assignment doubtful.] Fraser, 1957. *Eoc.,* USA (Wyoming).

Family STELEOPTERIDAE Handlirsch, 1906

[Steleopteridae Handlirsch, 1906b, p. 597]

Wings distinctly petiolate and slender; nodus only a short distance from level of arculus; several antenodals and numerous postnodals; only 2 rows of cells between veins RS1 and RS2; RS2, IRS2, RS3+4, MA, and CUP nearly parallel, with only a single row of cells between adjacent veins except at wing margin; CUP long, extending to level of proximal edge of pterostigma. *Jur.*

Steleopteron Handlirsch, 1906b, p. 598 [*S. deichmulleri*; OD]. RS2, IRS2, RS3+4, MA, and CUP not equally spaced over wing. Prytykina, 1968. *Jur.,* Europe (Germany).——Fig. 53,2. *S. deichmulleri*; fore wing, ×2 (Handlirsch, 1906b).

Auliella Prytykina, 1968, p. 35 [*A. crucigera*; OD]. Wings as in *Steleopteron*, but RS2, IRS2, R3+4, MA, and CUP equally spaced over wing. *Jur.,* USSR (Kazakh).——Fig. 53,3. *A. crucigera*; hind(?) wing, ×2 (Prytykina, 1968).

Family CALOPTERYGIDAE Selys-Longchamps, 1850

[Calopterygidae Selys-Longchamps in Selys-Longchamps & Hagen, 1850, p. 133]

Wings not petiolate; crossveins very numerous, forming a dense reticulation; antenodals numerous, extending to vein R, primaries not differentiated; RS1+2 arched toward R basally; pterostigma commonly obsolescent. *Eoc.–Holo.*

Calopteryx Leach, 1815, p. 137. Heer, 1847, 1849, 1853a; Hagen, 1848, 1861–1863; Scudder, 1890; Esaki & Asahina, 1957. *Oligo.,* Europe (Baltic), USA (Colorado); *Mio.,* Europe (Germany, Yugoslavia); *Pleist.,* Japan–*Holo.*
Eocalopteryx Cockerell, 1920c, p. 236 [*E. atavina*; OD]. Little-known genus, based on wing fragment. [Possibly a synonym of *Mnais* (recent).] Fraser, 1940. *Eoc.,* USA (Wyoming).
Eodichroma Cockerell, 1923c, p. 397 [*E. mirifica*; OD]. Little-known genus, based on fragment of broad wing, with 13 antenodals. *Eoc.,* USA (Texas).

Family EUPHAEIDAE Selys-Longchamps, 1853

[Euphaeidae Selys-Longchamps, 1853, p. 47] [=Zacallitidae Cockerell, 1928c, p. 297]

Wings subpetiolate or not petiolate; antenodals numerous and usually aligned with crossveins below; primary antenodals not differentiated; nodus nearly at level of midwing; discoidal cell short, its base connected to vein R by arculus; pterostigma strongly developed. DAVIES, 1981. *Eoc.–Holo.*

Euphaea SELYS-LONGCHAMPS, 1840, p. 200. *Holo.*
Epallagites COCKERELL, 1924a, p. 9 [*E. avus*; OD]. Little-known genus; arculus at almost a third of distance from wing base to nodus. [Family assignment doubtful.] FRASER, 1957. *Eoc.*, USA (Colorado).
Indophaea FRASER, 1929, p. 293. THÉOBALD, 1937a; FRASER, 1957. *Oligo.*, Europe (France)–*Holo.*
Litheuphaea FRASER, 1955a, p. 43 [*L. carpenteri*; OD]. Similar to *Euphaea* (recent), but nodus situated more proximad, pterostigma much larger, and anal vein extending beyond level of nodus. *Oligo.*, USA (Oregon).——FIG. 52,5. **L. carpenteri*; wing, ×1.5 (Fraser, 1955a).
Zacallites COCKERELL, 1928c, p. 298 [*Z. balli*; OD]. Wings subpetiolate; antenodals numerous, not aligned with crossveins below; pterostigma long; apical regions of both fore and hind wings conspicuously darkened. [Type of family Zacallitidae COCKERELL, 1928c.] FRASER, 1940, 1957. *Eoc.*, USA (Colorado).——FIG. 52,3. **Z. balli; a,* fore and *b,* hind wings, ×1.4 (Fraser, 1940).

Family UNCERTAIN

The following genera, apparently belonging to the order Odonata, suborder Zygoptera, are too poorly known to permit assignment to families.

Austrolestidion TILLYARD in TILLYARD & DUNSTAN, 1916, p. 45 [*A. duaringae*; OD]. Little-known genus, based on nymph. *Paleoc.–Plio.*, Australia (Queensland).
Daemhippus NAVÁS, 1927, p. 91 [*Platycnemis cineuneguli* COLLADO, 1926, p. 101; OD]. Little-known genus, based on wing fragment. *Oligo.*, Europe (Spain).
Eolestes COCKERELL, 1940a, p. 104 [*E. synthetica*; OD]. Little-known genus, based on wing fragment, with forking of 1A into 2 irregular branches. FRASER, 1945. *Eoc.*, USA (Colorado).——FIG. 53,1. **E. synthetica*; wing, ×2.5 (Fraser, 1945).
Eosagrion HANDLIRSCH, 1920, p. 184 [*E. risi*; OD]. Little-known genus, based on wing fragment. [Type of family Eosagrionidae HANDLIRSCH, 1920.] *Jur.*, Europe (Germany).
Eothaumatoneura PONGRÁCZ, 1935, p. 527 [*E. ptychoptera*; OD]. Little-known genus, based on small wing fragments. *Eoc.*, Europe (Germany).
Euphaeopsis HANDLIRSCH, 1906b, p. 596 [*Euphaea multinervis* HAGEN, 1862, p. 119; OD]. Little-known genus, based on wing fragment. *Jur.*, Europe (Germany).
Megasemum MANEVAL, 1936, p. 28 [*M. ronzonense*; OD]. Little-known genus, based on wing fragment. *Oligo.*, Europe (France).
Protothore COCKERELL, 1930, p. 50 [*P. explicata*; OD]. Little-known genus, based on wing fragment, possibly related to Polythoridae (recent). *Eoc.*, USA (California).
Pseudoeuphaea HANDLIRSCH, 1906b, p. 596 [*Euphaea areolata* HAGEN, 1862, p. 106; SD COWLEY, 1934b, p. 252]. Little-known genus, based on wing fragment. *Jur.*, Europe (Germany).

Suborder UNCERTAIN

The following genera, apparently belonging to the order Odonata, are too poorly known to permit assignment to suborders.

Family STENOPHLEBIIDAE Handlirsch, 1906

[Stenophlebiidae HANDLIRSCH, 1906b, p. 581]

Fore and hind wings similar in form, narrowed basally but not petiolate; primary antenodals absent; numerous secondary antenodals; discoidal cell closed in both pairs of wings, irregular in shape, and divided by 1 or 2 crossveins; vein CUP strongly bent at arculus; 1A well developed. *Jur.*

Stenophlebia HAGEN, 1866a, p. 79 [*Heterophlebia amphitrite* HAGEN, 1862, p. 105; SD CARPENTER, herein]. Nodus at midwing; 1A with 3 distinct terminal branches. [This peculiar genus has been placed in the Anisoptera by HAGEN (1866a) and NEEDHAM (1903); in the Anisozygoptera by FRASER (1957); and in the Zygoptera by PRITYKINA (1980a), who designated a new superfamily, Stenophlebioidea, for it.] PRITYKINA, 1980a. *Jur.*, Europe (Germany).——FIG. 54,2. *S. latreillei* (GERMAR); *a,* fore and *b,* hind wings, ×0.7 (Hagen, 1866a); *c,* fore wing, region of discoidal cell, ×5 (Carpenter, 1932a).

Family UNCERTAIN

The following genera, apparently belonging to the order Odonata, are too poorly known to permit assignment to families.

Antitaxineura TILLYARD, 1935b, p. 382 [*A. anomala*; OD]. Small wing fragment showing nodal area. [Ordinal assignment doubtful.] RIEK, 1956. *Trias.*, Australia (New South Wales).
Camptotaxineura TILLYARD, 1937a, p. 88 [*C. ephialtes*; OD]. Apical wing fragment. [Type of family Camptotaxineuridae TILLYARD.] *Perm.*, USA (Kansas).
Kaltanoneura ROHDENDORF, 1961a, p. 86 [*K. bar-

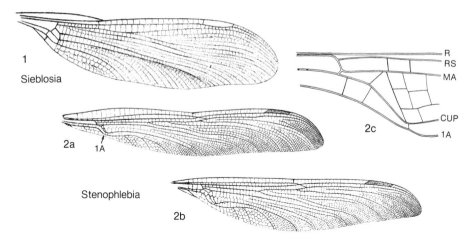

FIG. 54. Stenophlebiidae and Uncertain (p. 88–89).

tenevi; OD]. Wing fragment, with few crossveins and weak pterostigma. Probably protozygopterous. [Type of family Kaltanoneuridae ROHDENDORF.] *Perm.*, USSR (European RSFSR).

Kargalotypus ROHDENDORF, 1962a, p. 72 [**Megatypus kargalensis* MARTYNOV, 1932, p. 19; OD]. Little-known genus, based on apical wing fragment. SC extending about two-thirds wing length from base; pterostigma absent. PRITYKINA, 1981. *Perm.*, USSR (Asian RSFSR).

Magnasupplephlebia ZESSIN, 1982, p. 105 [**M. kallweita*; OD]. Little-known genus, based on apical wing fragment. *Jur.*, Europe (Germany).

Mesonetopsis PING, 1935, p. 112 [**M. zeni*; OD]. Little-known nymph, probably anisopterous. *Jur.*, China (Xinjiang).

Mesophlebia TILLYARD in TILLYARD & DUNSTAN, 1916, p. 24 [**M. antinodalis*; OD]. Little-known genus, based on small wing fragment, including nodal area. [Type of family Mesophlebiidae TILLYARD, 1916; originally placed in Anisoptera.] PRITYKINA, 1981. *Trias.*, Australia (New South Wales).

Orthaeschnites HAUPT, 1956, p. 31 [**O. primus*; OD]. Little-known genus, based on small wing fragment. *Eoc.*, Europe (Germany).

Palaeophlebia BRAUER, REDTENBACHER, & GANGLBAUER, 1889, p. 6 [**P. synlestoides*; OD]. Little-known wing fragment. *Jur.*, USSR (Asian RSFSR).

Samarura BRAUER, REDTENBACHER, & GANGLBAUER, 1889, p. 7 [**S. gigantea*; SD COWLEY, 1934b, p. 253]. Little-known nymphs. HANDLIRSCH, 1906b. *Jur.*, USSR (Asian RSFSR).

Sieblosia HANDLIRSCH, 1907, p. 896 [**Heterophlebia jucunda* HAGEN, 1858a, p. 121; OD]. Little-known genus, based on poorly preserved wing, with closed discoidal cell and weakly developed nodus much nearer to arculus than to pterostigma. [Type of family Sieblosiidae HANDLIRSCH, 1907; placed by HANDLIRSCH in the Anisozygoptera.] *Oligo.*, Europe (Germany).——FIG. 54,*1*. **S. jucunda* (HAGEN); fore(?) wing, ×1.8 (Hagen, 1858a).

Syrrhoe BODE, 1953, p. 69 [**S. commissa*; OD]. Wing fragment. *Jur.*, Europe (Germany).

Infraclass PALAEOPTERA
Order UNCERTAIN

The following genera, apparently belonging to the infraclass Palaeoptera, are too poorly known to permit assignment to orders.

Aedoeophasma SCUDDER, 1885g, p. 265 [**A. anglica*; OD]. Little-known genus, based on distal fragment of wing. [Placed by SCUDDER (1885g), HANDLIRSCH (1906b, 1919b), and BOLTON (1916, 1917b) in the Palaeodictyoptera, *incertae sedis*, but transferred to the Protodonata, *incertae sedis*, by HANDLIRSCH (1922).] *U. Carb.*, England.

Archaeoptilites HANDLIRSCH, 1919b, p. 534 [**Archaeoptilus lucasi* BRONGNIART, 1885a, p. 60; OD]. Little-known genus, based on very small fragment of wing. [Originally placed in the Palaeodictyoptera.] BRONGNIART, 1893; HANDLIRSCH, 1922. *U. Carb.*, Europe (France).

Archaeoptilus SCUDDER, 1881b, p. 295 [**A. ingens*; OD]. Little-known genus, based on small fragment of large wing. [Type of the family Archaeoptilidae HANDLIRSCH, 1906b. Originally considered by SCUDDER to be "neuropterous," this genus was subsequently (1883b) placed by him in the orthopteroid complex. However, HANDLIRSCH (1906b) and BOLTON (1925) were of the opinion that it was more likely a member of the Palaeodictyoptera.] *U. Carb.*, England.

Bardapteron ZALESSKY, 1944a, p. 342 [**B. ovale*;

OD]. Little-known genus, based on fragment of wing. [Type of the family Bardapteridae ZALESSKY, 1944a. Originally placed in a new order, Permodictyoptera, but transferred by ROHDENDORF (1962a) to the Palaeodictyoptera.] *Perm.*, USSR (European RSFSR).

Breyeriodes HANDLIRSCH, 1906b, p. 118 [*B. kliveri*; OD]. Little-known genus, based on small fragment of wing. [Originally placed in the Palaeodictyoptera, *incertae sedis.*] HANDLIRSCH, 1919b, 1922. *U. Carb.*, Europe, (Germany).

Campteroneura HANDLIRSCH, 1906a, p. 685 [*C. reticulata*; OD]. Little-known genus, based on small fragment of wing. HANDLIRSCH, 1906b, 1922. *U. Carb.*, USA (Alabama).

Campyloptera BRONGNIART, 1893, p. 406 [*C. eatoni*; OD]. Little-known genus, based on incomplete wing. [The generic name *Campyloptera* was first used in 1885 (BRONGNIART, 1885a), but no species was mentioned until 1893. Placed in the Megasecoptera by BRONGNIART (1885a), HANDLIRSCH (1906b), CARPENTER (1943b), and LAURENTIAUX (1953); in the Protodonata by BRONGNIART (1893) and TILLYARD (1928d); and in a new order, Campylopterodea, by ROHDENDORF (1962a).] *U. Carb.*, Europe (France).

Cryptovenia BOLTON, 1912, p. 315 [*C. moyseyi*; OD]. Little-known genus, based on wing fragment. [Originally placed in the Palaeodictyoptera. Type of the family Cryptoveniidae BOLTON, 1912]. LAMEERE, 1917b; HANDLIRSCH, 1919b, 1922. *U. Carb.*, England.

Dictyoneurella LAURENTIAUX, 1949b, p. 207 [*D. perfecta*; OD]. Little-known genus, based on incomplete wing. [Type of the family Dictyoneurellidae KUKALOVÁ-PECK, 1975. Placed in the Palaeodictyoptera by LAURENTIAUX (1949b); transferred to the Megasecoptera by KUKALOVÁ-PECK (1975).] *U. Carb.*, Europe (France).

Dyadentomum HANDLIRSCH, 1904b, p. 7 [*D. permense*; OD]. Little-known genus, based on a body fragment thought by HANDLIRSCH to be that of an ephemerid nymph. *Perm.*, USSR (European RSFSR).

Eohymen MARTYNOV, 1937b, p. 9 [*E. maculipennis*; OD]. Little-known genus, based on poorly preserved wing. [Type of the family Eohymenidae MARTYNOV, 1937b. Placed in the Megasecoptera (Protohymenoptera) by MARTYNOV (1937b), in the Palaeodictyoptera by ROHDENDORF (1962a), and in the Caloneurodea by RASNITSYN (1980b).] *Perm.*, USSR (European RSFSR).

Erasipterella BRAUCKMANN, 1983, p. 9 [*E. piesbergensis*; OD]. Little-known genus, based on fragments of fore and hind wings. [Almost certainly a member of the odonate complex, but order doubtful.] *U. Carb.*, Europe (Germany).

Erasipteron PRUVOST, 1933a, p. 151 [*E. larischi*; OD]. Little-known genus, based on incomplete wings. [Type of the family Erasipteridae CARPENTER, 1939. Placed in the Odonata by PRUVOST (1933a) and KUKALOVÁ (1964b); transferred to the Protodonata by CARPENTER (1939), LAURENTIAUX (1953), WHALLEY (1979), and PRITYKINA (1980b).] *U. Carb.*, Europe (Czechoslovakia), England.

Eurytaenia HANDLIRSCH, 1906a, p. 674 [*E. virginica*; OD]. Little-known genus, based on small fragment of wing. [Originally placed in the Palaeodictyoptera.] HANDLIRSCH, 1906b, 1922. *U. Carb.*, USA (West Virginia).

Frankenholzia GUTHÖRL, 1962c, p. 227 [*F. culmanni*; OD]. Little-known genus, based on wing fragment. [Originally placed in the Palaeodictyoptera, but transferred to the Megasecoptera by KUKALOVÁ-PECK (1975).] *U. Carb.*, Europe (Germany).

Gerephemera SCUDDER, 1880, p. 12 [*G. simplex*; OD]. Little-known genus, based on small fragment of wing. [Originally placed in the order Ephemeroptera by SCUDDER, but later (1890) transferred to the Orthoptera; assigned to the Odonata by HAGEN (1881a, 1881b, 1885) and to the Palaeodictyoptera by HANDLIRSCH (1906a, 1906b).] *U. Carb.*, Canada (Nova Scotia).

Hypermegethes HANDLIRSCH, 1906a, p. 672 [*H. schucherti*; OD]. Little-known genus, based on a small, proximal fragment of very large wing. [Type of the family Hypermegethidae HANDLIRSCH, 1906a. Placed in the Palaeodictyoptera by HANDLIRSCH (1906a, 1906b, 1922), but transferred to the Protohemiptera by LAMEERE (1917c).] *U. Carb.*, USA (Illinois).

Kuloja MARTYNOV, 1928b, p. 7 [*K. expansa*; OD]. Little-known genus, based on distal fragment of wing. [Type of the family Kulojidae MARTYNOV, 1928b. Originally placed in the Megasecoptera, but transferred to the Diaphanopterodea by ROHDENDORF (1962a).] MARTYNOV, 1932. *Perm.*, USSR (European RSFSR).

Leipsanon HANDLIRSCH, 1906b, p. 120 [*L. reticulatum*; OD]. Little-known genus, based on minute wing fragment. [Originally placed in the Palaeodictyoptera, *incertae sedis.*] HANDLIRSCH, 1919b. *U. Carb.*, Europe (Belgium).

Lithentomum SCUDDER, 1868c, p. 206 [*L. hartti*; OD]. Little-known genus, based on small fragment of wing. [Originally placed in the Palaeodictyoptera.] SCUDDER, 1880; HANDLIRSCH, 1906a, 1922. *U. Carb.*, Canada (New Brunswick).

Litoneura SCUDDER, 1885a, p. 169 [*Dictyoneura anthracophila* GOLDENBERG, 1854, p. 35; SD HANDLIRSCH, 1906b, p. 77]. Little-known genus, based on fragment of small wing. [Originally placed in the Palaeodictyoptera.] *U. Carb.*, Europe (Germany).

Litophlebia HUBBARD & RIEK, 1978, p. 260, *nom. subst. pro Xenophlebia* RIEK, 1976e, p. 150, *non*

Palaeoptera—Order Uncertain

DEMOULIN, 1968 [*Xenophlebia optata RIEK, 1976e, p. 150; OD]. Little-known genus, based on incomplete wing. [Type of the family Litophlebiidae HUBBARD & RIEK, 1978. Placed in the Ephemeroptera by RIEK (1976e) and HUBBARD & RIEK (1978) and in the Megasecoptera by HUBBARD & KUKALOVÁ-PECK (1980).] Trias., South Africa.

Megathentomum SCUDDER, 1868b, p. 570 [*M. pustulatum; OD]. Little-known genus, based on distal fragment of very large wing. SCUDDER, 1891; HANDLIRSCH, 1906a, 1906b, 1922. U. Carb., USA (Illinois).

Melanoblattula COCKERELL, 1927g, p. 415 [*M. nigressens; OD]. Little-known genus, based on fragment of small wing. [Originally placed in the Protorthoptera.] U. Carb., USA (Maryland).

Microblattina SCUDDER, 1895c, p. 57 [*M. perdita; OD]. Little-known genus, based on wing fragment. [Originally placed in the Blattaria, but transferred by HANDLIRSCH (1906a, 1906b) to the Protoblattoidea.] U. Carb., USA (Rhode Island).

Orthogonophora HANDLIRSCH, 1906a, p. 686 [*O. distincta; OD]. Little-known genus, based on distal fragment of wing. [Originally placed in the Palaeodictyoptera, incertae sedis.] HANDLIRSCH, 1906b, 1922. U. Carb., USA (West Virginia).

Palaeodictyopteron HANDLIRSCH, 1906a, p. 688 [collective group]. Little-known nymphal forms. HANDLIRSCH, 1906b; CARPENTER, 1948a. U. Carb., USA (Illinois, West Virginia), Europe (Germany).

Palaeopalara HANDLIRSCH, 1904a, p. 10 [*P. gracilis; OD]. Little-known genus, based on small fragment of wing. [Placed in the Megasecoptera by HANDLIRSCH (1906b) and KUKALOVÁ-PECK (1975).] U. Carb., Europe (Belgium).

Parapaolia HANDLIRSCH, 1906a, p. 687 [*Paolia superba SCUDDER, 1885a, p. 173; OD]. Little-known genus. [Placed, with some doubt, by HANDLIRSCH (1906b) in the Palaeodictyoptera.] U. Carb., USA (Illinois).

Perissophlebia TILLYARD, 1918c, p. 422 [*P. multiseriata; OD]. Little-known genus, based on small wing fragment. [Placed in the Odonata by TILLYARD (1918c) and PRITYKINA (1981).] Trias., Australia (Queensland).

Permoneura CARPENTER, 1931b, p. 124 [*P. lameerei; OD]. Little-known genus, based on complete hind wing. [Type of the family Permoneuridae CARPENTER, 1931b. Placed in the order Palaeodictyoptera by CARPENTER (1931b) and TILLYARD (1937); transferred to a new order, Permoneurodea (allied to the Palaeodictyoptera), by LAURENTIAUX (1953); and included in a new order, Archodonata (along with several other genera formerly in the Palaeodictyoptera), by ROHDENDORF (1962a). The ordinal name Archodonata was changed by SINITSHENKOVA (1980a, 1980b) to Permothemistida. CARPENTER (1976) proposed that the genus Permoneura be assigned to the Palaeoptera, incertae sedis.] Perm., USA (Kansas).

Piroutetia MEUNIER, 1907, p. 522 [*P. liassina; OD]. Little-known genus, based on wing fragment and placed in the Odonata. MEUNIER, 1908b. Jur., Europe (France).

Progonopteryx HANDLIRSCH, 1904a, p. 5 [*P. belgica; OD]. Little-known genus, based on wing fragment. [Originally placed in the Palaeodictyoptera (family Dictyoneuridae).] HANDLIRSCH, 1906b, 1919b. U. Carb., Europe (Belgium).

Protagrion BRONGNIART, 1893, p. 403 [*P. audouini; OD]. Little-known genus, based on incomplete wing. [Type of the family Protagrionidae HANDLIRSCH, 1906b. The generic name Protagrion was first used in 1885 (BRONGNIART, 1885a), but no species was mentioned until 1893. Placed in the Protodonata by BRONGNIART (1893), HANDLIRSCH (1906b), and MARTYNOV (1932); transferred to the Palaeodictyoptera by CARPENTER (1943b) and ROHDENDORF (1962a).] U. Carb., Europe (France).

Pseudohomothetus HANDLIRSCH, 1906a, p. 685 [*Homothetus erutus MATTHEW, 1895a, p. 95; OD]. Little-known genus, based on small wing fragment. [Originally placed in the Palaeodictyoptera, incertae sedis.] HANDLIRSCH, 1906b, 1919b. U. Carb., Canada (New Brunswick).

Pseudopalingenia HANDLIRSCH, 1906b, p. 124 [*Palingenia feistmanteli FRITSCH, 1882, p. 1; OD]. Little-known genus, based on part of body, including cerci. [Originally placed, with some uncertainty, in the Palaeodictyoptera, incertae sedis.] HANDLIRSCH, 1922. U. Carb., Europe (Czechoslovakia).

Pseudopaolia HANDLIRSCH, 1906a, p. 687 [*Paolia lacoana SCUDDER, 1885a, p. 173; OD]. Little-known genus. [Placed by HANDLIRSCH (1906b), with uncertainty, in the order Palaeodictyoptera, incertae sedis.] U. Carb., USA (Pennsylvania).

Rectineura BOLTON, 1934, p. 181 [*R. lineata; OD]. Little-known genus, based on poorly preserved wing fragment. [Originally placed in the Palaeodictyoptera.] U. Carb., England.

Reisia HANDLIRSCH, 1909c, p. 81, nom. subst. pro Handlirschia REIS, 1909, p. 693, non KOHL, 1896 [*Handlirschia gelasii REIS, 1909; OD]. Little-known genus, based on small fragment of wing. [Placed by HANDLIRSCH (1909c, 1920) and REIS (1909) in the Protodonata.] Trias., Europe (Germany).

Severinula PRUVOST, 1930, p. 151 [*S. leopoldi; OD]. Little-known genus, based on wing fragment. [Placed in the Palaeodictyoptera by PRUVOST (1930) and ROHDENDORF (1962a).] U. Carb., Europe (Belgium).

Sherborniella HANDLIRSCH, 1919b, p. 535

[*Palaeodictyopteron* (collective group) *higginsi* HANDLIRSCH, 1906b, p. 125; OD]. Little-known genus, based on small basal fragment of wing. [Originally placed in the Palaeodictyoptera, *incertae sedis*.] BOLTON, 1921, 1934. *U. Carb.*, England.

Sypharoptera HANDLIRSCH, 1911, p. 372 [*S. pneuma*; OD]. Little-known genus, based on incomplete wings. [Originally placed in the new order Sypharopteroidea by HANDLIRSCH (1911); transferred to order Diaphanopterodea by ROHDENDORF (1962a).] HANDLIRSCH, 1919b, 1922. *U. Carb.*, USA (Illinois).

Titanoptera BRONGNIART, 1893, p. 379 [*T. maculata*; OD]. Little-known genus, based on small fragment of wing; probably a palaeodictyopteron. HANDLIRSCH, 1906b; LAMEERE, 1917b. *U. Carb.*, Europe (France).

Triadologus RIEK, 1976b, p. 793 [*T. biseriatus*; OD]. Little-known genus, based on small fragment of wing. [Placed in the Protodonata by RIEK (1976b) and in the Odonata by PRITYKINA (1981).] *Trias.*, South Africa.

Wulasua T'AN, 1980, p. 159 [*W. maculata*; OD]. Little-known genus, based on a poorly preserved, small fragment of a wing. [Originally placed in the Diaphanopterodea.] *Perm.*, China (Inner Mongolia).

Xenoneura SCUDDER, 1868c, p. 206 [*X. antiquorum*; OD]. Little-known genus, based on wing fragment. [Type of the family Xenoneuridae SCUDDER, 1885b. Originally placed in the Palaeodictyoptera.] SCUDDER, 1880; HANDLIRSCH, 1906b, 1922. *U. Carb.*, Canada (New Brunswick).

Infraclass NEOPTERA
Martynov, 1923

[Neoptera MARTYNOV, 1923, p. 89]

Wings articulated to thorax by sclerotized plates (axillaries), not fused or rigidly connected; third axillary Y-shaped and attached to second axillary and posterior notal process, and connected by flexor muscle to thorax; venation basically as in Palaeoptera, but vein MA flat or nearly so or absent; cerci commonly present but vestigial or absent in higher orders. Immature stages very diverse in structure and development. *U. Carb.–Holo.*

This infraclass has been the predominant one since the Permian. It includes 25 existing orders and about 98 percent of the existing species of insects.

Division EXOPTERYGOTA
Sharp, 1899

[Exopterygota SHARP, 1899, p. 247]

Immature stages typically resembling the adults in general form, living in the same kind of environments, and having similar feeding habits; metamorphosis to adults gradual, wings developing within an externally visible cuticular sheath; pupal stage absent. *U. Carb.–Holo.*

Fifteen existing orders are generally recognized in this division, including about 11 percent of the existing species of insects. The orders are usually grouped into two categories, the orthopteroids and the hemipteroids, which have basic structural differences and which appear to represent two distinct lines of exopterygote evolution, although there is some doubt that either one is monophyletic (RICHARDS & DAVIES, 1977; I. M. MACKERRAS, 1970). The orthopteroids have mandibulate mouthparts; the fore wings are commonly tegminous or rarely elytroid; the hind wings commonly have a large fan-shaped anal area; cerci are present and are commonly well developed. These insects are known from the Upper Carboniferous to the present. Four very small, existing orders (Grylloblattodea, Zoraptera, Mallophaga, and Anoplura) belonging here are the only existing orders of insects absent from the geological record. They are discussed briefly below, within the Exopterygota.

The hemipteroids have haustellate mouthparts and feed on liquid food; the fore wings are diverse in structure, membranous or modified to hemelytra or elytra. The hind wings are broad, commonly with an anal fan in the more primitive families, but are small or very small in the more specialized families. Cerci are absent. These orders are known from the Permian to the present.

ORTHOPTEROID EXOPTERYGOTES

Order PERLARIA Latreille, 1802

[*nom. transl.* HANDLIRSCH, 1903, p. 733, *ex* Perlariae LATREILLE, 1802a, p. 292] [=Plecoptera BURMEISTER, 1838 in BURMEISTER, 1838–1839, p. 863] [Although the name Plecoptera is often used for this order, it has the distinct disadvantage of being easily confused with the ordinal name Plectoptera, occasionally used for mayflies. Perlaria is the older name.]

Fore wing membranous; costa marginal; vein SC usually extending to about midwing, rarely beyond, terminating on costa but connected distally to R; costal veinlets commonly few, even absent; R with several oblique veinlets leading to wing margin; RS arising at or before midwing, commonly near base, with 3 or 4 branches; M apparently dividing into MA and MP very near wing base; MA forked; MP obliquely or transversely directed toward CUA and anastomosed with it; MP+CUA with at least 2 terminal branches; CUA diverging from CUP near wing base; CUP, 1A, and 2A unbranched. Hind wing typically with expanded and folded anal area, reduced or absent in a few specialized genera; RS arising at or near wing base; M coalesced with base of RS; MA forked; MP diverging toward and anastomosing with CUA; MP+CUA and also CUP unbranched; anal veins varying in number and degree of development; crossveins usually few, highly variable, and in many genera restricted to certain areas of wing. Wings at rest held flat, not slanted, over abdomen. Mouthparts mandibulate, weak in recent species; antennae setaceous, long; body weakly sclerotized; cerci usually well developed, with numerous segments; ovipositor absent or vestigial. Nymphs similar to adults in general form but aquatic; tracheal gills on thorax, coxae, or other parts of body, including sides of abdominal segments (Eustheniidae). *Perm.–Holo.*

Difference of opinion exists about the homologies of some wing veins. The media (M) appears to be represented by a forked vein that has been generally interpreted as MA (HANDLIRSCH, 1906b, 1907, 1908a; ROHDENDORF, 1962a). SHAROV (1961b) concluded that MP is actually present at the wing base, diverging from MA and coalescing with CUA, as in certain families of Protorthoptera. The free, diverging part of MP is apparently more distinct in the hind wing than in the fore. This interpretation of MA, MP, and CUA is followed here.

The existing Perlaria are usually divided into several suborders, but opinions differ about the number of these and the structural bases for the divisions (cf. ILLIES, 1965, and RASNITSYN, 1980d). However, since almost none of the fossil specimens shows the morphological features used in the subordinal classification, these groups are omitted from the following account.

The recognition of two Permian genera, *Stenoperlidium* and *Palaeotaeniopteryx,* as members of living families (Eustheniidae and Taeniopterygidae, respectively) is necessarily dubious. Only part of the fore wings and none of the hind wings are known for *Stenoperlidium*; the fore wings and part of the hind are known for *Palaeotaeniopteryx* but not the anal area of the hind wing. Nevertheless, the geological record as now known suggests that the stone flies were well established at the ordinal level before the end of the Paleozoic Era and that relatively few modifications, such as the reduction of the anal area of the hind wing and of crossveins, have taken place in the adults subsequently.

Family PALAEOPERLIDAE Sharov, 1961

[Palaeoperlidae SHAROV, 1961e, p. 227]

Costal area of fore wing with 4 or 5 veinlets; 3 or 4 veinlets from vein R to margin; free basal piece of MA slightly oblique; MP+CUA with 3 terminal branches; anal area narrow. Hind wing and body unknown. *Perm.*

Palaeoperla SHAROV, 1961e, p. 227 [**P. exacta*; OD]. RS with 3 or 4 branches; crossveins numerous between MA and MP+CUA, few elsewhere. *Perm.*, USSR (Asian RSFSR).——Fig. 55,2a. **P. exacta*; fore wing, ×5.5 (Sharov, 1961e). ——Fig. 55,2b. *P.(?) prisca* Sharov; nymph, ×9 (Sharov, 1961e).

Family PERLOPSEIDAE Martynov, 1940

[*nom. correct.* Rohdendorf, 1957, p. 81, *pro* Perlopsididae Martynov, 1940, p. 31]

Fore wing as in Palaeoperlidae but with only 1 veinlet from vein R to margin, 2 branches on MP+CUA. Hind wing unknown. Body slender, legs long, tarsi with 3 segments. *Perm.*

Perlopsis Martynov, 1940, p. 31 [**P. filicornis*; OD]. Costal area very narrow; wing widest beyond middle. *Perm.*, USSR (Asian RSFSR).——Fig. 55,1. **P. filicornis*; *a*, fore wing, ×3.2 (Rohdendorf, 1962a); *b*, body, ×4.0 (Martynov, 1940).

Family SIBERIOPERLIDAE Sinitshenkova, 1983

[Siberioperlidae Sinitshenkova, 1983, p. 96]

Antennae long, moniliform, shorter than body. Wings of females of normal size, with branches of RS directed posteriorly and MA and MP unbranched; hind wings with enlarged anal area; anal veins branched. Males micropterous; costal region of the fore wing unusually wide. Legs short, femora wide; cerci shorter than body. Nymphs with body densely covered with short hairs; antennae and cerci relatively short; tracheal gills absent. Apparently related to the existing family Gripopterygidae. *Jur.*

Siberioperla Sinitshenkova, 1983, p. 96 [**S. lacunosa*; OD]. Posterior margin of head convex; first antennal segment short. Female with fore wing about three times as long as wide; RS with 4 or 5 branches; CUA with 2 or 3 branches. *Jur.*, USSR (Asian RSFSR).

Family EUSTHENIIDAE Tillyard, 1921

[Eustheniidae Tillyard, 1921d, p. 35]

Fore wing with few to many veinlets in costal area; vein RS with at least 3 branches; 3 anal veins; crossveins present over most of wing. Hind wing with prominent anal area

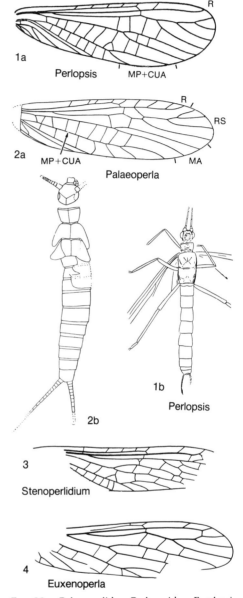

Fig. 55. Palaeoperlidae, Perlopseidae, Eustheniidae, and Uncertain (p. 94–97).

but without marginal indentation at end of CUP. Nymphs with 5 or 6 pairs of lateral abdominal gills. *Perm.–Holo.*

Eusthenia Westwood, 1832, p. 348. *Holo.*
Stenoperlidium Tillyard, 1935c, p. 386 [**S. permianum*; OD]. Similar to *Stenoperla* (recent) but

Perlaria

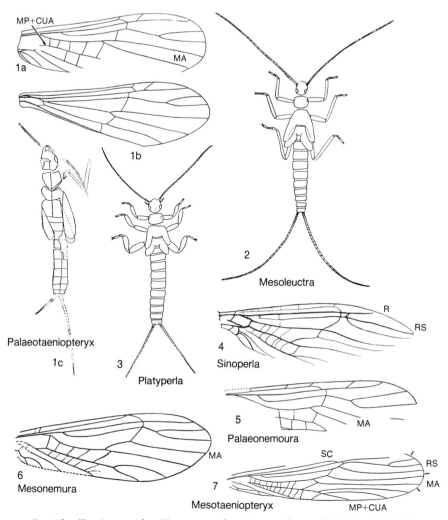

Fig. 56. Taeniopterygidae, Notonemouridae, Platyperlidae, and Perlidae (p. 95–97).

with fewer crossveins and with the broadest part of wing proximal to midwing. [Family assignment doubtful.] *Perm.*, Australia (New South Wales).——Fig. 55,*3*. **S. permianum*; fore wing, ×2 (Tillyard, 1935c).

Family TAENIOPTERYGIDAE
Klapálek, 1905

[Taeniopterygidae KLAPÁLEK, 1905, p. 30]

Costal area of fore wing with few veinlets; vein R with not more than single veinlet leading to margin; MA commonly with 2 branches; crossveins fewer than in Palaeoperlidae. *Perm.–Holo.*

Taeniopteryx PICTET, 1841, p. 343. Adults. HAGEN in PICTET & HAGEN, 1856; ILLIES, 1965. *Oligo.*, Europe (Baltic)–*Holo.*

Brachyptera NEWPORT, 1848, p. 388. ILLIES, 1967a. *Plio.*, Europe (Germany)–*Holo.*

Mesonemura BRAUER, REDTENBACHER, & GANGLBAUER, 1889, p. 11 [**M. maaki*; OD]. Crossveins from end of SC to MA forming continuously curved series. *Jur.*, USSR (Asian RSFSR).——Fig. 56,*6*. *M. turanica* MARTYNOV; fore wing, ×5.5 (Martynov, 1937a).

Mesotaeniopteryx MARTYNOV, 1937a, p. 81 [**M. elongata*; OD]. RS forked to level of end of SC; MP+CUA with 3 terminal branches, first 2 close together, curved, and directed anteriorly; crossveins numerous between MA and MP+CUA, more numerous between MP+CUA and CUP. *Jur.*, USSR (Tadzhik).——Fig. 56,7. *M. splendida* MARTYNOV; fore wing, ×2.5 (Martynov, 1937a).

Palaeonemoura SHAROV, 1961e, p. 233 [**P. clara*; OD]. Fore wing as in *Palaeotaeniopteryx* but with MA more strongly curved and with fewer crossveins between MA and MP+CUA. *Perm.*, USSR (Asian RSFSR).——Fig. 56,5. **P. clara*; fore wing, ×6.5 (Sharov, 1961e).

Palaeotaeniopteryx SHAROV, 1961e, p. 230 [**P. elegans*; OD]. MA forked before level of end of SC; several crossveins between MA and MP+CUA. *Perm.*, USSR (Asian RSFSR).——Fig. 56,1. **P. elegans*; *a*, fore and *b*, hind wings, ×8; *c*, nymph, ×10 (all Sharov, 1961e).

Perlariopsis PING, 1928, p. 31 [**P. peipiaoensis*; OD]. Little-known adult. [Family assignment doubtful.] ILLIES, 1965. *Cret.*, China.

Sinonemoura PING, 1928, p. 24 [**S. grabaui*; OD]. Little-known nymph. [Family position doubtful.] ILLIES, 1965. *Cret.*, China.

Family LEUCTRIDAE Klapálek, 1905

[Leuctridae KLAPÁLEK, 1905, p. 32]

Costal area with 1 or 2 veinlets; vein R with 1 veinlet to margin; RS forked to about half its length, branches parallel to those of MA; MP+CUA and CUP markedly divergent distally. *Eoc.–Holo.*

Leuctra STEPHENS, 1836, p. 144. Adults. PICTET & HAGEN, 1856; COCKERELL, 1922b; ILLIES, 1965; JARZEMBOWSKI, 1980. *Eoc./Oligo.*, England; *Oligo.*, Europe (Germany, Baltic), USA (Colorado)–*Holo*.

Megaleuctra NEAVE, 1934, p. 4. Adult female. RICKER, 1935; ILLIES, 1967b; ZWICK, 1973. *Oligo.*, Europe (Baltic)–*Holo*.

Family NOTONEMOURIDAE Ricker, 1950

[Notonemouridae RICKER, 1950, p. 201]

Similar to Nemouridae (recent). Adults with vein SC arching toward C. Nymphs small, lacking external gills; cerci multisegmented, almost as long as antennae. ZWICK, 1973. *Jur.–Holo.*

Notonemoura TILLYARD, 1923, p. 215. *Holo.*

Mesoleuctra BRAUER, REDTENBACHER, & GANGLBAUER, 1889, p. 9 [**M. gracilis*; OD]. Nymph: posterior margin of head rounded; basal antennal segment about as wide as long; mandibles slightly longer than wide, with many denticles; femora only slightly shorter and broader than tibiae; pronotum about as long as wide; posterior wing pads slightly broader than the fore pair; cerci very thin distally; body without hair covering. PING, 1928; ILLIES, 1965; SINITSHENKOVA, 1982. *Jur.*, USSR (Asian RSFSR); *Cret.*, China (Inner Mongolia).——Fig. 56,2. **M. gracilis*; restoration, dorsal view, ×2 (Sinitshenkova, 1982).

Family PERLODIDAE Klapálek, 1912

[Perlodidae KLAPÁLEK, 1912, p. 5]

Venation very diverse; costal space usually with several short veinlets; veinlets from vein R to margin commonly longer and more oblique than those of costal area; RS commonly with several branches; MP+CUA diverging anteriorly in distal region, appearing to coalesce with branches of MA. *Oligo.–Holo.*

Perlodes BANKS, 1903, p. 241. Adult. PICTET & HAGEN, 1856; ILLIES, 1965. *Oligo.*, Europe (Baltic)–*Holo*.

Isoperla BANKS, 1906, p. 175. Adult. PICTET & HAGEN, 1856; ILLIES, 1965. *Oligo.*, Europe (Baltic)–*Holo*.

Family PLATYPERLIDAE Sinitshenkova, 1982

[Platyperlidae SINITSHENKOVA, 1982, p. 118]

Nymph: head about as long as wide; antennae long, basal segment large; labrum transverse; mandibles with only a few denticles apically; pronotum transverse; femora and tibiae relatively short and broad; 2 basal segments of tarsi short and broad, their combined lengths less than that of third segment; tarsal claws prominent; fore wing pads relatively long, hind pair much shorter and broader; body covered with hair; external gills apparently absent. *Jur.*

Platyperla BRAUER, REDTENBACHER, & GANGLBAUER, 1889, p. 10 [**P. platypoda*; OD]. Posterior margin of head strongly convex; basal segment of antenna conical; pronotum with prominent posterior angles; posterior margin of terminal abdominal tergite with a short, broad median projection. PING, 1928; SINITSHENKOVA, 1982. *Jur.*, USSR (Asian RSFSR).——Fig. 56,3. **P. platypoda*; restoration of nymph, dorsal view, ×2 (Sinitshenkova, 1982).

Family PERLIDAE Latreille, 1802

[Perlidae LATREILLE, 1802a, p. 292]

Costal area with numerous, short veinlets, those between vein R and margin continuing series; RS usually with 3 or 4 branches; distal part of MP+CUA curving anteriorly. *Cret.—Holo.*

Perla GEOFFROY, 1762, p. 229. Adult. PICTET & HAGEN, 1856. *Oligo.,* Europe (Baltic)–*Holo.*
Sinoperla PING, 1928, p. 28 [**S. abdominalis*; OD]. Little-known wings, with 2 crossveins between R and RS near end of SC. ILLIES, 1965. *Cret.,* China.——FIG. 56,4. **S. abdominalis*; fore wing, ×5.3 (Ping, 1928).

Family UNCERTAIN

The following genera, apparently belonging to the order Perlaria, are too poorly known to permit assignment to families.

Euxenoperla RIEK, 1973, p. 531 [**E. simplex*; OD]. Little-known genus, possibly related to the Gripopterygidae. Vein RS of fore wing arched anteriorly at level of end of SC; RS with 3 or 4 branches; M deeply forked, branching before midwing; few crossveins. *Perm.,* South Africa; *Trias.,* Australia (Queensland).——FIG. 55,4. **E. simplex,* South Africa; fore wing, ×2.8 (Riek, 1973).
Euxenoperlella RIEK, 1976a, p. 770 [**E. jacquesi*; OD]. Similar to *Euxenoperla,* but RS with only 2 branches and M forking more distally. *Perm.,* South Africa.
Gondwanoperlidium PINTO & PURPER, 1978, p. 79 [**G. argentinarum*; OD]. Little-known genus, similar to *Euxenoperla,* but fore wings with many more crossveins. *Trias.,* South America (Argentina).
Mesonotoperla RIEK, 1954c, p. 167 [**M. sinuata*; OD]. Fore wing fragment, possibly of an eustheniid. *Trias.,* Australia (New South Wales).
Permoleuctropsis MARTYNOV, 1937b, p. 34 [**P. gracilis*; OD]. Little-known nymph. *Perm.,* USSR (European RSFSR).
Uralonympha ZALESSKY, 1939, p. 64 [**U. varica*; OD]. Little-known nymph. CARPENTER, 1969. [Ordinal assignment uncertain.] *Perm.,* USSR (Asian RSFSR); *Jur.,* Antarctica (Ohio Range).

Order PROTORTHOPTERA Handlirsch, 1906

[Protorthoptera HANDLIRSCH, 1906a, p. 695] [=Hadentomoidea HANDLIRSCH, 1906a, p. 692; Hapalopteroidea HANDLIRSCH, 1906a, p. 694; Protoblattoidea HANDLIRSCH, 1906a, p. 704; Reculoidea HANDLIRSCH, 1906b, p. 127; Protoperlaria TILLYARD, 1928b, p. 187; Cnemidolestoidea HANDLIRSCH, 1937, p. 63; Paraplecoptera MARTYNOV, 1938b, p. 98; Strephocladodea MARTYNOV, 1938b, p. 100; Protocicadida HAUPT (in part), 1941, p. 75; Protofulgorida HAUPT (in part), 1941, p. 75] [The ordinal name Protoblattoidea HANDLIRSCH (1906a) was changed to Protoblattodea by SHAROV (1962a) and has generally been accepted.]

Wings typically containing all main veins, including MA and MP, but without complete alternation of convexities and concavities; MA apparently absent in some families (see below). Main veins usually independent; in some families CUA anastomosed with part of M or MP and CUP; in a few families MA tending to coalesce with branches of RS. Fore and hind wings commonly similar in form and venation, anal area of hind wing rarely expanded to form prominent lobe. Fore wing membranous in more primitive groups, but slightly coriaceous or distinctly coriaceous in others; hair covering usually well developed on membranous wings, reduced or absent on coriaceous wings; prominent setae may be present on certain parts of wings; wing area between veins with archedictyon, resembling that of Palaeodictyoptera, or with coarse network of crossveins or more commonly with regular system of nearly straight crossveins; anal area in many families set off from remigium by strongly concave CUP. Hind wing membranous, venation of remigium usually slightly different from that of fore wing; RS arising nearer wing base, and stems of CUA and M coalesced; CUA much less developed; CUP setting off anal area, which includes several anal veins. Fore wings (and more rarely hind wings) may have conspicuous maculations or prominent cuticular thickenings.

Body structure: antennae prominent, usually long (e.g., Liomopteridae), with numerous segments; head (known in very few families) small, almost always hypognathous; prothorax commonly bearing pronotal disc (e.g., Liomopteridae) or slender, without such disc; some families (e.g., Geraridae) with elongate prothorax, which may bear prominent spines; prothorax very rarely (e.g., Lemmatophoridae) with pair of membranous paranotal lobes (Fig. 57, *Lemmatophora*), resembling those of certain Palaeodictyoptera; pterothorax with usual form; legs usually cursorial, but forelegs in some families apparently raptorial; in none, as the order is treated here, were hind legs modified for jumping (saltatorial); five tarsomeres (little known). Abdomen of moderate length; cerci usually prominent, long in some forms (e.g., Liomopteridae), but commonly small (e.g.,

Fig. 57. *Lemmatophora typa* Sellards, Lemmatophoridae, Permian of Kansas. Dorsal view of head and anterior portion of thorax, showing the reticulate, pronotal lobes and hair covering. Specimen MCZ 3539, ×40 (Carpenter, new).

Protembiidae) or modified (e.g., males, Chelopteridae).

Nymphal forms little known; antennae and cerci well developed; most nymphs clearly terrestrial (e.g., *Liomopterites, Kaltanympha*), others apparently modified for aquatic life, with tracheal gills along sides of abdomen (see Fig. 61,1b, *Lemmatophora*). U. Carb.–Trias.

The extinct order Protorthoptera was named by Handlirsch (1906a, 1906b) for a diverse assemblage of Paleozoic species with presumed orthopteroid affinities. He also named another extinct order, Protoblattoidea, for other species that he considered to be intermediate between the Palaeodictyoptera and the Blattodea, Manteodea, and Phasmatodea; at the same time he named a third order, "Protorthoptera vel Protoblattoidea," for species that were apparently intermediate between the Protorthoptera and the Protoblattodea. As more Paleozoic insects became known, a gradual diminution of the distinctions between the Protorthoptera and the Protoblattodea became apparent, the number of genera placed in the "Protorthoptera vel Protoblattodea" complex being nearly double that in the Protoblattodea (HANDLIRSCH, 1922). This classification proved unsatisfactory, and MARTYNOV subsequently (1937b, 1938b) proposed that the order Protorthoptera be restricted to species having saltatorial hind legs, like the true Orthoptera, and that the remaining species in that order be placed in another new, extinct order, Paraplecoptera. In the same year ZEUNER transferred the saltatorial species cited by MARTYNOV to the order Orthoptera, where they clearly belonged (Zeuner, 1937). The cursorial species were then distributed among the other three orders, the Protorthoptera, Protoblattodea, and Paraplecoptera. This arrangement was followed for many years. However, SHAROV (1961a, 1966a) was convinced that one family, Sthenaropodidae, which had slender, cursorial hind legs and which was previously included in the Protorthoptera, represented the actual stock from which the true Orthoptera were derived. Accordingly, he proposed that the order Protorthoptera be limited to that family. He placed the other families in the Protoblattodea and Paraplecoptera. CARPENTER (1966), objecting to SHAROV's concept of the Sthenaropodidae, proposed that the Protoblattodea and Paraplecoptera be merged with the Protorthoptera to form a single order until more is known of the morphology of the families involved. SHAROV (1968) agreed with CARPENTER that the Protoblattodea and Paraplecoptera were inseparable and should be combined into one order, the Protoblattodea, but insisted that the order Protorthoptera, with its single family, Sthenaropodidae, should be retained. Since then, there has been little consistency in the use of these ordinal names.

In a general review of the orthopteroids, RASNITSYN (1980c) transferred the Sthenaropodidae to the order Orthoptera, completely dropping the names Protorthoptera, Protoblattodea, and Paraplecoptera, and assigned most of the families previously included in those orders to the small existing order of flightless insects, the Grylloblattodea. The family Geraridae, formerly included in the Paraplecoptera, was placed in a new order, Gerarida. More recently, BURNHAM (1983), following her study of the types of orthopteroids in the Museum d'Histoire Naturelle in Paris, including those of *Sthenaropoda,* the type genus of Sthenaropodidae, placed *Sthenaropoda* in synonymy with *Gerarus,* the type genus of the Geraridae. It is clear that we need to know much more about the morphology of these extinct families before we can reach an acceptable conclusion about their relationships.

In the present work the order Protorthoptera is retained and includes the families formerly in the Protoblattodea and Paraplecoptera, as well as in the Protorthoptera itself. However, a substantial number of genera, based on fragments of wings, have been placed in the category of order Uncertain.

The division of the Protorthoptera into suborders seems virtually meaningless at present. Although a few groups of families can be recognized, most of the families remain isolated, mainly as a result of the lack of detailed knowledge of both fore and hind wings and the body. The lines of evolution within this Paleozoic complex have not yet been satisfactorily untangled. The assumption of most workers that these lines must lead to existing orders (e.g., Blattodea, Orthoptera, Perlaria) appears incorrect; more likely they radiated in diverse directions, only a very few leading to existing ordinal groups.

The homologies of the protorthopterous venation present no special difficulties, except for the media. In palaeopterous orders the anterior and posterior median veins (MA, MP) are readily recognizable as convex and concave, respectively. That both MA and MP exist in any of the orthopteroids (or in any of the Neoptera) is not certain; the loss of the convex vein in the median complex of the Neoptera has convinced some entomologists that the only element remaining is MP. On the other hand, even this vein is not always clearly concave, the coriaceous nature of the tegmen altering the thickness and general nature of the wing surface and membrane. Evidence for the presence of both MA and MP in the primitive Neoptera is indicated by the similarity between the fore wing venation of primitive Palaeodictyoptera (e.g., Dictyoneuridae) and that of the existing Orthoptera of the family Pneumoridae (RAGGE, 1955a).

In the following account of the Protorthoptera, the branches of the media are designated MA and MP only if the media divides before or near the middle line of the wing, and then only if there is no specific evidence against this interpretation, such as the presence of a media that is strongly concave entirely. In view of the uncertainties noted above, as well as the lack of knowledge of hind wings and body structures of most Protorthoptera, it is not possible to identify with confidence the most primitive families in the order. However, the Homoeodictyidae, Thoronysididae, and Paoliidae, which have an archedictyon as well as a concave MP in the fore wing, might well occupy that position. The venational specializations that have developed in the order have apparently involved the loss of the archedictyon and its replacement by a reticulation of crossveins and eventually by more regular crossveins; this has apparently taken place independently of the development of the fore wing as a tegmen. In contrast to most other orders of insects, only rarely has MA tended to anastomose with R and RS. On the other hand, in many families, MP and CUA show various degrees of anastomosis (Cacurgidae, Aenigmatodidae, Protokollariidae); in some of these, the stem of CUA is apparently anastomosed with the stem of M (or MP), CUP arising independently from the wing base. Also, in some of these genera, CUA, after diverging from MP, coalesces for a short distance with a branch of CUP. In all probability, these specializations of MP, CUA, and CUP, have been developed independently several times within the order. The fore wings commonly bear maculations (see Fig. 61,3, *Lisca*) or more elaborate markings (see Figs. 75 and 76, *Protodiamphipnoa*).

The hind wing is known in so few families of Protorthoptera that little can be said about its evolution. A well-developed anal area was present in many families (Lemmatophoridae, Liomopteridae), and this probably indicates a specialized condition of the hind wing. In others (Geraridae), the hind wing apparently had a very small anal area.

The general body structure is known in a very few families of Protorthoptera, and details of structure are known in even fewer (Lemmatophoridae, Chelopteridae, Probnidae, Liomopteridae, and Eucaenidae). The prothorax seems to show the greatest diversity of structure. In some families (e.g., Liomopteridae) the prothorax consists of a discrete pronotal plate surrounded by an oval or nearly oval disc, which in some genera may be covered with fine hairs. In the Lemmatophoridae the prothorax bears a pair of distinct paranotal lobes, which are membranous and covered with microtrichia, like those on the wings. In other families the prothorax is more slender, and in the Geraridae it is long and bears numerous long spines. The legs of the Protorthoptera also show various structural trends. In most families the three pairs of legs are similar, the third pair being slightly longer than the others. In a few families the forelegs are apparently adapted for raptorial purposes.

There are several basic features characteristic of the Protorthoptera in addition to the cursorial legs. The wings at rest, as far as is known, were folded flat over the abdomen, not slanted, as in the Orthoptera. The costal vein of the fore wing was usually marginal basally, but if it were submarginal the subcostal area was small and included only a few veinlets at most, in contrast to the Orthoptera, in which the precostal area tended to be long and replete with veinlets.

As treated here, the order Protorthoptera was almost exclusively Permo-Carboniferous, with most of the genera from the Permian. Two genera, not well known, are from the Triassic: *Tomia* MARTYNOV (family Tomiidae) and *Mesorthopteron* TILLYARD (family uncertain). Both of these are poorly known and may turn out to belong elsewhere.

Family HOMOEODICTYIDAE Martynov, 1937

[Homoeodictyidae MARTYNOV, 1937b, p. 26]

Fore wing slender, with fine archedictyon; vein SC terminating on costa; MA without definite convexity; MP concave; CUP branched. *Perm.*

Homoeodictyon MARTYNOV, 1937b, p. 26 [*H. elongatum*; OD]. Fore wing with broad costal area, traversed by several distinct veinlets, and with archedictyon. *Perm.*, USSR (European RSFSR).——FIG. 58,5. *H. elongatum*; fore wing, ×1 (Martynov, 1937b).

Family THORONYSIDIDAE Handlirsch, 1919

[Thoronysididae HANDLIRSCH, 1919b, p. 544]

Fore wing slender, crossveins forming irregular coarse network over entire wing, no anastomosis of main veins. Vein CUA extensively developed, its most distal branch terminating well beyond midwing. Hind wing unknown. *U. Carb.*

Thoronysis HANDLIRSCH, 1906b, p. 139 [*Oedischia ingbertensis* VON AMMON, 1903, p. 282; OD]. Fore wing with SC terminating on R near wing apex; M forking before midwing and after origin of RS. GUTHÖRL, 1934. *U. Carb.*, Europe (Germany).——FIG. 58,11. *T. ingbertensis* (VON AMMON); fore wing, ×0.9 (Guthörl, 1934).

Family PAOLIIDAE Handlirsch, 1906

[Paoliidae HANDLIRSCH, 1906a, p. 682]

Fore wing oval, slender, with broad costal area; fine network, resembling archedictyon, over entire wing; veinlets also present in costal area and some other parts of wing; vein SC terminating (usually on R) in distal fourth of wing; RS arising proximally of midwing; MA apparently absent; MP (concave) well developed; CU dividing very close to wing base; CUA branched; anal area weakly set off by marginal indentation at end of CUP. Hind wing triangular, with anal-posterior extension but no anal fan; venation basically as in fore wing. KUKALOVÁ, 1958a. *U. Carb.*

Paolia SMITH, 1871, p. 44 [*P. vetusta*; OD]. Hind wing with MP dividing before midwing but distally of origin of RS; proximal part of wing only slightly broader than distal half. LAURENTIAUX, 1950; KUKALOVÁ, 1958a. *U. Carb.*, USA (Indiana), Europe (The Netherlands).——FIG. 58,3. *P. vetusta*, Indiana; hind wing, ×8 (Smith, 1871).

Holasicia KUKALOVÁ, 1958a, p. 942 [*H. vetula*; OD]. Fore wing slender; MP forked at midwing; costal margin straight. *U. Carb.*, Europe (Czechoslovakia).——FIG. 58,1. *H. vetula*; fore wing, ×1.4 (Kukalová, 1958a).

Olinka KUKALOVÁ, 1958a, p. 944 [*O. modica*; OD]. Little-known genus; similar to *Holasicia*, but fore wing with convex costal margin and MP forking distally of midwing. *U. Carb.*, Europe (Czechoslovakia).——FIG. 58,9. *O. modica*; fore wing, ×1.5 (Kukalová, 1958a).

Paoliola HANDLIRSCH, 1919b, p. 533 [*P. gurleyi*; OD]. Hind wing similar to that of *Paolia* but with MP forked more deeply. MELANDER, 1903. *U. Carb.*, USA (Indiana).——FIG. 58,10. *P. gurleyi*; hind wing, ×1.4 (Melander, 1903).

Pseudofouquea HANDLIRSCH, 1906b, p. 125 [*Fouquea cambrensis* ALLEN, 1901, p. 68; OD]. Fore wing similar to *Olinka*, but costal space narrower and MP dividing before midwing. LAURENTIAUX, 1950. *U. Carb.*, Wales.——FIG. 58,12. *P. cambrensis* (ALLEN); fore wing, ×1 (Laurentiaux, 1950).

Sustaia KUKALOVÁ, 1958a, p. 946 [*S. impar*; OD]. Little-known fore wing, similar to *Olinka* but much larger; hind wing with branch of MP terminating on apical end of hind margin. *U. Carb.*, Europe (Czechoslovakia).——FIG. 58,7. *S. impar*; a, fore and b, hind wings, ×0.4 (Kukalová, 1958a).

Zdenekia KUKALOVÁ, 1958a, p. 937 [*Z. grandis*; OD]. Fore wing broader than in *Holasicia*; costal margin convex; MP forked at midwing. Hind wing much broader proximally than in distal half; branches of MP terminating along middle part of hind margin. *U. Carb.*, Europe (Czechoslovakia). —— FIG. 58,8. *Z. grandis; a,* fore wing, ×1; *b,* hind wing, ×0.8 (both Kukalová, 1958a).

Family STYGNIDAE Handlirsch, 1906

[Stygnidae HANDLIRSCH, 1906b, p. 115]

Related to Paoliidae, but crossveins distinct, though irregular, and vein MP less developed. *U. Carb.*

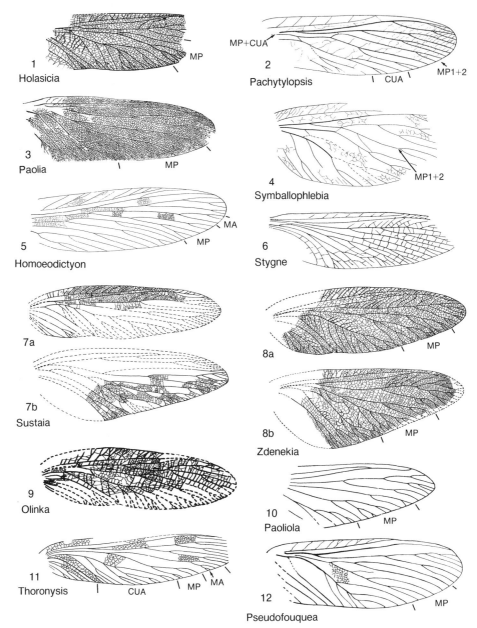

Fig. 58. Homoeodictyidae, Thoronysididae, Paoliidae, Stygnidae, and Pachytylopsidae (p. 101–103).

Stygne HANDLIRSCH, 1906b, p. 115 [*S. roemeri*; OD]. Fore wing with narrow costal area; RS arising very close to base of wing, its first branch at about midwing. SCHWARZBACH, 1939. *U. Carb.*, Poland (Upper Silesia).——FIG. 58,6. *S. roemeri*; fore wing, ×2.2 (Handlirsch, 1906b).

Family HAPALOPTERIDAE
Handlirsch, 1906

[Hapalopteridae HANDLIRSCH, 1906b, p. 304]

Fore wing similar to that of Cacurgidae but with fewer branches of main veins; vein

CUP forking further from wing base; cuticular spots absent from wings. *U. Carb.*

Hapaloptera HANDLIRSCH, 1906a, p. 694 [**H. gracilis*; OD]. Fore wing with SC extending nearly to wing apex; costal veinlets unbranched; RS with 4 branches; MP forked to about midwing; CUA with terminal fork only; crossveins numerous, weakly formed. CARPENTER, 1965. *U. Carb.*, USA (Pennsylvania).——FIG. 59,1. **H. gracilis*; fore wing, ×4 (Carpenter, 1965).

Family PACHYTYLOPSIDAE
Handlirsch, 1906

[Pachytylopsidae HANDLIRSCH, 1906b, p. 138] [=Anthraconeuridae LAURENTIAUX & LAURENTIAUX-VIEIRA, 1980, p. 407]

Fore wing with vein SC terminating on costal margin well before wing apex; MA apparently absent; CUA anastomosed with MP for a short interval basally; crossveins weak, apparently forming an irregular network over most of wing. Probably related to Paoliidae. *U. Carb.*

Pachytylopsis BORRE, 1875a, p. xl [*P. persenairei*; SD BORRE, 1875b, p. lvi] [=*Palorthopteron* HANDLIRSCH, 1904a, p. 3 (type, *P. melas*)]. Costal area of moderate width; MP1+2 directed anteriorly shortly after its separation from CUA1 and connected to RS by a short but stout crossvein; R with several terminal branches. HANDLIRSCH, 1906b; PRUVOST, 1930, 1933b; LAURENTIAUX & LAURENTIAUX-VIEIRA, 1981. *U. Carb.*, Europe (Belgium).——FIG. 58,2. **P. persenairei*; fore wing, ×1.4 (Handlirsch, 1904a).
Anthraconeura LAURENTIAUX & LAURENTIAUX-VIEIRA, 1980, p. 407 [*A. silvatica*; OD]. Similar to *Protopachytylopsis*, but costal area much narrower, especially basally; R without terminal branches; CUA with 2 long branches. [Type of family Anthraconeuridae LAURENTIAUX & LAURENTIAUX-VIEIRA.] *U. Carb.*, Europe (Belgium).
Protopachytylopsis LAURENTIAUX & LAURENTIAUX-VIEIRA, 1981, p. 83 [*P. leckwycki*; OD]. Similar to *Pachytylopsis*, but costal area broader basally; R without terminal branches; CUA with several short marginal branches. *U. Carb.*, Europe (Belgium).
Symballophlebia HANDLIRSCH, 1904a, p. 3 [*S. latipennis*; OD]. Similar to *Pachytylopsis,* but fore wing much broader; MP1+2 in short contact with RS. [Family assignment doubtful.] PRUVOST, 1930. *U. Carb.*, Europe (Belgium). ——FIG. 58,4. **S. latipennis*; fore wing, ×1.2 (Pruvost, 1930).

Family BLATTINOPSIDAE
Bolton, 1925

[Blattinopsidae BOLTON, 1925, p. 23] [=Oryctoblattinidae HANDLIRSCH, 1906b, p. 155]

Fore wing with vein SC terminating on costal margin well before apex; R usually sigmoidally curved, numerous oblique veinlets between R and costal margin beyond SC; RS with numerous branches; MA apparently absent; MP often with one or more branches anastomosed with R or RS; CUA anastomosed with basal portion of M, diverging away, and then fusing with CUA2; strong indentation at end of CUP; anal veins straight; crossveins numerous, commonly forming meshwork of cells. Hind wing unknown. *U. Carb.–Perm.*

The venation is highly variable within genera and species of this family. In addition, some specimens show a more or less distinct curving line near the middle of the wing and extending from R to the hind margin. This has led some workers to consider the Blattinopsidae to be Homoptera, related to the Fulgoridae. However, a similar line, present on the wings of some species of roaches, is apparently due to a pressure mark on the tegmina, resulting from the flexed position of the wings. It is commonly better developed on one tegmen than on the other and may be missing from one of them. Such an origin of the cross lines could explain why they are present in some blattinopsid fore wings but lacking in others. KUKALOVÁ, 1959b; CARPENTER, 1966.

Blattinopsis GIEBEL, 1867, p. 417 [*Blattina reticulata* GERMAR, 1851 in GERMAR, 1844–1853, p. 87; OD] [=*Oryctoblattina* SCUDDER, 1879b, p. 122, obj.; *Protociccus* BRONGNIART, 1885a, p. 67, nom. nud.; *Prisca* K. W. FRITSCH, 1900, p. 45 (type, *P. wittinensis*); *Oryctomylabris* HANDLIRSCH, 1906b, p. 346 (type, *Oryctoblattina oblonga* DEICHMÜLLER, 1882, p. 41); *Pseudofulgora* HANDLIRSCH, 1906b, p. 357 (type, *Fulgora ebersi* DOHRN, 1867, p. 131); *Blattinopsiella* MEUNIER, 1907, p. 523 (type, *B. pygmaea*); *Anadymenella* STRAND, 1929, p. 19, nom. subst. pro *Anadyomene* K. W. FRITSCH, 1900, p. 45, non GISTEL, 1848 (type, *A. huysseni*); *Palaeorincanites* HAUPT, 1941, p. 90 (type, *Blattinopsis anthracina* HANDLIRSCH,

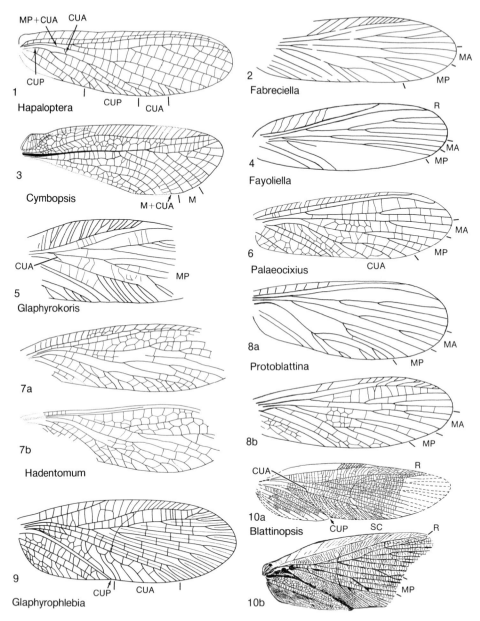

Fig. 59. Hapalopteridae, Blattinopsidae, Cymbopsidae, and Hadentomidae (p. 103–106).

1906a, p. 706)]. Fore wing with SC extending to midwing; crossveins very close together over most of wing; numerous oblique veinlets between CUA and CUP. KUKALOVÁ, 1959b; CARPENTER, 1966; MÜLLER, 1977. *U. Carb.*, Europe (Germany, France), USA (Ohio); *Perm.*, Europe (Germany, Czechoslovakia).——FIG. 59,*10a*. *B. angustai* KUKALOVÁ, Perm., Czechoslovakia; fore wing, ×2 (Kukalová, 1959b).——FIG. 59,*10b*. *B. reticulata* (GERMAR), U. Carb., Germany; fore wing, ×2 (Schlectendal, 1913).

Glaphyrokoris RICHARDSON, 1956, p. 38 [**G. mirandus*; OD]. Similar to *Glaphyrophlebia* but with SC and R longer. *U. Carb.*, USA (Illinois).——FIG. 59,*5*. **G. mirandus*; fore wing as preserved; ×2.6 (Richardson, 1956).

Glaphyrophlebia HANDLIRSCH, 1906a, p. 707 [*G. pusilla*; OD] [=*Pursa* SELLARDS, 1909, p. 153 (type, *P. ovata*); *Sindon* SELLARDS, 1909, p. 154 (type, *S. speciosa*)]. Similar to Blattinopsis but with 2 rows of cells proximally between CUA and CUP. BOLTON, 1934; KUKALOVÁ, 1965; CARPENTER, 1966. *U. Carb.*, USA (Pennsylvania), Wales; *Perm.*, USA (Kansas), Europe (Czechoslovakia).——FIG. 59,9. *G. speciosa*, (SELLARDS), Perm., Kansas; fore wing, ×7 (Carpenter, 1966).

Protoblattiniella MEUNIER, 1912d, p. 1194 [*P. minutissima*; OD]. Little-known genus, based on fragment of nymph. [Family assignment doubtful.] LAURENTIAUX, 1959b. *U. Carb.*, Europe (France).

Family CYMBOPSIDAE
Kukalová, 1965

[Cymbopsidae KUKALOVÁ, 1965, p. 86]

Little-known family of uncertain affinities. Fore wing tegminous; vein SC extending very nearly to wing apex; RS arising in distal third of wing, with several branches; M apparently coalesced with stem R to about midwing; crossveins numerous. [Placed in Protorthoptera by KUKALOVÁ, but ordinal assignment doubtful.] *Perm.*

Cymbopsis KUKALOVÁ, 1965, p. 86 [*C. excelsa*; OD]. SC sigmoidally curved; crossveins reticulate only in basal costal area and in area between SC and R+M. *Perm.*, Europe (Czechoslovakia). ——FIG. 59,3. *C. excelsa*; fore wing, ×5.7 (Kukalová, 1965).

Family EUCAENIDAE
Handlirsch, 1906

[Eucaenidae HANDLIRSCH, 1906a, p. 709] [=Teneopteridae RICHARDSON, 1956, p. 46]

Fore wing coriaceous, oval; costal space broad, with numerous veinlets; vein R with few distal branches; RS arising near wing base, with many branches; M well developed, with several branches leading to posterior border; CUP curved, well developed. Hind wing little known; costal area narrow; R with few terminal branches; anal area unknown. Head slender, long; antennae long, setaceous; mandibles dentate; maxillary palpi very long; prothorax long, broad posteriorly, narrowed anteriorly, with a constricted area adjoining the head; legs alike, all femora stout; tarsi with 5 segments; abdominal segments with posteriorly directed lateral lobes; cerci very short. Females with a short ovipositor. CARPENTER & RICHARDSON, 1976. *U. Carb.*

Eucaenus SCUDDER, 1885d, p. 325 [*E. ovalis*; OD] [=*Teneopteron* CARPENTER, 1944, p. 17 (type, *T. mirabile*)]. Fore wing: veinlets of costal area unbranched; RS with branches directed toward wing apex. MELANDER, 1903; HANDLIRSCH, 1922; CARPENTER & RICHARDSON, 1976. *U. Carb.*, USA (Illinois).——FIG. 60. *E. ovalis*; whole insect, ×2.8 (Carpenter & Richardson, 1976).

Family HADENTOMIDAE
Handlirsch, 1906

[Hadentomidae HANDLIRSCH, 1906b, p. 303] [=Fayoliellidae HANDLIRSCH, 1919b, p. 558; Palaeocixiidae HANDLIRSCH, 1919b, p. 539]

Fore wing with vein SC ending on costal margin well beyond level of midwing; costal area with a series of veinlets, mostly straight; RS arising well before midwing, with 2 to 4 long branches; M forking near or before midwing; CUA with several terminal branches. Hind wing with venation essentially as in fore wing, but costal area much narrower; anal area unknown. [Placed by HANDLIRSCH (1906a) in order Hadentomoidea.] CARPENTER, 1965. *U. Carb.*

Hadentomum HANDLIRSCH, 1906a, p. 693 [*H. americanum*; OD]. M forking at about midwing; RS with at least 3 branches in both wings. CARPENTER, 1965. *U. Carb.*, USA (Illinois).——FIG. 59,7. *H. americanum; a,* fore and *b,* hind wings, ×2.4 (Carpenter, 1965).

Fabreciella CARPENTER, 1934, p. 327 [*F. pennsylvanica*; OD]. Fore wing as in *Palaeocixius*, but costal area broader; M with 6 branches, RS with 4. [Family assignment doubtful.] *U. Carb.*, USA (Pennsylvania).——FIG. 59,2. *F. pennsylvanica*; fore wing, ×4.2 (Carpenter, 1934).

Fayoliella MEUNIER, 1908j, p. 247 [*F. elongata*; OD]. Costal area of fore wing broad, with irregular veinlets; RS and M with 4 branches. [Family assignment doubtful.] *U. Carb.*, Europe (France). ——FIG. 59,4. *F. elongata*; fore wing, ×2.5 (Carpenter, new).

Palaeocixius HANDLIRSCH, 1906b, p. 326 [*P. antiquus*; SD HANDLIRSCH, 1922, p. 74] [=*Palaeocixius* BRONGNIART, 1885a, p. 67, *nom. nud.*; *Fabrecia* MEUNIER, 1911a, p. 123 (type, *F. pygmaea*)]. Fore wing with fork of M at about level of origin of RS; RS and MA forking at about same level; RS with 2 branches, M with 5; a few large cells formed in region of M and

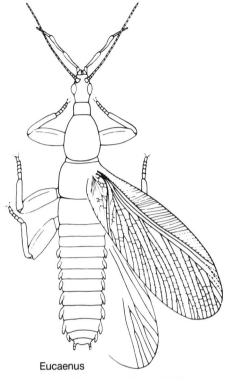

Eucaenus

Fig. 60. Eucaenidae (p. 105).

in anal area. *U. Carb.*, Europe (France).——Fig. 59,6. **P. antiquus*; fore wing, ×4 (Carpenter, new).

Protoblattina MEUNIER, 1909d, p. 48 [**P. bouvieri*; OD]. Fore wing as in *Palaeocixius*, but RS with 3 branches, M with 6. *U. Carb.*, Europe (France).——Fig. 59,8a. **P. bouvieri*; fore wing, ×2.8 (Carpenter, new).——Fig. 59,8b. *P. giardi* MEUNIER, 1912d; fore wing, ×2 (Carpenter, new).

Family LEMMATOPHORIDAE Sellards, 1909

[Lemmatophoridae SELLARDS, 1909, p. 162] [=Ortaidae SELLARDS, 1909, p. 167; Lecoriidae HANDLIRSCH, 1919b, p. 556]

Fore wing with costal area moderately broad; vein SC terminating on costa at least slightly beyond midwing; R unbranched; MA and MP present, but neither convex nor concave; proximal half of MP obsolescent; CUA very strongly developed, usually with 3 branches; CUP obsolescent, forming a straight vena dividens; 2 anals present. Hind wing shorter than fore wing but with expanded anal area, and with at least slight incision of wing margin at end of CUP; R unbranched; MA unbranched and coalesced to variable extent with RS; CUA well developed, CUP obsolescent; anal fan with four main veins. Venation highly variable among species. Antennae long, multisegmented; head small, hypognathous; eyes small; prothorax bearing pair of membranous paranota, with reticulated venation and covered with microtrichia; mesonotum and metanotum broad and flat; five tarsomeres; abdomen unspecialized but bearing on the first nine segments small lateral processes resembling vestigial gills; cerci about as long as abdomen; female with very short ovipositor. Nymphs apparently aquatic, with lateral gills on first nine abdominal segments. CARPENTER, 1935a, 1939. *Perm.*

Lemmatophora SELLARDS, 1909, p. 162 [**L. typa*; SD TILLYARD, 1928b, p. 189]. SC terminating just beyond midwing; RS unbranched; hind wing with deep incision at end of CUP. *Perm.*, USA (Kansas).——Fig. 57. **L. typa*; head, prothorax, prothoracic lobes, neotype, ×40 (Carpenter, new).——Fig. 61,1a. **L. typa*; restoration of adult, ×2.4 (Carpenter, 1935a).——Fig. 61,1b. *Lemmatophora* sp. (probably *typa*); nymph, ×2.4 (Carpenter, 1935a).

Artinska SELLARDS, 1909, p. 165 [**A. clara*; SD TILLYARD, 1928e, p. 321] [=*Estadia* SELLARDS, 1909, p. 166 (type, *E. elongata*); *Lectrum* SELLARDS, 1909, p. 167 (type, *L. anomalum*); *Orta* SELLARDS, 1909, p. 168 (type, *O. ovata*)]. SC extending well beyond midwing; RS with at least one fork. *Perm.*, USA (Kansas).——Fig. 62,1. **A. clara; a*, fore wing, ×6 (Carpenter, 1935a); *b*, hind wing, ×6 (Tillyard, 1928e).

Blania KUKALOVÁ, 1964c, p. 101 [**B. rotunda*; OD]. Fore wing as in *Artinska* but broader and more nearly oval; costal area relatively broad. *Perm.*, Europe (Czechoslovakia).——Fig. 62,4. **B. rotunda*; fore wing, ×7.5 (Kukalová, 1964c).

Lecorium SELLARDS, 1909, p. 167 [**L. elongatum*; OD] [=*Stemma* SELLARDS, 1909, p. 168 (type, *S. elegans*); *Sellardsia* TILLYARD, 1928e, p. 343 (type, *S. kansensis*); *Metalecorium* HANDLIRSCH, 1937, p. 96, *nom. nud.*; *Paralecorium* HANDLIRSCH, 1937, p. 96 (type, *Lecorium parvum*)]. Fore wing with costal area narrow, as in *Paraprisca*; CUA coalesced with M basally but diverging just before origin of MA. CARPENTER, 1935a, 1939. *Perm.*, USA (Kansas).——Fig. 61,2. **L. elongatum*; adult, ×6 (Carpenter, 1935a).

Lisca SELLARDS, 1909, p. 163 [**L. minuta*; OD].

Protorthoptera

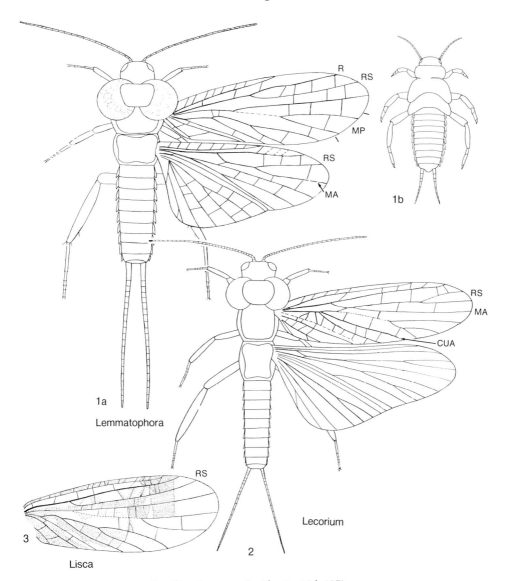

Fig. 61. Lemmatophoridae (p. 106–107).

Fore wing with costal area much narrower than in *Lemmatophora*; RS unbranched; RS arising much nearer wing base than in *Lemmatophora*. *Perm.*, USA (Kansas).——FIG. 61,3. **L. minuta*; fore wing, ×8 (Tillyard, 1928e).

Maculopterum KUKALOVÁ, 1964c, p. 107 [**M. maculatum*; OD]. Little-known genus. Fore wing as in *Torrentopterum* but with numerous maculations. [Probably a synonym of *Torrentopterum*.] *Perm.*, Europe (Czechoslovakia).

Oborella KUKALOVÁ, 1964c, p. 93 [**O. matura*; OD]. Fore wing as in *Artinska*, but costal area much broader; CU usually anastomosed with M before separating into CUA and CUP. Central disc of pronotum oval. *Perm.*, Europe (Czechoslovakia).

Paraprisca HANDLIRSCH, 1919b, p. 555, *nom. subst. pro Prisca* SELLARDS, 1909, p. 167, *non* K. W. FRITSCH, 1900 [**Prisca fragilis* SELLARDS, 1909, p. 167]. Fore wing more slender than in *Lemmatophora*; R straight; CUA not anastomosed with M; hind wing with only slight incision at

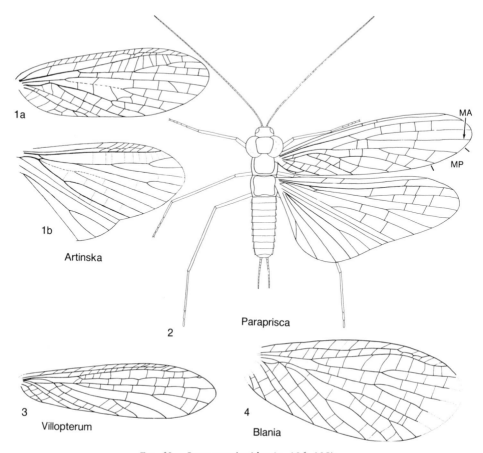

Fig. 62. Lemmatophoridae (p. 106–108).

CUP. Antennae longer and legs much longer than in *Lemmatophora*. ZALESSKY, 1952. *Perm.*, USA (Kansas), USSR (Asian RSFSR).——FIG. 62,*2*. **P. fragilis* (SELLARDS), Kansas; adult, ×6 (Carpenter, 1935a).

Quecopterum KUKALOVÁ, 1964c, p. 98 [**Q. decussatum*; OD]. Little-known genus. Fore wing as in *Oborella*, but central disc of pronotum irregular in shape. *Perm.*, Europe (Czechoslovakia).

Sharovipterum KUKALOVÁ, 1964c, p. 100 [**S. alatum*; OD]. Little-known genus. Fore wing apparently broadly oval; costal area narrow; RS and MA apparently unbranched. Pronotum subtriangular. *Perm.*, Europe (Czechoslovakia).

Torrentopterum KUKALOVÁ, 1964c, p. 105 [**T. pallidum*; OD]. Fore wing as in *Lecorium*, but crossveins apparently more numerous. [Probably a synonym of *Lecorium*.] *Perm.*, Europe (Czechoslovakia).

Villopterum KUKALOVÁ, 1964c, p. 108 [**V. villosum*; OD]. Fore wing as in *Lecorium* but more elongate and slightly broader distally; RS apparently unbranched; CUA1 with a deep fork. *Perm.*, Europe (Czechoslovakia).——FIG. 62,*3*. **V. villosum*; fore wing, ×8 (Kukalová, 1964c).

Family LIOMOPTERIDAE Sellards, 1909

[Liomopteridae SELLARDS, 1909, p. 157] [=Khosaridae MARTYNOV, 1937b, p. 29]

Fore wing membranous, usually with microtrichia well developed on at least part of wing; vein SC terminating on costa beyond midwing; costal area with numerous, slanted veinlets, not forming regular cells; R extending nearly to wing apex; RS with at least two branches; M forked at about level of origin of RS or slightly toward base; MA and MP not anastomosed with other veins; CUA typically diverging anteriorly shortly after its origin and forking into CUA1 and CUA2;

CUP and 1A unbranched. Hind wing with costal area narrow; R extending nearly to apex; RS arising much nearer base than in fore wing; M and CU fused at base; CUA strong and deeply forked; 1A close to CUP; anal area forming lobe containing numerous anal veins. Antennae long, multisegmented; head hypognathous; eyes prominent; prothorax usually with paranotal expansions continuing anteriorly and posteriorly around pronotum itself; reticulation not visible, but paranotals commonly membranous and covered with microtrichia; legs slender, hind legs longer than others, not modified for jumping; 5 tarsomeres; cerci long. CARPENTER, 1950. *Perm.*

Liomopterum SELLARDS, 1909, p. 157 [*L. ornatum*; SD CARPENTER, 1950, p. 189] [=*Horates* SELLARDS, 1909, p. 158 (type, *H. elongatus*)]. Costal space moderately broad; fork of M proximal to origin of RS; cells (when present) almost exclusively confined to area of CUA and CUP. *Perm.*, USA (Kansas).——FIG. 63,7. **L. ornatum*; adult, ×3.5 (Carpenter, 1950).

Abashevia SHAROV, 1961d, p. 194 [**A. suchovi*; OD]. Similar to *Parapermula*, but costal area narrower and branches of MA arising pectinately. *Perm.*, USSR (Asian RSFSR).——FIG. 63,3. **A. suchovi*; fore wing, ×1.8 (Sharov, 1961d).

Alicula SCHLECHTENDAL, 1913, pl. 2 [**A. lebachensis*; OD] [=*Permula* HANDLIRSCH, 1919b, p. 542, obj.]. Little-known genus, with fore wing apparently similar to that of *Liomopterum*, but crossveins numerous and forming a fine reticulation; CUA with several additional branches. [HANDLIRSCH (1919b) considered the names *Alicula lebachensis* to be nomina nuda, but under article 12, section 7, of the ICZN (p. 35, 1985 ed.) both names are available.] KUKALOVÁ, 1964c. *Perm.*, Europe (Germany, Czechoslovakia). ——FIG. 64,1. *A. acra* (KUKALOVÁ); fore wing as preserved, ×3 (after Kukalová, 1964c).

Cerasopterum KUKALOVÁ, 1964c, p. 60 [**C. gracile*; OD]. Fore wing as in *Tapopterum*, but RS with only 3 branches. *Perm.*, Europe (Czechoslovakia).——FIG. 64,4. **C. gracile*; fore wing, ×6 (Kukalová, 1964c).

Climaconeurites SHAROV, 1961d, p. 195 [**C. asiaticus*; OD]. MA branching dichotomously; anterior branch of MA anastomosed for short distance with RS. *Perm.*, USSR (Asian RSFSR). ——FIG. 63,1. **C. asiaticus*; a, fore wing; b, hind wing, ×2.6 (Sharov, 1961d).

Depressopterum KUKALOVÁ, 1964c, p. 48 [**D. senior*; OD]. Little-known genus. Fore wing as in *Parapermula* but more elongate and with less convex anterior margin. *Perm.*, Europe (Czechoslovakia).——FIG. 64,2. **D. senior*; fore wing as preserved, ×4 (Kukalová, 1964c).

Donopterum KUKALOVÁ, 1964c, p. 54 [**D. carpenteri*; OD]. Fore wing as in *Turbopteron* but broader; costal area narrower; RS and MA with more branches. *Perm.*, Europe (Czechoslovakia). ——FIG. 64,5. **D. carpenteri*; fore wing, ×2 (Kukalová, 1964c).

Drahania KUKALOVÁ, 1964c, p. 51 [**D. avia*; OD]. Similar to *Depressopterum*, but fore wing more slender; CUA with only 2 main branches. *Perm.*, Europe (Czechsolovakia).——FIG. 64,3. **D. avia*; fore wing, ×5 (Kukalová, 1964c).

Fumopterum KUKALOVÁ, 1964c, p. 59 [**F. largum*; OD]. Little-known genus, based on distal half of wing. Venation as in *Donopterum*, but wing much more slender. *Perm.*, Europe (Czechoslovakia).

Ideliopsis CARPENTER, 1948b, p. 101 [**I. ovalis*; OD]. Costal margin only slightly curved; MP coalesced proximally with CUA1. Crossveins numerous, regular; no reticulation in apical part of wing. [Family assignment doubtful.] *Perm.*, USA (Texas).——FIG. 63,4. **I. ovalis*; fore wing, ×1.8 (Carpenter, 1948b).

Kaltanella SHAROV, 1961d, p. 206 [**K. lata*; OD]. Fore wing broadly oval, with almost no cells; MA with 2 main stems arising before level of origin of RS. *Perm.*, USSR (Asian RSFSR). ——FIG. 63,8. **K. lata*; a, fore and b, hind wings, ×2.6 (Sharov, 1961d).

Kaltanympha SHAROV, 1961d, p. 220 [**K. thysanuriformis*; OD]. Nymph with long, slender cerci; apparently terrestrial. *Perm.*, USSR (Asian RSFSR). —— FIG. 63,2. **K. thysanuriformis*; nymph, ×4 (Sharov, 1961d).

Kazanella MARTYNOV, 1930d, p. 1116 [**K. rotundipennis*; OD]. Little-known fore wing, with broad costal margin. *Perm.*, USSR (European RSFSR).——FIG. 63,6. **K. rotundipennis*; fore wing, ×4 (Sharov, 1962b).

Khosara MARTYNOV, 1937b, p. 30 [**K. permiakovae*; OD]. Apex of fore wing rounded but markedly asymmetrical; no cells; MA with long branches. *Perm.*, USSR (European RSFSR).—— FIG. 63,5. **K. permiakovae*; fore wing, ×2.2 (Sharov, 1962c).

Lioma KUKALOVÁ, 1964c, p. 56 [**L. moravica*; OD]. Fore wing as in *Donopterum* but more slender and with a longer SC. *Perm.*, Europe (Czechoslovakia).

Liomopterella SHAROV, 1961d, p. 202 [**L. vulgaris*; OD]. Similar to *Abashevia*, but M forking well before origin of RS and MA dichotomously branched. *Perm.*, USSR (Asian RSFSR).——FIG. 65,1. **L. vulgaris*; a, fore and b, hind wings, ×2.6 (Sharov, 1961d).

Liomopterina RIEK, 1973, p. 518 [**L. clara*; OD]. Little-known genus, based on proximal fragment

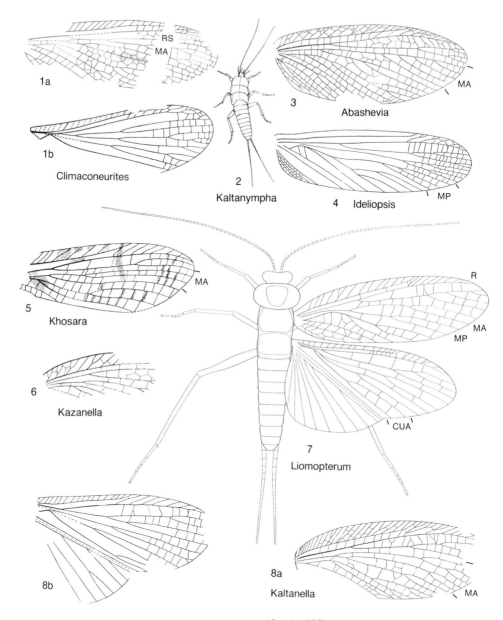

FIG. 63. Liomopteridae (p. 109).

of wing. [Family assignment doubtful.] *Perm.*, South Africa (Natal).

Liomopterites SHAROV, 1961d, p. 207 [**L. expletus*; OD]. Fore wing similar to that of *Liomopterum* but with curvature of CUA less pronounced and with fewer crossveins; in hind wing, MA anastomosed with RS. Nymph slender, apparently terrestrial. *Perm.*, USSR (Asian RSFSR).——FIG. 65,*2a*. **L. expletus*; fore wing, ×4.5 (Sharov, 1961e).——FIG. 65,*2b*. *L. comans* SHAROV; hind wing, ×4.5 (Sharov, 1961e).——FIG. 65,*2c*. *L.(?) gracilis* SHAROV; nymph, ×6 (Sharov, 1961d).

Liomoptoides RIEK, 1973, p. 515 [**L. similis*; OD]. Little-known genus, based on small, apical fragment of wing. [Family assignment doubtful.] *Perm.*, South Africa (Natal).

Mioloptera RIEK, 1973, p. 515 [**M. stuckenbergi*;

OD]. Little-known genus, apparently similar to *Parapermula*. RIEK, 1976a. *Perm.,* South Africa (Natal).

Mioloperina RIEK, 1976a, p. 762 [**M. tenuipennis*; OD]. Little-known genus, based on small fragment of wing. *Perm.,* South Africa (Natal).

Mioloptoides RIEK, 1976a, p. 761 [**M. andrei*; OD]. Little-known genus, based on wing fragment; similar to *Mioloptera*. *Perm.,* South Africa (Natal).

Neoliomopterum RIEK, 1976a, p. 762 [**N. picturatum*; OD]. Little-known genus, based on apical fragment of wing. [Family assignment doubtful.] *Perm.,* South Africa (Natal).

Ornaticosta SHAROV, 1961d, p. 197 [**O. magna*; OD]. Apex of fore wing acute; costal area with dark pigmentation extending nearly to apex of wing. *Perm.,* USSR (Asian RSFSR).——FIG. 65,3. **O. magna*; fore wing, ×1 (Sharov, 1961d).

Paraliomopterum SHAROV, 1961d, p. 218 [**P. paulum*; OD]. Similar to *Liomopterum*, but SC extending much further towards apex. *Perm.,* USSR (Asian RSFSR).——FIG. 65,4. **P. paulum*; fore wing, ×2.4 (Sharov, 1961d).

Parapermula SHAROV, 1961d, p. 191 [**P. sibirica*; OD]. Fore wing oval, with very broad costal space; RS with numerous terminal branches; MA dichotomously branched; at least a few cells between most main veins. *Perm.,* USSR (Asian RSFSR).——FIG. 65,5. **P. sibirica; a,* fore and *b,* hind wings, ×2.5 (Sharov, 1961d).

Sarbalopterum SHAROV, 1961d, p. 217 [**S. ignorabile*; OD]. Little-known fore wing, with broad costal area and no cells. *Perm.,* USSR (Asian RSFSR).——FIG. 65,6. **S. ignorabile*; fore wing, ×6.6 (Sharov, 1961d).

Semopterum CARPENTER, 1950, p. 197 [**S. venosum*; OD]. Fore wing similar to that of *Liomopterum* but with more numerous crossveins and with several additional anal veins. *Perm.,* USA (Kansas).——FIG. 65,7. **S. venosum*; fore wing, ×1.8 (Carpenter, 1950).

Sibirella SHAROV, 1961d, p. 215 [**S. paucinervis*; OD]. Subcostal area nearly as wide as costal area; few crossveins and branches of main veins. *Perm.,* USSR (Asian RSFSR).——FIG. 65,8. **S. paucinervis*; fore wing, ×3.4 (Sharov, 1961d).

Tapopterum CARPENTER, 1950, p. 195 [**T. celsum*; OD]. Costal space narrower than in *Liomopterum*; crossveins more numerous, with at least a few cells between most main veins. *Perm.,* USA (Kansas).——FIG. 65,9. **T. celsum*; fore wing, ×2.5 (Carpenter, 1950).

Turbopterum KUKALOVÁ, 1964c, p. 52 [**T. finum*; OD]. Fore wing as in *Drahania,* but costal area broader, SC much shorter, and MA with a short fork. *Perm.,* Europe (Czechoslovakia).

Tyrannopterum KUKALOVÁ, 1964c, p. 70 [**T. minimum*; OD]. Similar to *Cerasopterum* but much smaller; fore wing with branches of RS directed anteriorly, away from M. *Perm.,* Europe (Czechoslovakia).

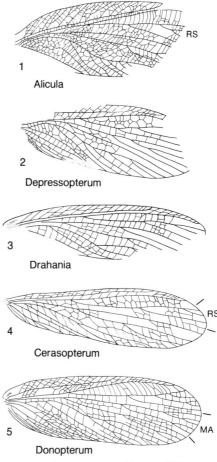

FIG. 64. Liomopteridae (p. 109).

Family PHENOPTERIDAE
Carpenter, 1950

[Phenopteridae CARPENTER, 1950, p. 204]

Related to Liomopteridae. Fore wing membranous, delicate; vein SC terminating on margin well beyond midwing; costal area with numerous, oblique veinlets; RS arising before midwing, with a few branches; M forked at about level of origin of RS, rarely with three branches; CUA with a basal branch (CUA2) and a distal branch dividing near wing margin; crossveins numerous, irregular,

Hexapoda

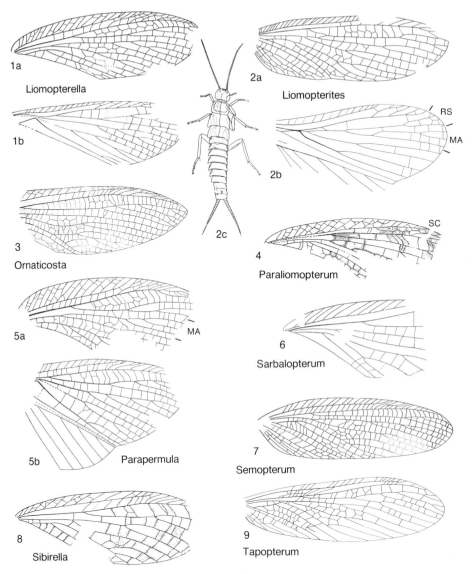

FIG. 65. Liomopteridae (p. 109–111).

forming irregular reticulation over most of wing. Hind wing with RS, M, and CUA arising from single stem near wing base; CUA branched; anal area well developed. Body structure unknown. *Perm.*

Phenopterum CARPENTER, 1950, p. 205, *nom. subst. pro Lepium* SELLARDS, 1909, p. 156, *non* ENDERLEIN, 1906 [**Lepium elongatum* SELLARDS, 1909, p. 156; OD]. RS with 3 branches; fork of M slightly basal of origin of RS. *Perm.*, USA (Kansas).——FIG. 66,*3*. **P. elongatum* (SELLARDS); *a*, fore and *b*, hind wings, ×4 (Carpenter, 1950).

Brunia KUKALOVÁ, 1964c, p. 72 [**B. raketa*; OD]. Similar to *Phenopterum*, but wings more slender and RS with only 2 branches; costal area very narrow. *Perm.*, Europe (Czechoslovakia).——FIG. 66,*2*. **B. raketa*; fore wing, ×4 (Kukalová, 1964c).

Chlumia KUKALOVÁ, 1964c, p. 77 [**C. parva*; OD]. Fore wing as in *Brunia* but much broader. *Perm.*, Europe (Czechoslovakia).

Protorthoptera

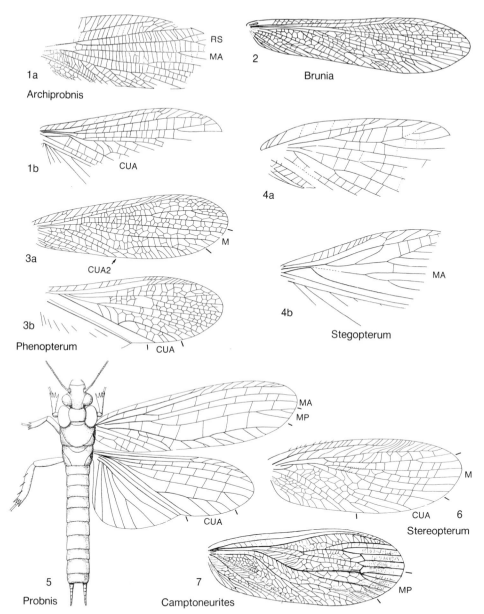

Fig. 66. Phenopteridae, Probnidae, Stegopteridae, Stereopteridae, Camptoneuritidae, and Archiprobnidae (p. 112–115).

Family HAVLATIIDAE
Kukalová, 1964

[Havlatiidae KUKALOVÁ, 1964c, p. 83]

Similar to Liomopteridae, but fore wing markedly broader distally and crossveins less numerous. *Perm.*

Havlatia KUKALOVÁ, 1964c, p. 84 [*H. annae*; OD]. Costal and subcostal areas very narrow; SC extending nearly to wing apex. *Perm.*, Europe (Czechoslovakia). —— FIG. 67,3. *H. annae*; fore wing, ×8 (Kukalová, 1964c).

Ventopterum KUKALOVÁ, 1964c, p. 87 [*V. rapidum*; OD]. Little-known genus. Fore wing as in *Zephyropterum*, but subcostal area broader; cross-

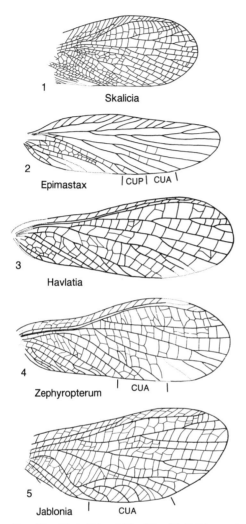

Fig. 67. Havlatiidae, Jabloniidae, Skaliciidae, and Epimastacidae (p. 113–115).

veins more irregular. *Perm.,* Europe (Czechoslovakia).
Zephyropterum KUKALOVÁ, 1964c, p. 85 [*Z. lentum*; OD]. Fore wing as in *Havlatia,* but costal margin more concave before midwing; CUA with 2 long branches. *Perm.,* Europe (Czechoslovakia).——FIG. 67,4. *Z. lentum*; fore wing, ×6.5 (Kukalová, 1964c).

Family JABLONIIDAE Kukalová, 1964

[Jabloniidae KUKALOVÁ, 1964c, p. 81]

Small species, related to the Phenopteridae, but fore wing very broad, with strongly convex hind margin; vein CUA markedly sigmoidal, its terminal branches almost semicircular. *Perm.*

Jablonia KUKALOVÁ, 1964c, p. 82 [*J. aestiva*; OD]. Anterior wing margin almost straight; both RS and MA with 3 branches. *Perm.,* Europe (Czechoslovakia).——FIG. 67,5. *J. aestiva*; fore wing, ×10 (Kukalová, 1964c).

Family SKALICIIDAE Kukalová, 1964

[Skaliciidae KUKALOVÁ, 1964c, p. 88]

Fore wing slightly coriaceous, with covering of fine hairs and with prominent setae distally; vein SC terinating well before apex; RS arising near midwing, with several terminal branches; MA and MP terminating near wing apex; CUA2 with long fork; crossveins reticulate in distal part of wing only. *Perm.*

Skalicia KUKALOVÁ, 1964c, p. 89 [*S. rara*; OD]. Fore wing very broad distally; apex rounded. *Perm.,* Europe (Czechoslovakia).——FIG. 67,1. *S. rara*; fore wing, ×3.5 (Kukalová, 1964c).
Doubravia KUKALOVÁ, 1964c, p. 90 [*D. annosa*; OD]. Little-known genus. Fore wing apparently long and slender. [Family assignment doubtful.] *Perm.,* Europe (Czechoslovakia).

Family PROBNIDAE Sellards, 1909

[*nom. correct.* CARPENTER, herein, *pro* Probnisidae SELLARDS, 1909, p. 159]

Related to Lemmatophoridae. Fore wing coriaceous, granular in texture; costal area narrow; vein R extending nearly to wing apex; RS unbranched; MA and MP separating at about level of origin of RS, their branching very variable; CUA1 producing variable number of arched branches along posterior margin of wing; CUA2 usually unbranched; 1A unbranched. Hind wing membranous but with wrinkles; costal space narrower than in fore wing; RS arising almost at base of wing, unbranched; CUA1 as in fore wing but with longer branches; deep indentation of hind margin at end of CUP; anal fan well developed, with 6 anal veins. Head small, eyes prominent; antennae rather short and robust; prothorax with small lateral lobes, without venation; legs of moderate length; tarsi three-segmented; abdomen robust; cerci short; ovipositor apparently small. CARPENTER, 1943a. *Perm.*

Probnis SELLARDS, 1909, p. 159 [*P. speciosa*; SD TILLYARD, 1937c, p. 415] [=*Espira* SELLARDS, 1909, p. 160 (type, *E. obscura*); *Stoichus* SELLARDS, 1909, p. 160 (type, *S. elegans*); *Stinus* SELLARDS, 1909, p. 161 (type, *S. breve-cubitalis*)]. Fore wing: SC terminating at midwing; R with several veinlets from R to costal margin beyond SC; crossveins straight, widely separated over most of wing; CUA1 extending nearly to wing apex. *Perm.*, USA (Kansas).——FIG. 66,5. **P. speciosa*; whole insect, X4 (Carpenter, 1943a).

Family STEGOPTERIDAE
Sharov, 1961

[Stegopteridae SHAROV, 1961d, p. 220]

Similar to Liomopteridae, but fore wing more coriaceous and rough. *Perm.*

Stegopterum SHAROV, 1961d, p. 221 [*S. hirtum*; OD]. Fore wing with few crossveins; no cells. Hind wing with MA free from RS. *Perm.*, USSR (Asian RSFSR).——FIG. 66,4. **S. hirtum; a,* fore and *b,* hind wings, X4.4 (Sharov, 1961d).

Family STEREOPTERIDAE
Carpenter, 1950

[Stereopteridae CARPENTER, 1950, p. 201]

Related to Liomopteridae. Fore wing slightly coriaceous, with few patches of conspicuous setae but without covering of microtrichia; vein SC terminating on margin beyond midwing; costal area narrow, with numerous oblique veinlets; RS arising before midwing; CUA anastomosed with stem of M for short distance; crossveins numerous, irregular. Hind wing and body structure little known. CARPENTER, 1966. *Perm.*

Stereopterum CARPENTER, 1950, p. 202 [**S. rotundum*; OD]. M forking at level of origin of RS; row of stout setae along basal third of costal margin; smaller setae or branches of M near midwing. CARPENTER, 1966. *Perm.*, USA (Kansas).——FIG. 66,6. **S. rotundum;* fore wing, X4 (Carpenter, 1950).

Family EPIMASTACIDAE
Martynov, 1928

[Epimastacidae MARTYNOV, 1928b, p. 63]

Fore wing narrowed beyond midwing; vein SC remote from wing margin and terminating on costal margin near midwing; RS arising before midwing and with several long branches; CUA anastomosed for a short interval with M basally before forking. *Perm.*

Epimastax MARTYNOV, 1928b, p. 63 [**E. parvulus*; OD]. Fore wing: R with several branches to the costal margin of wing; RS with 5 terminal branches. *Perm.*, USSR (European RSFSR), Europe (Czechoslovakia).——FIG. 67,2. *E. celer* KUKALOVÁ, 1965; fore wing, X8 (Kukalová, 1965).

Family CAMPTONEURITIDAE
Martynov, 1931

[Camptoneuritidae MARTYNOV, 1931a, p. 98, *nom. subst. pro* Camptoneuridae MARTYNOV, 1928b, p. 53]

Related to Phenopteridae. Fore wing with narrow costal area; vein RS arising before midwing; crossveins forming strong, irregular network; distal branches of RS, MA, and MP straight and parallel, without crossveins. Hind wing unknown. *Perm.*

Camptoneurites MARTYNOV, 1931a, p. 98, *nom. subst. pro Camptoneura* MARTYNOV, 1928b, p. 53, *non* AGASSIZ, 1846 [**Camptoneura reticulata* MARTYNOV, 1928b, p. 35; OD]. Fore wing with costal margin slightly concave; 2 rows of irregular cells between MP and CUA. *Perm.*, USSR (European RSFSR).——FIG. 66,7. **C. reticulata* (MARTYNOV); fore wing, X3.5 (Martynov, 1928b).

Family ARCHIPROBNIDAE
Sharov, 1961

[*nom. correct.* CARPENTER, herein, *pro* Archiprobnisidae SHAROV, 1961d, p. 185]

Fore wing with main veins more widely spaced than in Ideliidae; crossveins irregular but not forming distinct reticulation except in and near anal areas; veins RS and MA arising at same level close to base of wing. Hind wing little known; CUA sharply bent near base, as in some Liomopteridae. *Perm.*

Archiprobnis SHAROV, 1961d, p. 186 [**A. repens*; OD]. Fore wing broadly oval with rounded apex. *Perm.*, USSR (Asian RSFSR).——FIG. 66,1. **A. repens; a,* fore and *b,* hind wings, X2.5 (Sharov, 1961d).

Family PROTEMBIIDAE
Tillyard, 1937

[Protembiidae TILLYARD, 1937b, p. 243] [=Telactinopterygidae CARPENTER, 1943a, p. 78]

Related to Phenopteridae. Fore wing slightly coriaceous; distal parts of veins RS, MA, MP bordered by delicate lines on each side; SC terminating on costal margin; costal space very narrow; RS arising at midwing,

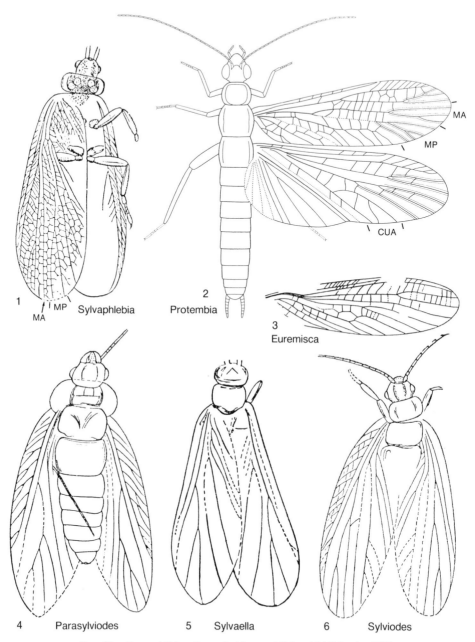

FIG. 68. Protembiidae, Euremiscidae, and Sylvaphlebiidae (p. 117).

branched; M forked before origin of RS; CUA without definite CUA2, having several distal branches; crossveins few, unequally distributed. Hind wing with RS arising near wing base; M forking slightly beyond midwing; CUA forked distally; anal lobe well developed. Antennae long; head small, with large eyes; prothorax with small disc; hind legs longer than others, all with five tarsomeres; cerci short but distinct. CARPENTER, 1950. *Perm.*

Protembia TILLYARD, 1937b, p. 245 [*P. permiana*; OD] [=*Telactinopteryx* TILLYARD, 1937c, p. 422 (type, *T. striatipennis*)]. Fore wing: RS with 3 terminal branches; several veinlets between R and margin beyond end of SC; a few cells formed between R and RS. *Perm.,* USA (Kansas).——FIG. 68,2. *P. permiana;* whole insect, ×6.5 (Carpenter, 1950).

Family EUREMISCIDAE
Zalessky, 1951

[Euremiscidae ZALESSKY, 1951b, p. 81]

Related to Phenopteridae, but fore wing without network of crossveins. *Perm.*

Euremisca ZALESSKY, 1951b, p. 82 [*E. splendens*; OD]. Slender fore wing, with very narrow costal space. *Perm.,* USSR (Asian RSFSR).——FIG. 68,3. *E. splendens;* fore wing, ×3.5 (Sharov, 1962c).

Family SYLVAPHLEBIIDAE
Martynov, 1940

[Sylvaphlebiidae MARTYNOV, 1940, p. 18] [=Sylvaelidae MARTYNOV, 1940, p. 26; Sylviodidae MARTYNOV, 1940, p. 23]

Related to Phenopteridae; fore wing coriaceous, without hairs; veins MA and MP long, with few branches; prothorax short and broad, with small or large membranous lobes. *Perm.*

Sylvaphlebia MARTYNOV, 1940, p. 18 [*S. tuberculata*; OD] [=*Biarmopteron* ZALESSKY, 1953c, p. 42 (type, *B. protoblattoides*)]. Fore wing little known; 2 rows of cells between MA and MP. *Perm.,* USSR (Asian RSFSR).——FIG. 68,1. *S. tuberculata;* fore wing and part of body, ×3.2 (Sharov, 1962c).
Parasylviodes MARTYNOV, 1940, p. 23 [*P. tetracladus*; OD]. Fore wing with very broad costal area; RS arising near midwing; large lobes on prothorax. *Perm.,* USSR (Asian RSFSR).——FIG. 68,4. *P. tetracladus;* fore wings and part of body, ×3.4 (Sharov, 1962c).
Sylvaella MARTYNOV, 1940, p. 55 [*S. paurovenosa;* OD]. Fore wing little known; RS arising at midwing but with fewer branches than in *Parasylviodes;* costal area narrow. *Perm.,* USSR (Asian RSFSR).——FIG. 68,5. *S. paurovenosa;* fore wings and part of body, ×6 (Sharov, 1962c).
Sylviodes MARTYNOV, 1940, p. 23 [*S. perloides;* OD] [=*Biarmopterites* ZALESSKY, 1953c, p. 45 (type, *B. reticulatus*)]. Fore wing little known; costal space much broader than in *Sylvaphlebia*. *Perm.,* USSR (Asian RSFSR).——FIG. 68,6. *S. perloides;* fore wing and part of body, ×1.6 (Sharov, 1962c).

Family CHELOPTERIDAE
Carpenter, 1950

[Chelopteridae CARPENTER, 1950, p. 198]

Related to the Liomopteridae. Fore wing membranous or only slightly coriaceous; hairs absent; costal area broad; subcostal area very narrow in proximal region; crossveins numerous, forming coarse reticulation between veins CUA1 and CUA2 and in anal area; crossveins between R and RS very slanted and parallel. Hind wing with RS arising nearer base than in Liomopteridae; CUA unbranched; crossveins forming coarse network in distal and cubital areas of wing. Antennae long, but with fewer segments than in Liomopteridae; head broad; pronotum with flat, membranous marginal area, lacking hairs; tarsi five-segmented; cerci of male modified to form forceps; female with prominent ovipostior. *Perm.*

Chelopterum CARPENTER, 1950, p. 199 [*C. peregrinum;* OD]. Fore wing with SC approaching costal margin at about midwing; main fork of M just proximal to origin of RS; MA and MP about equally developed. *Perm.,* USA (Kansas). ——FIG. 69,4. *C. peregrinum;* complete insect, male, ×3.4 (Carpenter, 1950).

Family DEMOPTERIDAE
Carpenter, 1950

[Demopteridae CARPENTER, 1950, p. 203]

Related to Liomopteridae. Fore wing with membrane strongly coriaceous; costal margin slightly concave; costal area narrow, about as wide as subcostal; CUA branched only in its distal half. Hind wing unknown. *Perm.*

Demopterum CARPENTER, 1950, p. 203 [*D. gracile;* OD]. Fore wing slender; SC with series of stout spines along its proximal part; MP much more extensively developed than MA. *Perm.,* USA (Kansas).——FIG. 69,5. *D. gracile;* fore wing, ×2.7 (Carpenter, 1950).

Family ATACTOPHLEBIIDAE Martynov, 1930

[Atactophlebiidae Martynov, 1930c, p. 952]

Fore wing with costal area slightly broader than subcostal; crossveins tending to be irregularly shaped; two rows of irregular cells between veins R and anterior branch of RS. Hind wing little known, with broad costal area and very narrow subcostal. Branching of veins of both wings highly variable. Ovipositor small but distinct. *Perm.*

Atactophlebia Martynov, 1928b, p. 51 [*A. termitoides*; OD]. Fore wing with area of RS narrow; RS with few branches. *Perm.*, USSR (Asian RSFSR).——Fig. 69,3. *A. termitoides*; *a*, fore and *b*, hind wings, ×1.4 (Martynov, 1930c).

Family MEGAKHOSARIDAE Sharov, 1961

[Megakhosaridae Sharov, 1961d, p. 178]

Fore wing long, slender; costal area very narrow; both veins MA and MP apparently present; series of strong crossveins between CUP and most posterior branch of CUA, more basal ones being abruptly curved at junction with CUP; crossveins over rest of wing numerous and irregular. Hind wing with MA anastomosed for short distance with RS. *Perm.*

Megakhosara Martynov, 1937b, p. 31 [*M. fasciipennis*; OD] [=*Syndesmophora* Martynov, 1937b, p. 41 (type, *S. composita*)]. Fore wing with RS dichotomously branched; no anastomosis between main veins; both fore and hind wings with irregular crossveins. *Perm.*, USSR (Asian RSFSR).——Fig. 69,*1a*. *M. dilucida* Sharov; fore wing, ×1.8 (Sharov, 1961e).——Fig. 69,*1b*. *M. fasciipennis*; hind wing, ×1.5 (Martynov, 1937b).

Megakhosarella Sharov, 1961d, p. 182 [*M. regressa*; OD]. Little-known fore wing; MA and RS anastomosed for very short distance. *Perm.*, USSR (Asian RSFSR).——Fig. 69,*2*. *M. regressa*; fore wing, ×3 (Sharov, 1961d).

Family IDELIIDAE Zalessky, 1929

[Ideliidae M. D. Zalessky, 1929, p. 21] [=Rachimentomidae G. M. Zalessky, 1939, p. 55]

Fore wing with broad costal area having numerous slanting veinlets, usually forming reticulation; no anastomosis of veins MA and RS; RS usually with more than three branches; stem CU formed as in Liomopteridae but CUA more elaborately branched; CUP not so strongly developed as in Liomopteridae; crossveins numerous, usually forming reticulation. Hind wing little known, apparently similar to that of Liomopteridae. Antennae prominent; pronotum with broad, coriaceous expansions; cerci probably well developed; long ovipositor present. *U. Carb.–Perm.*

Stenaropodites Martynov, 1928b, p. 47 [*S. reticulata*; OD] [=*Idelia* Zalessky, 1929, p. 4 (type, *I. permiakovi*)]. Fore wing with fine network of cells, resembling archedictyon; costal margin weakly curved; CUA2 strongly sigmoidal. *Perm.*, USSR (Asian RSFSR).——Fig. 70,*6*. *S. permiakovi* (Zalessky); fore wing, ×1.2 (Zalessky, 1929).

Aenigmidelia Sharov, 1961d, p. 175 [*A. incredibilis*; OD]. Fore wing oval, with strongly curved costal margin; SC with basal branch, resembling submarginal costa; main branch of SC coalesced with R basally; crossveins as in *Archidelia*; M forking at level of origin of RS. *Perm.*, USSR (Asian RSFSR).——Fig. 70,*8*. *A. incredibilis*; fore wing, ×1 (Sharov, 1961d).

Archidelia Sharov, 1961d, p. 172 [*A. elongata*; OD]. Fore wing with strongly convex costal margin; crossveins forming irregular reticulation, much finer than that in *Kortshakolia*, but no archedictyon; costal veinlets branched; M forking before origin of RS. *Perm.*, USSR (Asian RSFSR). —— Fig. 70,*2*. *A. elongata*; *a*, fore and *b*, hind wings, ×1 (Sharov, 1961d).

Kortshakolia Sharov, 1961d, p. 171 [*K. ideliformis*; OD]. Little-known genus, based on fragment of fore wing, with costal margin shaped as in *Stenaropodites*; RS with at least 4 branches; MA directed anteriorly at its origin toward R before curving distally. *U. Carb.*, USSR (Asian RSFSR).——Fig. 70,*1*. *K. ideliformis*; fore wing, ×1.6 (Sharov, 1961d).

Metidelia Martynov, 1937b, p. 23 [*M. kargalensis*; OD]. Fore wing with costal area narrower than in *Stenaropodites*; crossveins forming nearly regular network, not so fine as archedictyon. *Perm.*, USSR (Asian RSFSR).——Fig. 70,*3*. *M. kargalensis*; fore wing, ×1.8 (Martynov, 1937b).

Paridelia Sharov, 1961d, p. 175 [*P. pusilla*; OD]. Fore wing with costal margin as in *Stenaropodites*; RS arising near midwing, with 2 branches; MA with 2 branches, MP with 3. Sharov, 1962c. *Perm.*, USSR (Asian RSFSR).——Fig. 70,*9*. *P. pusilla*; fore wing, ×1.8 (Sharov, 1961d).

Rachimentomon Zalessky, 1939, p. 56 [*R. reticulatum*; OD]. Little-known genus. Costal margin of fore wing nearly straight; fine archedictyon present; venation little known; pronotal disc large; ovipositor well developed, nearly half as long as

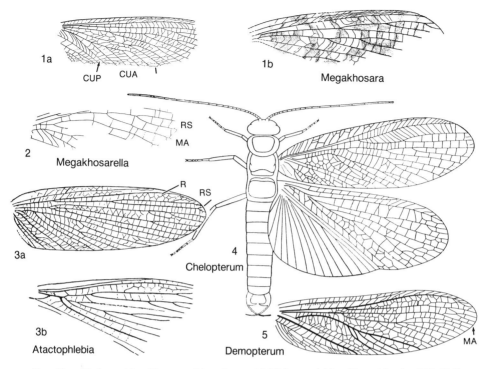

FIG. 69. Chelopteridae, Demopteridae, Atactophlebiidae, and Megakhosaridae (p. 117–118).

abdomen. *Perm.*, USSR (Asian RSFSR).——FIG. 70,5. **R. reticulatum*; whole specimen, ×2 (Zalessky, 1939).

Sylvidelia MARTYNOV, 1940, p. 19 [**S. latipennis*; OD]. Fore wing with archedictyon as in *Stenaropodites* but with more strongly curved costal margin, more branches to RS, and without sigmoidal CUA2. *Perm.*, USSR (Asian RSFSR).——FIG. 70,7. **S. latipennis*; fore wing, ×1.4 (Martynov, 1940).

Family EURYPTILONIDAE Martynov, 1940

[Euryptilonidae MARTYNOV, 1940, p. 16]

Fore wing oval, with narrow costal area; vein RS arising near midwing; CUA arising from stem of CU at base of wing and coalescing with M for short distance; CUA with several long, parallel branches. Pronotal disc well developed; legs adapted for running, spinous. *Perm.*

Euryptilon MARTYNOV, 1940, p. 16 [**E. blattoides*; OD]. Fore wing with subcostal space much broader than costal; M sigmoidally curved. *Perm.*, USSR (Asian RSFSR).——FIG. 70,10. **E. blattoides*; fore wing, ×5.5 (Martynov, 1940).

Family NARKEMIDAE Handlirsch, 1911

[Narkemidae HANDLIRSCH, 1911, p. 321]

Little-known family. Vein SC of fore wing terminating on R at level of midwing; RS with numerous parallel branches ending on wing apex; M apparently with a single distal fork; CUA apparently extensively branched; anal area unknown. *U. Carb.*

Narkema HANDLIRSCH, 1911, p. 322 [**N. taeniatum*; OD]. Little-known genus, based on incomplete fore wing. RS with at least 5 terminal branches. Wing with 7 narrow, dark transverse bands. SHAROV, 1961e; PINTO & ORNELLAS, 1978c. *U. Carb.*

Family HERBSTIALIDAE Schmidt, 1953

[Herbstialidae SCHMIDT, 1953, p. 165]

Related to Cacurgidae (probably synonymous). Fore wing with reticulation of cross-

120   *Hexapoda*

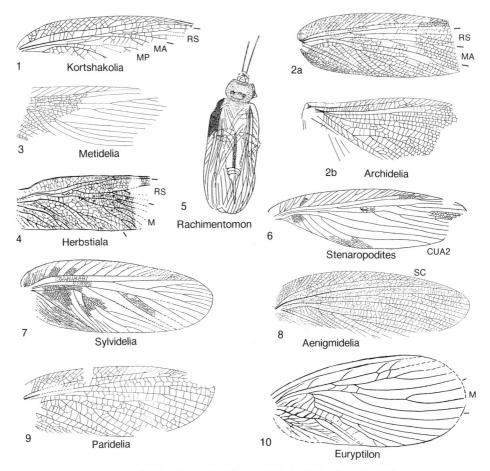

Fig. 70. Ideliidae, Euryptilonidae, and Herbstialidae (p. 118–120).

veins over entire wing surface; cuticular spots as in Cacurgidae. *U. Carb.*

Herbstiala SCHMIDT, 1953, p. 153 [**H. herbsti*; OD]. Origin of RS at about same level as first fork of M. [Placed in order Protocicadida.] *U. Carb.*, Europe (Germany).——FIG. 70,4. **H. herbsti*; fore wing, ×1.5 (Schmidt, 1953).

Family CACURGIDAE
Handlirsch, 1911

[Cacurgidae HANDLIRSCH, 1911, p. 324]

Apparently related to Omalidae. Fore wing oval, apex broadly rounded; vein SC nearly straight, terminating on wing margin beyond midwing; RS arising well before midwing; MA apparently absent; CUA anastomosed with MP basally, diverging before level of origin of RS, then anastomosed with CUP1; crossveins numerous. Hind wing unknown. *U. Carb.*

Cacurgus HANDLIRSCH, 1911, p. 324 [**C. spilopterus*; OD]. Little-known genus. Fore wing broadest at level of midwing; R with several oblique veinlets leading to fore margin of wing; crossveins forming a coarse network over most of wing except costal area; wing membrane with many circular thickenings. *U. Carb.*, USA (Illinois).——FIG. 71,6. **C. spilopterus*; fore wing, ×1 (Handlirsch, 1911).

Heterologus CARPENTER, 1944, p. 14 [**H. langfordorum*; OD]. Fore wing as in *Cacurgus,* but costal area narrower and more tapering and no network of crossveins. *U. Carb.*, USA (Illinois). ——FIG. 71,8. **H. langfordorum*; fore wing, ×3.5 (Carpenter, 1944).

Protodictyon MELANDER, 1903, p. 196 [**P. pulchripenne*; OD]. Similar to *Heterologus,* but crossveins of fore wing forming a coarse reticulation in several areas of the wing; RS remote from M

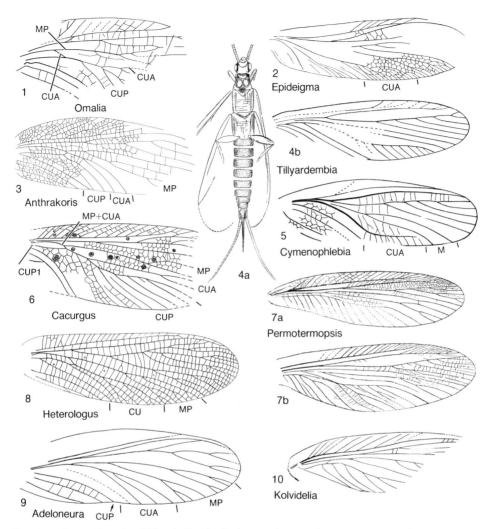

FIG. 71. Cacurgidae, Omaliidae, Tillyardembiidae, Epideigmatidae, Permotermopsidae, and Adeloneuridae (p. 120–124).

basally. [Placed by MELANDER in Hemeristina group of Palaeodictyoptera.] *U. Carb.*, USA (Illinois).

Spilomastax HANDLIRSCH, 1911, p. 326 [**S. oligoneurus*; OD]. Apparently similar to *Cacurgus*, but M forked before level of origin of RS. *U. Carb.*, USA (Illinois).

Family OMALIIDAE Handlirsch, 1906

[Omaliidae HANDLIRSCH, 1906b, p. 145]

Related to Cacurgidae, but fore wing without circular thickenings and with vein RS less developed than MP; CUA coalesced with MP at base. Hind wing unknown. *U. Carb.*

Omalia VAN BENEDEN & COEMANS, 1867, p. 384 [**O. macroptera*; OD] [=*Palaeomastax* HANDLIRSCH, 1904a, p. 16 (type, *P. carbonis*)]. Little-known fore wing; costal margin strongly curved and costal area broad; first fork of CUP beyond anastomosis with CUA. *U. Carb.*, Europe (Belgium). —— FIG. 71,*1*. **O. macroptera*; fore wing, ×1 (Pruvost, 1930).

Anthrakoris RICHARDSON, 1956, p. 36 [**A. aetherius*; OD]. Related to *Omalia*, but costal margin of fore wing more strongly curved and CUA with only 2 branches. *U. Carb.*, USA (Illinois).——

FIG. 71,3. **A. aetherius*; fore wing, ×5 (Carpenter, new).

Coselia BOLTON, 1922, p. 81 [**C. palmiformis*; OD]. Little-known genus, based on small fragment of wing. [Type of family Coseliidae BOLTON.] PRUVOST, 1930. *U. Carb.,* England.

Family GERARIDAE Scudder, 1885

[*nom. correct.* HANDLIRSCH, 1906a, p. 701, *pro* Gerarina SCUDDER, 1885b, p. 762] [=Sthenaropodidae HANDLIRSCH, 1906b, p. 141; Genopterygidae RICHARDSON, 1956, p. 41]

Fore wing membranous; costal area of uniform width for most of its length, with many crossveins, mostly unbranched; vein SC terminating on C; R without branches; RS commonly anastomosed with M for a short distance or connected to it by a short crossvein; CUA strongly developed, arising from the combined bases of R and M; CUP forked. Hind wing with the costal area more narrow than in the fore wing; RS arising very near the wing base; anal area little known but apparently not enlarged. Head relatively small, with long, filamentous antennae; prothorax long, slender anteriorly but broad posteriorly, bearing prominent spines; legs cursorial, slender, with five tarsal segments. Abdomen very little known. BURNHAM, 1983. *U. Carb.*

Gerarus SCUDDER, 1885d, p. 344 [**G. vetus*; OD] [=*Genopteryx* SCUDDER, 1885d, p. 327 (type, *G. constricta*); *Sthenaropoda* BRONGNIART, 1885a, p. 59 (type, *S. fischeri*); *Archaeacridites* MEUNIER, 1909c, p. 39 (type, *A. bruesi*); *Rossites* RICHARDSON, 1956, p. 44 (type, *R. inopinus*)]. Moderately large species. Fore wing with RS branched 2 or 3 times; M with 4 or 5 branches and either anastomosed for a short interval with RS or connected to it by a strong crossvein. BURNHAM, 1983. *U. Carb.,* USA (Illinois), Europe (France).——FIG. 72,3a. *G. bruesi* (MEUNIER); fore wing, ×1 (Burnham, 1983).——FIG. 72,3b. *G. danielsi*; reconstruction, based on many specimens, ×0.7 (Burnham, 1983).

Anepitedius HANDLIRSCH, 1911, p. 318 [**A. giraffa*; OD]. Little-known genus, based on wing and body fragments. BURNHAM, 1983. *U. Carb.,* USA (Illinois).

Genentomum SCUDDER, 1885d, p. 329 [**G. validum*; OD]. Similar to *Gerarus,* but branches of M straight and parallel in fore wing; first fork of CUP very close to wing base. BURNHAM, 1983. *U. Carb.,* USA (Illinois).

Gerarulus HANDLIRSCH, 1911, p. 316 [**G. radialis*; OD]. Little-known genus, based on wing fragments; RS with not more than 4 branches. BURNHAM, 1983. *U. Carb.,* USA (Illinois).

Nacekomia RICHARDSON, 1956, p. 33 [**N. rossae*; OD]. Fore wing similar to that of *Gerarus* but more slender; M not connected to RS by a thickened crossvein. BURNHAM, 1983. *U. Carb.,* USA (Illinois).——FIG. 72,2. **N. rossae*; fore wing, ×1.4 (Carpenter, new).

Progenentomum HANDLIRSCH, 1906a, p. 701 [**P. carbonis*; OD]. Fore wings as in *Gerarus* but more pointed; SC shorter; RS with 4 main branches; branches of CUA nearly parallel. BURNHAM, 1983. *U. Carb.,* USA (Illinois).

Family SPANIODERIDAE Handlirsch, 1906

[Spanioderidae HANDLIRSCH, 1906a, p. 695]

Fore wing coriaceous, with granular surface resembling that of Probnidae; costal margin only slightly curved; costal area narrow, with regular series of oblique, simple crossveins; vein SC terminating on R beyond midwing; RS arising near base; M flat or slightly concave, with several terminal branches; stem of CUA apparently anastomosed with base of M, diverging from M at about level of origin of RS, commonly with a series of long branches; CUP nearly straight; anal area with several veins; crossveins numerous, unbranched. Hind wing little known; remigium as in fore wing, but CUA strongly diverging away from M, its branches shorter than in fore wing; anal area unknown. Head small; prothorax elongate, without spines; legs slender, cursorial; abdomen little known; cerci unknown; ovipositor long. *U. Carb.*

Propteticus SCUDDER, 1885d, p. 334 [**P. infernus*; OD] [=*Petromartus* MELANDER, 1903, p. 191 (type, *P. indistinctus*); *Spaniodera* HANDLIRSCH, 1906a, p. 696 (type, *S. ambulans*); *Camptophlebia* HANDLIRSCH, 1906a, p. 698 (type, *Dictyoneura clarinervis* MELANDER, 1903); *Paracheliphlebia* HANDLIRSCH, 1906a, p. 699 (type, *Cheliphlebia extensa* MELANDER, 1903); *Metryia* HANDLIRSCH, 1906a, p. 700 (type, *M. analis*)]. Fore wing with vein R terminating just before wing apex; M with basal fork at about level of origin of RS, anterior branch with at least one fork; CUA with at least 4 to 7 branches. Hind wing narrower than fore wing. BURNHAM, 1986. *U. Carb.,* USA (Illinois).——FIG. 72,1. **P. infernus*; reconstruction, based on type and sev-

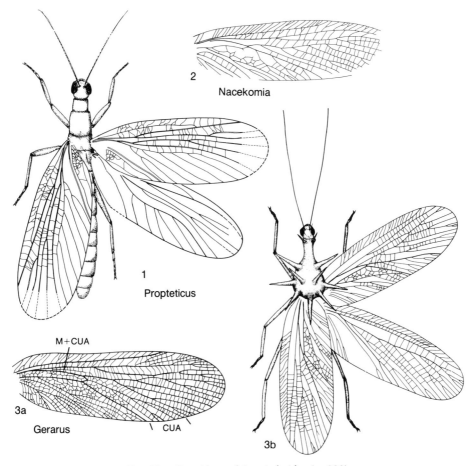

Fig. 72. Geraridae and Spanioderidae (p. 000).

eral additional specimens, ×1.3 (Burnham, 1986).

Dieconeura SCUDDER, 1885d, p. 336 [*D. arcuata; SD HANDLIRSCH, 1906a, p. 699]. Little-known genus, with fore wing more slender than in *Propteticus*; M with only one fork, anterior branch briefly anastomosed with branch of RS. [Family assignment doubtful.] *U. Carb.*, USA (Illinois).

Family APITHANIDAE
Handlirsch, 1911

[Apithanidae HANDLIRSCH, 1911, p. 320]

Related to Spanioderidae. Fore wing with vein R1 extending nearly to apex, a series of oblique veinlets between R1 and the costal margin beyond SC; RS arising in distal third of wing; prothorax shorter than in Spanioderidae. *U. Carb.*

Apithanus HANDLIRSCH, 1911, p. 320 [*A. jocularis; OD]. Fore wing little known; M apparently forking at about midwing. *U. Carb.*, USA (Illinois).

Family TILLYARDEMBIIDAE
Zalessky, 1938

[Tillyardembiidae ZALESSKY, 1938, p. 64] [=Permocapniidae MARTYNOV, 1940, p. 52]

Fore wing little known; vein SC terminating on costa; RS pectinately branched; CUA with at least 4 parallel branches to apical and hind margins. Body slender, with prominent cerci and ovipositor. SHAROV, 1962c. *Perm.*

Tillyardembia ZALESSKY, 1937d, p. 847 [*T. biarmica; OD] [=*Permocapnia* MARTYNOV, 1940, p.

52 (type, *P. brevipes*)]. RS with about 6 branches; M with deep fork. *Perm.,* USSR (Asian RSFSR).——FIG. 71,*4*. *T. brevipes* (MARTYNOV); *a,* complete specimen, ×4.5 (Martynov, 1940); *b,* fore wing, ×4 (Sharov, 1962c).

Family EPIDEIGMATIDAE
Handlirsch, 1911

[Epideigmatidae HANDLIRSCH, 1911, p. 356] [=Cymenophlebiidae PRUVOST, 1919, p. 128]

Fore wing coriaceous; costal area of moderate width; vein RS arising in basal third of wing, with several parallel branches; M apparently independent of RS; CUA branching only distally, forming many terminal branches along posterior border; anal area sharply marked by curved suture and very small; irregular reticulation over most of wing. Hind wing unknown. Pronotum elongate, oval. *U. Carb.*

Epideigma HANDLIRSCH, 1911, p. 357 [*E. elegans*; OD]. Fore wing slender, length almost 4 times width; SC terminating on R. *U. Carb.,* USA (Illinois).——FIG. 71,*2*. *E. elegans*; fore wing, ×2 (Handlirsch, 1911).

Cymenophlebia PRUVOST, 1919, p. 128 [*C. carpentieri*; OD]. Fore wing as in *Epideigma,* but costal area wider; SC terminating on costa. *U. Carb.,* Europe (France).——FIG. 71,*5*. *C. carpentieri*; fore wing, ×3 (Pruvost, 1919).

Family PERMOTERMOPSIDAE
Martynov, 1937

[Permotermopsidae MARTYNOV, 1937b, p. 84]

Fore wing similar to that of Ideliidae, but basal part narrowed and vein CUA more remote distally from wing margin. Hind wing unknown. *Perm.*

Permotermopsis MARTYNOV, 1937b, p. 84 [*P. roseni*; OD]. Costal veinlets simple; crossveins forming delicate, irregular network. *Perm.,* USSR (European RSFSR).——FIG. 71,*7a*. *P. roseni*; fore wing, ×1.0 (Martynov, 1937b).——FIG. 71,*7b*. *P. pectinata* MARTYNOV; fore wing, ×1.2 (Martynov, 1937b).

Kolvidelia ZALESSKY, 1956a, p. 282 [*K. curta*; OD]. Little-known fore wing, with costal area broader than in *Permotermopsis.* [Family assignment doubtful.] *Perm.,* USSR (Asian RSFSR).——FIG. 71,*10*. *K. curta*; fore wing, ×2 (Zalessky, 1956a).

Family ADELONEURIDAE
Carpenter, 1938

[Adeloneuridae CARPENTER, 1938, p. 450]

Fore wing with very broad costal area having long, oblique veinlets; vein MA apparently absent; CUA anastomosed with MP proximally; distinct marginal indentation at end of CUP. *U. Carb.*

Adeloneura CARPENTER, 1938, p. 450 [*A. thompsoni*; OD]. Little-known fore wing, with very narrow subcostal space; CUA and MP separating at about level of origin of RS. *U. Carb.,* USA (Illinois).——FIG. 71,*9*. *A. thompsoni*; fore wing, ×1.6 (Carpenter, 1938).

Family AENIGMATODIDAE
Handlirsch, 1906

[Aenigmatodidae HANDLIRSCH, 1906a, p. 683]

Crossveins forming reticulated network over most of fore wing; vein MA apparently absent; CUA anastomosed with stem of MP, unbranched. *U. Carb.*

Aenigmatodes HANDLIRSCH, 1906a, p. 683 [*A. danielsi*; OD]. Little-known genus, based on fragment of fore wing; RS and MP with 3 distinct branches. *U. Carb.,* USA (Illinois).

Family STREPHOCLADIDAE
Martynov, 1938

[Strephocladidae MARTYNOV, 1938b, p. 100]

Fore wing coriaceous; precostal area absent; vein SC well developed, extending to midwing or beyond, with several branches; RS arising before midwing; R ending well before apex, with several oblique branches to wing margin; RS with several long branches; M forked before origin of RS, anterior branch commonly touching RS or connected to it by a crossvein; CUA longitudinal, with several long branches; branches of RS, M, and CUA parallel and slightly sigmoidal; distinct furrow posterior to CUA; 1A close and parallel to CUP. Crossveins numerous and regularly arranged, an irregular network in costal area and between CUA and CUP and anal veins. Wing membrane with fine microtrichia between veins; prominent setae on most veins. Hind wing and body unknown. [The relationships of this family within the Protor-

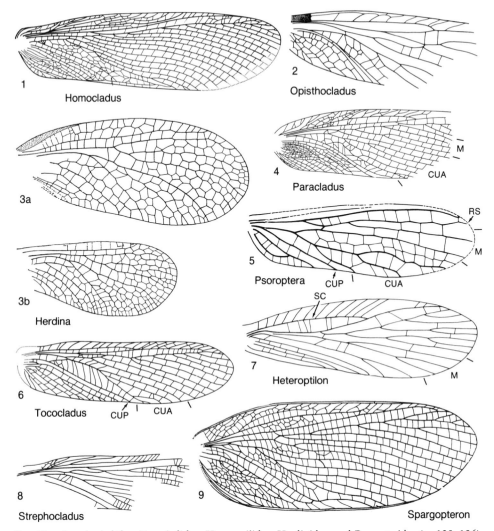

Fig. 73. Strephocladidae, Tococladidae, Heteroptilidae, Herdinidae, and Psoropteridae (p. 125–126).

thoptera are uncertain; MARTYNOV (1938a) placed it in a new order, Strephocladodea.] CARPENTER, 1966. *U. Carb.–Perm.*

Strephocladus SCUDDER, 1885d, p. 337 [**Petrablattina subtilus* KLIVER, 1883, p. 251; OD]. Little-known genus. Costal area with straight, oblique veinlets near level of midwing; branches of CUA and RS dichotomously formed. KUKALOVÁ, 1965; CARPENTER, 1966. *U. Carb.*, Europe (Germany).——FIG. 73,8. **S. subtilus*; fore wing as preserved (holotype), ×2 (Carpenter, 1966).

Homocladus CARPENTER, 1966, p. 60 [**H. grandis*; OD]. Fore wing slender; costal area much as in *Spargopteron*; branches of RS arising pectinately, those of CUA dichotomously. *Perm.*, USA (Kansas). —— FIG. 73,1. **H. grandis*; fore wing, ×1.3 (Carpenter, 1966).

Paracladus CARPENTER, 1966, p. 62 [**P. retardatus*; OD]. Fore wing as in *Homocladus,* but branches of RS, M, and CUA nearly straight. *Perm.*, USA (Kansas).——FIG. 73,4. **P. retardatus*; fore wing, ×3 (Carpenter, 1966).

Spargopteron KUKALOVÁ, 1965, p. 89 [**S. latericius*; OD]. Fore wing much broader than in *Strephocladus*; all veinlets in costal area branched or irregular; branches of RS dichotomous, those of CUA pectinate. CARPENTER, 1966. *Perm.*, Europe (Czechoslovakia).——FIG. 73,9. **S. latericius*; fore wing, ×1.8 (Kukalová, 1965).

Family TOCOCLADIDAE
Carpenter, 1966

[Tococladidae CARPENTER, 1966, p. 77]

Similar to Protokollaridae, but fore wing with anterior branch of vein M anastomosed for short interval with stem of RS; SC ending on R, just beyond midwing; crossveins not reticulate. Body little known (*Opisthocladus*); head relatively large; pronota of thoracic segments large and nearly circular. *Perm.*

Tococladus CARPENTER, 1966, p. 77 [*T. rallus*; OD]. Area between CUA and CUP very broad, traversed by long crossveins, not reticulate. *Perm.*, USA (Kansas).——FIG. 73,6. *T. rallus*; fore wing, ×2.3 (Carpenter, 1966).
Opisthocladus CARPENTER, 1976, p. 342 [*O. arcuatus*; OD]. Fore wing as in *Tococladus,* but costal veinlets looped and RS arising more distally; basal part of costal area thick and strongly sclerotized. *Perm.*, USA (Kansas).——FIG. 73,2. *O. arcuatus*; fore wing as preserved, ×4 (Carpenter, 1976).

Family HETEROPTILIDAE
Carpenter, 1976

[Heteroptilidae CARPENTER, 1976, p. 346]

Insects of moderate size; affinities uncertain within the Protorthoptera. Fore wing oval, anterior margin strongly curved; vein SC unusually remote from wing margin, ending on R near midwing; SC curving posteriorly near midwing; RS with several branches, M with few; CUA nearly straight and ending on hind margin about three-fourths wing length from base; CUP and anal veins close together and straight. *Perm.*

Heteroptilon CARPENTER, 1976, p. 346 [*H. costale*; OD]. Fore wing broadest beyond midwing; RS with 8 terminal branches; CUA with a short, distal fork. *Perm.*, USA (Kansas).——FIG. 73,7. *H. costale*; fore wing, ×3.7 (Carpenter, 1976).

Family HERDINIDAE
Carpenter & Richardson, 1971

[Herdinidae CARPENTER & RICHARDSON, 1971, p. 287]

Apparently related to Cacurgidae. Wings very short; venation strongly developed; small tubercles on all main veins and crossveins of fore and hind wings. Fore wing with base of costal area strongly sclerotized; vein SC ending on costal margin well before apex of wing; RS arising at level of midwing, with 3 main branches; M independent of R basally and with two main branches. Crossveins numerous, forming a coarse network over the wing. Hind wing much smaller than fore wing; costal area narrow; venation apparently as in fore wing. Body little known; pronotum large. *U. Carb.*

Herdina CARPENTER & RICHARDSON, 1971, p. 291 [*H. mirificus*; OD]. CUA apparently unbranched; CUP forked, one branch directed toward hind margin of wing and very irregular. [It has been suggested by some workers that the specimen on which this genus is based is in fact a nymph, not an adult. However, the wings are well sclerotized, have thick veins, and are covered with tubercles. Two additional specimens, with similarly reduced wings, have more recently been found in the same deposit.] *U. Carb.*, USA (Illinois).——FIG. 73,3. *H. mirificus*; *a*, fore and *b*, hind wings, ×7 (Carpenter & Richardson, 1971).

Family PSOROPTERIDAE
Carpenter, 1976

[Psoropteridae CARPENTER, 1976, p. 345]

Small insects of uncertain affinities. Fore wing membrane coriaceous and rugose, with hair covering; veins M and CU coalesced near base of wing; M with 2 branches, CUA with 3; longitudinal veins thick; crossveins weak. Hind wing and body unknown. *Perm.*

Psoroptera CARPENTER, 1976, p. 345 [*P. cubitalia*; OD]. Fore wing with R extending almost to wing apex; RS arising at about level of fork of CUA; 2 rows of cells between M3+4 and CUA. *Perm.*, USA (Kansas).——FIG. 73,5. *P. cubitalia*; fore wing, ×9 (Carpenter, 1976).

Family STREPHONEURIDAE
Martynov, 1940

[Strephoneuridae MARTYNOV, 1940, p. 14]

Fore wing with costal area of moderate width; vein SC with branched veinlets and terminating on costa; crossveins numerous, without reticulation; R with series of close veinlets to costal margin distally. *Perm.*

Strephoneura MARTYNOV, 1940, p. 14 [*S. robusta*; OD]. Subcostal area very narrow; MP anastomosed with CUA proximally. *Perm.*, USSR (Asian RSFSR).——FIG. 74,1. *S. robusta*; fore wing, ×1 (Sharov, 1962c).

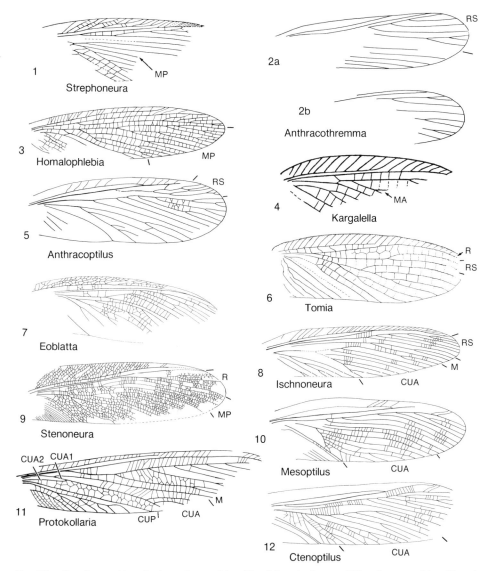

Fig. 74. Strephoneuridae, Anthracothremmidae, Tomiidae, Anthracoptilidae, Stenoneuridae, Homalophlebiidae, Ischnoneuridae, and Protokollariidae (p. 126–130).

Family ANTHRACOTHREMMIDAE Handlirsch, 1906

[Anthracothremmidae HANDLIRSCH, 1906a, p. 712]

Little-known insects. Fore wing with branches of vein RS forming series of nearly parallel veins leading to apical area of wing. Hind wing with remigium shaped as in fore wing and venation similar; anal area unknown. *U. Carb.*

Anthracothremma SCUDDER, 1885d, p. 327 [*A. robusta*; OD]. Wings with rounded apex; RS of fore wing with at least 6 straight, main branches, some forked. *U. Carb.*, USA (Illinois).——FIG. 74,2. *A. robusta*; *a*, fore and *b*, hind wings, ×2.5 (Carpenter, new).

Melinophlebia HANDLIRSCH, 1911, p. 364 [*M. analis*; OD]. Little-known genus, based on fragment of fore wing. RS with only 3 or 4 main branches. *U. Carb.*, USA (Illinois).

Pericalyphe HANDLIRSCH, 1911, p. 363 [*P. longa*; OD]. Similar to *Anthracothremma* but larger; RS with simple branches. [Probably a synonym of *Anthracothremma*.] *U. Carb.*, USA (Illinois).

Silphion HANDLIRSCH, 1911, p. 365 [**S. latipenne*; OD]. Fore wing as in *Pericalyphe* but broader. *U. Carb.*, USA (Illinois).

Family TOMIIDAE Martynov, 1936

[Tomiidae MARTYNOV, 1936, p. 1254]

Costal area wider than in Atactophlebiidae, with series of evenly spaced veinlets; vein SC terminating slightly beyond midwing; R with series of costal veinlets continuing series of veinlets from SC; crossveins of wing more regular than in Atactophlebiidae. [Ordinal assignment uncertain.] *Perm.–Trias.*

Tomia MARTYNOV, 1936, p. 1255 [**T. costalis*; OD]. Double row of cells between RS and R proximally; other crossveins simple; MA with 4 branches. *Trias.*, USSR (Asian RSFSR).——FIG. 74,6. **T. costalis*; fore wing, ×3.5 (Martynov, 1936).

Kargalella MARTYNOV, 1937b, p. 32 [**K. subcostilis*; OD]. Little-known genus, based on fragment of fore wing. Costal area broader than in *Tomia*; single row of cells between all veins. RS anastomosed with MA basally. [Family position doubtful.] *Perm.*, USSR (European RSFSR).

——FIG. 74,4. **K. subcostilis*; fore wing, ×3.5 (Martynov, 1937b).

Family ANTHRACOPTILIDAE Handlirsch, 1922

[Anthracoptilidae HANDLIRSCH, 1922, p. 98]

Related to Ischnoneuridae. Fore wing with vein SC terminating on R well before apex; RS apparently originating about midwing; CUA extensively developed with dichotomous branching; wing membrane granular, as in Probnidae; crossveins numerous, irregular, and weak. *U. Carb.*

Anthracoptilus LAMEERE, 1917b, p. 180 [**Homalophlebia perrieri* MEUNIER, 1909d, p. 46; OD] [=*Prostenoneura* HANDLIRSCH, 1919b, p. 559, obj.]. All branches of RS directed anteriorly; RS with fewer branches than M. *U. Carb.*, Europe (France).——FIG. 74,5. **A. perrieri* (MEUNIER); fore wing, ×1.5 (Carpenter, new).

Family CNEMIDOLESTIDAE Handlirsch, 1906

[Cnemidolestidae HANDLIRSCH, 1906b, p. 135]

Fore wing similar to that of Ischnoneuridae, but RS arising near midwing and cross-

Protodiamphipnoa

FIG. 75. Cnemidolestidae (p. 129).

veins fewer, more widely separated. Hind wing unknown. Prothorax small, about as long as wide; head small, slenderly oval; antennae long and filamentous; forelegs long and apparently raptorial. *U. Carb.*

Protodiamphipnoa BRONGNIART, 1893, p. 410 [*P. tertrini*; OD] [=*Cnemidolestes* HANDLIRSCH, 1906b, p. 135 (type, *Protophasma woodwardi* BRONGNIART, 1893, p. 427)]. Fore wing with M and CUA dividing at level of origin of RS; wings with conspicuous "eye spot." HANDLIRSCH, 1937. *U. Carb.*, Europe (France).——FIG. 75. **P. tertrini*; fore wings, ×1.5 (Carpenter, new).——FIG. 76. *P. gaudryi* (BRONGNIART); fore wings and part of body, holotype, ×1.7 (Carpenter, new).

Family STENONEURITIDAE Lameere, 1917

[Stenoneuritidae LAMEERE, 1917b, p. 197]

Fore wing similar to that of Stenoneuridae, but RS arising near midwing and MP much less developed, branching only in its distal half near posterior margin. *U. Carb.*

Stenoneurites HANDLIRSCH, 1906b, p. 153 [**Stenoneura maximi* BRONGNIART, 1893, p. 430; OD]. MP forking before the origin of RS. *U. Carb.*, Europe (France).

Family STENONEURIDAE Handlirsch, 1906

[Stenoneuridae HANDLIRSCH, 1906b, p. 152] [=Eoblattidae HANDLIRSCH, 1906b, p. 155]

Fore wing slightly coriaceous; vein SC extending nearly to wing apex, terminating on R; RS arising at wing base, parallel and close to R up to at least midwing; RS with several branches; MA apparently absent; MP dividing at about midwing, the most anterior branch directed anteriorly towards RS; CUA forking well before midwing; each main branch subdividing; CUP marking off the anal area, which contains numerous anal veins, very close together and nearly parallel. Hind wing little known, apparently similar to fore wing except that M is reduced and the anal area is expanded. Prothorax broad, distinctly broader than long; prothoracic legs robust, moderately long. *U. Carb.*

Stenoneura BRONGNIART, 1893, p. 429 [**S. fayoli*; OD]. Fore wing with numerous crossveins forming very irregular, fine reticulation over most of wing, except anal area. *U. Carb.*, Europe (France).

Protodiamphipnoa

FIG. 76. Cnemidolestidae (p. 129).

——FIG. 74,9. **S. fayoli*; fore wing, ×1.2 (Carpenter, new).

Eoblatta HANDLIRSCH, 1906b, p. 155 [**Stenoneura robusta* BRONGNIART, 1893, p. 429; OD]. Fore wing with fewer crossveins than in *Stenoneura*, forming coarse reticulation only in costal area and a few small regions of wing. *U. Carb.*, Europe (France).——FIG. 74,7. **E. robusta* (BRONGNIART); fore wing, ×0.9 (Carpenter, new).

Family HOMALOPHLEBIIDAE Handlirsch, 1906

[Homalophlebiidae HANDLIRSCH, 1906b, p. 136]

Related to Stygnidae, but fore wing with very extensive vein MP; MA apparently absent; CUA forked at margin. Hind wing unknown. *U. Carb.*

Homalophlebia BRONGNIART, 1893, p. 437 [**H. finoti*; SD HANDLIRSCH, 1906b, p. 136]. Fore wing with RS arising about one-third of wing length from base, with several branches; first fork of MP at same level. *U. Carb.*, Europe (France).——FIG. 74,3. **H. finoti*; fore wing, ×0.8 (Carpenter, new).

Parahomalophlebia HANDLIRSCH, 1906b, p. 137 [**Homalophlebia courtini* BRONGNIART, 1893, p. 438; OD]. Similar to *Homalophlebia*, but RS with single fork. U. Carb., Europe (France).

Family ISCHNONEURIDAE
Handlirsch, 1906

[Ischnoneuridae HANDLIRSCH, 1906b, p. 133] [=Stenoneurellidae HANDLIRSCH, 1919b, p. 560]

Related to Stenoneuridae. Fore wing with vein SC terminating on R well before apex; RS arising close to wing base with several long branches; CUA very extensively developed; crossveins numerous, mostly straight, not forming reticulation. Hind wing unknown. Prothorax and legs long. U. Carb.

Ischnoneura BRONGNIART, 1893, p. 433, *nom. subst.* pro *Leptoneura* BRONGNIART, 1885a, p. 62, *non* WALLENGREN, 1857 [*Leptoneura oustaleti* BRONGNIART, 1885a, p. 62; SD HANDLIRSCH, 1922, p. 76] [=*Ischnoneurilla* HANDLIRSCH, 1919b, p. 556 (type, *Ischnoneura elongata* BRONGNIART, 1893, p. 433); *Ischnoneurona* HANDLIRSCH, 1919b, p. 557 (type, *Ischnoneura delicatula* BRONGNIART, 1893, p. 433)]. Branches of CUA close together and parallel. U. Carb., Europe (France).——FIG. 74,8. *I. oustaleti* (BRONGNIART); fore wing, ×0.6 (Carpenter, new).
Ctenoptilus LAMEERE, 1917b, p. 180 [*Homalophlebia trouessarti* MEUNIER, 1911a, p. 127; OD]. Similar to *Ischnoneura*, but branches of CUA more widely separated and divergent. U. Carb., Europe (France).——FIG. 74,12. *C. trouessarti* (MEUNIER); fore wing, ×1.4 (Carpenter, new).
Mesoptilus LAMEERE, 1917b, p. 174 [*M. dolloi*; OD] [=*Pseudooedischia* HANDLIRSCH, 1919b, p. 557 (type, *P. berthaudi*); *Stenoneurella* HANDLIRSCH, 1919b, p. 559 (type, *S. fayoliana*)]. Similar to *Ischnoneura*, but first branch of RS arising well before midwing; posterior branch of CUA more oblique than in *Ischnoneura*. U. Carb., Europe (France).——FIG. 74,10. *M. dolloi*; fore wing, ×1.2 (Carpenter, new).

Family PROTOKOLLARIIDAE
Handlirsch, 1906

[Protokollariidae HANDLIRSCH, 1906b, p. 137] [=Sthenaroceridae HANDLIRSCH, 1906b, p. 149; Laspeyresiellidae SCHLECHTENDAL, 1913, p. 96]

Fore wing very long, slender; vein SC terminating not far beyond midwing; RS arising near wing base, with several long branches. M and CUA1 anastomosed at wing base, separating before level of origin of RS; M apparently unbranched; CUA2 arising independently of CUA1 but anastomosed with it shortly after separation of CUA1 from M; crossveins numerous. Hind wing unknown. Head small; antennae thick at base; prothorax narrow; front legs long. U. Carb.

Protokollaria BRONGNIART, 1893, p. 409 [*P. ingens*; OD]. CUA with 2 branches arising distally and curved; crossveins between CUA and CUP forming a coarse reticulation. U. Carb., Europe (France).——FIG. 74,11. *P. ingens*; fore wing, ×1.5 (Carpenter, new).
Laspeyresiella SCHLECTENDAL, 1913, p. 96, *nom. subst.* pro *Laspeyresia* HANDLIRSCH, 1906b, p. 140, *non* HÜBNER, 1825 [*Laspeyresia wettinensis* HANDLIRSCH, 1906b, p. 140; OD] [=*Laspeyresiella* KRAUSSE, 1922, p. 132, obj. synonym & homonym]. Little-known genus, with wings and body shaped as in *Protokollaria*. U. Carb., Europe (Germany).
Sthenarocera BRONGNIART, 1885a, p. 59 [*S. pachytyloides*; OD]. Similar to *Protokollaria*, but fore wing more slender; crossveins not forming a reticulation. U. Carb., Europe (France).

Family PROTOPHASMATIDAE
Brongniart, 1885

[*nom. correct.* CARPENTER, herein, *pro* Protophasmida BRONGNIART, 1885a, p. 59]

Little-known family, apparently related to Geraridae. Fore wing with small but distinct precostal area; several veinlets arising from costa; crossveins forming network; costal space much broader than subcostal; vein RS arising nearer wing base than in Geraridae, with several branches. U. Carb.

Protophasma BRONGNIART, 1878, p. 57 [*P. dumasi*; OD]. Fore and hind wings with several transverse rows of maculations. U. Carb., Europe (France).

Family UNCERTAIN

The following genera, apparently belonging to the order Protorthoptera, are too poorly known to permit assignment to families.

Acridites GERMAR, 1842, p. 93 [*A. carbonarius*; OD]. Little-known genus, based on poorly preserved fore wing with narrow costal area. [Probably related to Geraridae.] U. Carb., Europe (Germany).
Adiphlebia SCUDDER, 1885d, p. 345 [*A. lacoana*; OD]. Based on little-known insect, with short oval wings and robust body. [Type of Adiphlebiidae HANDLIRSCH, 1906a.] U. Carb., USA (Illinois).
Aenigmatella SHAROV, 1961c, p. 159 [*A. com-

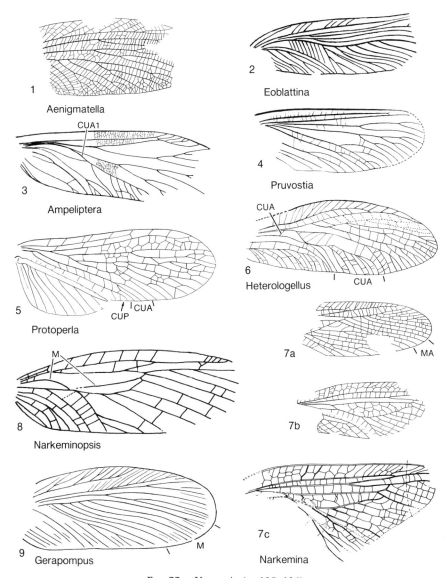

Fig. 77. Uncertain (p. 130–134).

parabilis; OD]. Costal area broad, with numerous veinlets; RS and M dividing at about same level; CUA very extensively developed. *U. Carb.*, USSR (Asian RSFSR).——Fig. 77,*1*. **A. comparabilis*; fore wing, ×1.4 (Sharov, 1961c).

Aetophlebia Scudder, 1885d, p. 338 [**A. singularis*; OD]. Based on fragments of wing. [Type of Aetophlebiidae Handlirsch, 1906a, p. 708.] *U. Carb.*, USA (Illinois).

Aetophlebiopsis Zalessky, 1955b, p. 347 [**A. fusca*; OD]. Based on fragments of wings. *Perm.*, USSR (Asian RSFSR).

Agogoblattina Handlirsch, 1906a, p. 714 [**Oryctoblattina occidua* Scudder, 1885e, p. 37; OD]. Based on fragments of wings and body. *U. Carb.*, USA (Illinois).

Ampeliptera Pruvost, 1927, p. 76 [**A. limburgica*; OD]. Little-known genus, based on incomplete fore wing with fine archedictyon; CUA1 apparently anastomosed with M near wing base and diverging away before the level of the origin of RS, as in some Permian Orthoptera. [Placed in Hapalopteridae (order Hapaloptera) by Pruvost (1927); in Palaeocixiidae (order Protorthoptera) by HANDLIRSCH (1937); in family uncertain (order Paraplecoptera) by MARTYNOV (1938b); in new family Ampelipteridae (order Protocicadida) by HAUPT (1941); and in Paoliidae (order Protorthoptera) by KUKALOVÁ (1958b).] *U. Carb.*, Europe (Holland).——FIG. 77,3. **A. limburgica*; fore wing, holotype, ×5 (Kukalová, 1958b).

Anegertus HANDLIRSCH, 1911, p. 353 [**A. cubitalis*; OD]. Based on wing fragments. *U. Carb.*, USA (Illinois).

Anthracomastax HANDLIRSCH, 1904a, p. 17 [**A. furcifer*; OD]. Based on wing fragment. *U. Carb.*, Europe (Belgium).

Archaeologus HANDLIRSCH, 1906a, p. 807 [**A. falcatus*; OD]. Based on fore and hind wing fragments. *U. Carb.*, USA (Illinois).

Archimastax HANDLIRSCH, 1906a, p. 806 [**A. americanus*; OD]. Based on wing fragment. *U. Carb.*, USA (Arkansas).

Asiopompus SHAROV, 1961c, p. 160 [**A. tomicus*; OD]. Based on fragment of fore wing. [Type of Asiopompidae SHAROV, 1961d.] *U. Carb.*, USSR (Asian RSFSR).

Atava SELLARDS, 1909, p. 157 [**A. ovata*; OD]. Based on fragment of hind wing. *Perm.*, USA (Kansas).

Axiologus HANDLIRSCH, 1906a, p. 805 [**A. thoracicus*; OD]. Based on fragments of body and wings. *U. Carb.*, USA (Illinois).

Balduriella MEUNIER, 1925, p. 105 [**B. latissima*; OD]. Based on wing fragment. *U. Carb.*, Europe (Germany).

Boutakovia PRUVOST, 1934, p. 1 [**B. saleei*; OD]. Based on wing fragment. [Placed in Homalophlebiidae by PRUVOST.] *U. Carb./Perm.*, Africa (Zaire).

Cacurgellus PRUVOST, 1919, p. 125 [**C. barryi*; OD]. Based on wing fragment. *U. Carb.* Europe (France).

Cheliphlebia SCUDDER, 1885d, p. 328 [**C. carbonaria*; OD]. Little-known genus, based on fragments of fore wings. [Type of Cheliphlebiidae HANDLIRSCH, 1906a, p. 709.] *U. Carb.*, USA (Illinois).

Chrestotes SCUDDER, 1868b, p. 567 [**C. lapidea*; OD]. Based on fragments of wings. *U. Carb.*, USA (Illinois).

Commentrya LAMEERE, 1917b, p. 176 [**Oedischia maximae* BRONGNIART, 1893, p. 437; OD]. Little-known genus, based on body fragment. *U. Carb.*, Europe (France).

Danielsiella COCKERELL, 1916c, p. 100 [**D. priscula*; OD]. Based on fragments of wings and body. *U. Carb.*, USA (Illinois).

Didymophleps SCUDDER, 1885d, p. 330 [**Termes contusa* SCUDDER, 1878c, p. 300; OD]. Based on small fragment of wing. *U. Carb.*, USA (Illinois).

Dieconeurites HANDLIRSCH, 1906a, p. 699 [**Dieconeura rigida* SCUDDER, 1885d, p. 336; OD]. Based on wing fragment. *U. Carb.*, USA (Pennsylvania).

Distasis HANDLIRSCH, 1904a, p. 17 [**D. rhipophora*; OD]. Based on distal fragment of hind wing. *U. Carb.*, Europe (Belgium).

Endoiasmus HANDLIRSCH, 1906a, p. 805 [**E. reticulatus*; OD]. Based on wing fragment. *U. Carb.*, USA (Illinois).

Eoblattina BOLTON, 1925, p. 19 [**E. complexa*; OD]. Little-known genus, based on fragments of wings and body; fore wing with extensive, sigmoidal CUA having several short, straight branches leading to CUP. *U. Carb.*, Europe (France). —— FIG. 77,2. **E. complexa*; fore wing, ×1.3 (Bolton, 1925).

Gerapompus SCUDDER, 1885d, p. 326 [**G. blattinoides*; SD HANDLIRSCH, 1906a, p. 711]. Little-known genus, based on poorly preserved fore wing; R and RS apparently with several long branches distally; M extensively branched. [Type of Gerapompidae HANDLIRSCH, 1906a, p. 711.] *U. Carb.*, USA (Illinois).——FIG. 77,9. *G. schucherti* HANDLIRSCH; fore wing, ×2 (Handlirsch, 1906a).

Gerarianus HANDLIRSCH, 1919b, p. 551 [**G. commentryanus*; OD]. Based on wing fragments. *U. Carb.*, Europe (France).

Gerarites HANDLIRSCH, 1919b, p. 551 [**Gerarus commentryi* BRONGNIART, 1893, pl. 24, fig. 10; OD]. Based on wing fragment. *U. Carb.*, Europe (France).

Geraroides HANDLIRSCH, 1906a, p. 704 [**Dieconeura maxima* MELANDER, 1903, p. 193; OD]. Little-known genus, based on fragments of wings and body. HANDLIRSCH, 1922. *U. Carb.*, USA (Illinois).

Germanoprisca ZEUNER, 1936a, p. 267 [**F. zimmermanni*; OD]. Little-known insect with prominent, slender cerci; fore wing fragment. [Type of Germanopriscidae ZEUNER, 1936a.] MÜLLER, 1978b. *Perm.*, Europe (Germany).

Gyrophlebia HANDLIRSCH, 1906a, p. 697 [**G. longicollis*; OD]. Little-known genus, based on poorly preserved fore wing and body. *U. Carb.*, USA (Illinois).

Hadentomoides RIEK, 1974a, p. 15 [**H. dwykensis*; OD]. Little-known genus, based on apical fragment of wing. [Originally placed in family Hadentomidae.] *Perm.*, South Africa (Natal).

Haplopterum MARTYNOV, 1928b, p. 84 [**H. majus*; OD]. Based on hind wing fragment. *Perm.*, USSR (European RSFSR).

Hemeristia DANA, 1864, p. 35 [**H. occidentalis*; OD] [=*Hemeristica* GOLDENBERG, 1877, p. 15, obj.]. Based on wing fragments. *U. Carb.*, USA (Illinois).

Heterologellus SCHMIDT, 1962, p. 843 [**H. teichmuellerae*; OD]. Apparently similar to *Omalia*, but fore wing with SC terminating on R well before wing apex; MP dividing distally, near wing margin; CUA only just touching CUP, not coalesced with it; MP not as extensively branched as CUA. [Placed in Omaliidae by SCHMIDT.] *U. Carb.,* Europe (Germany).——FIG. 77,6. **H. teichmuellerae;* fore wing, ×3 (Schmidt, 1962).

Heterologopsis BRAUCKMANN & KOCH, 1982, p. 18 [**H. ruhrensis;* OD]. Little-known genus, apparently related to the Cacurgidae, but SC much shorter and terminating on R. *U. Carb.,* Europe (Germany).

Kaltanopterodes SHAROV, 1961d, p. 223 [**K. vanus;* OD]. Based on hind wing of nymph. *Perm.,* USSR (Asian RSFSR).

Kargalodes MARTYNOV, 1937b, p. 33 [**K. incerta;* OD]. Based on wing fragment. *Perm.,* USSR (European RSFSR).

Kelleropteron BRAUCKMANN & HAHN, 1980, p. 308 [**K. kaelberbesgense;* OD]. Little-known genus, based on small fragment of wing. *U. Carb.,* Europe (Germany).

Khosarophlebia MARTYNOV, 1940, p. 24 [**K. sylvaensis;* OD]. Based on hind wing fragment. *Perm.,* USSR (Asian RSFSR).

Klebsiella MEUNIER, 1908c, p. 242 [**K. exstincta;* OD]. Based on fragments of fore and hind wings. [Type of Klebsiellidae HANDLIRSCH, 1919b, p. 552.] *U. Carb.,* Europe (France).

Lecopterum SELLARDS, 1909, p. 161 [**L. delicosum;* OD]. Based on wing fragment. *Perm.,* USA (Kansas).

Limburgina LAURENTIAUX, 1950, p. 14 [**L. antiqua;* OD]. Based on wing fragment, with suggestions of convex MA. [Ordinal position doubtful.] *U. Carb.,* Europe (Netherlands).

Macrophlebium GOLDENBERG, 1869, p. 164 [**M. hollebeni;* OD]. Based on wing fragment. *U. Carb.,* Europe (Germany).

Megalometer HANDLIRSCH, 1906a, p. 713 [**M. lata;* OD]. Based on fragments of wings. *U. Carb.,* USA (Illinois).

Mesorthopteron TILLYARD in TILLYARD & DUNSTAN, 1916, p. 14 [**M. locustoides;* OD]. Little-known genus, based on fragments of fore wing. Fore wing elongate-oval, with fine archedictyon; costal area broad, with numerous, parallel veinlets; SC straight and close to R and RS, with several distal branches; M weakly formed; CUA with numerous long, pectinate branches; anal area small. [Type of family Mesorthopteridae TILLYARD, 1922b.] RIEK, 1956. *Trias.,* Australia (New South Wales).

Metacheliphlebia HANDLIRSCH, 1906a, p. 698 [**Cheliphlebia elongata* SCUDDER, 1885d, p. 328; OD]. Little-known genus, based on small fragments of wings. HANDLIRSCH, 1906b. *U. Carb.,* USA (Illinois).

Miamia DANA, 1864, p. 34 [**M. bronsoni;* OD]. Little-known genus, based on wing fragments. [Possibly belonging to the Spanioderidae.] *U. Carb.,* USA (Illinois).

Mitinovia SHAROV, 1961d, p. 223 [**M. dubia;* OD]. Based on hind wing with extensive veinlets from R to costa beyond SC. *Perm.,* USSR (Asian RSFSR).

Narkemina MARTYNOV, 1931a, p. 81 [**N. angustata;* OD]. Fore wing elongate-oval; venation with some resemblance to that of the Narkemidae. RS forking near midwing; M dividing near base, MA continuing in a straight line and branching beyond origin of RS; MP diverging posteriorly and coalescing with CUA for a considerable interval before diverging anteriorly, its terminal branches joining some of those of MA; CUA with a few short, terminal branches; CUP unbranched. Hind wing very broad, with an enlarged anal area, but with the venation of the remigium much as in the fore wing. [PINTO & ORNELLAS (1978c) correctly recognized that the genus *Narkemina,* formerly placed in the Narkemidae, required a separate family, but they proposed the invalid family name, Narkemocarcurgidae, for the type genus *Narkemina.*] SHAROV, 1961e; PINTO & ORNELLAS, 1978c; LEWIS, 1979; RASNITSYN, 1980c. *U. Carb.,* USSR (Asian RSFSR), Brazil (Parana Basin), USA (Missouri).——FIG. 77,7a. **N. angustata;* fore wing, ×2.6 (Sharov, 1961e).——FIG. 77,7b. *N. angustiformis* SHAROV; fore wing, ×2 (Sharov, 1961e).——FIG. 77,7c. *N. rodendorfi* PINTO & ORNELLAS; hind wing, ×1.2 (Pinto & Ornellas, 1978c).

Narkeminopsis WHALLEY, 1979, p. 87 [**N. eddi;* OD]. Little-known genus. Apparently similar to *Narkemina,* but fore wing with M diverging posteriorly near wing base, then anastomosing briefly with CUA before diverging anteriorly and joining RS at level of end of SC; few costal veinlets; archedictyon present in costal and CUA areas. *U. Carb.,* England.——FIG. 77,8. **N. eddi;* fore wing, ×2.5 (Whalley, 1979).

Ochetopteron COCKERELL, 1927g, p. 414 [**O. canaliculatum;* OD]. Little-known genus, based on wing fragment. *U. Carb.,* USA (Maryland).

Orthoneurites MARTYNOV, 1928b, p. 49 [**O. regularis;* OD]. Based on distal wing fragment. *Perm.,* USSR (European RSFSR).

Palaeocarria COCKERELL, 1917e, p. 80 [**P. ornata;* OD]. Based on fragment of wing. *U. Carb.,* USA (Illinois).

Palaeoedischia MEUNIER, 1914d, p. 364 [**P. boulei;* OD]. Based on fragment of fore wing. *U. Carb.,* Europe (France).

Palaeomantopsis MARTYNOV, 1928b, p. 83 [**P. furcatella;* OD]. Based on distal wing fragment. *Perm.,* USSR (European RSFSR).

Paolekia RIEK, 1976a, p. 764 [**P. perditae;* OD]. Little-known genus, based on small apical fragment of wing. [Placed originally in Paoliidae.] *Perm.,* South Africa (Natal).

Paranarkemina PINTO & ORNELLAS, 1980a, p. 288

[*P. kurtzi; OD]. Little-known genus, based on incomplete wing. SC ending on R beyond level of midwing; RS arising basally and forking at about level of end of SC, with numerous branches; M forking before origin of RS; MA with 2 distal branches; CUA anastomosed briefly with MP before terminating in many branches. *U. Carb.,* Argentina (San Luis).

Polyernus SCUDDER, 1885d, p. 343 [*D. complanatus; OD]. Based on fragments of wings and body. *U. Carb.,* USA (Illinois).

Polyetes HANDLIRSCH, 1906a, p. 715 [*P. furcifer; OD]. Based on small wing fragment. *U. Carb.,* USA (Illinois).

Protoperla BRONGNIART, 1893, p. 407 [*P. westwoodi; OD]. Little-known genus, based on hind wing. SC ending on costal margin near midwing; M apparently coalesced with R and RS basally; RS with 2 main branches; M with numerous irregular branches; CUA with 3 very short terminal branches; CUP unbranched. Anal area enlarged, with a series of long pectinate branches from 2A. [Placed in the family Protoperlidae by BRONGNIART.] LAMEERE, 1917b. *U. Carb.,* Europe (France).——FIG. 77,5. *P. westwoodi;* hind wing, ×4 (Carpenter, new).

Prototettix GIEBEL, 1856, p. 306 [*Gryllacris lithanthraca GOLDENBERG, 1854, p. 24; OD]. Based on fore wing fragment. [Type of Prototettigidae HANDLIRSCH, 1906b, p. 135.] *U. Carb.,* Europe (Germany).

Pruvostia BOLTON, 1921, p. 48 [*P. spectabilis; OD]. Little-known wing (probably hind) with basal origin of RS. *U. Carb.,* Europe (England). ——FIG. 77,4. *P. spectabilis;* wing, ×0.9 (Bolton, 1921).

Pseudetoblattina HANDLIRSCH, 1906a, p. 714 [*Etoblattina reliqua SCUDDER, 1893b, p. 18; OD]. Based on wing fragment. *U. Carb.,* USA (Rhode Island).

Pseudogerarus HANDLIRSCH, 1906a, p. 804 [*P. scudderi; OD]. Based on small fragments of wings. *U. Carb.,* USA (Illinois).

Pseudopolyernus HANDLIRSCH, 1906a, p. 803 [*Polyernus laminarum SCUDDER, 1885d, p. 343; OD]. Little-known genus, based on wing fragments. *U. Carb.,* USA (Pennsylvania).

Ptenodera BOLTON, 1922, p. 90 [*P. dubius; OD]. Based on distal wing fragment. *U. Carb.,* England.

Rhipidioptera BRONGNIART, 1893, p. 447 [*R. elegans; OD]. Little-known genus, based on small fragment of wing. *U. Carb.,* Europe (France).

Roomeria MEUNIER, 1914e, p. 388 [*R. carbonaria; OD]. Based on little-known fore wing. [Type of Roomeriidae HANDLIRSCH, 1919.] *U. Carb.,* Europe (France).

Schuchertiella HANDLIRSCH, 1911, p. 311 [*S. gracilis; OD]. Little-known genus, based on small wing fragment. [Type of Schuchertiellidae HANDLIRSCH, 1911.] *U. Carb.,* USA (Illinois).

Sellardsiopsis ZALESSKY, 1939, p. 51 [*S. conspicua; OD]. Little-known fore wing. *Perm.,* USSR (Asian RSFSR).

Sharovia PINTO & ORNELLAS, 1978b, p. 100, *junior homonym, Sharovia* SINITSHENKOVA, 1977 [*S. permiafricana; OD]. Little-known genus, based on wing fragment. [Originally placed in Lemmatophoridae.] *Perm.,* South Africa (Cape of Good Hope).

Sindonopsis MARTYNOV, 1928b, p. 61 [*S. subcostalis; SD SHAROV, 1962c, p. 117]. Little-known wing with short SC. *Perm.,* USSR (European RSFSR).

Thaumatophora RIEK, 1976d, p. 147 [*T. pronotalis; OD]. Little-known genus, based on nymph with lateral abdominal gills. *Perm.,* South Africa (Natal).

Order BLATTARIA
Latreille, 1810

[Blattaria LATREILLE, 1810, p. 246] [=Blattodea BRUNNER, 1882, p. 26; Protofulgorida HAUPT (in part), 1941, p. 75]

Exopterygotes with dorsoventrally compressed bodies; head free, commonly hypognathous or opisthognathous, rarely prognathous; antennae filiform, multisegmented; compound eyes of moderate size; mandibles well developed; pronotum large, commonly covering head and extending laterally (Fig. 78); legs cursorial, spinous, with 5 tarsal segments; wings typically well developed, aptery not uncommon; fore wings tegminous, broadly oval, commonly as broad basally as at midwing; hind wings membranous, with an expanded anal fan, at least as large as remigium and containing radiating veins; abdomen with tenth tergite enlarged, forming a conspicuous supra-anal plate; cerci typically multisegmented, commonly of moderate length; external ovipositors absent in existing species but well developed in Paleozoic and many Mesozoic species. Most existing Blattaria nocturnal, omnivorous, commonly occurring in warm, moist environments. *U. Carb.–Holo.*

These are primitive orthopteroids, probably most closely related to the Isoptera among existing orders (MCKITTRICK, 1965). The order is now a relatively small one, containing less than 4,000 species (M. J. MAC-

KERRAS, 1970), but the geological record indicates that it was one of the largest orders of insects during the late Paleozoic.

The venational pattern of the cockroaches is typically orthopteroid (Fig. 78,*1,2,*). In the fore wing, however, the costa is completely marginal, there being no precostal area. Veins RS, M, and CUA are well developed, and CUP is strongly concave and curved. Crossveins are numerous but weak in existing species; in most Paleozoic species they are much stronger or commonly form a fine network (archedictyon). The venational pattern of the remigium of the hind wing is like that of the fore wing except that RS, M, and CUA have fewer branches.

The basic venational pattern of the fore wing is unusually constant throughout the order, with very few exceptions. On the other hand, the detailed branching of the veins is extremely variable within all taxonomic levels. Early attempts at family classification, in which wing venation was used (REHN, 1951), were very controversial, but the one proposed by MCKITTRICK (1964) has been generally accepted. This classification bases the families on the genitalic structures of both sexes, the nature of the proventriculus, egg-laying behavior, and the structure of certain appendages. MCKITTRICK recognized five existing families: Blattidae, Cryptoceridae, Polyphagidae, Blattellidae, and Blaberidae. The existing genera are usually based on the more detailed structure of the genitalia, hind wings, legs, and male tergal glands.

Unfortunately, such details of structure are rarely preserved in fossil roaches, with the exception of those in amber. The vast majority of fossil Blattaria, close to 90 percent, consist of isolated wings or wing fragments. Furthermore, most of the specimens with bodies preserved have the two pairs of wings folded back over the body in the usual resting position, obscuring most of the body structures that are preserved (Fig. 78). Study of extensive series of Paleozoic roaches has shown that their venational variability was at least as great as that of existing species (SCHNEIDER, 1977, 1978a, 1978b). The tendency in

FIG. 78. Blattaria; dorsal view of an archimylacrid roach from the Upper Carboniferous of Illinois in its normal resting posture, ×3.4 (Carpenter, new).

publications on these fossils has been to place emphasis on slight differences in venation, resulting in many families and genera. At least 25 extinct families and 370 extinct genera have been named from Paleozoic and early Mesozoic deposits, and fully half of these are based on single specimens.

The fossil record shows only a few obvious trends in the evolution of the fore wings of the Blattaria. In the most primitive and largest extinct family of the order, the Archimylacridae, the subcosta arises as a separate vein, isolated from R and giving rise to a series of branches toward the costa (Fig. 79,*1,*). Also, R arises as a distinct branch of stem R and has several branches. In most specialized species, as in the existing family Blattidae (Fig. 79,*2*), SC, R, and RS arise from a single stem.

Quite apart from the wings, the geological record has provided some interesting data bearing on the reproduction of the Blattaria.

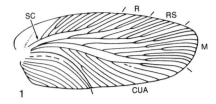

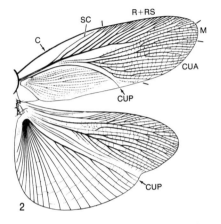

FIG. 79. Blattaria; typical venation.——*1*. Tegmen of *Phyloblatta manebachensis* GOLDENBERG, Upper Carboniferous of Germany, Archimylacridae, ×2.2 (Schneider, 1978b).——*2*. Tegmen and hind wing of *Periplaneta brunnea,* Holocene of Australia, Blattidae, ×2.3 (CSIRO, 1970).

The females of existing roaches lack a true ovipositor but commonly possess short internal valves that serve only to guide the fertilized eggs (several to many) into the genital atrium, where the ootheca is formed. In some species the oothecae are deposited within a few days, the embryos developing outside the body of the female. In others, the oothecae are first extruded, then retracted, and finally deposited in the brood sac, where the embryos continue their development. When they are mature, the ootheca is extruded again and the nymphs emerge from it (ROTH, 1967; M. J. MACKERRAS, 1970). Several small and obscure fossils, presumed to be oothecae, have been reported from Paleozoic deposits, but these are not very convincing as oothecae, and all are now considered to be fragments of other organisms (BROWN, 1957; VISHNIAKOVA, 1968). Furthermore, numerous specimens of female Blattaria with long, external ovipositors are known from Upper Carboniferous and Permian deposits of Europe, Asia, and North America (BRONGNIART, 1889; SELLARDS, 1904; ZALESSKY, 1939, 1940, 1953b). In more recent years Dr. V. N. VISHNIAKOVA of the Paleontological Institute in Moscow has described similar ovipositors on specimens from Triassic and Jurassic deposits of the USSR (VISHNIAKOVA, 1965, 1968, 1973) (Fig. 80,*1–3*). From her detailed study of these remarkable fossils, Dr. VISHNIAKOVA concluded that these ovipositors were derived from the eighth and ninth abdominal sternites and that they were therefore homologous with the ovipositors of the Orthoptera but not with the short internal valves of the existing Blattaria, which are derived from the seventh sternite (NEL, 1929; SHAROV, 1966b). It is noteworthy that in some of the Jurassic species the ovipositor is very short (Fig. 80,*1*). The gradual shortening and ultimate loss of the long external ovipositors apparently took place toward the end of the Mesozoic.

The systematics of the fossil Blattaria has been in need of a thorough revision for many years, especially considering the additional information acquired during the past twenty years. Recognizing the unsatisfactory state of the classification of the extinct species, in 1977, Dr. Jörg SCHNEIDER, of the Department of Geological Sciences, Bergakademie Freiberg, Germany, began a long-range study of type specimens, as well as new material, from the Paleozoic and Mesozoic, with full recognition of the variability of the wing venation. Up to the present time seven papers in this series have been published (SCHNEIDER, 1977, 1978a, 1978b, 1980, 1982, 1983, 1984), and others are in preparation. Since Dr. SCHNEIDER's studies are continuing, it would be presumptuous and futile for me to attempt to present here a systematic treatment of the extinct genera of the order.

FIG. 80. Blattaria; wings and body of several female specimens from the family Mesoblattinidae.——1. *Artitocoblatta asiatica* VISHNIAKOVA, Jurassic of the USSR, ×3.5.——2. *Karatavoblatta longicaudata* VISHNIAKOVA, Jurassic of the USSR, ×1.——3. *Rhipidoblatta brevivalvata* VISHNIAKOVA, Jurassic of the USSR, ×2 (all Vishniakova, 1968).

Order ISOPTERA Brullé, 1832

[Isoptera BRULLÉ, 1832, p. 66]

Wings membranous, usually very similar, held flat over abdomen at rest, and possessing a transverse humeral or basal suture; veins in anterior part of wings more strongly sclerotized than in remainder; crossveins very weakly developed, commonly forming delicate reticulation covering all or greater part of wing surface; vein C marginal; SC simple or branched, in some species very short or

Fig. 81. Isoptera; holotype specimen of *Cretatermes carpenteri,* Hodotermitidae, Cretaceous of Labrador, ×16.5 (Carpenter, new).

completely absent as distinct vein; radial system usually consisting of distinct R of variable length, equally distinct RS1+2 arising from R very near base of wing, and more highly developed RS3+4 arising about at origin of RS1+2 and forming a series of branches extending to apical or subapical region of wing (with much variation in degree of development of these radial veins); M weak but often extensively branched, no indication of division into MA and MP; CUA also weak, tending to be extensively branched along posterior margin; CUP usually weak and commonly short; anal fold formed along CUP in some; 1A usually short and reduced. Mouthparts mandibulate; antennae moniliform; cerci distinct in all castes. All species social and polymorphic, their communities composed of reproductive forms and numerous workers and soldiers. *Cret.–Holo.*

The Isoptera, commonly known as termites, are clearly related to the orthopteroids and show closest affinities with the Blattodea (MCKITTRICK, 1964, 1965). The known range of the existing family Mastotermitidae, universally regarded as the most primitive family, is only from the late Oligocene or early Miocene. The earliest records of the order, however, consist of two genera from the Cretaceous (EMERSON, 1967; JARZEMBOWSKI, 1981). Both belong to the existing family Hodotermitidae, which, although less specialized than the Termitidae, is distinctly more advanced than the Mastotermitidae with respect to both morphology and social behavior. It seems almost certain, therefore, that some species closely related to the Mastotermitidae will eventually be found in Jurassic or even Triassic deposits.

The brief geological record of the Mastotermitidae is, in fact, of much significance. The family includes only one living species, *Mastotermes darwiniensis,* restricted to tropical Australia. The Tertiary record of the family, however, contains representatives from all other continents except Africa, suggesting a wide dispersal during the Mesozoic (EMERSON, 1965). Also, the presence of specimens of all castes in mid-Tertiary amber from Mexico shows that by that time the family had achieved as complicated a social structure as now exists in *M. darwiniensis* (KRISHNA & EMERSON, 1983).

That some degree of social behavior was present among the Isoptera as far back as the Cretaceous is also apparent. In all existing termites the main part of the wing is separated from its base by a line of fracture, the humeral or basal suture (Fig. 81, *Cretatermes;* see also Fig. 82,*2a, Mastotermes,* and Fig. 83,*7a, Proelectrotermes*); shortly after the adult's flight from the parent colony, the wings break off at the suture, leaving a stub or scale. The dropping of the wings after nuptial flight is obviously related to the founding of a new colony, the wings no longer being useful. The presence of humeral sutures in the Cretaceous specimens is therefore convincing evidence that colony founding had already been developed in the family Hodotermitidae by the Early Cretaceous.

Five of the six families of Isoptera generally recognized (EMERSON & KRISHNA, 1975) have records extending at least into the Tertiary: the Mastotermitidae and Kalotermitidae from the Eocene, the Hodotermitidae from the Cretaceous, the Rhinotermitidae from the Oligocene, and the Termitidae from the Miocene. The family Serritermitidae, which is based on a single genus, has no known geological record.

In the course of their evolution the Isoptera have tended toward a secondarily homonomous condition of the wings. The primitive hind wing of the Mastotermitidae has a small but distinct anal lobe, which does not occur in any other family. In general, also, the tendency has been for reduction of the wing veins, with R and SC losing their identity as the anterior veins become compressed toward the anterior margin. These are relatively minor changes, however, in comparison with the differentiation of castes and the development of social behavior, which reach extraordinary levels of complexity in the Termitidae.

The Isoptera is one of the very few orders of insects of which the extinct forms have received careful study by specialists on recent species. SNYDER's catalogue of the Isoptera of the world, including the extinct species (1949); EMERSON's review of the Termopsinae (1933), his account of the geographic origins of termite genera (1955), and his reviews of the Mastotermitidae (1965), Kalotermitidae (1969), and Rhinotermitidae (1971); and KRISHNA's earlier revisional study of the Kalotermitidae (1961) and his joint paper with EMERSON on *Mastotermes* (1983) cover almost completely the record of the fossil Isoptera.

Family MASTOTERMITIDAE
Desneux, 1904

[Mastotermitidae DESNEUX, 1904a, p. 284]

Hind wing with distinct anal lobe; tarsi clearly with 5 segments; left mandible with 2 marginal teeth. EMERSON, 1965. *Eoc.–Holo.*

Mastotermes FROGGATT, 1896, p. 517 [=*Pliotermes* PONGRÁCZ, 1926, p. 26 (type, *P. hungaricus*)]. EMERSON, 1965; JARZEMBOWSKI, 1980; KRISHNA & EMERSON, 1983. *Eoc.*, England; *Oligo.*, Europe (Germany), England; *Oligo./Mio.*, Mexico (Chiapas)–*Holo.*——FIG. 82,*2*. *M. darwiniensis* FROGGATT, recent; *a*, fore and *b*, hind wings (humeral suture absent), ×2 (CSIRO, 1970).
Blattotermes RIEK, 1952b, p. 17 [*B. neoxenus*; OD]. Similar to *Mastotermes* but with less consolidation of RS. COLLINS, 1925; EMERSON, 1965. ?*Eoc.*, Australia (Queensland); *Eoc.*, USA (Tennessee).——FIG. 82,*3a*. *B. neoxenus*, ?*Eoc.*, Australia; fore wing, ×2.4 (Riek, 1952a).——FIG. 82,*3b*. *B. wheeleri* (COLLINS), *Eoc.*, Tennessee; wing, ×2.0 (Collins, 1925).
Miotermes VON ROSEN, 1913, p. 325 [*M. procerus* HEER; OD]. Wing venation as in *Mastotermes* but with more extensively developed M. [Family assignment doubtful.] *Mio.*, Europe (Germany, Yugoslavia).——FIG. 82,*1*. *M. procerus* (HEER); hind wing, ×1.8 (Pongrácz, 1926).
Spargotermes EMERSON, 1965, p. 19 [*S. costalimai*; OD]. Hind wings: RS diffuse, with several main branches forking to form additional branches reaching to wing tip. *Mio./Plio.*, Brazil.——FIG. 82,*4*. *S. costalimai*; hind wing with anal area folded under rest of wing, ×3.4 (Emerson, 1965).

Family KALOTERMITIDAE
Froggatt, 1896

[Kalotermitidae FROGGATT, 1896, p. 516]

Wing membrane reticulate; vein R short and almost always unbranched; pronotum as wide as head or nearly so; 4 tarsal segments. KRISHNA, 1961; EMERSON, 1969. *Eoc.–Holo.*

Kalotermes HAGEN, 1853, p. 479. HAGEN, 1861–1863; HANDLIRSCH, 1907; COCKERELL, 1917a; SNYDER, 1949; KRISHNA, 1961; EMERSON, 1969; JARZEMBOWSKI, 1980. *Eoc.*, Europe (France); *Oligo.*, England, Europe (Baltic, Germany); *Mio.*, Europe (Germany, Italy), Asia (Burma)–*Holo.*
Calcaritermes SNYDER, 1925, p. 155. EMERSON, 1969. *Oligo./Mio.*, Mexico (Chiapas)–*Holo.*
Cryptotermes BANKS, 1906, p. 336. [Generic position of fossil uncertain.] PIERCE, 1958. *Mio.*, USA (California)–*Holo.*
Electrotermes VON ROSEN, 1913, p. 331 [*Termes*

Hexapoda

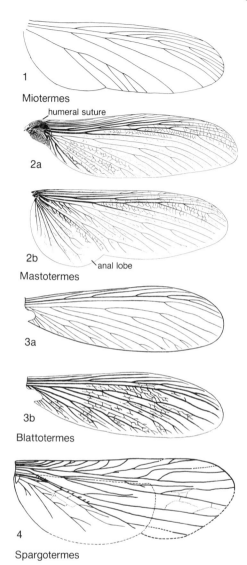

FIG. 82. Mastotermitidae (p. 139).

affinis HAGEN in PICTET & HAGEN, 1856, p. 49; OD]. Similar to *Kalotermes,* but middle tibiae with 2 outer spines distally. KRISHNA, 1961; EMERSON, 1969. *Oligo.,* Europe (Baltic).——FIG. 83,6. *E. affinis* (HAGEN); *a,* fore wing, ×7.0; *b,* right middle leg, ×1.2 (both Krishna, 1961).

Eotermes STATZ, 1939–1940, p. 13 [*E. grandaeva*; OD]. Similar to *Proelectrotermes* but much larger; lateral spines on middle tibiae relatively shorter; M weak, its main stem close and parallel to RS. EMERSON, 1969. *Oligo.,* Europe (Germany).——FIG. 83,8. *E. grandaeva*; fore wing, ×1.5 (Emerson, 1969).

Incisitermes KRISHNA, 1961, p. 353. EMERSON, 1969. *Oligo./Mio.,* Mexico (Chiapas)–*Holo.*

Neotermes HOLMGREN, 1911, p. 53. PITON, 1940a; EMERSON, 1969. *Eoc.,* Europe (France)–*Holo.*

Proelectrotermes VON ROSEN, 1913, p. 331 [*Kalotermes berendtii* PICTET in PICTET & HAGEN, 1856, p. 49; OD]. Similar to *Kalotermes,* but middle tibiae with a single inner-lateral spine and 2 outer-lateral spines; fore wing with a very short SC; branches of RS directed anteriorly and terminating on anterior margin; M slightly nearer to RS than to CU at midwing. SCUDDER, 1883a; KRISHNA, 1961; EMERSON, 1969. *Oligo.,* Europe (Baltic).——FIG. 83,7. **P. berendtii* (PICTET); *a,* fore wing as preserved, ×5.5; *b,* right middle leg, ×6.0 (both Krishna, 1961).

Prokalotermes EMERSON, 1933, p. 189 [*Parotermes hageni* SCUDDER, 1883a, p. 139; OD]. Similar to *Proelectrotermes* but with 24 to 26 antennal segments. EMERSON, 1969; LEWIS, 1977a. *Oligo.,* USA (Colorado, Montana).

Family HODOTERMITIDAE Desneux, 1904

[Hodotermitidae DESNEUX, 1904b, p. 14]

Wings with vein CU well developed; short anal vein present in hind wing; ocelli absent; left mandible with 3 marginal teeth; pronotum usually much narrower than head; tarsi with 4 segments. *Cret.–Holo.*

Hodotermes HAGEN, 1853, p. 480. *Holo.*

Archotermopsis DESNEUX, 1904b, p. 13. VON ROSEN, 1913. *Oligo.,* Europe (Baltic)–*Holo.*

Cretatermes EMERSON, 1967, p. 284 [**C. carpenteri*; OD]. Fore wing small, humeral suture evenly curved; RS area gradually widened from base to apical quarter of wing; M about midway between RS and CU; CU short, not reaching beyond basal half of posterior margin of wing. *Cret.,* Canada (Labrador).——FIG. 81. **C. carpenteri*; holotype, ×16.5 (Carpenter, new).——FIG. 83,5. **C. carpenteri*; venation of fore wing, ×8.0 (Emerson, 1967).

Parotermes SCUDDER, 1883a, p. 135 [**P. insignis*; OD]. Second marginal tooth of left mandible slightly shorter than first marginal tooth; posterior edge of second marginal tooth and anterior edge of third marginal tooth not symmetrical. *Oligo.,* USA (Colorado).——FIG. 83,4. **P. insignis*; outline of left mandible, ×26 (Emerson, 1933).

Termopsis HEER, 1849, p. 23 [**T. bremii*; SD HAGEN, 1858c, p. 74] [=*Xestotermopsis* VON ROSEN, 1913, p. 330, obj.]. Similar to *Zootermopsis* but having 5 hind tarsal segments visible above and below; humeral suture in fore wing only slightly curved. EMERSON, 1933. *Oligo.,*

Isoptera

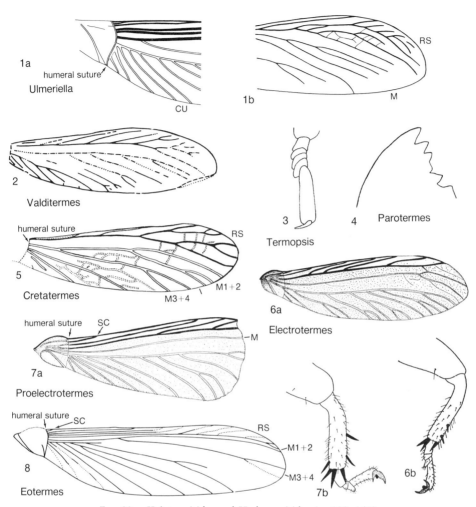

FIG. 83. Kalotermitidae and Hodotermitidae (p. 139–141).

Europe (Baltic).——FIG. 83,*3*. **T. bremii*; hind tarsus, ×7 (Emerson, 1933).

Ulmeriella MEUNIER, 1920a, p. 728 [**U. bauckhorni*; OD] [=*Diatermes* MARTYNOV, 1929, p. 178 (type, *D. cockerelli*)]. Vein RS with several branches directed posteriorly and terminating on hind margin. MARTYNOV, 1929; ZEUNER, 1938; STATZ, 1939–1940; SNYDER, 1949; WEIDNER, 1967, 1968b, 1971; EMERSON, 1968. *Oligo.,* Europe (Germany), USSR (Kazakh); *Mio.,* Europe (Germany), USA (Washington); *Plio.,* Europe (Germany).——FIG. 83,*1*. **U. bauckhorni*; *a*, base of fore wing, ×8; *b*, distal half of fore wing, ×6 (both Emerson, 1968).

Valditermes JARZEMBOWSKI, 1981, p. 92 [**V. brenanae*; OD]. Similar to *Cretotermes* but larger; fore wing more elongate; branching of RS and CU more complex. *Cret.,* England.——FIG. 83,*2*. **V. brenanae*; fore wing, ×5.5 (Jarzembowski, 1981).

Zootermopsis EMERSON, 1933, p. 182. [Generic assignment doubtful.] SCUDDER, 1890. *Oligo.,* USA (Colorado)–*Holo.*

Family RHINOTERMITIDAE Froggatt, 1896

[Rhinotermitidae FROGGATT, 1896, p. 518]

Wings commonly reticulate; vein R much reduced; RS unbranched; M usually approaching very close to CU or coalesced with it; frontal gland always present; left mandible with 3 marginal teeth; ocelli present; tarsi with 4 segments, cerci with 2. *Oligo.–Holo.*

Rhinotermes HAGEN, 1858, p. 233. *Holo.*

Coptotermes WASMANN, 1896, p. 629. SNYDER, 1960; EMERSON, 1971. *Oligo./Mio.,* Dominican Republic, Mexico (Chiapas)–*Holo.*

Heterotermes FROGGATT, 1896, p. 518. Adult. SNYDER, 1960. *Mio.,* Mexico (Chiapas)–*Holo.*

Parastylotermes SNYDER & EMERSON in SNYDER, 1949, p. 378 [**Stylotermes washingtonensis* SNYDER, 1931, p. 317; OD]. Similar to *Reticulitermes*, but wing membrane and veins almost without hairs; eyes relatively larger than in *Reticulitermes*; ocelli distinct; stump of fore wing (basal scale) proportionately large. SNYDER, 1950, 1955; PIERCE, 1958. *Oligo.,* Europe (Baltic); *Mio.,* USA (Washington, California).

Reticulitermes HOLMGREN, 1913, p. 60. ARMBRUSTER, 1941; WEIDNER, 1955, 1971; PIERCE, 1958; EMERSON, 1971. *Oligo.,* Europe (Baltic, Germany), USA (Colorado); *Mio.,* USA (California), Europe (Germany); *Plio.,* Europe (Germany)–*Holo.*

Rhinotermites ARMBRUSTER, 1941, p. 21 [**R. dzierzoni*; OD]. Little-known genus, based on wing and body fragments. EMERSON, 1971. *Mio.,* Europe (Germany).

Family TERMITIDAE
Westwood, 1840

[Termitidae WESTWOOD, 1840, p. 11]

Wings not conspicuously reticulate; vein R greatly reduced or absent; left mandible usually with 2 marginal teeth; frontal gland well developed; basal scale of fore wing always proportionately small. *Mio.–Holo.*

Termes LINNÉ, 1758, p. 609. *Holo.*

Gnathamitermes LIGHT, 1932, p. 390. PIERCE, 1958. *Mio.,* USA (California)–*Holo.*

Macrotermes HOLMGREN, 1909, p. 193. [Generic assignment of fossil doubtful.] CHARPENTIER, 1843; SNYDER, 1949. *Mio.,* Europe (Yugoslavia)–*Holo.*

Family UNCERTAIN

The following genera, apparently belonging to the order Isoptera, are too poorly known to permit family assignment.

Architermes HAUPT, 1956, p. 28 [**A. simplex*; OD]. Little-known wing. *Eoc.,* Europe (Germany).

Mastotermites ARMBRUSTER, 1941, p. 13 [**M. stuttgartensis*; OD]. Little-known genus, possibly a synonym of *Miotermes*. EMERSON, 1971. *Plio.,* Europe (Germany).

Metatermites ARMBRUSTER, 1941, p. 26 [**M. statzi*; OD]. Little-known genus. EMERSON, 1971. *Mio.,* Europe (Germany).

Order MANTEODEA
Burmeister, 1838

[Manteodea BURMEISTER, 1838 in BURMEISTER, 1838–1839, p. 517, as Mantodea]

Fore wings usually tegminous, strongly so in most, more rarely membranous; costa marginal, no precostal space; vein SC distinct, long, extending well beyond midwing; R strongly developed, terminating nearly at wing apex; RS arising distally, consisting of 1 or several distal branches, or commonly absent as distinct vein; M well developed, typically dividing near base into 2 main branches, which may represent MA and MP (SHAROV, 1962a); CUA apparently anastomosed with stem of posterior branch of M; CUP separating from CUA at wing base, nearly straight, unbranched; anal veins at least slightly sigmoidal; posterior part of anal area commonly expanded to form small, prominent lobe containing distal parts of several anal veins. Hind wings with slender remigium, anal area greatly expanded; RS unbranched, arising near wing base; M fused basally with stem of R; MP apparently diverging from R and anastomosing with CUA; MA continuing nearly straight, unbranched; CUA extensively developed, with several branches; CUP and 1A unbranched, nearly straight; several radiating anal veins. Antennae of moderate length, multisegmented; mouthparts mandibulate; forelegs raptorial, others cursorial; tarsi typically with 5 segments; pronotum not usually extending over head; prothorax commonly (but not invariably) elongate, forelegs attached near anterior end; ovipositor not usually developed externally but rarely protruding slightly; cerci usually conspicuous, multisegmented. *Oligo.–Holo.*

The Manteodea, although clearly related to the Orthoptera, are less specialized in some respects (e.g., five-segmented tarsi, segmented cerci). In all probability they are even more closely related to the Blattaria (McKITTRICK, 1964, 1965) but appear to have been derived independently from a protorthopterous stock and to have evolved

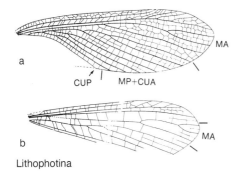

FIG. 84. Chaeteessidae (p. 143).

entirely as predators; the raptorial forelegs, present in all known species, show several types of modification. Although the oldest known Manteodea are from the Baltic amber (Oligocene), the group probably existed in the Mesozoic and even in the Permian. Some of the Late Carboniferous Protorthoptera are known to have possessed raptorial forelegs, but they do not appear to be related closely to the Manteodea.

The venation of the Manteodea is clearly orthopteroid but is characterized by such distinct features as the differences in RS in the fore and hind wings and the apparent anastomosis of MP and CUA in both wings. Convexities and concavities, as in the Orthoptera, are not distinct for all veins. Veins SC and CUP are markedly concave and R and CUA similarly convex; RS and M show no definite topography. That both MA and MP are present in the hind wing is suggested by the basal separation of veins that appear to be main branches of M, although other interpretations of these branches are possible. The evidence for the presence of MP in the fore wing is virtually nonexistent in the Manteodea and rests upon the similarity of that part of the fore wing to the corresponding part of the hind wing.

Family CHAETEESSIDAE Handlirsch, 1920

[Chaeteessidae HANDLIRSCH, 1920, p. 497]

Fore wing having vein R with several distinct, anteriorly pectinate branches distally; fore tarsus attached to distal end of tibia; tibia lacking terminal projecting hook. *Oligo.– Holo.*

Chaeteessa BURMEISTER, 1838, p. 527. [Generic assignment very dubious.] GIEBEL, 1862. *Oligo.*, Europe (Baltic)–*Holo.*

Lithophotina COCKERELL, 1908s, p. 343 [*L. floccosa*; OD]. Similar to *Chaeteessa* (recent) but with more pectinate branches on R in fore wing. SHAROV, 1962a. *Oligo.*, USA (Colorado).——FIG. 84. **L. floccosa; a,* fore wing and *b,* remigium of hind wing, ×2.5 (Cockerell, 1908s).

Family MANTEIDAE Saussure, 1859

[nom. correct. ROBERTS, 1941, p. 15, pro Mantidae SAUSSURE, 1859, p. 59]

Fore wing having R with 2 or fewer anteriorly pectinate branches distally; fore tibia extending beyond point of tarsal attachment, forming curved, projecting hook. *Oligo.– Holo.*

Mantis LINNÉ, 1758, p. 425. ZEUNER, 1931. *Mio.*, Europe (Germany)–*Holo.*

Eobruneria COCKERELL, 1913b, p. 343 [*E. tessellata*; OD]. Little-known genus, based on fragment of fore wing with broad costal area. [Possibly related to *Stagmomantis* (recent).] *Oligo.*, USA (Colorado).

Order PROTELYTROPTERA Tillyard, 1931

[Protelytroptera TILLYARD, 1931, p. 234] [=Protocoleoptera TILLYARD, 1924b, p. 434]

Small to medium-sized insects, related to the orthopteroids. Head small, eyes conspicuous; antennae prominent, moderately long, stout, multisegmented; pronotum broad, flattened, commonly with microtrichia laterally; legs robust, spiny, with 5 tarsal segments. Fore wings typically forming convex elytra (only rarely flat) with distinct venation in primitive forms and weak venation in specialized species; costal area expanded at base of wing, forming prominent, flattened lobe (costal expansion); veins SC, R, RS, M, CUA, CUP, and 3 anal veins present in more generalized forms; in more specialized species only basal parts of SC, RS, and CUP discernible; most species with submarginal thickening (sutural margin) parallel to posterior margin of elytron; cluster of small setae commonly present near subcosta and another along basal part of sutural margin; in Megelytridae microtrichia covering entire elytron. Hind wings typically longer and much broader than elytra; anal area expanded, with longitudinal and, in some families, transverse folding. Abdomen broad, terminating in short but prominent cerci about as long as last 3 abdominal segments. Females with short external ovipositor. Immature stages unknown. CARPENTER & KUKALOVÁ, 1964. *Perm.–Cret.*

The fore wings in this extinct order of elytrophorous insects resemble those of the Coleoptera, but the general nature of the venation of both fore and hind wings and the presence of prominent cerci indicate relationship with the orthopteroids, especially the Blattaria and Dermaptera. Although few species of Protelytroptera are known at present, their diversity suggests that the order was a large and varied group, at least during the Permian. The relatively recent discovery of the family Umenocoleidae in a Cretaceous deposit (see below) indicates that the order may have continued to diversify throughout the Mesozoic.

In the more primitive species the fore wings tended to be tegminous and almost flat, with the costal expansion small, the sutural margin absent or weakly developed, and the crossveins numerous (e.g., Archelytridae, Apachelytridae). In the more specialized forms, in which the fore wings were convex and heavily sclerotized, the main veins were reduced or obsolescent distally, crossveins were virtually absent, and the surfaces of the elytra were granulate or rugose (e.g., Protelytridae, Permelytridae, Planelytridae, Umenocoleidae).

The hind wings, which are known in four families (Archelytridae, Protelytridae, Permelytridae, Apachelytridae), had a narrow remigium and a well-developed anal fan. However, there were substantial differences in the structure of the wings in these families. Those of the Protelytridae could fold at rest transversely as well as longitudinally (TILLYARD, 1931; CARPENTER, 1933a), but those of the Permelytridae and Apachelytridae could apparently fold only longitudinally (CARPENTER & KUKALOVÁ, 1964).

The body structure is known, very incompletely, in the families Protelytridae, Permelytridae, and Apachelytridae.

Family ELYTRONEURIDAE Carpenter, 1933

[Elytroneuridae CARPENTER, 1933a, p. 478]

Elytron nearly flat, not convex; sutural margin apparently absent; vein SC branched; M and CUA fused for considerable distance beyond wing base. *Perm.*

Elytroneura CARPENTER, 1933a, p. 478 [*E. permiana*; OD]. Costal margin convex, costal expansion very prominent; SC remote from anterior wing margin. *Perm.*, USA (Kansas).——FIG. 85,4. *E. permiana*; elytron, ×4 (Carpenter, 1933a).

Family ARCHELYTRIDAE Carpenter, 1933

[Archelytridae CARPENTER, 1933a, p. 477]

Elytron slightly convex; costal expansion weakly developed; stems of main veins independent; vein SC long, terminating about two-thirds of wing length from base; SC, R, M, CUP, and 1A unbranched; CUP strongly concave; sutural margin well developed, terminating before apex; weak crossveins over entire wing. *Perm.*

Archelytron CARPENTER, 1933a, p. 477 [*A. superbum*; OD]. Costal margin strongly arched. Vein SC weakly developed at base but strong distally; RS arising at level of termination of SC; CUA and CUP diverging about one-fifth wing length from base; CUA with 3 terminal branches; most crossveins unbranched. [The generic name *Archelytron* was subsequently proposed by HAUPT (1952, p. 248) for a Permian species, *priscus*, that he assigned to the Coleoptera. However, Dr. Jörg SCHNEIDER of the Bergakademie Freiberg, Germany, who recently examined the unique specimen for me, found that it is a plant fragment, not an insect. No homonymy is, therefore, involved.] CARPENTER & KUKALOVÁ, 1964. *Perm.*, USA (Kansas).——FIG. 85,2. *A. superbum*; fore wing, ×8 (Carpenter & Kukalová, 1964).

Ortelytron KUKALOVÁ, 1965, p. 66 [*O. europaeum*; OD]. Similar to *Archelytron,* but fore wing with much shorter sutural margin and with crossveins

Protelytroptera

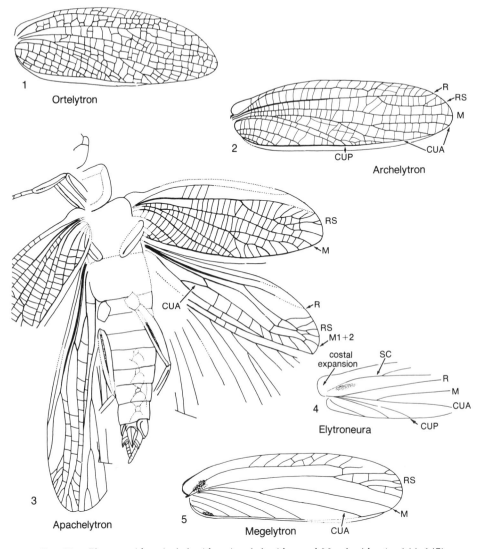

FIG. 85. Elytroneuridae, Archelytridae, Apachelytridae, and Megelytridae (p. 144–147).

more irregular. Hind wing with CUP and 1A close together and parallel from wing base to distal margin. *Perm.*, Europe (Czechoslovakia).——FIG. 85,*1*. **O. europaeum*; fore wing, ×14 (Kukalová, 1965).

Family APACHELYTRIDAE
Carpenter & Kukalová, 1964

[Apachelytridae CARPENTER & KUKALOVÁ, 1964, p. 185]

Related to Archelytridae. Fore wing weakly tegminous, similar to that of Archelytridae, but costal expansion larger and crossveins more numerous and virtually all unbranched. Hind wing with expanded anal area as in Protelytridae; stem of vein M independent of R; M forked near midwing; CUP coalesced with 1A. Body little known; eyes relatively small; pronotum rectangular; forelegs short, hind legs much longer, with well-developed femora; tarsi short, segmented. *Perm.*

Apachelytron CARPENTER & KUKALOVÁ, 1964, p. 187 [**A. transversum*; OD]. Costal margin of fore wing arched; RS arising slightly beyond

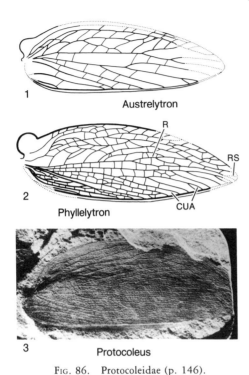

Fig. 86. Protocoleidae (p. 146).

midwing. *Perm.*, Europe (Moravia).——Fig. 85,*3*. **A. transversum*; holotype as preserved, ×9 (Carpenter & Kukalová, 1964).

Family PROTOCOLEIDAE
Tillyard, 1924

[Protocoleidae Tillyard, 1924b, p. 434]

Fore wing tegminous, only slightly convex; anterior margin strongly arched; sutural margin nearly straight and bordering entire posterior margin; wing surface granulate and with tubercles at least in some areas; setae commonly present in subcostal area; costal expansion large; vein SC long with several to many branches; RS arising near midwing; M and CUA with several branches, variable in form. Hind wing and body unknown. *Perm.*

Protocoleus Tillyard, 1924b, p. 434 [**P. mitcheli*; OD]. Fore wing uniformly covered with flat tubercles; main veins and their branches parallel with longitudinal axis of wing; RS arising near midwing. Kukalová, 1966. *Perm.*, Australia (New South Wales).——Fig. 86,*3*. **P. mitcheli*; fore wing, holotype, ×3.2 (Tillyard, 1924b).

Austrelytron Kukalová, 1966, p. 96 [**A. tillyardi*; OD]. Similar to *Protocoleus* but with fewer branches of main veins and crossveins; tubercles few and pointed. *Perm.*, Australia (New South Wales).——Fig. 86,*1*. **A. tillyardi*; fore wing, ×5 (Kukalová, 1966).

Phyllelytron Kukalová, 1966, p. 94 [**P. folium*; OD]. Similar to *Protocoleus,* but main veins and their branches very irregular, not aligned or parallel with longitudinal axis of wing; granulation of wing surface coarse. *Perm.*, Australia (New South Wales).——Fig. 86,*2*. **P. folium*; fore wing, ×2.3 (Kukalová, 1966).

Family PROTELYTRIDAE
Tillyard, 1931

[Protelytridae Tillyard, 1931, p. 235]

Elytron convex, anterior margin strongly arched; vein SC short, not extending beyond midwing; venation and sutural margin well developed; RS and CUP unbranched; M free from CUA or coalesced with it basally for a short interval; 3 or 4 anal veins. Hind wing: anal area with about 10 anal veins; stem of M coalesced with R. Body apparently flattened; antenna well developed, with short, thick segments. *Perm.*

Protelytron Tillyard, 1931, p. 239 [**P. permianum*; OD]. Vein CUA unbranched; patches of setae along SC and basal part of sutural margin. *Perm.*, USA (Kansas).——Fig. 87,*3a*. *P. furcatum* Carpenter; elytron, ×8.8 (Carpenter, 1933a).——Fig. 87,*3b*. **P. permianum*; reconstruction, ×5.4 (Carpenter, 1933a).

Permelytropsis Carpenter, 1933a, p. 474 [**P. cubitalis*; OD]. CU unbranched. *Perm.*, USA (Kansas).——Fig. 87,*1*. **P. cubitalis*; elytron, ×10 (Carpenter, 1933a).

Uralelytron Rohdendorf, 1939, p. 506 [**U. martynovi*; OD]. Little-known elytron and body fragments. [Family assignment uncertain.] *Perm.*, USSR (Asian RSFSR).——Fig. 87,*2*. **U. martynovi*; elytron, ×8 (Rohdendorf, 1939).

Family MEGELYTRIDAE
Carpenter, 1933

[Megelytridae Carpenter, 1933a, p. 476]

Fore wing flat except for basal part of costal area; costal expansion very small; vein R very strong; RS arising in distal part of wing, branched; CUA and M coalesced basally, both unbranched; sutural margin complete; several oblique crossveins in costal area; entire wing with a dense covering of fine hair. Hind wing unknown. *Perm.*

Megelytron TILLYARD, 1931, p. 247 [*M. robustum*; OD]. Vein RS with 4 terminal branches and several twigs; one cluster of setae at base of subcosta and another near inner margin of anal area. CARPENTER & KUKALOVÁ, 1964. *Perm.*, USA (Kansas).——FIG. 85,5. *M. robustum*; fore wing, ×5 (Carpenter & Kukalová, 1964).

Family PLANELYTRIDAE Kukalová, 1965

[Planelytridae KUKALOVÁ, 1965, p. 75]

Fore wing almost flat; anterior margin strongly arched; costal expansion well developed; vein SC extending about two-thirds of wing length from base; veins M and CUA coalesced for about one-third wing length from base; sutural margin well developed. *Perm.*

Planelytron KUKALOVÁ, 1965, p. 75 [*P. planum*; OD]. SC strongly arched in costal area; weak crossveins in subcostal area. *Perm.*, Europe (Czechoslovakia).——FIG. 88,2. *P. planum*; fore wing, ×8 (Kukalová, 1965).

Family PERMELYTRIDAE Tillyard, 1931

[Permelytridae TILLYARD, 1931, p. 246] [=Blattelytridae TILLYARD, 1931, p. 249; Acosmelytridae TILLYARD, 1931, p. 252]

Fore wing convex; costal margin arched; veins weakly developed, commonly obsolescent distally; vein RS absent; MA commonly free from CU, rarely coalesced with it; sutural margin normally developed. Hind wing apparently like that of *Protelytron*, but SC much longer and terminating on RS. Body little known; head smaller than in Protelytridae and Apachelytridae; antenna, as preserved, about as long as abdomen; cerci short, segmented. *Perm.*

Permelytron TILLYARD, 1931, p. 246 [*P. schucherti*; OD]. Vein M of fore wing not coalesced with CUA. CARPENTER & KUKALOVÁ, 1964. *Perm.*, USA (Kansas).——FIG. 88,4. *P. schucherti*; elytron, ×6.7 (Carpenter, 1939).

Blattelytron TILLYARD, 1931, p. 250 [*B. permianum*; OD]. Little-known genus, based on fragment of elytron. *Perm.*, USA (Kansas).—— FIG. 88,5. *B. permianum*; elytron, ×5 (Tillyard, 1931).

Parablattelytron TILLYARD, 1931, p. 251 [*P. sub-

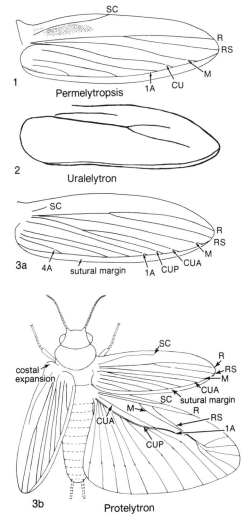

FIG. 87. Protelytridae (p. 146).

incisum; OD] [=*Acosmelytron* TILLYARD, 1931, p. 253 (type, *A. elongatum*)]. Vein CUA coalesced with M basally; main veins commonly obsolescent in distal half of wing. *Perm.*, USA (Kansas).——FIG. 88,6. *P. subincisum*; dorsal view as preserved, ×7 (Carpenter & Kukalová, 1964).

Family PERMOPHILIDAE Tillyard, 1924

[Permophilidae TILLYARD, 1924b, p. 430]

Fore wing tegminous, slightly convex; sutural margin distinct but narrow; wing surface with granulation and tubercles; costal

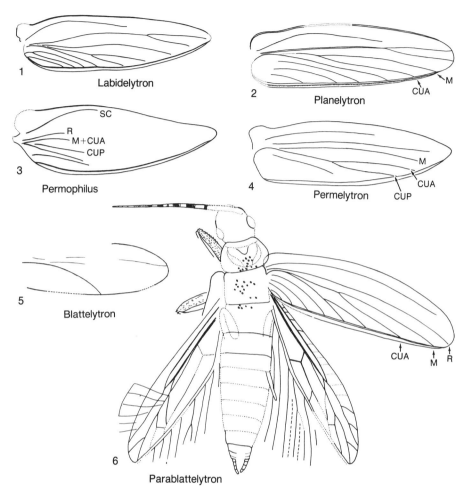

Fig. 88. Planelytridae, Permelytridae, Permophilidae, and Labidelytridae (p. 147–148).

expansion prominent; costal margin arched but asymmetrical; apex acute; main veins present in basal half of wing only. [Originally placed in Coleoptera.] *Perm.*

Permophilus TILLYARD, 1924b, p. 430 [*P. pincombei*; OD]. Fore wing with very narrow sutural margin; wing surface with dense granulation and indistinct tubercles; wing strongly narrowed in distal half. KUKALOVÁ, 1966. *Perm.,* Australia (New South Wales).——FIG. 88,*3. P. hirtus* KUKALOVÁ; fore wing, ×8 (Kukalová, 1966).

Elytrathrix KUKALOVÁ, 1966, p. 102 [*E. hirsuta*; OD]. Similar to *Permophilus* but with conspicuous tubercles and setae in basal half of wing, including costal expansion. *Perm.,* Australia (New South Wales).

Family LABIDELYTRIDAE Kukalová-Peck, 1988

[Labidelytridae KUKALOVÁ-PECK, 1988, p. 339, *nom. subst. pro* Stenelytridae KUKALOVÁ, 1966, p. 102]

Fore wing tegminous, nearly flat, long and slender; apex pointed; surface finely granulate; costal expansion large; venation as in Protelytridae. *Perm.*

Labidelytron KUKALOVÁ-PECK, 1988, p. 339, *nom. subst. pro Stenelytron* KUKALOVÁ, 1966, p. 102, *non* HANDLIRSCH, 1906 [*Stenelytron enervatum* KUKALOVÁ, 1966; OD]. Vein M of fore wing unbranched, not coalesced with CU basally. *Perm.,* Australia (New South Wales).——FIG. 88,*1. *L. enervatum*; fore wing, ×4 (Kukalová, 1966).

Xenelytron KUKALOVÁ, 1966, p. 105 [*X. ligula*; OD]. Similar to *Stenelytron,* but M coalesced with CU in basal half of wing. *Perm.,* Australia (New South Wales).

Family DERMELYTRIDAE Kukalová, 1966

[Dermelytridae KUKALOVÁ, 1966, p. 105]

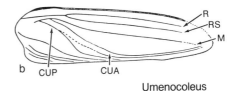

Umenocoleus

Fore wing convex, apparently weakly sclerotized; anterior margin convex; sutural margin well developed; venation much reduced, at most with only basal parts of veins R, CU, and A present. *Perm.*

Dermelytron KUKALOVÁ, 1966, p. 106 [*D. conservativum*; OD]. Fore wing oval, apex directed posteriorly; costal expansion small. *Perm.,* Australia (New South Wales).

Chanoselytron KUKALOVÁ, 1966, p. 108 [*C. gingiva*; OD]. Similar to *Dermelytron,* but costal expansion much larger. *Perm.,* Australia (New South Wales).

Psychelytron KUKALOVÁ, 1966, p. 108 [*P. progressivum*; OD]. Fore wing as in *Dermelytron,* but apex directed anterolaterally. *Perm.,* Australia (New South Wales).

Family UMENOCOLEIDAE Chen & T'an, 1973

[Umenocoleidae CHEN & T'AN, 1973, p. 174]

Elytron apparently only slightly convex, elongate, with well-developed longitudinal veins; vein SC very close to and paralleling anterior margin of wing; R nearly parallel to SC; RS arising about one-fifth of wing length from base; stems of M and CU coalesced; M diverging from common stem just before level of origin of RS, parallel to RS; CUA diverging posteriorly as far as midwing, then continuing parallel to posterior margin of wing; CUP apparently forked, with an anterior branch directed toward posterior margin of wing, then continuing parallel to CUA; 4 short anal veins; crossveins apparently absent, but surface of wing finely granulate. Hind wing little known, extending a short distance beyond end of fore wing. Body little known; antennae filiform, with at least 16 segments; pronotum broader than long, coarsely granulate. *Cret.*

The remarkable fossil on which this genus

FIG. 89. Umenocoleidae (p. 150).

and family is based was placed by its authors in the order Coleoptera. However, the general structure of the elytra and of their venation in particular is so much like that of the Protelytroptera that I transferred the insect to that order. The filiform and segmented nature of the antennae and the peculiar vena-

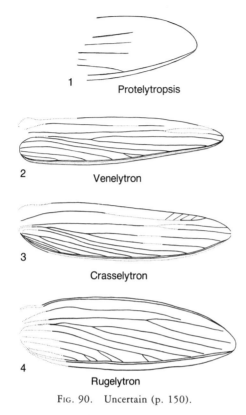

FIG. 90. Uncertain (p. 150).

tional pattern are not at all like those of the Coleoptera, as noted by the authors. Of course, since no Protelytroptera have been reported from deposits later than the Permian, this Cretaceous fossil extends the range of the order at least a hundred million years. It also suggests the possibility that some of the isolated and poorly preserved elytra found in Triassic or Jurassic deposits and identified as Coleoptera may actually be Protelytroptera.

Umenocoleus CHEN & T'AN, 1973, p. 169 [*U. sinuatus; OD]. RS and CUA unbranched. Cret., China (Kansu).——FIG. 89. *U. sinuatus; a, dorsal view of holotype, ×7.0 (Chen & T'an, 1973); b, elytron, ×4.5 (after Chen & T'an, 1973).

Family UNCERTAIN

The following genera, apparently belonging to the order Protelytroptera, are too poorly known to permit assignment to families.

Artocoleus MARTYNOV, 1933b, p. 78 [*A. ivensis; OD]. Little-known elytron. [Ordinal assignment doubtful.] Perm., USSR (European RSFSR).

Crasselytron KUKALOVÁ, 1965, p. 70 [*C. convexum; OD]. Fore wing very convex, slender and long; anterior margin nearly straight; SC extending nearly to wing apex; M coalesced with CUA for about one-third of wing length from base; wing surface granulate; sutural margin very narrow. Perm., Europe (Czechoslovakia).——FIG. 90,3. *C. convexum; fore wing, ×7 (Kukalová, 1965).

Glabelytron KUKALOVÁ, 1965, p. 77 [*G. lativenosum; OD]. Fore wing flat; anterior margin arched; SC sigmoidal, unbranched, extending about two-thirds of wing length. Perm., Europe (Czechoslovakia).

Protelytropsis TILLYARD, 1931, p. 245 [*P. grandis; OD]. Distal fragment of large elytron. Perm., USA (Kansas).——FIG. 90,1. *P. grandis; elytron, ×7 (Tillyard, 1931).

Rugelytron KUKALOVÁ, 1965, p. 72 [*R. fuscum; OD]. Fore wing convex, relatively long; anterior margin arched; sutural margin well developed, extending to wing apex; M coalesced with CUA for about one-third of wing length from base; wing surface granulate. Perm., Europe (Czechoslovakia).——FIG. 90,4. *R. fuscum; fore wing, ×7 (Kukalová, 1965).

Venelytron KUKALOVÁ, 1965, p. 73 [*V. tuberculatum; OD]. Fore wing long and narrow; anterior margin nearly staight; SC long, extending nearly to apex; M and CUA coalesced for about one-third wing length from base; posterior margin of wing slightly concave distally; wing surface granulate. Perm., Europe (Czechoslovakia).——FIG. 90,2. *V. tuberculatum; fore wing, ×4.5 (Kukalová, 1965).

Order DERMAPTERA de Geer, 1773

[Dermaptera DE GEER, 1773, p. 399]

Head broad, with mandibulate mouthparts and conspicuous antennae, consisting of at least 10 segments, usually many more; compound eyes commonly very large; ocelli absent in recent species; fore wings forming short, convex tegmina or elytra, typically lacking veins; hind wings semicircular, with greatly expanded anal area; remigium much reduced, with at most vestiges of veins SC, R, M, and CU; at rest, hind wings folded radially and also transversely beneath tegmina; hind wings commonly absent; abdomen usually broad, first tergum fused with metathorax; in typical species (suborder Forficulina) cerci forming pair of heavily sclerotized, unsegmented forceps; ovipositor present in primitive species, absent in others. Immature stages similar to adults in general characteristics; cerci usually styliform. Subsocial habits in several genera. Most species omnivorous. *Jur.–Holo.*

The Dermaptera share many features of the Orthoptera and Blattaria, and they almost certainly arose from related stock. However, their peculiarly modified wings and thorax indicate that they belong to a widely divergent line.

The most distinctive characteristics of the Dermaptera are found in the modifications of their wings and cerci. Although the tegmina of existing species lack veins, those of most Jurassic species (suborder Archidermaptera) have veins that are apparently homologous with R, RS, M, CUA, and CUP of other insects (VISHNIAKOVA, 1980a). Also, two Jurassic genera belonging to the existing family Pygidicranidae, considered to be the most primitive of existing families, have several simple veins in the tegmina. The hind wings are membranous and when expanded are large and semicircular. The anterior half of the remigium is at least slightly sclerotized, forming a leathery scale; several of the main veins seem to have been lost in the sclerotization, only their basal parts persisting (Fig. 91). The rest of the hind wing is supported by a series of radiating veins, which appear to arise from a fulcrum at the distal end of the scale (Fig. 91). The complicated folding of these hind wings has been described in detail by MARTYNOV (1938b) and VERHOEFF (1917).

The heavily sclerotized forceps, which are modified cerci, show much diversity of form and size in recent Dermaptera. Segmentation of the forceps is not visible in adults of recent Dermaptera but is clearly indicated in the nymphs of some of the Pygidicranidae. The cerci of the adults of the Archidermaptera, all Jurassic, are very diverse in form, some being long and setaceous, with as many as 40 segments (VISHNIAKOVA, 1980a). In oth-

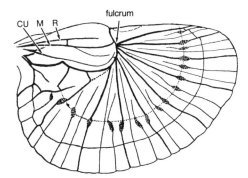

FIG. 91. Dermaptera; expanded hind wing of *Forficula auricularia* LINNÉ, Holocene (Bey-Bienko, 1936).

ers they are shorter, with a tendency for the basal segments to be more sclerotized and coalesced, forming incipient forceps (see Fig. 92,3).

The order Dermaptera is generally considered to consist of four suborders: Archidermaptera, Forficulina, Arixeniina, and Diploglossata. The last two, which include only a very few species, have no geological record; they are apterous, with short, styliform, unsegmented cerci, and are associated with bats (Arixeniina) or are ectoparasites of rodents (Diploglossata). The Archidermaptera include the most generalized members of the order and are known at present only from the Jurassic. The Forficulina, consisting of several families of typical earwigs, extend back to the Jurassic.

Suborder ARCHIDERMAPTERA
Bey-Bienko, 1936

[Archidermaptera BEY-BIENKO, 1936, p. 215]

Tarsi with 4 or 5 segments; tegmina with distinct but much reduced venation; cerci commonly long, slender, multisegmented, rarely short. Hind wings unknown. *Jur.*

Family PROTODIPLATYIDAE
Martynov, 1925

[Protodiplatyidae MARTYNOV, 1925b, p. 573] [=Protodiplatidae ROHDENDORF, 1957, p. 78, unjustified emendation]

Antenna filiform, with 17 to 23 segments, the first segment enlarged and second at least

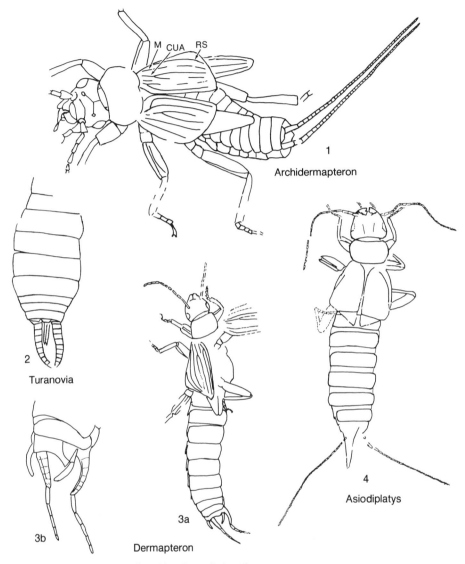

FIG. 92. Protodiplatyidae (p. 152–153).

as long as third; ocelli present; pronotum transverse; tegmina relatively long, apically dilated, and with veins RS and M unbranched, CUA and A forked, and CUP reduced; tarsi long; fore and middle tarsi with 4 segments, hind tarsi with 5; pretarsi with 2 claws and arolium; abdomen with 10 visible segments. Females with prominent, external ovipositor. *Jur.*

Protodiplatys MARTYNOV, 1925b, p. 573 [**P. fortis*; OD]. Head small; antenna with 17 to 19 segments; tegmina broad, not extending beyond second abdominal segment; cerci about half as long as body, with no more than 40 segments. VISHNIAKOVA, 1980a. *Jur.,* USSR (Kazakh).

Archidermapteron VISHNIAKOVA, 1980a, p. 82 [**A. martynovi*; OD]. Similar to *Protodiplatys* but with much larger and longer tegmina; cerci nearly as long as body, with more than 40 segments. *Jur.,* USSR (Kazakh).——FIG. 92,*1*. **A. martynovi*; lateral view of holotype as preserved, ×2.5 (after Vishniakova, 1980a).

Asiodiplatys VISHNIAKOVA, 1980a, p. 85 [**P. speciosus*; OD]. Similar to *Protodiplatys,* but head larger; antenna with 22 segments; tegmina apically truncate in straight line; cerci about half length of body. *Jur.*, USSR (Kazakh).——FIG. 92,4. **A. speciosus*; dorsal view of holotype as preserved, ×5.5 (after Vishniakova, 1980a).

Dermapteron MARTYNOV, 1925b, p. 575 [**D. incerta*; OD]. Similar to *Protodiplatys*; anterior margin of pronotum concave, posterior margin slightly convex; femora without spines; cerci short, with only 6 segments and only about one-fifth length of body; basal segment of cerci enlarged and falciform, with vestigial segmentation; ovipositor short. VISHNIAKOVA, 1980a. *Jur.*, USSR (Kazakh).——FIG. 92,3. **D. incerta; a*, paratype as preserved, ×3; *b*, apex of abdomen, ×6 (both after Vishniakova, 1980a).

Microdiplatys VISHNIAKOVA, 1980a, p. 85 [**V. campodeiformis*; OD]. Similar to *Protodiplatys* but smaller; antenna with 19 segments; pronotum with lateral margins almost parallel; cerci as long as body. *Jur.*, USSR (Kazakh).

Turanovia VISHNIAKOVA, 1980a, p. 88 [**T. incompleta*; OD]. Similar to *Dermapteron*, but anterior and posterior margins of pronotum nearly parallel; cerci short, about one-sixth length of body, weakly curved and converging, consisting of 9 more or less coalesced segments. *Jur.*, USSR (Kazakh).——FIG. 92,2. **T. incompleta*; distal part of abdomen and cerci, ×7 (after Vishniakova, 1980a).

Suborder FORFICULINA Newman, 1835

[Forficulina NEWMAN, 1835, p. 424]

Tarsi with 3 segments; cerci forming heavy forceps, without segmentation in adults; eyes well developed. *Jur.–Holo.*

Family PYGIDICRANIDAE Verhoeff, 1902

[Pygidicranidae VERHOEFF, 1902, p. 188]

Head depressed, truncate, concave, not emarginate posteriorly; femora commonly compressed and carinulate; body typically pubescent; tarsi commonly simple. Cerci of nymphs of two subfamilies segmented. [A diverse family generally considered to be the most primitive of the existing families of the order.] *Jur.–Holo.*

Pygidicrana SERVILLE, 1831, p. 30. *Holo.*
Semenoviola MARTYNOV, 1925c, p. 74 [**S. obliquotruncata*; OD]. Head large, transverse, posterior margin concave; antenna with 11 moniliform segments, first 2 segments of nearly identical length; ocelli present; tegmina with unbranched veins RS, M, and CUA; vein A forked; cerci short, strongly curved; ovipositor short, external. VISHNIAKOVA, 1980a. *Jur.*, USSR (Kazakh).

Semenovioloides VISHNIAKOVA, 1980a, p. 92 [**S. capitatus*; OD]. Similar to *Semenoviola* but larger; anterior margin of pronotum concave. *Jur.*, USSR (Kazakh).

Turanoderma VISHNIAKOVA, 1980a, p. 92 [**T. sepultum*; OD]. Similar to *Semenoviola*, but tegmina widened distally and truncate apically in a straight line; antenna with 12 segments; cerci short, strongly falciform. *Jur.*, USSR (Kazakh). ——FIG. 93,2. **T. sepultum*; dorsal view of holotype as preserved, ×3.7 (after Vishniakova, 1980a).

Family LABIDURIDAE Verhoeff, 1902

[Labiduridae VERHOEFF, 1902, p. 189]

Body usually convex; femora not flattened or carinulate; cerci of nymphs not segmented. *Paleoc.–Holo.*

Labidura LEACH, 1815, p. 118. COCKERELL, 1920e; ZEUNER, 1962b. *Mio.*, Asia (Burma); *Pleist.*, St. Helena–*Holo.*

Carcinophora SCUDDER, 1876b, p. 291. COCKERELL, 1925e; BOGACHEV, 1940. *Paleoc.–Plio.*, Argentina; *Mio.*, Europe (Germany)–*Holo.*

Labiduromma SCUDDER, 1890, p. 203 [**L. avia*; SD TOWNES, 1945, p. 350]. First segment of anterior tarsus stout and swollen; forceps very broad. [Family assignment doubtful.] COCKERELL, 1924a; BROWN, 1984. *Oligo.*, USA (Colorado). ——FIG. 93,1 **L. avia*; whole insect, ×2 (Scudder, 1890).

Family FORFICULIDAE Verhoeff, 1902

[Forficulidae VERHOEFF, 1902, p. 190]

Body usually moderately flattened; antenna with 12 to 15 cylindrical or subcylindrical segments; elytra commonly present; legs short, flattened; second tarsal segment dilated on each side; abdomen with parallel sides, rarely tapering or dilated; forceps flattened or cylindrical. *Eoc.–Holo.*

Forficula LINNÉ, 1758, p. 423. BOGACHEV, 1940. *Eoc.*, Europe (Italy); *Mio.*, Europe (Germany) –*Holo.*

Suborder UNCERTAIN

The following genera, apparently belonging to the order Dermaptera, are too poorly known to permit assignment to suborders.

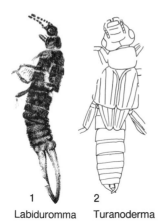

1 Labiduromma 2 Turanoderma

FIG. 93. Pygidicranidae and Labiduridae (p. 153).

Mesoforficula PING, 1935, p. 107 [**M. sinkianensis*; OD]. Little-known insect with short antennae and elytra. *Jur.,* China (Xinjiang).

Sinolabia ZHOU & CHEN, 1983, p. 62 [**S. longyouensis*; OD]. Little-known genus, based on poorly preserved specimen, apparently lacking tegmina. *Cret.,* China (Zhejing).

Order ORTHOPTERA
Olivier, 1789

[*nom. transl. et correct.* OLIVIER, 1811, p. 550, *ex* Orthopteres OLIVIER, 1789a, p. 12] [=*Pruvostitoptera* ZALESSKY, 1928b, p. 381]

Fore wings typically tegminous, rarely membranous; costa submarginal, precostal area usually prominent; vein SC generally extending beyond midwing, with series of oblique veinlets directed to anterior margin beyond C; R with oblique veinlets or definite terminal branches; RS arising from R, usually having several branches; M typically with at least 2 main branches, which in some families may represent MA and MP; CUA well developed, commonly anastomosed with M or its branches; CUP and 1A nearly straight, unbranched. Hind wings membranous, with slender remigium and expanded anal lobe; costa reduced, usually marginal; anal lobe including several to many radiating anal veins. Crossveins usually well developed and numerous, in many forming reticulation, which may develop into series of weak intercalary veins. Fore wings or both fore and hind wings may be reduced or completely absent. Mouthparts mandibulate; antennae well developed, commonly long; prothorax prominent; tarsi usually with from 3 to 5 segments; hind legs modified for jumping; female usually with ovipositor; cerci small, usually inconspicuous and unsegmented. Stridulatory organs usually present (generally alary or femoroalary) at least in males; tympanal organs on either abdomen or fore tibiae. *U. Carb.–Holo.*

As usually treated and as presented here, the order Orthoptera includes only the saltatorial orthopteroids. Lack of knowledge of the leg structure of a few Permian orthopteroids has made their ordinal positions uncertain. In these cases I have accepted SHAROV'S conclusions, as given in his detailed account of the phylogeny of the orthopteroids (1968).

The venation of the Orthoptera has some controversial features. The topography (i.e., convexity or concavity) of veins C, SC, R, and CUP is retained, but the branches of RS and M, as well as of CUA, are usually flat or neutral in the fore wings. In certain extinct families, such as the Oedischiidae, however, MP is clearly preserved as a strongly concave vein and CUA as convex. MA does not occur in any known Orthoptera as a distinctly convex vein; its presence, as in the Protorthoptera, can only be assumed on the basis of the proximal position of the first fork of M. One area of controversy is the relationship between M and CUA. In the fore wings of most Orthoptera there is some kind of connection between CUA and M (Fig. 94, *Oedischia williamsoni*; see also Fig. 95,*4b*). In others, CUA curves anteriorly from the stem of CU near the base of the wing and anastomoses with the stem of M for a brief interval before diverging posteriorly (see Fig. 110,*2, Mesoedischia madygenica*). In most families, CUA has merged with M at the very base of the wing and is usually no longer visible as a distinct vein (see Fig. 103,*1, Hagla gracilis*). Also, as noted by SHAROV (1968), some crossveins have tended to become relatively thick, functioning as struts (as in the Odonata), especially in the fore wings of the males,

Fig. 94. Orthoptera; fore wing and hind leg of *Oedischia williamsoni*, Upper Carboniferous of France, ×1.6 (Carpenter, new).

in which the venation has been much modified by the development of the stridulatory apparatus.

The evolution of the Orthoptera has apparently involved (1) increasing specializations of the fore wings as tegmina or wing covers, (2) development of stridulatory organs on the fore wings at least of the males, (3) expansion of the anal lobes of the hind wings, and (4) development of tympanal organs. The ability of the Orthoptera to jump and to stridulate has placed them among the most conspicuous of the existing insects. The saltatorial legs were well developed in the Oedischiidae of the Upper Carboniferous (Fig. 94). Stridulatory organs were thought by SHAROV (1968) to have been present on the wings of some oedischiids; they were obviously well developed on the wings of several Triassic genera of the related, existing family Haglidae (see Figs. 104,*1, Archihagla* and 106,*2, Protshorkuphlebia*).

The order Orthoptera is here divided into two suborders, Ensifera and Caelifera.

Suborder ENSIFERA
Chopard, 1920

[Ensifera CHOPARD, 1920, p. 56]

Antennae long, filiform, commonly longer than body and consisting of at least 30 segments; tympanal (auditory) organs, when present, located on fore tibiae; stridulatory structures (if present) on the overlapping, horizontal part of the fore wings in resting position; ovipositor, when present, sword-shaped. *U. Carb.–Holo.*

The morphological features included in the diagnosis of the Ensifera are only rarely preserved in fossils. However, there are other

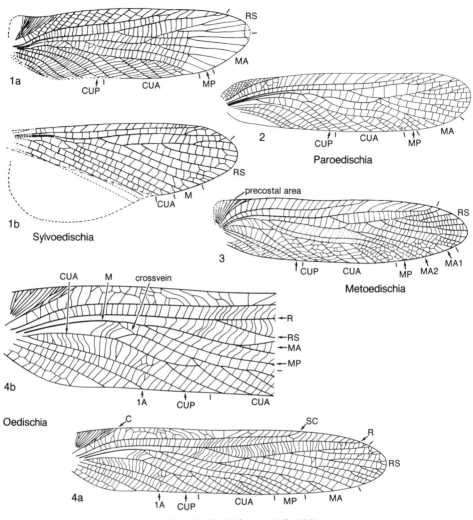

Fig. 95. Oedischiidae (p. 157–158).

characteristics, especially in the venational patterns, that are sufficiently associated with the body structures mentioned to enable suborder classification in most instances.

Family OEDISCHIIDAE Handlirsch, 1906

[Oedischiidae Handlirsch, 1906a, p. 700] [=Anhomalophlebiidae Handlirsch, 1919b, p. 547; Pruvostitidae Zalessky, 1928b, p. 381]

Fore wing weakly coriaceous; precostal area well developed; vein SC extending well beyond midwing; RS arising at about midwing, with several branches; stem of M independent basally, dividing into MA and MP after a short interval; MA with a distinct anterior branch (MA1) diverging toward RS and commonly at least touching it; MP commonly unbranched; CUA and CUP separating close to wing base; CUA1 directed toward stem of M and typically connected to it by a thickened crossvein, just before origin of MP; CUA with several long branches; CUP and anal veins unbranched. Hind wing incompletely known (in *Sylvoedischia* and *Macroedischia*); remigium about same size as in fore wing; MA not anastomosed with

RS; anal area unknown. Body known only in *Oedischia* and *Sylvoedischia*; head hypognathous; pronotum long; legs long, posterior pair with enlarged femora (Fig. 94), tarsi with 5 segments; cerci and ovipositor short. *U. Carb.–Perm.*

The oedischiids are generally considered to be the most primitive of the known Orthoptera. At present thirteen genera are known, two from the Upper Carboniferous and the rest from Permian strata. It should be noted here that I have excluded from the Oedischiidae the Triassic genus *Mesoedischia* SHAROV, which seems to me to represent a distinct family. It is here placed in the category of family Uncertain.

Oedischia BRONGNIART, 1885a, p. 58 [*O. williamsoni*; OD]. Fore wing with SC terminating at about one-quarter wing length from apex; MA1 touching or nearly touching RS; MP branched. ZEUNER, 1939; CARPENTER, 1966; SHAROV, 1968. *U. Carb.*, Europe (France).——FIG. 94. *O. williamsoni*; fore wing and hind leg, ×1.6 (Carpenter, new).——FIG. 95,4. *O. williamsoni*; a, fore wing, ×1.0; b, proximal half of fore wing, ×1.7 (both Carpenter, new).

Anhomalophlebia HANDLIRSCH, 1919b, p. 547 [*Homalophlebia couloni* MEUNIER, 1911a, p. 128; OD]. Fore wing as in *Oedischia* but relatively shorter and broader; MP unbranched; MA1 not quite reaching RS. [Type of family Anhomalophlebiidae HANDLIRSCH, 1919b.] *U. Carb.*, Europe (France).——FIG. 96,1. *A. couloni* (MEUNIER); fore wing, ×1.6 (Carpenter, new).

Jasvia ZALESSKY, 1934, p. 150 [*J. reticulata*; OD]. Little-known genus, apparently similar to *Oedischia*, but crossveins forming a dense reticulation over most of wing; MP unbranched. *Perm.*, USSR (Asian RSFSR).——FIG. 96,2. *J. reticulata*; fore wing, ×1.0 (Zalessky, 1934).

Macroedischia SHAROV, 1968, p. 159 [*M. elongata*; OD]. Fore wing as in *Jasvia*, but precostal area longer and more pointed; crossveins not forming a dense reticulation apically; anal area longer. *Perm.*, USSR (Asian RSFSR).——FIG. 96,6. *M. elongata*; fore wing, ×1.0 (Sharov, 1968).

Metoedischia MARTYNOV, 1928b, p. 45 [*M. magnifica*; OD]. Fore wing as in *Jasvia*, but relatively broader; MA1 anastomosed with RS for longer interval; crossveins between branches of RS more nearly straight. [The small wing fragment of an oedischiid from the Permian of Portugal and described by LAURENTIAUX & TEIXEIRA (1958b, p. 212) as *Metoedischia lusitanica* obviously does not belong to this genus.] SHAROV, 1968. *Perm.*,

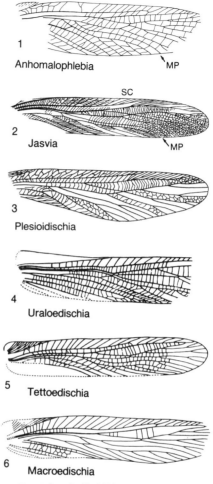

FIG. 96. Oedischiidae (p. 157–158).

USSR (European RSFSR).——FIG. 95,3. *M. magnifica*; fore wing, ×1.8 (Martynov, 1928b).

Paroedischia CARPENTER, 1966, p. 79 [*P. recta*; OD]. Similar to *Metoedischia*, but precostal area very long; SC long; crossveins not reticulate. *Perm.*, USA (Kansas).——FIG. 95,2. *P. recta*; fore wing, ×1.8 (Carpenter, 1966).

Permoedischia KUKALOVÁ, 1955a, p. 542 [*P. moravica*; OD]. Little-known fore wing; precostal area more extensive than in *Oedischia*; MP unbranched. SHAROV, 1968. *Perm.*, Europe (Czechoslovakia).

Plesioidischia HANDLIRSCH, 1906b, p. 346 [*P. baentschi*; OD]. Fore wing as in *Oedischia* but markedly widened near middle; crossveins forming reticulation in region of RS. *Perm.*, Europe (Germany).——FIG. 96,3. *P. baentschi*; fore wing, ×1.0 (Guthörl, 1934).

Pruvostites ZALESSKY, 1928b, p. 381 [*P. takhtachurensis*; OD]. Little-known genus, based on wing fragment with broad costal area. [Type of family Pruvostitidae and order Pruvostitoptera.] SHAROV, 1968. *Perm.*, USSR (European RSFSR).

Rimnosentomon ZALESSKY, 1955b, p. 349 [*R. grande*; OD]. Little-known genus, based on distal fragment of fore wing. SHAROV, 1962c. *Perm.*, USSR (Asian RSFSR).

Sylvoedischia SHAROV, 1968, p. 158 [*S. uralica*; OD]. Fore wing with large precostal area, nearly as long as in *Macroedischia*; costal veinlets connected by crossveins; crossveins very dense over most of wing. Hind wing with SC nearly straight. *Perm.*, USSR (Asian RSFSR).——FIG. 95,*1*. *S. uralica*; *a*, fore and *b*, hind wings, both ×1.8 (Sharov, 1968).

Tettoedischia SHAROV, 1968, p. 159 [*T. minuta*; OD]. Fore wing slender; precostal area large, tapering; costal veinlets not connected by crossveins. *Perm.*, USSR (Asian RSFSR).——FIG. 96,*5*. *T. minuta*; fore wing, ×1.8 (Sharov, 1968).

Uraloedischia SHAROV, 1968, p. 157 [*U. permiensis*; OD]. Little-known genus, based on proximal fragment of fore wing. Precostal area long and narrow, extending about halfway to origin of RS; subcostal veinlets not reticulate. *Perm.*, USSR (Asian RSFSR).——FIG. 96,*4*. *U. permiensis*; fore wing, ×1.7 (Sharov, 1968).

Family TCHOLMANVISSIIDAE Zalessky, 1934

[Tcholmanvissiidae ZALESSKY, 1934, p. 153] [=Tillyardiellidae HANDLIRSCH, 1937, p. 82]

Fore and hind wings similar to those of the Oedischiidae. Fore wing with vein MA1 not anastomosed with RS and without a sharp bend toward RS; branches of CUA nearly parallel and close together. Hind wing with 1A forked distally; anal area with about 12 radiating veins. Body as in the Oedischiidae; ovipositor well developed and bearing many small spines. SHAROV, 1968. *Perm.*

Pinegia MARTYNOV, 1928b, p. 47 [*P. oknowae*; OD] [=*Thnetodes* MARTYNOV, 1928b, p. 5 (type, *T. craticus*); *Tcholmanvissia* ZALESSKY, 1929, p. 19 (type, *T. noinskii*); *Kamaites* ZALESSKY, 1929, p. 21 (type, *K. mirabilis*); *Tillyardiella* MARTYNOV, 1930a, p. 76 (type, *T. distincta*)]. Crossveins very numerous and close together, forming a reticulation only in distal part of wings; posterior margin of fore wing concave. SHAROV, 1962c, 1968. *Perm.*, USSR (Asian and European RSFSR).——FIG. 97,*1*. *P. longipes* (MARTYNOV); *a*, fore and *b*, hind wings, both ×1.0 (Sharov, 1968).

Jubilaeus SHAROV, 1968, p. 161 [*J. beybienkoi*; OD]. Fore wing as in *Pinegia*, but subcostal area broader; precostal area bulging; posterior margin of wing straight or slightly convex. *Perm.*, USSR (Asian RSFSR).——FIG. 97,*2*. *J. beybienkoi*; *a*, fore and *b*, hind wings, both ×0.8 (Sharov, 1968).

Family PERMELCANIDAE Sharov, 1962

[Permelcanidae SHAROV, 1962b, p. 112]

Fore wing more membranous than in Oedischiidae; vein SC extending at least to midwing; RS arising near midwing, with very short, oblique stem and anastomosed with MA for a considerable interval; MP diverging from MA before level of origin of RS; CUA separating from CUP near wing base; CUA forking before level of main fork of M; CUA diverging toward M and typically anastomosed with M and MP for a short interval; CUP arising from the common stem CU; several anal veins. Hind wing little known; costa submarginal; SC terminating near midwing; anal area apparently well developed; anal veins unknown. Body (known only in *Permelcana*): antennae long, filiform; legs slender, hind femora thick basally; tarsi with 4 segments. SHAROV, 1968. *Perm.*

Permelcana SHAROV, 1962b, p. 114 [*P. sojanense*; OD]. MA in fore wing anastomosed with RS for an interval almost equal to length of free part of MA. *Perm.*, USSR (European RSFSR).——FIG. 98,*1*. *P. kukalovae* SHAROV; *a*, fore and *b*, hind wings, both ×5.0 (Sharov, 1968).

Meselcana SHAROV, 1968, p. 162 [*P. madygenica*; OD]. Similar to *Permelcana*, but SC much longer; branches of CUA1 in fore wing strongly curved. *Perm.*, USSR (Asian RSFSR).——FIG. 98,*4*. *M. madygenica*; *a*, fore and *b*, hind wings, both ×5.3 (Sharov, 1968).

Proelcana SHAROV, 1962b, p. 113 [*P. uralica*; OD]. Little-known genus, based on fragment of fore wing. MA anastomosed with RS for a very short interval. *Perm.*, USSR (Asian RSFSR).

Promartynovia TILLYARD, 1937a, p. 99 [*P. venicosta*; OD]. Similar to *Permelcana*, but fore wing more broadly rounded distally; RS with only 2 terminal branches. Hind wing unknown. *Perm.*, USA (Kansas).——FIG. 98,*3*. *P. venicosta*; fore wing, ×7.3 (Carpenter, 1966).

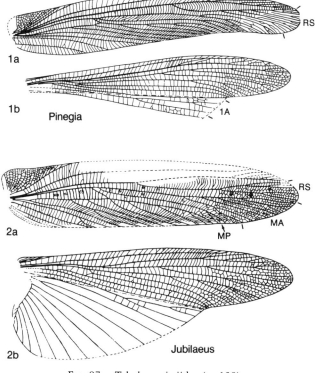

FIG. 97. Tcholmanvissiidae (p. 158).

Family ANELCANIDAE
Carpenter, 1987

[Anelcanidae CARPENTER, 1987, p. 375, *nom. subst. pro* Parelcanidae CARPENTER, 1966, p. 84]

Fore wing as in Oedischiidae but more coriaceous; precostal area very large and very acute distally, extending about one-third wing length from wing base; vein SC remote from costal margin, terminating beyond midwing. Hind wing unknown. *Perm.*

Anelcana CARPENTER, 1987, p. 375, *nom. subst. pro Parelcana* CARPENTER, 1966, p. 84, *non* HANDLIRSCH, 1906b [**Parelcana dilatata*; OD]. Costal area about as wide as area between SC and R at midwing; crossveins numerous but not branched. *Perm.*, USA (Kansas).——FIG. 98,*2*. **A. dilatata* (CARPENTER); fore wing as preserved, ×2.8 (Carpenter, 1966).
Petrelcana CARPENTER, 1966, p. 85 [**P. elongata*; OD]. Fore wing elongate, with many irregular veinlets; precostal area not so long or so broad as in *Anelcana*; costal area much wider than area between SC and R; R with several irregular, oblique veinlets distally; basal piece of MA very long; crossveins forming a coarse reticulation in several areas of wing. *Perm.*, USA (Kansas).——FIG. 98,*5*. **P. elongata*; fore wing, ×2.2 (Carpenter, 1966).

Family PERMORAPHIDIIDAE
Tillyard, 1932

[Permoraphidiidae TILLYARD, 1932a, p. 5]

Similar to Permelcanidae. Fore wing apparently lacking precostal area; vein SC with an anterior basal branch, connected to costal margin of wing by crossveins; MA anastomosed for a short interval with RS; crossveins numerous. Hind wing with MA anastomosed with RS; CUA1 anastomosed with M; anal area unknown. *Perm.*

Permoraphidia TILLYARD, 1932a, p. 6 [**P. americana*; OD]. Costal area of fore wing much broader than subcostal area; CUP extending

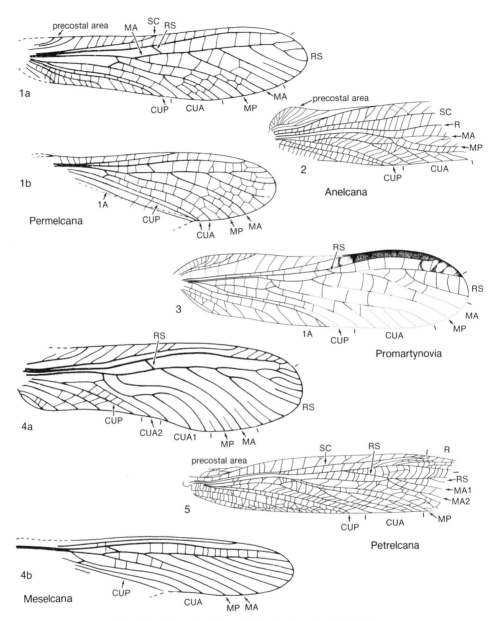

Fig. 98. Permelcanidae and Anelcanidae (p. 158–159).

beyond midwing. CARPENTER, 1943a; SHAROV, 1968. *Perm.*, USA (Kansas).——FIG. 99,*1*. **P. americana; a,* fore and *b,* hind wings, ×8.0 (Carpenter, 1943a).

Family ELCANIDAE Handlirsch, 1906

[Elcanidae HANDLIRSCH, 1906b, p. 412]

Similar to Permelcanidae, but fore wing with vein CUA forking much nearer wing base; RS and MA in brief contact. SHAROV, 1968. *Jur.–Cret.*

Panorpidium WESTWOOD, 1854, p. 394 [**P. tessellatum*; OD] [=*Elcana* GIEBEL, 1856, p. 259, obj.; *Rapha* GIEBEL, 1856, p. 290 (type, *R. liassina*); *Clathrotermes* HEER, 1865, p. 85 (type, *C. signatus*); *Parelcana* HANDLIRSCH, 1906b, p. 420 (type, *P. tenuis*)]. Fore wing: submarginal part

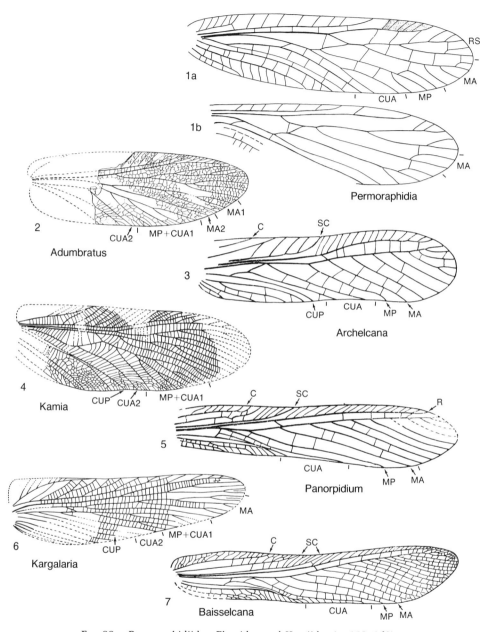

FIG. 99. Permoraphidiidae, Elcanidae, and Kamiidae (p. 159–162).

of C with several branches leading to front margin of wing; apex of wing rounded. ZEUNER, 1942d; BODE, 1953. *Jur.,* England, Europe (Germany, Switzerland), USSR (Kazakh).——FIG. 99,5. *P. karatavica* (SHAROV), USSR; fore wing, ×4.0 (Sharov, 1968).

Archelcana SHAROV, 1968, p. 163 [**Elcana britannica* HANDLIRSCH, 1906b, p. 414; OD]. Fore wing as in *Panorpidium,* but submarginal part of C without distinct branches. *Jur.,* England, USSR (Kirghiz).——FIG. 99,3. *A. britannica* (HANDLIRSCH), England; fore wing, ×4.5 (Sharov, 1968).

Baisselcana SHAROV, 1968, p. 164 [*B. sibirica*; OD]. Fore wing: precostal area long, with many veinlets; apex of wing pointed. *Cret.,* USSR (Asian RSFSR).——FIG. 99,7. *B. sibirica*; fore wing, ×2.5 (Sharov, 1968).

Family KAMIIDAE Sharov, 1968

[Kamiidae Sharov, 1968, p. 165]

Related to Oedischiidae. Costal area of fore wing very broad; area between veins SC and R very narrow; MA not anastomosed with RS or even curved toward it and with 3 or more branches; crossveins very numerous. *Perm.*

Kamia Martynov, 1928b, p. 4 [*K. angustovenosa*; OD] [=*Spongoneura* Martynov, 1928b, p. 6 (type, *S. incerta*); *Permacridites* Martynov, 1931c, p. 156 (type, *P. maximus*)]. Fore wing broad; crossveins mostly straight, not branched. Sharov, 1968. *Perm.*, USSR (European and Asian RSFSR).——Fig. 99,*4*. *K. angustovenosa*; fore wing as preserved, ×1.3 (Sharov, 1968).

Adumbratus Sharov, 1961f, p. 246 [*A. extentus*; OD]. Similar to *Kamia*, but crossveins forming a dense reticulation over most of fore wing. Sharov, 1968. *Perm.*, USSR (Asian RSFSR).——Fig. 99,*2*. *A. extentus*; fore wing as preserved, ×3.0 (Sharov, 1961f).

Kargalaria Sharov, 1968, p. 165 [*K. maculata*; OD]. Fore wing as in *Kamia* but more slender; crossveins not so dense; main veins and their branches nearly parallel. *Perm.*, USSR (Asian RSFSR).——Fig. 99,*6*. *K. maculata*; fore wing as preserved, ×1.0 (Sharov, 1968).

Family VITIMIIDAE Sharov, 1968

[Vitimiidae Sharov, 1968, p. 152]

Similar to Kamiidae, but anastomosis of MP and CUA in fore wing apparently occurring at base of wing; hind wing with large anal area. *Trias.–Cret.*

Vitimia Sharov, 1968, p. 167 [*V. evoluta*; OD]. Crossveins mostly simple, unbranched. *Cret.*, USSR (Asian RSFSR).——Fig. 100,*4*. *V. evoluta*; fore wing as preserved, ×3.5 (Sharov, 1968).

Fergania Sharov, 1968, p. 167 [*F. reducta*; OD]. Fore wing similar to that of *Vitimia*, but precostal area much larger and CUP diverging from CU more proximally. Hind wing with a concave anterior margin. *Trias.*, USSR (Kirghiz).——Fig. 100,*2*. *F. reducta*; *a*, fore and *b*, hind wings, both ×2.8 (Sharov, 1968).

Provitimia Sharov, 1968, p. 166 [*P. pectinata*; OD]. Little-known genus, similar to *Vitimia*, but SC with many forked branches. *Trias.*, USSR (Kirghiz).

Family BINTONIELLIDAE Handlirsch, 1939

[Bintoniellidae Handlirsch, 1939, p. 55]

Related to Vitimiidae. Fore wing with prominent precostal area; veinlets in costal area commonly unbranched, without connecting crossveins; vein MA not anastomosed with RS and remote from it; M with 3 main, parallel branches. *Trias.–Jur.*

Bintoniella Handlirsch, 1939, p. 55 [*B. brodiei*; OD]. Costal area of fore wing narrow, nearly as narrow as subcostal area; branches of vein RS curved and parallel. Remigium of hind wing relatively slender and pointed; crossveins very dense distally. Fore wings showing some sexual dimorphism, those of males being more sclerotized than those of females, as well as slightly larger. Whalley, 1982. *Jur.*, England.——Fig. 100,*1*. *B. brodiei*; *a*, fore and *b*, hind wings, both ×1.8 (Sharov, 1968).

Probintoniella Sharov, 1968, p. 168 [*P. triassica*; OD]. Fore wing as in *Bintoniella*, but costal area very broad, about twice as wide as subcostal area; branches of RS divergent and irregular. Remigium of hind wing broad, with rounded apex. *Trias.*, USSR (Kirghiz).——Fig. 100,*3*. *P. triassica*; *a*, fore and *b*, hind wings, both ×2.8 (Sharov, 1968).

Family TRIASSOMANTEIDAE Tillyard, 1922

[nom. correct. Brues, Melander, & Carpenter, 1954, p. 809, ex Triassomantidae Tillyard, 1922b, p. 449, nom. imperf.] [=Xenopteridae Riek, 1955, p. 687]

Apparently related to the Oedischiidae. Precostal area of fore wing small and narrow; costal area very broad, its veinlets unbranched; subcostal area narrow; vein M much reduced; crossveins unbranched, except in anal area. Hind wing with large anal area, with CUA extending farther distally than in fore wing. [I have followed Sharov (1968) in placing his genera *Ferganopterus*, *Ferganopterodes*, and *Triassomanteodes* in the family Triassomanteidae. However, because of the fragmentary nature of the unique specimen of the type genus, that placement is uncertain. The family diagnosis above is based in part on Sharov's genera.] *Trias.*

Triassomantis Tillyard, 1922b, p. 450 [*T. pygmaeus*; OD]. Little-known genus, based on fragment of fore wing. *Trias.*, Australia (Queensland).

Ferganopterodes Sharov, 1968, p. 169 [*F. reductus*; OD]. Fore wing as in *Ferganopterus*, but MA not anastomosed with RS. *Trias.*, USSR (Kirghiz).——Fig. 101,*1*. *F. reductus*; fore wing, ×4.0 (Sharov, 1968).

Ferganopterus Sharov, 1968, p. 169 [*F. clarus*; OD]. Fore wing with long and tapering costal area; SC extending well beyond midwing; MA

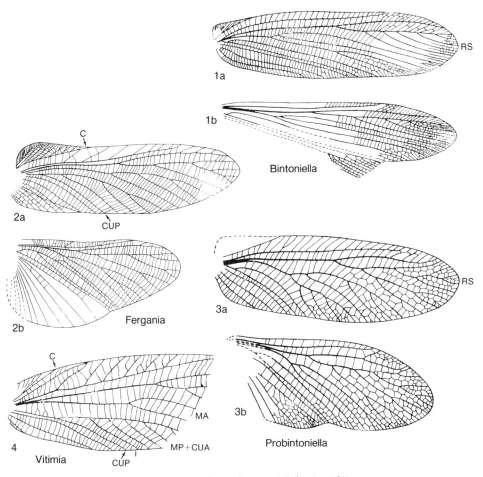

FIG. 100. Vitimiidae and Bintoniellidae (p. 162).

anastomosed with RS; crossveins between CUP and 1A very close together. Hind wing with MA and CUA close together and parallel. *Trias.*, USSR (Kirghiz).——FIG. 101,*4*. **F. clarus; a,* fore and *b,* hind wings, both ×3.0 (Sharov, 1968).

Triassomanteodes SHAROV, 1968, p. 170 [**T. madygenicus;* OD]. Similar to *Ferganopterodes,* but SC terminating at about midwing, MA anastomosed with RS, and stem of CUA much shorter. *Trias.*, USSR (Kirghiz).——FIG. 101,*3*. **T. madygenicus;* fore wing, ×7.0 (Sharov, 1968).

Xenopterum RIEK, 1955, p. 678 [**X. crosbyi;* OD]. Little-known genus, based on wing fragments. [Type of family Xenopteridae RIEK.] SHAROV, 1968. *Trias.*, Australia (Queensland).

Family TETTAVIDAE Sharov, 1968

[Tettavidae SHAROV, 1968, p. 171]

Wing venation similar to that of Oedischiidae; veins RS and MA anastomosed in hind wing but not in fore wing; R in fore wing with several branches distally. *Perm.–Trias.*

Tettavus SHAROV, 1968, p. 171 [**Pinegia fenestrata* MARTYNOV, 1931c, p. 208; OD]. Fore wing (presumably of female): distal branches of R very long; RS with pectinate branching; crossveins in radial field forming a reticulation distally. *Perm.*, USSR (Asian RSFSR).——FIG. 101,*5*. **T. fenestrata; a,* fore wing as preserved, *b,* hind wing, both ×2.0 (Sharov, 1968).

Madygenia SHAROV, 1968, p. 171 [**M. orientalis;* OD]. Female: fore wing as in *Tettavus,* but RS branched dichotomously; crossveins reticulate over most of wing; hind wing with branches of RS nearly parallel to MA, MP, and CUA. Male much smaller than female; venation little known. *Trias.*, USSR (Kirghiz).——FIG. 101,*2*. **M. orientalis; a,* fore and *b,* hind wings of female, *c,* fore wing of male as preserved, all ×2.0 (Sharov, 1968).

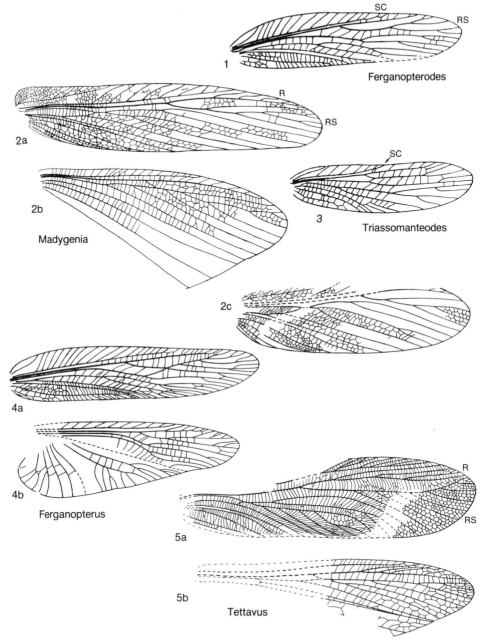

Fig. 101. Triassomanteidae and Tettavidae (p. 162–163).

Family TETTIGONIIDAE
Krauss, 1902

[Tettigoniidae KRAUSS, 1902, p. 538]

Similar in appearance to Oedischiidae and Tettavidae. Fore wings tegminous, their apices folded together at rest; vein RS only rarely fused with MA, even for a short interval; CUA with an anterior branch coalescing with MP; stridulatory structures commonly well developed; right tegmen with a membranous area between CUP and branches of CUA; left tegmen with a similar structure, but CUP larger and serrulate. Hind wings much as in Oedischiidae, but MA commonly anastomosed for a short interval; MP typically fused with an anterior branch of CUA. Body laterally compressed; antennae long, filiform; fore and middle legs relatively short; hind femora skittle-shaped; tarsi with 4 segments; ovipositor laterally flattened; auditory organs commonly present, situated on fore tibiae. *Eoc.–Holo.*

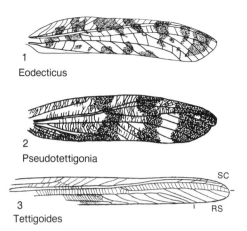

FIG. 102. Tettigoniidae (p. 165).

Tettigonia LINNÉ, 1758, p. 425. *Holo.* [A wing from the Miocene/Pliocene of France has been identified (BRICE & LAURENTIAUX, 1964) as belonging to a female of the existing species *Tettigonia viridissima* LINNÉ, but SHAROV (1968) reports that the wing is from a male and that the taxonomic identification is erroneous.]
Anabrus HALDEMANN, 1852, p. 372. Ovipositor only. [Generic position doubtful.] COCKERELL, 1908p. *Oligo.*, USA (Colorado)–*Holo.*
Arctolocusta ZEUNER, 1937, p. 157 [**Locusta groenlandica* HEER, 1883, p. 146; OD]. Little-known fore wing, female. SHAROV, 1968. *Eoc.*, Greenland.
Eodecticus PONGRÁCZ, 1928, p. 128 [**E. maculatus*; OD]. Fore wing similar to that of *Decticus* (recent) but with more branches of SC to costal margin and with irregular maculations. ZEUNER, 1939. *Mio.*, Europe (Austria).——FIG. 102,*1*. **E. maculatus*; fore wing, ×1.0 (Zeuner, 1939).
Eomortoniellus ZEUNER, 1936b, p. 291 [**E. handlirschi*; OD]. Similar to *Mortoniellus* (recent) but pronotum smaller. ZEUNER, 1944b. *Oligo.*, Europe (Baltic).
Lipotactes BRUNNER, 1898, p. 274. Nymphal male. ZEUNER, 1936b. *Oligo.*, Europe (Baltic)–*Holo.*
Lithymnetes SCUDDER, 1878a, p. 532 [**L. guttatus*; OD]. Little-known genus, based on fore wing. THÉOBALD, 1937a; ZEUNER, 1939; SHAROV, 1968; KEVAN & WIGHTON, 1983. *Oligo.*, USA (Colorado).
Nymphomorpha HENRIKSEN, 1922b, p. 13 [**N. medialis*; OD]. Little-known hind wing. ZEUNER, 1939. *Eoc.*, Europe (Denmark).
Orchelimum SERVILLE, 1839, p. 522. Complete male and female. SCUDDER, 1890. *Oligo.*, USA (Colorado)–*Holo.*
Orphania FISCHER, 1853, p. 222. Hind tibia. CHARPENTIER, 1825; ZEUNER, 1929, 1939. *Pleist.*, Europe (Poland)–*Holo.*
Platycleis FIEBER, 1852, p. 2. Fore wings, hind wings, and parts of body. ZEUNER, 1929, 1939. *Mio.*, Europe (Austria, Germany)–*Holo.*
Pseudotettigonia ZEUNER, 1937, p. 157 [**Tettigonia amoena* HENRIKSEN, 1929, p. 317; OD]. Similar to *Tettigonia*, but crossveins of fore wing more regular. SHAROV, 1968. *Eoc.*, Europe (Denmark).——FIG. 102,*2*. **P. amoena* (HENRIKSEN); fore wing, ×0.8 (Zeuner, 1939).
Rammea ZEUNER, 1931, p. 253 [**R. laticeps*; OD]. Similar to *Decticus* but with a conelike process between meso- and metasternum. ZEUNER, 1939. *Mio.*, Europe (Germany).
Tettigoides RIEK, 1952b, p. 20 [**T. pectinata*; OD]. Elytron very narrow, with long SC; RS arising at midwing; M nearly parallel to R. [Family position uncertain.] *?Eoc.*, Australia (Queensland).——FIG. 102,*3*. **T. pectinata*; elytron, ×2.0 (Riek, 1952b).

Family HAGLIDAE Handlirsch, 1906

[Haglidae HANDLIRSCH, 1906b, p. 425] [=Eospilopteronidae COCKERELL, 1915, p. 472; Pamphagopsidae MARTYNOV, 1925b, p. 577; Aboilidae MARTYNOV, 1925b, p. 581; Prophalangopsidae HANDLIRSCH, 1929, p. 724; Isfaropteridae MARTYNOV, 1937a, p. 61; Tshorkuphlebiidae MARTYNOV, 1937a, p. 72]

Related to Oedischiidae but with the venation of the fore wings conspicuously different in males and females; fore wings of males with a more or less elaborate stridulatory organ. Fore wing, male: costal area very broad, at least basally, with numerous veinlets and crossveins; costal vein submarginal in specialized species, forming a long precostal area; R typically with several terminal branches; RS with numerous branches in generalized genera, but few in genera having elaborate stridulatory organs; CUA coalesced with M basally but diverging posteriorly before M divides into MA and MP; MA not anastomosed with RS; MA and MP commonly unbranched but strongly curved in specialized genera; CUP unbranched. Fore wing, female: MA not directed toward RS or in contact with it; MA and MP much as in males but without curves. Hind wing known in only a few species; venation as in Oedischiidae but with an enlarged anal area in some specialized genera. Antennae long, multisegmented; legs relatively short; 4 tarsal segments; auditory organs on fore tibiae; ovipositor broad and well developed. ZEUNER, 1939; SHAROV, 1968. *Trias.–Holo.*

SHAROV's account (1968) of the fossil Haglidae shows that the family was very large and diverse during the early half of the Mesozoic. In contrast, there are only two recent genera, *Prophalangopsis* WALKER (India) and *Cyphoderis* UHLER (western North America), neither of which is represented in the fossil record.

Hagla GIEBEL, 1856, p. 265 [**H. gracilis*; SD ZEUNER, 1939, p. 139] [=*Haglodes* HANDLIRSCH, 1906b, p. 425 (type, *Hagla similis* GIEBEL, 1856, p. 265)]. Male: fore wing broadest before midwing. Female: fore wing with 5 to 6 terminal branches on RS. Hind wing unknown. ZEUNER, 1939; SHAROV, 1968. *Jur.*, England.——FIG. 103,*1*. **H. gracilis*; fore wings of *a*, male and *b*, female, ×2.0 (Zeuner, 1939).

Aboilus MARTYNOV, 1925b, p. 581 [**A. fasciatus*; OD] [=*Pamphagopsis* MARTYNOV, 1925b, p. 578 (type, *P. maculata*); *Syndesmophyllum* MARTYNOV, 1934, p. 1004, *nom. nud.*]. Male: fore wing broadly oval; precostal area large; costal area very broad; RS arising just before midwing; MA and MP smoothly curved and parallel; hind wing with narrow costal area; RS arising nearer wing base than in fore wing; anal area unknown. Female: fore wing with nearly straight anterior margin, M forking before origin of RS. [Type of family Aboilidae MARTYNOV, 1925b.] ZEUNER, 1939; SHAROV, 1968. *Jur.*, USSR (Kazan).——FIG. 103,*2*. *A. columnatus* MARTYNOV; *a*, fore and *b*, hind wings of male, both ×1.7 (Rohdendorf, 1962a); *c*, fore wing of female, ×0.2 (Sharov, 1968).

Albertoilus KEVAN & WIGHTON, 1981, p. 1825 [**A. cervirufi*; OD]. Little-known genus, based on fragments of fore and hind wings, apparently related to *Aboilus*. KEVAN & WIGHTON, 1983. *Paleoc.*, Canada (Alberta).

Alloma HONG, 1982a, p. 79 [**A. facialata*; OD]. Female: fore wing as in *Hebeihagla* but broader near midwing. HONG, 1982b. *Jur.–Cret.*, China (Liaoning Province).

Archaboilus MARTYNOV, 1937a, p. 51 [**A. kisylkiensis*; OD]. Male: fore wing as in *Aboilus*, but costal veinlets more numerous and closer together; precostal area much smaller. ZEUNER, 1939; SHAROV, 1968. *Jur.*, USSR (Kirghiz, Tadzhik).——FIG. 104,*6*. *A. shurabicus* MARTYNOV; fore wing of male, ×0.8 (Rohdendorf, 1962a).

Archaeohagla LIN, 1965, p. 364 [**A. sinensis*; OD]. Little-known genus, based on proximal fragment of fore wing of female. [Probably a synonym of *Sinohagla*.] *Jur.*, China (Inner Mongolia).

Archihagla SHAROV, 1968, p. 175 [**A. zeuneri*; OD]. Male: fore wing very similar to that of *Hagla*, but front margin of fore wing smoothly curved and branching of R more extensive; subcostal veinlets mostly forked. Female unknown. *Trias.*, USSR (Kirghiz).——FIG. 104,*1*. **A. zeuneri*; fore wing, ×3.2 (Sharov, 1968).

Cyrtophyllites OPPENHEIM, 1888, p. 223 [**C. rogeri*; OD]. Little-known genus. Male: fore wing broadly oval, with very wide costal area; RS with 3 or 4 long branches; crossveins close together and forming cellules distally. SHAROV, 1968. *Jur.*, Europe (Germany).——FIG. 104,*2*. **C. rogeri*; fore wing, ×1.2 (Zeuner, 1939).

Eospilopteron COCKERELL, 1915, p. 472 [**E. ornatum*; OD]. Little-known genus, based on apical fragment of wing. [Family assignment doubtful. Type of family Eospilopteronidae.] SHAROV, 1968. *Jur.*, England.

Hebeihagla HONG, 1982b, p. 1121 [**H. songyingziensis*; OD]. Female: fore wing similar to that of *Parahagla*, but RS apparently with a few more branches; wing broadest basally. [Probably a synonym of *Parahagla*.] *Jur.*, China (Hebei Province).

Isfaroptera MARTYNOV, 1937a, p. 61 [**I. grylliformis*; OD]. Similar to *Cyrtophyllites*. Male: fore wing about as wide as long, almost circular; RS with only 1 or 2 branches. [Type of family Isfaropteridae MARTYNOV, 1937a.] ZEUNER, 1939; SHAROV, 1968. *Jur.*, USSR (Kirghiz).——FIG. 104,*3*. **I. grylliformis*; fore wing of male, ×2.0 (Martynov, 1937a).

Jurassobatea ZEUNER, 1937, p. 154 [**J. gryllacroides*; OD]. Little-known genus, based on a poorly preserved specimen. Large species, similar to *Aboilus*. [Probably a synonym of *Pycnophlebia*. Placed by ZEUNER (1937, 1939) in Gryllacrididae; by SHAROV (1968) in Haglidae.] KEVAN & WIGHTON, 1981, 1983. *Jur.*, Europe (Germany).

Liassophyllum ZEUNER, 1935, p. 106 [**L. abbreviatum*; OD]. Little-known genus, based on distal fragment of fore wing of male; apex pointed; RS arising much nearer apex than in *Cyrtophyllites*; MA strongly arched anteriorly. [Family assignment doubtful.] ZEUNER, 1939; SHAROV, 1968. *Jur.*, England.——FIG. 104,*5*. **L. abbreviatum*; fore wing as preserved, ×0.8 (Zeuner, 1935).

Mesogryllus HANDLIRSCH, 1906b, p. 523 [**Blattidium achelous* WESTWOOD, 1854, p. 390; OD]. Little-known genus, based on poorly preserved

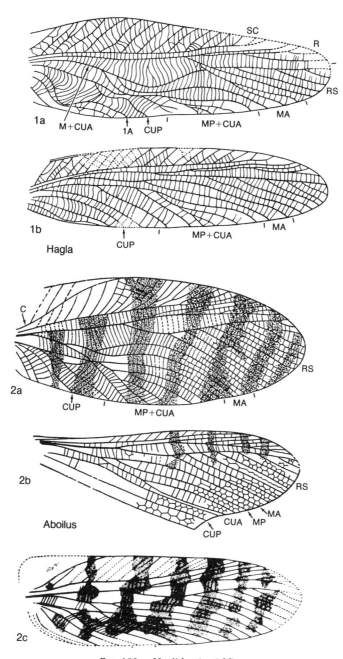

Fig. 103. Haglidae (p. 166).

wing. [Placed in Gryllidae by Zeuner (1939) and in Haglidae by Sharov (1968).] *Jur.*, England.

Neohagla Riek, 1955, p. 683 [**N. sinuata*; OD]. Male: venation of fore wing as preserved similar to that of *Hagla*, but with fewer branches on R; crossveins between R and RS more sinuous. Sharov, 1968. *Trias.*, Australia (Queensland).

Nipponohagla Fujiyama, 1978, p. 183 [**N. kaga*; OD]. Female: fore and hind wings apparently as

Hexapoda

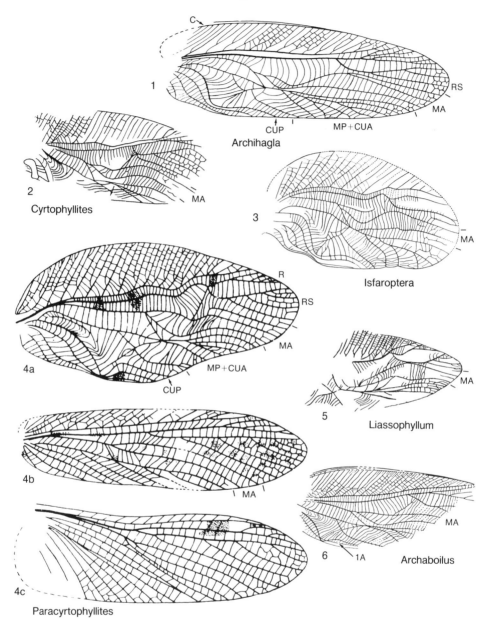

Fig. 104. Haglidae (p. 166–168).

in *Aboilus*, but fore wing much broader, with convex posterior margin. *Cret.*, Japan.

Notopamphagopsis CABRERA, 1928, p. 371 [*N. bolivari*; OD]. Little-known genus, based on apical fragment of wing. SHAROV, 1968. *Trias.*, Argentina (Mendoza).

Palaeorehnia COCKERELL, 1908t, p. 126 [*P. maculata*; OD] [=*Cymatomera maculata* SCUDDER, 1890, p. 230]. Little-known genus, based on small proximal fragment of fore wing of female. ZEUNER, 1939; SHAROV, 1968; KEVAN & WIGHTON, 1983. *Paleog.*, Scotland; *Oligo.*, USA (Colorado).

Paracyrtophyllites SHAROV, 1968, p. 177 [*P. undulatus*; OD]. Similar to *Cyrtophyllites*, but male with stem MA of fore wing more remote from MP and deeply forked. Fore wing of female with normal costal area and shape. Hind wing

Orthoptera—Ensifera

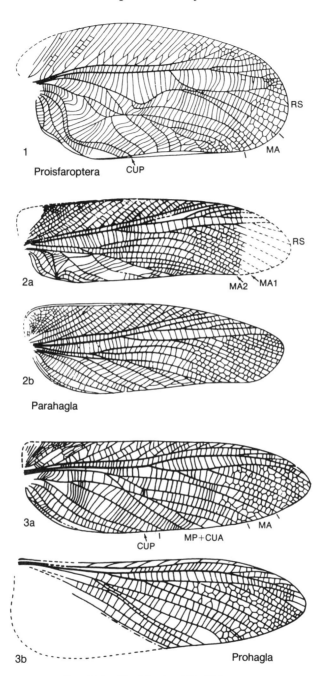

Fig. 105. Haglidae (p. 169–170).

with enlarged anal area. *Jur.*, USSR (Kazakh).
——Fig. 104,4. **P. undulatus; a,* fore wing of male, ×4.0; *b,* fore wing of female, ×3.2; *c,* hind wing of female, ×4.0 (Sharov, 1968).

Parahagla SHAROV, 1968, p. 178 [**P. sibirica;* OD]. Similar to *Hagla,* but fore wing of male with MA1 and MA2 almost parallel and with a strong bend at midwing. *Cret.,* USSR (Asian RSFSR).

———FIG. 105,2. *P. sibirica; a, fore wing of male, ×2.3; b, fore wing of female, ×1.8 (Sharov, 1968).

Paratshorkuphlebia SHAROV, 1968, p. 176 [*P. multivenosa; OD]. Similar to *Tshorkuphlebia*. Male: costal area of fore wing very broad; MA curved anteriorly near midwing. *Jur.*, USSR (Kirghiz).——FIG. 106,3. *P. multivenosa; fore wing of male as preserved, ×1.6 (Sharov, 1968).

Procyrtophyllites ZEUNER, 1935, p. 106 [*P. britannicus; OD]. Little-known genus, based on small wing fragment of male; similar to *Archaboilus*, but crossveins between RS and M more sigmoidal. ZEUNER, 1939; SHAROV, 1968. *Jur.*, England.

Prohagla RIEK, 1954c, p. 164 [*P. superba; OD]. Female: fore wing relatively broad, with prominent precostal area; oblique costal veinlets as in *Protshorkuphlebia*. RS with only 5 terminal branches; vein 2A serrate. RIEK, 1955; SHAROV, 1968. *Trias.*, Australia (New South Wales, Queensland).——FIG. 105,3. *P. superba; a, fore and b, hind wings of female, ×2.0 (Sharov, 1968).

Proisfaroptera SHAROV, 1968, p. 173 [*P. martynovi; OD]. Male: fore wing broadly oval, with wide costal area; MA strongly curved; stridulatory organ well developed. Female: fore wing similar to that of *Turkestania*, but branches of R much shorter. *Trias.*, USSR (Kirghiz).——FIG. 105,1. *P. martynovi; fore wing of male, ×3.0 (Sharov, 1968).

Prophalangopseides SHAROV, 1968, p. 178 [*P. vitimicus; OD]. Similar to *Prophalangopsis* (recent), but fore wing of male with branching of RS more extensive and that of M more reduced. *Cret.*, USSR (Asian RSFSR).——FIG. 106,7. *P. vitimicus; fore wing of male, ×2.5 (Sharov, 1968).

Protohagla ZEUNER, 1962a, p. 165 [*P. langi; OD]. Little-known genus, based on proximal fragment of fore wing of male. Precostal area well developed; subcostal area with dense, parallel crossveins; longitudinal veins straight, except CUP. SHAROV, 1968. *Jur.*, England.

Protshorkuphlebia SHAROV, 1968, p. 174 [*P. triassica; OD]. Male: fore wing elongate, pointed; R and RS branching at same level, almost symmetrically. Female: fore wing very similar to that of *Turkestania*. *Trias.*, USSR (Kirghiz).——FIG. 106,2. *P. triassica; fore wing of male, ×2.0 (Sharov, 1968).

Pseudohagla SHAROV 1962g, p. 152 [*Hagla pospelovi MARTYNOVA, 1949b, p. 923; OD]. Female: fore wing broader than in *Hagla* and precostal area longer; crossveins numerous but not reticulate. SHAROV, 1968. *Jur.*, USSR (Asian RSFSR).——FIG. 106,6. *P. pospelovi (MARTYNOVA); fore wing of female, ×1.4 (Sharov, 1962g).

Pycnophlebia DEICHMÜLLER, 1886, p. 20 [*Locusta speciosa* GERMAR, 1839, p. 198; OD]. Little-known genus, based on numerous but poorly preserved specimens; apparently similar to *Aboilus*. Large species; remigium of fore wing subtriangular; RS with about 10 parallel branches. Tympanal organ on fore tibia. [Placed by MARTYNOV (1937a) in Haglidae, by ZEUNER (1939) in Ensifera, *incertae sedis*, and by SHAROV (1968) in Haglidae.] *Jur.*, Europe (Germany).

Sinohagla LIN, 1965, p. 363 [*S. anthoides; OD]. Little-known genus, based on distal portion of fore wing of female. Venation as in *Aboilus*, with some reticulate crossveins distally. SHAROV, 1968. *Jur.*, China (Inner Mongolia).

Sunoprophalangopsis HONG, 1982b, p. 1124 [*S. elegantis; OD]. Similar to *Aboilus*, but fore wing with CUA more abruptly curved. [Probably a synonym of *Aboilus*.] *Jur.*, China (Hebei Province).

Tshorkuphlebia MARTYNOV, 1937a, p. 154 [*T. compressa; OD]. Related to *Hagla*, but with fewer branches on RS. Male: costal area very broad basally; precostal area wide but short; stridulatory organ as in *Hagla*. Female: fore wing with branches of M and CUA parallel. [Type of family Tshorkuphlebiidae.] SHAROV, 1968. *Jur.*, USSR (Tadzhik).——FIG. 106,1. *T. shurabica* SHAROV; a, fore wing of male as preserved; b, fore wing of female, ×1.5 (Sharov, 1968).

Turkestania SHAROV, 1968, p. 173 [*T. deviata; OD]. Female: fore wing as in *Zeunerophlebia*, but CUP nearly straight basally; crossveins very dense over entire wing. *Trias.*, USSR (Kirghiz).——FIG. 106,4. *T. deviata; fore wing of female, ×0.75 (Sharov, 1968).

Zalmona GIEBEL, 1856, p. 266 [*Z. brodiei; OD]. Little-known genus, based on small fragment of fore wing of female, probably related to *Paracyrtophyllites*. ZEUNER, 1939; SHAROV, 1968. *Jur.*, England.

Zalmonites HANDLIRSCH, 1906b, p. 422 [*Z. geinitzi; OD]. Little-known genus, based on small distal fragment of wing. [Placed in Locustidae by HANDLIRSCH (1906b) and in Haglidae by SHAROV (1968).] *Jur.*, Europe (Germany).

Zeunerophlebia SHAROV, 1968, p. 172 [*Z. gigas; OD]. Male: costal area of fore wing very broad; RS with at least 7 branches; CUA with at least 6 branches; stridulatory organs present. Female: costal area of fore wing less broad; CUP curved basally, serrate. *Trias.*, USSR (Kirghiz).——FIG. 106,5. *Z. gigas; a, fore wing of male, ×1.0; b, fore wing of female, ×0.7 (Sharov, 1968).

Family PHASMOMIMIDAE Sharov, 1968

[Phasmomimidae SHAROV, 1968, p. 179]

Related to the Haglidae. Fore wing elongate; precostal area small or absent; costal area narrow, with few veinlets; vein RS typ-

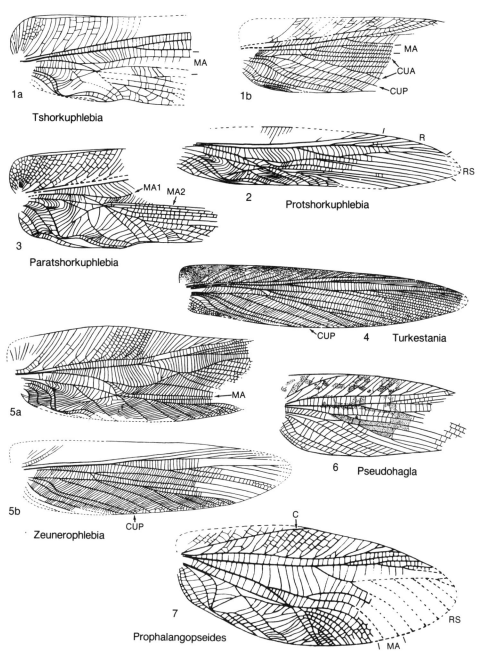

FIG. 106. Haglidae (p. 170).

ically with 2 or 3 terminal branches; MA and MP commonly unbranched; CUA not attached to stem of M. Hind wing much as in Haglidae but with a large anal fan. Ovipositor long, curved. *Jur.–Paleoc.*

Phasmomima SHAROV, 1968, p. 179 [*P. maculomarginata*; OD]. Fore wing: R and RS branched distally only; MA and MP ending almost at wing apex. *Jur.,* USSR (Kazakh).——FIG. 107,2. *P. maculomarginata; a,* fore and *b,* hind wings, both ×2.0 (Sharov, 1968).

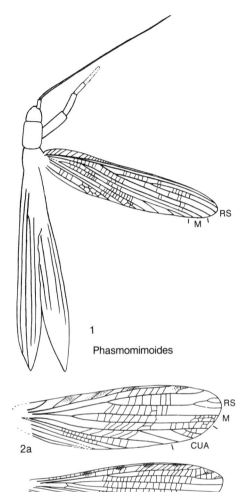

Fig. 107. Phasmomimidae (p. 171–172).

Palaeopteron RICE, 1969, p. 7 [*P. complexum*; OD]. Little-known genus, based on a fore wing fragment; similar to *Phasmomimella*, with veins SC, R, and RS1 convergent apically. [Originally placed in order Perlaria.] KEVAN & WIGHTON, 1983. *Cret.*, Canada (Labrador).

Phasmomimella KEVAN & WIGHTON, 1981, p. 1826 [*P. paskapoensis*; OD]. Little-known genus, based on fragments of fore and hind wings. Fore wing apparently as in *Phasmomimoides*, but stem of RS much longer; M apparently unbranched. *Paleoc.*, Canada (Alberta).

Phasmomimoides SHAROV, 1968, p. 180 [*P. lineatus*; OD]. Fore wing as in *Phasmomima*, but RS deeply forked and MA and MP ending below wing apex. *Jur.*, USSR (Kazakh); *Cret.*, USSR (Asian RSFSR).——FIG. 107,*1*. **P. lineatus*, Jur.; fore wing, part of body, and antenna, ×1 (Sharov, 1968).

Phasmomimula KEVAN & WIGHTON, 1981, p. 1828 [*P. enigma*; OD]. Little-known genus, based on fragments of fore and hind wings. Apparently similar to *Phasmomimella*, but RS unbranched in fore wing. *Paleoc.*, Canada (Alberta).

Family GRYLLACRIDIDAE Stål, 1874

[Gryllacrididae STÅL, 1874, p. 4]

Related to the Haglidae. Fore wing with precostal area very long, submarginal costa extending typically beyond midwing; branches of main veins more or less parallel; few branches on vein RS. Hind wing with M apparently unbranched; anal fan larger than remigium. Stridulatory organs absent from both pairs of wings, but tympanal organs present on fore tibiae of some existing species. Tarsi with 4 segments; ovipositor as in Haglidae. *Paleoc.–Holo.*

Gryllacris SERVILLE, 1831, p. 138. *Holo.*
Macrelcana KARNY, 1932, p. 67 [*Gryllacris ungeri* HEER, 1849, p. 8; OD]. Little-known genus, based on poorly preserved fore wing and hind legs; apparently related to *Gryllacris* (recent), but spines in hind femora broadened to form flat plates. ZEUNER, 1939, SHAROV, 1968. *Mio.*, Europe (Germany).
Prorhaphidophora CHOPARD, 1936a, p. 163 [*P. antiqua*; OD]. Similar to *Rhaphidophora* (recent), but fore and middle femora armed only with a small geniculate spine. ZEUNER, 1939. *Oligo.*, Europe (Baltic).
Zeuneroptera SHAROV 1962g, p. 153 [*Palaeorehnia scotica* ZEUNER, 1939, p. 126; OD]. Little-known genus, based on fore wing fragment. Vein C terminating slightly before midwing. SHAROV, 1968; KEVAN & WIGHTON, 1983. *Paleoc.*, Scotland.

Family GRYLLIDAE Latreille, 1802a

[Gryllidae LATREILLE, 1802a, p 274]

Related to the Haglidae. Male: fore wing tegminous, each tegmen with a longitudinal line of folding; tegmina at rest forming a boxlike cover over meso- and metanotum, hind wings, and abdomen; posterior branches of vein CUA modified as part of stridulatory

Orthoptera—Ensifera 173

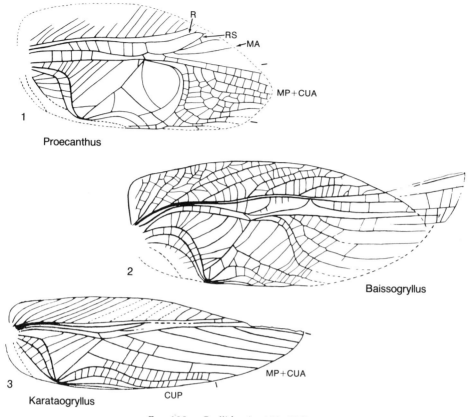

FIG. 108. Gryllidae (p. 173-174).

organ, extending almost to wing apex; stridulatory organ expanded distally and anteriorly, commonly occupying at least half of wing area, with an oblique secondary vein extending diagonally across central part of wing; CUP curving posteriorly, serrulate. Hind wing typically with an enlarged anal fan, as in Haglidae. Female: fore wing as in male in most respects, but lacking modifications of veins associated with stridulatory organ; branching of R and RS reduced, with loss of RS in highly specialized genera. Tarsi with 3 segments; cerci long, flexible; ovipositor cylindrical. ZEUNER, 1939, SHAROV, 1968. *Trias.–Holo.*

Gryllus LINNÉ, 1758, p. 425. Specimens of fore wings only. COCKERELL, 1925e; THÉOBALD, 1937a; ZEUNER, 1937, 1939; SHAROV, 1968. *Paleoc.-Plio.*, Argentina (Jujuy); *Oligo.*, England, Europe (France, Germany)–*Holo.*

Allopterites COCKERELL, 1920a, p. 275 [*A. multilineatus*; OD]. Little-known genus; similar to *Gryllus*, but M with more branches. *Oligo.*, England.

Baissogryllus SHAROV, 1968, p. 183 [*B. sibiricus*; OD]. Male: fore wing nearly oval in shape; costal area long and nearly of uniform width; branches of SC long, oblique. *Jur.*, USSR (Asian RSFSR).
——FIG. 108,2. *B. sibiricus*; fore wing of male, ×4 (Sharov, 1968).

Eneopterotrypus ZEUNER, 1937, p. 156 [*E. chopardi*; OD]. Little-known genus, based on fragment of fore wing of male. ZEUNER, 1939. *Oligo.*, England.

Gryllavus SHAROV, 1968, p. 181 [*G. madygenicus*; OD]. Male: fore wing relatively slender; R with a few, short terminal branches; RS with a shallow fork; MA and MP parallel and nearly straight, terminating at wing apex; CUA with 8 terminal branches. Female: fore wing similar to that of male except for details in region of stridulatory organ; RS unbranched. *Trias.*, USSR (Kirghiz).
——FIG. 109,1. *G. madygenicus*; fore wings of *a*, male and *b*, female, both ×4 (Sharov, 1968).

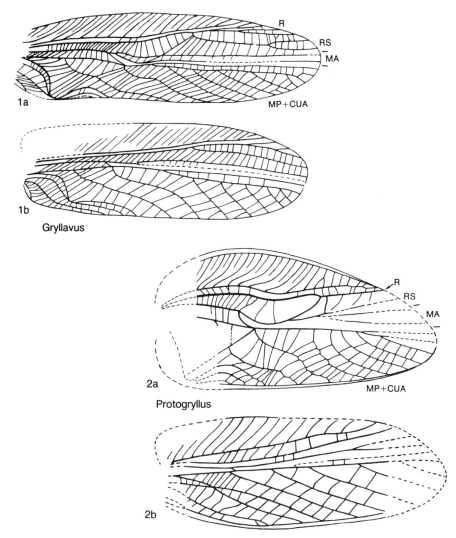

Fig. 109. Gryllidae (p. 173–175).

Heterotrypus Saussure, 1878, p. 537. Whole adult male. Chopard, 1936b; Weidner, 1956. *Oligo.*, Europe (Baltic)–*Holo.*

Karataogryllus Sharov, 1968, p. 182 [*K. gryllotalpiformis*; OD]. Male: fore wing as in *Protogryllus* but more elongate; anterior margin not so convex; branches of M and CUA parallel. *Jur.*, USSR (Kazakh).——Fig. 108,3. *K. gryllotalpiformis*; fore wing of male, ×4 (Sharov, 1968).

Liassogrylloides Bode, 1953, p. 105 [*L. basifastigatus*; OD]. Little-known fore wing. [Family assignment doubtful.] *Jur.*, Europe (Germany).

Lithogryllites Cockerell, 1908p, p. 64 [*L. lutzii*; OD]. Little-known genus, based on apterous male, probably immature. Zeuner, 1939. *Oligo.*, USA (Colorado).

Madasumma Walker, 1869, p. 64. Female adult. Chopard, 1936b; Zeuner, 1939, 1944b. *Oligo.*, Europe (Baltic)–*Holo.*

Paroecanthus Sharov, 1968, p. 184 [*P. caucasicus*; OD]. Little-known genus, based on fragment of fore wing of male. Similar to *Proecanthus*, but costal area much narrower. *Mio.*, USSR (Asian RSFSR).

Proecanthus Sharov, 1968, p. 183 [*P. anatolicus*; OD]. Little-known genus, based on incomplete fore wing of male. Similar to *Baissogryllus* but with fewer crossveins in the stridulatory area. *Cret.*, USSR (Kazakh).——Fig. 108,1. *P. anatolicus*; fore wing of male as preserved, ×5 (Sharov, 1968).

Pronemobius Scudder, 1890, p. 234 [*Nemobius*

tertiarius SCUDDER, 1878b, p. 774; SD ZEUNER, 1939, p. 227]. Related to *Nemobius* (recent), but hind tibiae without spines. ZEUNER, 1939. *Eoc.,* USA (Wyoming, Colorado).

Protogryllus HANDLIRSCH, 1906b, p. 424 [*Gryllus dobbertinensis* GEINITZ, 1880, p. 523; OD] [=*Achaetites* HANDLIRSCH, 1906b, p. 523 (type, *Acheta sedgwicki* BRODIE, 1845, p. 32); *Archaegryllodes* HAUGHTON, 1924, p. 336 (type, *A. stormburgensis*)]. Male: fore wing very broad, about half as wide as long; anterior border of wing very convex; R unbranched; distal branches of CUA more curved than in *Gryllavus*. ZEUNER, 1939; SHAROV, 1968. *Trias.,* South Africa; *Jur.,* USSR (Kirghiz, Kazakh), England, Europe (Germany).——FIG. 109,2. *P. asiaticus* SHAROV, USSR; fore wings of *a*, male and *b*, female, both ×6 (Sharov, 1968).

Stenogryllodes CHOPARD, 1936b, p. 382 [*S. brevipalpis*; OD]. Little-known genus, based on a male nymph. Similar to *Stenogryllus,* but spines of hind tibiae less numerous but more closely arranged. ZEUNER, 1939. *Oligo.,* Europe (Baltic).

Trichogryllus CHOPARD, 1936b, p. 378 [*Gryllus macrocercus* GERMAR & BERENDT, 1856, p. 36; OD]. Related to the recent *Pteroplistes*. Posterior tibiae with 4 widely separated spines on each side. ZEUNER, 1939. *Oligo.,* Europe (Baltic).

Family GRYLLOTALPIDAE
Leach, 1815

[Gryllotalpidae LEACH, 1815, p. 119]

Fore wing short, with a simple stridulatory organ on that of male; hind wing typically functional, but apterous species not uncommon. Forelegs well developed, adapted for digging; femora and tibiae broad and compressed; middle legs small; hind legs relatively short, with prominent femora; tarsi with 3 segments; external ovipositor absent. *Oligo.–Holo.*

Gryllotalpa LATREILLE, 1802, p. 275. Fore wing and whole specimen. COCKERELL, 1921d; ZEUNER, 1931, 1939; WEIDNER, 1968a. *Oligo.,* England; *Mio.,* Europe (Germany); *Plio.,* Europe (Germany)–*Holo.*

Neocurtilla KIRBY, 1906, p. 2. Foreleg only. ZEUNER, 1937, 1939. *Mio.,* Europe (Germany)–*Holo.*

Family PROPARAGRYLLACRIDIDAE
Riek, 1956

[*nom. transl.* SHAROV, 1968, p. 185, *ex* Proparagryllacridinae RIEK, 1956, p. 106]

Male: fore wing with vein C submarginal; precostal area long, as in Oedischiidae; R with several branches, some arising close to origin of RS; RS with only 3 terminal branches; branches of RS, MA, MP, and CUA close together and parallel; stridulatory organ absent. Female: fore wing as in male, but RS with several long branches. Antennae long, filiform; prothorax of moderate length; hind legs long, with femora only slightly thickened; tarsi with 5 segments; arolium present; ovipositor broad, serrate; cerci long and thin. SHAROV, 1968. *Trias.*

The relationships of this family with others in the Ensifera are uncertain. The type genus is known only by small fragments. The preceding family diagnosis is based on SHAROV'S account (1968) of the family, which was based mainly on two other genera that he placed in the family, *Mesogryllacris* RIEK and *Gryllacrimima* SHAROV.

Proparagryllacris RIEK, 1956, p. 106 [*P. crassifemur*; OD]. Little-known genus, based on fragments of wings and body. Venation apparently similar to that of *Gryllacrimima*. SHAROV, 1968. *Trias.,* Australia (Queensland).

Gryllacrimima SHAROV, 1968, p. 185 [*G. perfecta*; OD]. Similar to *Proparagryllacris*. Both MA and MP+CUA forked in fore wing of male. Body structure discussed under family. *Trias.,* USSR (Kirghiz).——FIG. 110,3. *G. perfecta*; fore wing of male, ×2.0 (Sharov, 1968).

Mesogryllacris RIEK, 1955, p. 685 [*M. giganteus*; OD]. Large species. Costal area of fore wing broader than in *Proparagryllacris*; M forked in basal half of wing; MA and MP apparently not forked. SHAROV, 1968. *Trias.,* Australia (Queensland).

Family UNCERTAIN

The following genera, apparently belonging to the Orthoptera, suborder Ensifera, are too poorly known to permit assignment to families.

Huabeius HONG, 1982b, p. 1128 [*H. suni*; OD]. Little-known genus, possibly related to the Haglidae. Cubital and anal areas of fore wing of female very narrow. *Cret.,* China (Hebei Province).

Lithymnetoides KEVAN & WIGHTON, 1983, p. 220, footnote [*Lithymnetes laurenti* THÉOBALD, 1937a, p. 113; OD]. Little-known genus, based on poorly preserved specimens. *Oligo.,* Europe (France).

Mesoedischia SHAROV, 1968, p. 160 [*M. madygenica*; OD]. Fore wing of male: precostal area very short; costal veinlets widely spaced, not reticulate; area between SC and R very narrow; crossveins numerous but reticulate only near wing

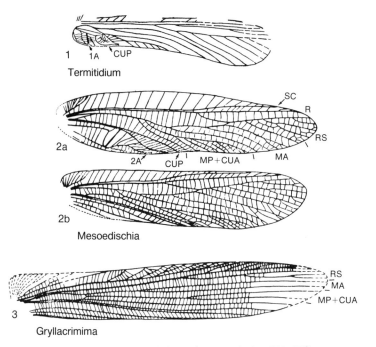

Fig. 110. Proparagryllacrididae and Uncertain (p. 175–176).

apex; CUA diverging basally from CUP and anastomosed with M for a considerable interval, before diverging off, usually in combination with MP; CUA with several branches; stridulatory apparatus well develpd at base of wing. Fore wing of female similar to that of male but without local specializations near wing base. [This genus was placed by Sharov (1968) in the Oedischiidae. The anastomosis of CUA with the stem of M, clearly shown at the base of the wing of the type specimen, and the presence of an advanced stridulatory apparatus at the base of the wing of the male seem to me to require a separate family. With regard to the structures mentioned, the genus seems more closely related to the Haglidae than to the Oedischiidae. So far as I am aware, however, this is the only extinct genus that shows the actual anastomosis of CUA1 with M at the base of the wing.] *Trias.*, USSR (Tadzhik).——Fig. 110,2. *M. madygenica*; fore wings of *a*, male and *b*, female, ×2.5 (Sharov, 1968).

Proedischia Pinto & Ornellas, 1978c, p. 309 [*P. mezzalirai*; OD]. Little-known genus, possibly related to the Oedischiidae. Fore wing elongate; costal area slender; precostal area very small or absent, the main veins not curved posteriorly near wing base; anal area unknown. [Type of family Proedischiidae Pinto & Ornellas]. *U. Carb.*, Brazil (São Paulo).

Protachaeta Handlirsch, 1939, p. 55 [*P. masculina*; OD]. Little-known genus, based on poorly preserved fore wing. Zeuner, 1939. *Jur.*, Europe (Germany).

Pseudohumbertiella Handlirsch, 1906b, p. 522 [*Humbertiella grandis* Brauer, Redtenbacher, & Ganglbauer, 1889, p. 13; OD]. Little-known genus, based on distal fragment of wing. [Placed in Locustidae by Handlirsch (1906b), and in Haglidae by Sharov (1968).] *Jur.*, USSR (Asian RSFSR).

Termitidium Westwood, 1854, p. 394 [*T. ignotum*; OD]. Fore wing with CUP remote from 1A. [Placed in Tettigoniidae by Zeuner (1939) and in Haglidae by Sharov (1968).] *Jur.*, England.——Fig. 110,*1*. *T. ignotum*; fore wing, ×1.5 (Zeuner, 1939).

Thuringopteryx Kuhn, 1937, p. 191 [*T. gimmi*; OD]. Little-known genus, based on hind wing fragment, possibly belonging to Haglidae. Zeuner, 1939. *Trias.*, Europe (Germany).

Zhemengia Hong, 1982b, p. 1123 [*Z. sinica*; OD]. Little-known genus, based on wing fragment. *Jur.*, China (Inner Mongolia).

Suborder CAELIFERA
Ander, 1936

[Caelifera Ander, 1936, p. 93]

Antennae shorter than body, with not more than 30 segments; tympanal organs, when present, on first abdominal segment; stridulatory mechanism diverse (rarely absent), hind tibiae or femora commonly scraped across ridges on abdomen or fore wings; ovipositor, when present, typically short. *Trias.–Holo.*

Family LOCUSTOPSEIDAE Handlirsch, 1906

[nom. correct ROHDENDORF, 1957, p. 83, pro Locustopsidae HANDLIRSCH, 1906b, p. 421]

Related to Acrididae. Fore wings long, commonly twice as long as body, typically broader distally than proximally; apex rounded; vein SC terminating near apex; M forking into MA and MP near level of origin of RS; MA typically with 2 long, parallel branches; CUA commonly with 3 terminal branches; CUP unbranched; crossveins not so numerous as in Acrididae. Stridulatory apparatus apparently absent. Hind wing as in Acrididae; M anastomosed with R basally. Body structures apparently similar to those of the Acrididae. ZESSIN, 1983a. *Trias.–Cret.*

Locustopsis HANDLIRSCH, 1906b, p. 421 [*L. elegans*; SD COCKERELL, 1915, p. 473] [=*Brodiana* ZEUNER, 1942d, p. 13 (type, *B. cubitalis*)]. Fore wing: RS with 3 to 5 branches; M with 3, rarely 2; crossveins forming an irregular network over at least part of wing. SHAROV, 1968; ZESSIN, 1983a. *Jur.,* England, Europe (Germany), USSR (Kazakh, Tadzhik, Kirghiz, Asian RSFSR).——FIG. 111,*1*. *L. karatavica* SHAROV; *a*, fore wing, *b*, hind wing, both ×3.5; *c*, entire specimen, ×2.5 (Sharov, 1968).

Conocephalella STRAND, 1926, p. 46, *nom. subst. pro Conocephalites* HANDLIRSCH, 1906b, p. 518, *non* BARRANDE, 1852 [*Conocephalus capito* DEICHMÜLLER, 1886, p. 24; OD] [=*Conocephalopsis* HANDLIRSCH, 1939, p. 154, obj.]. Large species; RS of fore wing with 6 or 7 branches. Strong spines on hind tibiae. ZEUNER, 1942d; SHAROV, 1968. *Jur.,* Europe (Germany).

Parapleurites BRAUER, REDTENBACHER, & GANGLBAUER, 1889, p. 13 [*P. gracilis*; OD]. Fore wing as in *Locustopsis* but with a double row of cells between M and CU basally. ZEUNER, 1942d; SHAROV, 1968. *Jur.,* USSR (Asian RSFSR).——FIG. 111,*4*. *P. sibirica* SHAROV; fore wing, ×4.0 (Sharov, 1968).

Praelocustopsis SHAROV, 1968, p. 187 [*P. mirabilis*; OD]. Fore wing: SC short; precostal area with very few veinlets and narrower than costal area below it; crossveins widely spaced distally. *Trias.,* USSR (Asian RSFSR).——FIG. 111,*2*. *P. mirabilis*; fore wing, ×9.0 (Sharov, 1968).

Schwinzia ZESSIN, 1983a, p. 180 [*S. sola*; OD]. Fore wing similar to that of *Triassolocusta,* but CUA with 4 terminal branches. *Jur.,* Europe (Germany).

Triassolocusta TILLYARD, 1922b, p. 451 [*T. leptoptera*; OD]. Little-known genus, based on part of a fore wing; M with 4 terminal branches, CUA with 2. SHAROV, 1968. *Trias.,* Australia (Queensland).

Zeunerella SHAROV, 1968, p. 189 [*Z. arborea*; OD]. Fore wing as in *Locustopsis* but C longer, extending almost to level of origin of RS; C strongly curved anteriorly near wing base. KEVAN & WIGHTON, 1983. *Cret.,* USSR (Kazakh).——FIG. 111,*3*. *Z. arborea*; fore wing, ×4.5 (Sharov, 1968).

Family LOCUSTAVIDAE Sharov, 1968

[Locustavidae SHAROV, 1968, p. 185]

Fore wing as in Locustopseidae, but vein CUA with 4 or 5 terminal branches. Body unknown. [Provisionally placed in Caelifera by SHAROV; lack of knowledge of body structure prevents definite assignment to either Caelifera or Ensifera.] *Trias.*

Locustavus SHAROV, 1968, p. 186 [*L. madygensis*; OD]. RS arising near midwing; forking of M at level of end of C. *Trias.,* USSR (Kirghiz).——FIG. 112,*2*. *L. madygensis*; fore wing, ×2.5 (Sharov, 1968).

Ferganopsis SHAROV, 1968, p. 186 [*F. lanceolatus*; OD]. Fore wing as in *Locustavus,* but RS arising much nearer wing base; M forking well before origin of RS. *Trias.,* USSR (Kirghiz).——FIG. 112,*4*. *F. lanceolatus*; *a*, fore and *b*, hind wings, both ×3.0 (Sharov, 1968).

Family EUMASTACIDAE Burr, 1899

[Eumastacidae BURR, 1899, p. 75]

Small species, with great diversity of structure. Fore and hind wings commonly reduced or absent; stridulatory structures and tympanal organs apparently absent; fore wing (when fully formed) with an unbranched CUA. *Cret.–Holo.*

Eumastax BURR, 1899, p. 94. *Holo.*

Archaeomastax SHAROV, 1968, p. 189 [*A. jurassicus*; OD]. Similar to *Erucius* (recent), but subcostal area broader and branches of M shorter. *Cret.,* USSR (Kazakh).——FIG. 112,*1*. *A. jurassicus*; fore wing, ×8.0 (Sharov, 1968).

Taphacris SCUDDER, 1890, p. 226 [*T. reliquata*; OD] [=*Eobanksia* COCKERELL, 1909h, p. 384 (type, *E. bittaciformis*)]. Little-known genus, based on fragments of wings and body. [Family

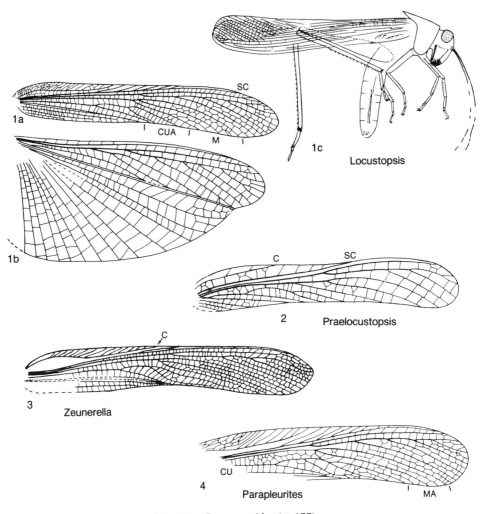

Fig. 111. Locustopseidae (p. 177).

assignment uncertain.] ZEUNER, 1944b. *Oligo.*, USA (Colorado).

Family PROMASTACIDAE Kevan & Wighton, 1981

[Promastacidae KEVAN & WIGHTON, 1981, p. 1834]

Apparently similar to Erucidae (recent), but fore wings broader and less constricted near middle; stems of veins SC, R, RS, and M more widely separated from each other. *Paleoc.–Eoc.*

Promastax HANDLIRSCH, 1910b, p. 97 [*P. archaicus*; OD]. Little-known genus, based on distal fragment of fore wing. Branches of RS arising in distal third of wing; RS with 3 terminal branches; MA unbranched. KEVAN & WIGHTON, 1981. *Eoc.*, Canada (British Columbia).——FIG. 112,3. *P. archaicus*; fore wing, ×2.6 (Handlirsch, 1910b).

Promastacoides KEVAN & WIGHTON, 1981, p. 1834 [*P. albertae*; OD]. Little-known genus, based on poorly preserved fore wing. Similar to *Promastax*, but branches of RS arising before midwing; RS with 5 terminal branches; MA forked. *Paleoc.*, Canada (Alberta).

Family TETRIGIDAE Rambur, 1838

[Tetrigidae RAMBUR, 1838, p. 64]

Small species. Fore wings short and thick, commonly scalelike or absent in existing species. Hind wings of moderate size, if present.

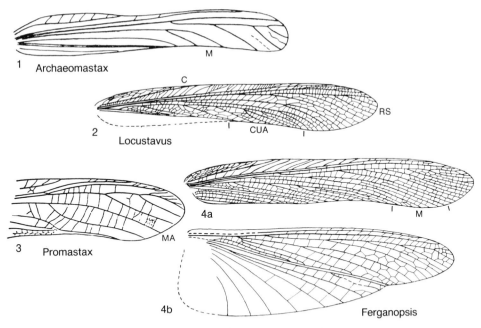

Fig. 112. Locustavidae, Eumastacidae, and Promastacidae (p. 177–178).

Pronotum projecting posteriorly at least to base of abdomen and commonly to or beyond apex of abdomen. Tympanal and stridulatory structures absent. *Cret.–Holo.*

Tetrix LATREILLE, 1802, p. 284. Adult male. ZEUNER, 1937. *Oligo.*, Europe (Baltic)–*Holo.*

Archaeotetrix SHAROV, 1968, p. 190 [*A. locustopseiformis*; OD]. Fore and hind wings fully formed. Fore wing tegminous, unusually thick; crossveins forming a coarse reticulation between longitudinal veins; precostal and costal areas nearly equally broad, the two combined about half width of wing; SC and stem of R very close together and parallel; M and CUA unbranched. Hind wing with MA and MP present. *Cret.*, USSR (Asian RSFSR).——FIG. 113,*3*. **A. locustopseiformis*; *a*, fore and *b*, hind wings, both ×6.5 (Sharov, 1968).

Prototetrix SHAROV, 1968, p. 190 [*P. reductus*; OD]. Fore wing with precostal area forming a small basal lobe; SC, R, and RS distinct; M and CU much reduced; CUA absent. *Cret.*, USSR (Asian RSFSR).——FIG. 113,*1*. **P. reductus*; fore wing, ×13.0 (Sharov, 1968).

Succinotettix PITON, 1938, p. 227 [*S. chopardi*; OD]. Little-known genus, apparently related to *Paratettix* (recent). Antennae with 19 segments; pronotum extending posteriorly slightly beyond end of abdomen. *Oligo.*, Europe (Baltic).

Tettigidea SCUDDER, 1862, p. 476. Whole insect. [Generic assignment doubtful.] HEER, 1865; SCUDDER, 1890. *Mio.*, Europe (Germany)–*Holo.*

Family TRIDACTYLIDAE Brunner, 1882

[Tridactylidae BRUNNER, 1882, p. 453]

Small, highly specialized species. Fore wing tegminous and short, commonly not reaching apex of abdomen; venation in existing species reduced to 2 or 3 veins (SC, R, 1A). Hind wing with remigium reduced to narrow strip; all veins unbranched; anal fan very large. Hind femora greatly enlarged; hind tibiae of recent species with a pair of articulated plates. *Cret.–Holo.*

Tridactylus OLIVIER, 1789, p. 26. *Holo.*

Monodactyloides SHAROV, 1968, p. 191 [*M. curtipennis*; OD]. Similar to *Monodactylus*, but fore wings short, extending only to middle of abdomen. SC of fore wing with short branches. *Cret.*, USSR (Asian RSFSR).

Monodactylus SHAROV, 1968, p. 191 [*M. dolichopterus*; OD]. Fore wing well developed and long, with apices reaching end of abdomen; SC extending about three-fourths wing length from base, with several long branches; M and CUA unbranched. Pronotum with broad lateral lobes.

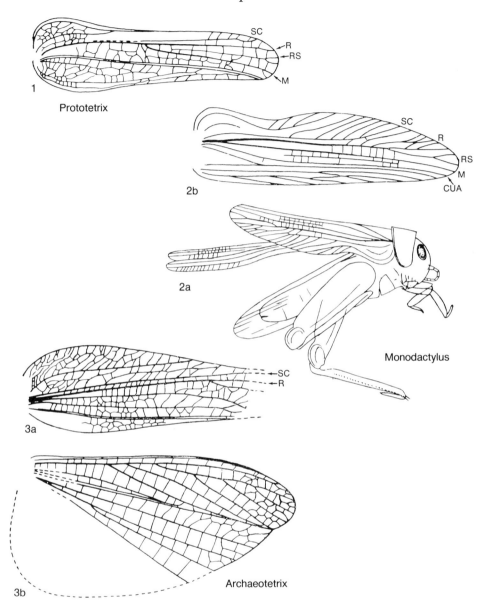

Fig. 113. Tetrigidae and Tridactylidae (p. 178–180).

Cret., USSR (Asian RSFSR).——Fig. 113,2. **M. dolichopterus; a,* entire specimen as preserved, ×4.5; *b,* fore wing, ×7.3 (Sharov, 1968).

Family ACRIDIDAE Latreille, 1825

[Acrididae LATREILLE, 1825, p. 414]

Both pairs of wings typically well developed. Fore wing with basal parts of veins SC, R, and M very close together; RS with numerous branches. Hind wing with R and M anastomosed basally. Tympanal organs on first abdominal segment. Stridulation by rubbing hind femora across posterior margin of fore wings, or by snapping hind wings in flight. *Eoc.–Holo.*

Acrida LINNÉ, 1758, p 425. *Holo.*

Bryodema FIEBER, 1853, p. 129. Parts of wings and body. PONGRÁCZ, 1928; ZEUNER, 1942a. *Mio.,* Europe (Hungary)–*Holo.*

Heeracris ZEUNER, 1937, p. 159 [**Acridium oeningense* SCUDDER, 1895a, p. 118; OD]. Little-known genus, based on part of fore wing. [Possibly related to *Catanops* (recent).] *Mio.,* Europe (Germany).

Mentacridium PITON, 1936b, p. 78 [**M. eocenicum*; OD]. Little-known genus, based on poorly preserved fore wing. [Family assignment doubtful.] ZEUNER, 1944b. *Eoc.,* Europe (France).

Miocaenacris ZEUNER, 1931, p. 275 [**M. soergeli*; OD]. Little-known genus, based on general body form; probably related to *Catanops* (recent). ZEUNER, 1941b. *Mio.,* Europe (Germany).

Nanthacia SCUDDER, 1890, p. 224 [**N. torpida*; OD]. Little-known genus, based on single hind wing. ZEUNER, 1944b. *Oligo.,* USA (Colorado).

Oedipoda LATREILLE, 1829, p. 188. Fore wings and body. [Generic assignment of fossils uncertain.] ZEUNER, 1942a. *Mio.,* Europe (Hungary)–*Holo.*

Proschistocerca ZEUNER, 1937, p. 158 [**P. oligocaenica*; OD]. Similar to *Schistocerca* (recent), but costal area of fore wing abruptly narrowed just before first branch of RS. *Oligo.,* England.

Protocatanops ZEUNER, 1931, p. 262 [**P. gracilis*; OD]. Little-known genus, based on head and prothorax. [Probably synonymous with *Catanops* (recent).] *Mio.,* Europe (Germany).

Taeniopodites COCKERELL, 1909q, p. 283 [**T. pardalis*; OD]. Little-known genus, based on fragment of fore wing; probably related to *Catanops* (recent). ZEUNER, 1941b. *Oligo.,* USA (Colorado).

Tyrbula SCUDDER, 1885b, p. 768 [**T. russelli*; OD]. Little-known genus, based on body only. ZEUNER, 1944b. *Eoc.,* USA (Wyoming); *Oligo.,* USA (Colorado).

Family UNCERTAIN

The following genera, apparently belonging to the Orthoptera, suborder Caelifera, are too poorly known to permit assignment to families.

Chresmoda GERMAR, 1839, p. 201 [**C. obscura*; OD] [=*Locusta prisca* GERMAR, 1839, p. 200]. Little-known genus, probably related to Acrididae. [Type of family Chresmodidae HAASE, 1890a, p. 11. Placed in the Orthoptera by GERMAR and in the Hemiptera (Hydrometridae or Gerridae) by most entomologists before 1900; placed by HANDLIRSCH (1906b) and almost everyone else since then in the Phasmatodea. Frequently confused with *Pygolampis gigantea* (GERMAR). See HAGEN, 1862; VISHNIAKOVA, 1980b, p. 173, footnote; and CARPENTER, 1992.] *Jur.,* Europe (Germany).

Miopyrgomorpha KEVAN in KEVAN & AKBAR, 1964, p. 1526, footnote [**Oedipoda fischeri* HEER, 1865, p. 367; OD]. Little-known genus, based on poorly preserved specimen. ZEUNER, 1944b; KEVAN, 1965. *Mio.,* Europe (Germany).

Suborder UNCERTAIN

The following genera, apparently belonging to the order Orthoptera, are too poorly known to permit assignment to suborders.

Locustopsites THÉOBALD, 1937a, p. 116 [**L. gigantea*; OD]. Little-known genus, based on fore wing fragment. ZEUNER, 1942d. *Oligo.,* Europe (France).

Phaneropterites HANDLIRSCH, 1906b, p. 519 [**Phaneroptera germari* MÜNSTER in GERMAR, 1842, p. 81; OD]. Little-known genus, based on a poorly preserved specimen. ZEUNER, 1942d. *Jur.,* Europe (Germany).

Order GRYLLOBLATTODEA
Brues & Melander, 1915

[*nom. correct.* BRUES & MELANDER, 1932, *pro* Grylloblattoidea BRUES & MELANDER, 1915, p. 13]

Wingless; antennae long, multisegmented; compound eyes absent or small; ocelli absent; legs cursorial; tarsi five-segmented; cerci well developed, segmented; ovipositor conspicuous. *Holo.*

Order TITANOPTERA
Sharov, 1968

[Titanoptera SHAROV, 1968, p. 122]

Orthopteroid insects of moderate to large size. Fore wing with or without a precostal area; spaces between veins RS, MA, and MP commonly wide, in some genera much enlarged and forming a stridulatory area, apparently in both sexes; CUP commonly branched; 2A with pectinate branching for its entire length, directed posteriorly. Hind wing with MP+CUA1 branched; 2A branched much as in fore wing; anal area large and forming a lobe in some genera but relatively small in others. Antennae very long, slender, and filiform; head hypognathous, with long, serrate mandibles; prothorax relatively short; forelegs prehensile, spinose; hind

Mesotitan

Fig. 114. Mesotitanidae (p. 182).

legs cursorial, relatively short; all tarsi with 5 segments; arolium present; cerci short, unsegmented; ovipositor also short. Wings at rest folded flat over abdomen, not inclined as in the Orthoptera (Saltatoria). Nymphs unknown. *Trias.*

Tillyard (in Tillyard & Dunstan, 1916) originally placed the genus *Mesotitan* in the Protorthoptera, but he later (Tillyard, 1925c) transferred it to the Protohemiptera, where it was also placed by McKeown (1937), who had much better specimens for study (Fig. 114). Zeuner (1939) transferred the genus and its family to the Orthoptera, in which they were also later placed by Riek (1954c). Sharov (1962b) assigned the Mesotitanidae to the Paraplecoptera, but in 1968, after study of a very extensive collection of Mesotitanidae and related families from the Triassic of the USSR, he designated the new order Titanoptera for their reception. More recently, Rasnitsyn (1980c) treated the Titanoptera as a suborder of the Orthoptera. However, since in the present work the order Orthoptera is restricted to the saltatorial orthopteroids, the Titanoptera have ordinal status.

Family MESOTITANIDAE Tillyard, 1925

[Mesotitanidae Tillyard, 1925c, p. 376] [=Clathrotitanidae Riek, 1954c, p. 165]

Large insects. Fore wing with areas between veins RS and MA1, MA1 and MA2, MA2 and MP+CUA1 commonly broad; crossveins in those areas straight, unbranched, alternately convex and concave, and forming a stridulatory apparatus; RS arising from R about one-third wing length from base and very close to R for most of its length; MA forking at about level of origin of RS. Venation of hind wing similar to that of fore wing except for the stridulatory area. *Trias.*

Mesotitan Tillyard in Tillyard & Dunstan, 1916, p. 40 [*M. giganteus*; OD] [=*Clatrotitan* McKeown, 1937, p. 32 (type, *C. andersoni*, =*M. scullyi* Tillyard, 1925c, p. 376)]. Fore wing broadest at level of midwing; precostal area apparently absent; stridulatory area about half the width of entire wing. Sharov, 1968. *Trias.,* Australia (New South Wales).——Fig. 114. *M. scullyi* Tillyard (type specimen of *C. andersoni* McKeown); fore wing, ×1 (McKeown, 1937).

Mesotitanodes Sharov, 1968, p. 197 [*M. tillyardi*; OD]. Fore wing very broad, especially medially; precostal area present; area between MA1 and MA2 about twice as wide as that

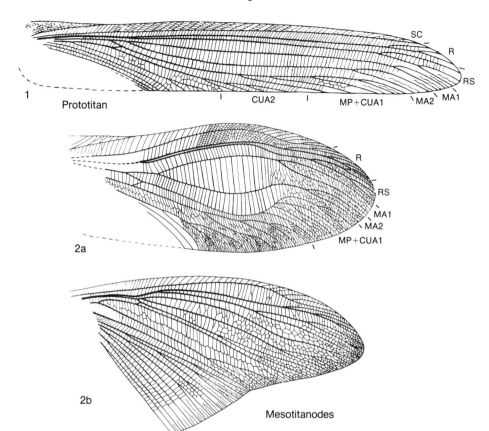

Fig. 115. Mesotitanidae (p. 182–183).

between RS and MA1. *Trias.*, USSR (Kirghiz).——Fig. 115,2. **M. tillyardi*; *a*, fore and *b*, hind wings, ×1 (Sharov, 1968).

Prototitan Sharov, 1968, p. 196 [**P. primitivus*; OD]. Fore wing nearly uniform in width, not increased medially; precostal area present; areas between RS and MP+CUA1 only slightly widened. *Trias.*, USSR (Kirghiz).——Fig. 115,*1*. **P. primitivus*; fore wing, ×1.5 (Sharov, 1968).

Ultratitan Sharov, 1968, p. 198 [**U. superior*; OD]. Little-known genus, based on distal fragment of wing. Stridulatory area extending nearly to wing apex. *Trias.*, USSR (Kirghiz).

Family PARATITANIDAE Sharov, 1968

[Paratitanidae Sharov, 1968, p. 198]

Fore wing with anterior margin uniformly curved; space between veins MA2 and MP+CUA1 very narrow; precostal area present; base of M anastomosed with R; RS arising from R in distal third of wing. Hind wing with enlarged anal lobe. *Trias.*

Paratitan Sharov, 1968, p. 199 [**P. libelluloides*; OD]. Fore wing with subcostal area nearly as broad as costal area; M branching from R nearer to forking of M than to wing base; MA2 slightly sigmoidal in distal half. *Trias.*, USSR (Kirghiz).——Fig. 116,*a,b*. **P. libelluloides*; *a*, fore and *b*, hind wings, ×1.6 (Sharov, 1968).——Fig. 116,*c*. *P. ovalis* Sharov; fore wing, ×1.5 (Sharov, 1968).

Family GIGATITANIDAE Sharov, 1968

[Gigatitanidae Sharov, 1968, p. 199]

Very large species. Fore wing with precostal area present; vein RS arising basally or near midwing; area between MA2 and

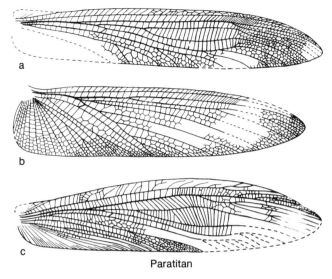

Fig. 116. Paratitanidae (p. 183).

MP+CUA1 the only broad space in the stridulatory area of the wing. *Trias.*

Gigatitan SHAROV, 1968, p. 200 [**G. vulgaris*; OD]. Fore wing: SC with basal branch extending about one-third wing length from base; RS arising from R near wing base. *Trias.*, USSR (Kirghiz).——FIG. 117,*3*. **G. vulgaris; a,* reconstruction, ×0.8; *b,* fore wing, ×0.6; *c,* hind wing, ×0.6; *d,* foreleg, ×2.0; *e,* head, antennae, pronotum, ×1 (all Sharov, 1968).

Nanotitan SHAROV, 1968, p. 202 [**N. magnificus*; OD]. Fore wing with large precostal area; RS arising near wing base; MA1 and MA2 unbranched; MA2 and MP+CUA1 anastomosed at base of wing. *Trias.*, USSR (Kirghiz).——FIG. 117,*1*. *N. extensus* SHAROV; fore wing, ×1 (Sharov, 1968).

Ootitan SHAROV, 1968, p. 201 [**O. curtis*; OD]. Fore wing very short and broad; RS arising at about midwing. *Trias.*, USSR (Kirghiz).——FIG. 117,*2*. **O. curtis;* fore wing, ×1 (Sharov, 1968).

Order PHASMATODEA Brunner, 1893

[*nom. correct.* BRUES, MELANDER, & CARPENTER, 1954, p. 102, *pro* Phasmodea BRUNNER, 1893, p. 76] [=Aeroplanoptera TILLYARD, 1923b, p. 481]

Moderate-sized to large insects, with much diversity of wing and body form. Mouthparts mandibulate, mandibles strong; eyes small; antennae typically long, slender, and multisegmented, less commonly short, with few to many segments; legs gressorial, long, and diversely modified; 5 tarsomeres. Fore wings typically reduced or absent in existing species but normally developed in some Mesozoic families. Hind wings of existing species rarely absent, commonly large; remigium tegminous; anal area greatly enlarged, fan-shaped. Venation of fore wings and of remigium of hind wings of existing species reduced, with dense reticulation, and a series of strong, parallel longitudinal veins, with very few branches. Abdomen long, slender, and cylindrical in Phasmatidae and most Mesozoic species, shorter and dorsoventrally compressed in Phylliidae; cerci unsegmented, typically short. Existing species foliage-feeders. Eggs deposited in ground litter or more rarely inserted in soil. *Trias.–Holo.*

For many years the Phasmatodea were considered to be a family within the order Orthoptera. However, their gressorial hind legs, five-segmented tarsi, unsegmented cerci, as well as several venational features, support the conclusion of BEIER (1967) that they represent a separate order within the orthopteroid complex. The order is now relatively small, with only two families, Phasmatidae and Phylliidae, generally recognized. The

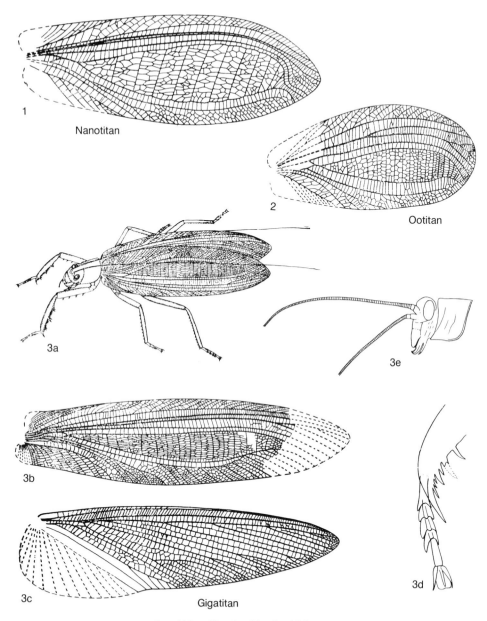

Fig. 117. Gigatitanidae (p. 184).

phasmatids, with long and cylindrical bodies, resemble twigs and small sticks; the phylliids, with dorsoventrally compressed bodies and flat extensions of the legs, resemble leaves.

Sharov (1968) has made the most significant contribution to our knowledge of the geological record of the order, and in particular to what appears to have been its early stages, with a study of a series of Mesozoic specimens. The geological record of the two existing families is limited to nymphs of one species of each family in Baltic amber. The early Mesozoic members of the order bear little resemblance, in general form, to the

Phasmatidae or Phylliidae. They do, however, have very long and narrow fore wings, with the longitudinal veins straight and parallel as in the phasmatids. Similarly, the remigium of the hind wing is long, narrow, and tegminous, with the longitudinal veins as in the fore wings. The compound vein, MP+CUA1, is present in the Mesozoic species and similar in form to that in both Phasmatidae and Phylliidae. The cerci in the Mesozoic fossils are unsegmented, as in the existing Phasmatodea. These features are indeed indicative of relationship, but more structural details are probably needed for conviction. At present, the Mesozoic record includes only five small families of these insects.

Family XIPHOPTERIDAE
Sharov, 1968

[Xiphopteridae SHAROV, 1968, p. 192]

Little-known family. Fore wing membranous; precostal area well developed; vein C with many fine branches; RS arising from R in distal half of wing, branched; MA forked; MP+CUA1 with a comb of branches directed posteriorly. Hind wing unknown. [Ordinal assignment doubtful.] *Trias.*

Xiphopterum SHAROV, 1968, p. 192 [*X. curvatum*; OD]. Fore wing broadest distally and curved posteriorly in distal area; precostal area forming a prominent bulge basally; MA forking in proximal part of wing. *Trias.*, USSR (Kirghiz).——FIG. 118,*2*. *X. curvatum*; fore wing, ×1.3 (Sharov, 1968).

Family AEROPLANIDAE
Tillyard, 1918

[Aeroplanidae TILLYARD, 1918c, p. 425]

Fore wing: precostal area long, but vein C without branches; longitudinal veins mostly parallel; crossveins unbranched, except in the anal area; RS arising from R near wing base; MP+CUA1 forked near level of origin of RS, with at most 5 branches. Hind wing unknown. MARTYNOV, 1928a; SHAROV, 1968; VISHNIAKOVA, 1980b. *Trias.*

Aeroplana TILLYARD, 1918c, p. 426 [*A. mirabilis*; OD]. Little-known genus, based on basal half of fore wing; CUA2, CUP, and 1A sigmoidal. [Originally placed in Protodonata, later in new order Aeroplanoptera (TILLYARD, 1923b).] *Trias.*, Australia (New South Wales).——FIG. 118,*1*. *A. mirabilis*; fore wing, ×2 (Sharov, 1968; after Tillyard, 1923b).

Paraplana SHAROV, 1968, p. 193 [*P. affinis*; OD]. Fore wing similar to that of *Aeroplana*, but the short, oblique base of CUA1 absent; MP+CUA1 with 3 branches. *Trias.*, USSR (Kirghiz).—— FIG. 118,*4*. *P. affinis*; fore wing, ×1.5 (Sharov, 1968).

Family AEROPHASMATIDAE
Martynov, 1928

[*nom. correct.* BRUES, MELANDER, & CARPENTER, 1954, p. 809, *pro* Aerophasmidae MARTYNOV, 1928a, p. 320]

Similar to Aeroplanidae, but fore wing lacking precostal area; vein MP+CUA1 with only 3 branches. Hind wing with RS and M anastomosed near wing base. *Jur.*

Aerophasma MARTYNOV, 1928a, p. 320 [*A. prynadai*; OD]. Fore wing with dense covering of hair; MA with 2 branches. SHAROV, 1968; VISHNIAKOVA, 1980b. *Jur.*, USSR (Kazakh).——FIG. 118,*3*. *A. prynadai*; *a*, fore and *b*, hind wings, ×1.3 (Sharov, 1968).

Family PROCHRESMODIDAE
Vishniakova, 1980

[Prochresmodidae VISHNIAKOVA, 1980b, p. 173, footnote]

Apparently related to Aeroplanidae. Antennae long and filiform. Fore wing: precostal area broad, with fine archedictyon; vein SC extending to wing apex; MP+CUA1 and branches of MA curved; 2A ending well before midwing. Hind wing: anal area very broad, branches of 2A directed to posterior margin of wing. Legs very long; male apparently with broad and spiny hind femora. *Trias.*

Prochresmoda SHAROV, 1968, p. 194 [*P. longipoda*; OD]. Precostal area extending nearly to midwing; crossveins very numerous, mostly straight, rarely branched. VISHNIAKOVA, 1980b. *Trias.*, USSR (Kirghiz).——FIG. 119. *P. longipoda*; *a*, fore and *b*, hind wings, ×2; *c*, whole insect, ×1 (Sharov, 1968).

Family CRETOPHASMATIDAE
Sharov, 1968

[Cretophasmatidae SHAROV, 1968, p. 193]

Fore wing much as in Aeroplanidae, but precostal area extending nearly to level of midwing; archedictyon present between veins

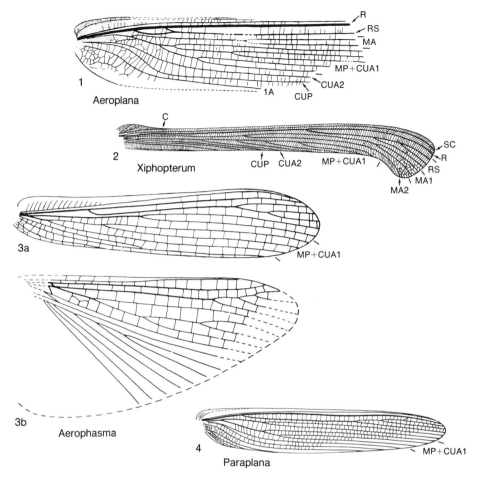

FIG. 118. Xiphopteridae, Aeroplanidae, and Aerophasmatidae (p. 186).

C and SC; RS unbranched; MA with 3 branches. *Cret.*

Cretophasma SHAROV, 1968, p. 193 [*C. raggei*; OD]. RS arising from R near base of wing. Anterior margin of wing slightly concave. *Cret.*, USSR (Kazakh).——FIG. 120,*4*. *C. raggei*; fore wing, ×2 (Sharov, 1968).

Family PHASMATIDAE Leach, 1815

[*nom. correct.* ROBERTS, 1941, p. 16, *pro* Phasmidae LEACH, 1815, p. 119]

Antennae commonly short and slender; fore wings coriaceous, commonly reduced or absent; hind wing typically well developed, with small coriaceous remigium and large anal fan, folded over the abdomen at rest; fore wing venation reduced, with a few, weakly developed, parallel veins; remigium of hind wings with a series of well-developed, parallel veins and numerous crossveins. Body commonly elongate; legs typically long, often spinose; middle and hind tibiae with a ventral, triangular, areolate area distally. *Oligo.– Holo.*

Phasma LICHTENSTEIN, 1796, p. 49. *Holo.*
Pseudoperla PICTET, 1854, p. 364 [*P. gracilipes*; OD]. Nymph, with small wing pads. Mesothorax slightly longer than pronotum. BACHOFEN-ECHT, 1949. *Oligo.*, Europe (Baltic).——FIG. 120,*2*. *P. gracilipes*; whole insect, ×2.5 (Germar & Berendt, 1856).

Family PHYLLIIDAE Brunner, 1893

[Phylliidae BRUNNER, 1893, p. 101]

Similar to Phasmatidae, but body flattened dorsoventrally; legs and abdominal

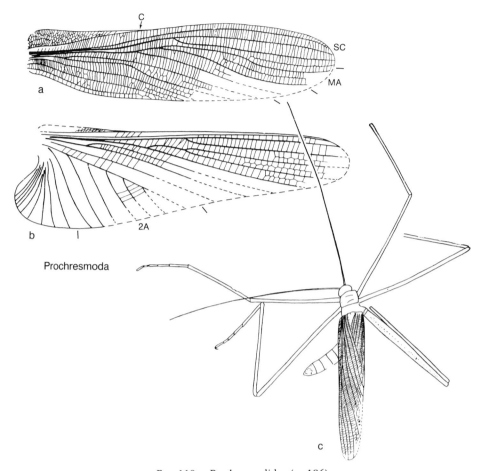

Fig. 119. Prochresmodidae (p. 186).

segments with lamellate extensions; fore wings reduced; hind wings commonly well developed, resting flat, not folded, over abdomen. *Oligo.,* Europe (Baltic)–*Holo.*

Phyllium ILLIGER, 1798, p. 499. *Holo.*
Electrobaculum SHAROV, 1968, p. 195 [**E. gracile*; OD]. Little-known genus, based on nymph. Ovipositor and cerci longer than those of other genera in family. *Oligo.,* Europe (Baltic).——FIG. 120,*3.* **E. gracile*; whole insect, ×3 (Sharov, 1968).

Family UNCERTAIN

The following genera, apparently belonging to the order Phasmatodea, are too poorly known to permit assignment to families.

Chresmodella BODE, 1953, p. 108 [**C. integra*; OD]. Little-known genus, based on fore wings, but details not clear; probably related to Halometridae (recent). SHAROV, 1968; VISHNIAKOVA, 1980b. *Jur.,* Europe (Germany).——FIG. 120,*1.* **C. integra*; fore wing, ×2.5 (after Bode, 1953).
Coniphasma BIRKET-SMITH, 1981, p. 245 [**C. rosenkrantzi*; OD]. Little-known genus, based on incomplete fore wing. R, RS, MA, MP, CUA, and CUP nearly straight and parallel; costa marginal; SC ending just beyond midwing; MA, MP, CUA, and CUP unbranched. [Ordinal assignment doubtful.] *Cret.,* Greenland.
Propygolampis WEYENBERGH, 1874, p. 84 [**P. bronni*; OD] [=*Halometra* OPPENHEIM, 1888, p. 230 (type, *Pygolampis gigantea* GERMAR, 1839, p. 207); *Sternarthron* HAASE, 1890b, p. 655 (type, *S. zitteli*)]. Little-known genus, similar to *Prochresmoda,* but antennae very short; longitudinal veins of fore wing straight and close together, much as in Aeroplanidae. Hind wing unknown. [*Propygolampis* and *Halometra* were originally placed in the order Hemiptera and have been confused with *Chresmoda* (Orthoptera). *Sternarthron* was placed by HAASE in the Araneae, but

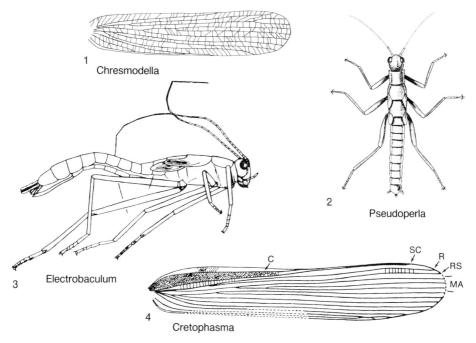

Fig. 120. Cretophasmatidae, Phasmatidae, Phylliidae, and Uncertain (p. 187–188).

as noted by HANDLIRSCH (1906b) the type specimens were insects; they were probably nymphs of *Propygolampis*.] HAGEN, 1862; HANDLIRSCH, 1906b, 1920; MARTYNOV, 1928a; CARPENTER, 1932a, 1992; ESAKI, 1949; SHAROV, 1968; VISHNIAKOVA, 1980b. *Jur.,* Europe (Germany); *Cret.,* China (Inner Mongolia).——FIG. 121. **Propygolampis gigantea* (GERMAR), *Jur.,* Germany; ventral view of whole insect, including antennae and cerci, specimen in Museum of Comparative Zoology, MCZ 6105, ×0.9 (Carpenter, new).

Order EMBIOPTERA Shipley, 1904

[Embioptera SHIPLEY, 1904, p. 261]

Small, subsocial insects with mandibulate mouthparts; tarsi with 3 segments, first segment of fore tarsi containing silk glands and much enlarged; hind femora enlarged; females apterous; males commonly winged; wings homonomous; veins except R and CUP usually weak; R and CUP thickened; SC short and not reaching midwing; RS typically forked before midwing; M simple or branched; CU usually with weak anterior branch and stronger CUP; anal lobes absent; cerci typically with 2 segments, generally asymmetrical in males. Ross, 1970. *Oligo.– Holo.*

The Embioptera are orthopteroids, apparently closely related to the Isoptera, although their precise ancestry is far from clear. Their morphological specializations, such as the slender body, short legs, and tendency for aptery, are adaptations to living in galleries. Lined with silk, produced by glands in the fore tarsi, the galleries are made on irregular surfaces of trees, rocks, moss, and even termite mounds. All existing Embioptera are subsocial, the female guarding the eggs and newly hatched nymphs in her galleries. The homonomous condition of the fore and hind wings is obviously secondary and the venation is much reduced.

The geological history of the Embioptera is poorly known. Two Permian genera, *Protembia* TILLYARD (1937b) and *Tillyardembia* ZALESSKY (1937d), originally placed in the order Embioptera, have been shown to be members of the Protorthoptera (MARTYNOV, 1940; CARPENTER, 1950; Ross, 1956); and another Permian genus, *Sheimia* MARTYNOVA

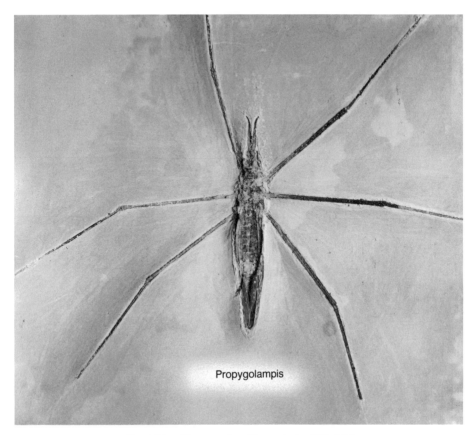

Fig. 121. Phasmatodea, Uncertain (p. 188–189).

(1958), also placed in the Embioptera, is known only by a poorly preserved wing that does not show sufficient structural features to justify assignment to the order (Carpenter, 1976; Ross, personal communication, 1977). The earliest record of the Embioptera, in the Baltic amber, consists of two apterous males belonging to the genus *Electroembia*. The genus is extinct but it is more specialized than some living genera. In all probability Embioptera will eventually be found in Cretaceous deposits.

Family EMBIIDAE Burmeister, 1838

[Embiidae Burmeister, 1838 in Burmeister, 1838–1839, p. 768]

Male wing (if present) with vein RS3+4 forked; hind tarsi with 1 or 2 ventral papillae (sole-bladders) on first segment; left cercus with 2 segments. *Oligo.–Holo.*

Embia Latreille, 1829, p. 257. *Holo.*
Electroembia Ross, 1956, p. 77 [*Embia antiqua Pictet, 1854, p. 370; OD]. Male apterous; basal segment of left cercus spiculate; hind basitarsus elongate and having 2 ventral papillae. *Oligo.,* Europe (Baltic).——Fig. 122,2. **E. antiqua* (Pictet); *a,* abdominal terminalia, *b,* hind basitarsus, ×30 (Ross, 1956).
Lithembia Ross, 1984, p. 83 [*Embia florissantensis Cockerell, 1908e, p. 230; OD]. Little-known genus, based on relatively large male, with typical embiid venation. Davis, 1939b. *Oligo.,* USA (Colorado).

Family NOTOLIGOTOMIDAE Davis, 1940

[Notoligotomidae Davis, 1940a, p. 681]

Male left cercus with 1 segment. *Mio.– Holo.*

Notoligotoma Davis, 1936, p. 244. *Holo.*
Burmitembia Cockerell, 1919d, p. 194 [*B. ve-

nosa; OD]. Male winged; veins strong; RS3+4 and M single; several strong oblique crossveins. DAVIS, 1939b, 1940a. *Mio.*, Burma.——FIG. 122,*1*. **B. venosa; a,* right fore wing, ×20; *b,* hind legs and abdomen from below, ×16 (both Davis, 1939b).

Order PSOCOPTERA
Shipley, 1904

[Psocoptera SHIPLEY, 1904, p. 261]

Small or minute insects, with short body; head relatively large; eyes large, ocelli usually present; antennae slender and commonly long, with numerous segments (12 to 50); mandibles well developed; laciniae forming sclerotized rods partially sunk into head and moved by muscles; labial palps vestigial in recent species; prothorax ordinarily small; meso- and metathorax usually distinct but may be fused to form compact unit; wings commonly present, reduced or absent in some, membranous and transparent, with distinct pterostigma. Fore wing commonly bearing conspicuous setae or scales; vein SC usually very short or absent; R and RS strongly developed, R enclosing distal end of pterostigma; RS usually forked; M arising from base, and in existing genera commonly coalesced with CUA basally and in contact with stem of R for short distance, terminating in several branches distally; CUA ordinarily forking near wing margin to form prominent cell, areola postica; CUP weakly developed, unbranched; usually only 1 anal vein. Hind wing generally smaller than fore wing, markedly so in some; in recent species, M, CUA, and CUP generally unbranched. Legs mostly homonomous, cursorial; tarsi in recent forms with 2 or 3 segments, in Permian Psocidiidae with 5 segments; abdomen with 9 or 10 distinct segments; cerci absent in recent forms, obsolescent in some Permian species. *Perm.–Holo.*

The Psocoptera constitute a very distinct and homogeneous group at present. Numerous recent families are now usually recognized, based on wing venation as well as tarsal and antennal segmentation, but much difference of opinion exists about character-

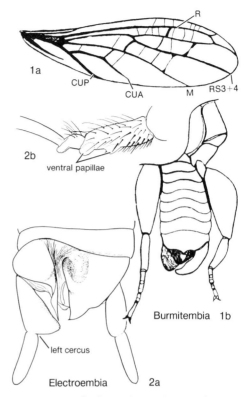

FIG. 122. Embiidae and Notoligotomidae (p. 190-191).

istics of the families. The classification used here is based on the systems of ROESLER (1944) and SMITHERS (1965, 1972). The families of Psocoptera are usually grouped into three or four suborders; but since there is little agreement among specialists regarding characteristics of the suborders or about the families included, the subordinal grouping is omitted here. In any case, most characteristics on which suborders are based are actually known in very few of the extinct genera.

The venation of the Psocoptera presents no problems in homology, except for the median system, which shows neither convexity nor concavity and which is accordingly designated here as the media (M). The evolution of the Psocoptera has involved extensive modifications of the wing venation; these have included varying degrees of anastomosis of RS and M, the branching of the media, and especially the form of the areola postica and its relationship to the media. Similar patterns have been developed independently in various lines of evolution within the order. Moreover, the Permian species seem to have evolved along lines very unlike those of recent families; none of the Permian groups seems to suggest an approach to any living family.

The discovery of Psocoptera in Permian deposits of Kansas in 1926 (TILLYARD, 1926f) was followed by finds of a similar nature in Australian and Russian beds. These fossils have revealed a fauna that is surprisingly diverse for Permian age and have indicated some lines of psocopterous evolution. On the basis of all evidence now available, the Permian Psocidiidae seem to be the most primitive members of the order known; they had homonomous wings, a relatively generalized venational pattern, five-segmented tarsi, and short but distinct cerci. Little is known of Mesozoic Psocoptera, but the order is well represented in the Oligocene of the Baltic amber. Fortunately, also, the amber species have been studied by several major workers on recent Psocoptera (HAGEN, ENDERLEIN, ROESLER), with the result that the systematics of the amber fauna is on as firm a basis as that of the recent one. It is clear that the Psocoptera in Baltic amber, although including several extinct genera, are remarkably modern, some of the species being as highly specialized as the most extensively modified recent species.

The Psocoptera are usually regarded as more closely related to the Hemiptera than to any other order. The mouthparts, although of a chewing type, are modified in several respects that are suggestive of the hemipterous pattern. In this connection, the prolongation of the head in *Dichentomum* (BECKER-MIGDISOVA, 1962a, p. 103) is especially interesting; it suggests the possibility that in these Permian species there was a tendency for the formation of a beaklike extension of the head. Relationship to the Hemiptera is also suggested by the close resemblance between the venation of the Psocidiidae and that of the Permian Archescytinidae (Hemiptera). In any event, although the Psocoptera as now known were almost certainly not ancestral to the Hemiptera, in all probability these two orders did arise from common ancestral stock.

Family PSOCIDIIDAE Tillyard, 1926

[Psocidiidae TILLYARD, 1926f, p. 319] [=Dichentomidae CARPENTER, 1932b, p. 6]

Fore wing usually slender; vein SC terminating on R; pterostigma commonly distinct; RS with 2 or 3 branches; M with at least 4 branches; length of areola postica about 3 times as long as its height. Hind wing similar to fore wing, usually about same size. Body structure known in *Dichentomum* only; head relatively large; antennae long, filamentous, with about 50 segments; head forming short rostrum; maxillary palpi long, with 3 segments; labial palpi shorter; fore tarsi with 4 segments; ovipositor prominent. *Perm.*

Dichentomum TILLYARD, 1926f, p. 320 [*D. tinctum*; OD] [=*Psocidium* TILLYARD, 1926f, p. 321 (type, *P. permianum*); *Chaetopsocidium* TILLYARD, 1926f, p. 331 (type, *C. sellardsi*); *Metapsocidium* TILLYARD, 1926f, p. 333 (type, *M. loxineurum*); *Pentapsocidium* TILLYARD, 1926f, p. 334 (type, *P. indistinctum*); *Permentomum* TILLYARD, 1926f, p. 335 (type, *P. tenuiforme*); *Parapsocidium* ZALESSKY, 1937d, p. 847 (type, *P. uralicum*)]. Pterostigma oval; M with 4 branches. *Perm.*, USA (Kansas), USSR (European RSFSR), Australia (New South Wales).——FIG. 123,2. *D. tinctum*, Perm., Kansas; *a*, whole insect, ×7 (Laurentiaux, 1953); *b*, fore wing, ×12 (Carpenter, 1933a).

Austropsocidium TILLYARD, 1935a, p. 267 [*A. pincombei*; OD]. R more remote from costa than in *Dichentomum*; pterostigma triangular. *Perm.*, Australia (New South Wales).——FIG. 123,1. *A. pincombei*; *a*, fore and *b*, hind wings, ×6 (Tillyard, 1935a).

Megapsocidium TILLYARD, 1935a, p. 269 [*M. australe*; OD]. Little-known wing apex; crossvein between base of pterostigma and RS. [Family assignment doubtful.] *Perm.*, Australia (New South Wales).

Stenopsocidium TILLYARD, 1935a, p. 270 [*S. elongatum*; OD]. Similar to *Dichentomum*, but M with 5 branches; pterostigma small. *Perm.*, Australia (New South Wales).——FIG. 123,5. *S. elongatum*; fore wing, ×9 (Tillyard, 1935a).

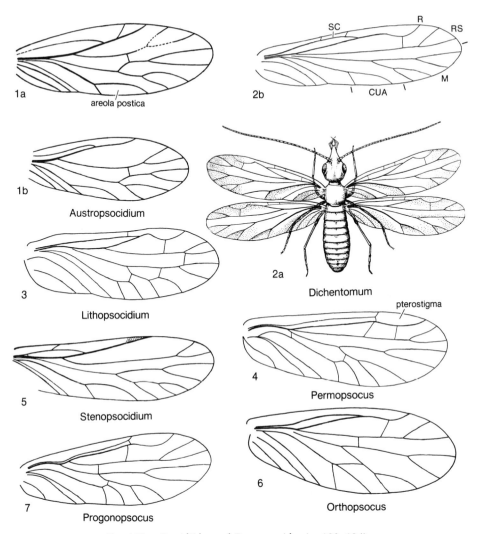

Fig. 123. Psocidiidae and Permopsocidae (p. 192–194).

Family PERMOPSOCIDAE Tillyard, 1926

[Permopsocidae Tillyard, 1926f, p. 340]

Fore wing broader than in Psocidiidae; veins SC and R close together and parallel; pterostigma well developed; fork of CUA high, with crossvein between CUA1 and M3+4. Hind wing similar to fore wing; fork of CUA slightly higher. Body structure little known; antennae much shorter than in Psocidiidae. *Perm.*

Permopsocus Tillyard, 1926f, p. 339 [**P. latipennis*; OD] [=*Ancylopsocus* Tillyard, 1926f, p. 344 (type, *A. insolitus*)]. End of SC connected to base of pterostigma by short crossvein; RS and stem of M divergent; fore wing with small sclerotized lobe near base of posterior margin. *Perm.*, USA (Kansas).——Fig. 123,*4*. **P. latipennis*; hind wing, ×11 (Carpenter, 1932b).

Lithopsocidium Carpenter, 1932b, p. 14 [**L. permianum*; OD]. Hind wing: SC clearly terminating on R near pterostigma; RS arising near middle of wing. *Perm.*, USA (Kansas). —— Fig. 123,*3*. **L. permianum*; hind wing, ×17 (Carpenter, 1933a).

Orthopsocus CARPENTER, 1932b, p. 15 [*O. singularis*; OD]. Hind wing: SC terminating on R near pterostigma; RS arising nearer wing base than in *Lithopsocidium*; fork of CUA triangular. *Perm.*, USA (Kansas).——FIG. 123,6. *O. singularis*; hind wing, ×12 (Carpenter, 1933a).

Progonopsocus TILLYARD, 1926f, p. 337 [*P. permianus*; OD]. Very similar to *Permopsocus*, but RS and stem of M parallel. *Perm.*, USA (Kansas).——FIG. 123,7. *P. permianus*; hind wing, ×14 (Carpenter, 1933a).

Family MARTYNOPSOCIDAE
Karny, 1930

[Martynopsocidae KARNY, 1930, p. 446, nom. subst. pro Dinopsocidae MARTYNOV, 1928b, p. 40]

Fore wing with vein SC terminating on R; pterostigma very slender; RS and M each with 3 branches. Hind wing and body unknown. *Perm.*

Martynopsocus KARNY, 1930, p. 446, nom. subst. pro *Dinopsocus* MARTYNOV, 1928b, p. 39, non BANKS, 1920 [*Dinopsocus arcuatus* MARTYNOV, 1928b, p. 40; OD] [=*Idelopsocus* ZALESSKY, 1929, p. 17 (type, *I. tartaricus*)]. RS with RS1, RS2, and RS3+4; fork of CUA shallow. *Perm.*, USSR (European RSFSR).——FIG. 124,1. *M. arcuatus* (MARTYNOV); fore wing, ×6 (Martynov, 1928b).

Family ARCHIPSYLLIDAE
Handlirsch, 1906

[Archipsyllidae HANDLIRSCH, 1906b, p. 502]

Antennae filiform, with at least 13 segments; mandibles elongate; fore and hind wings slender, very similar in shape and venation; vein SC short, coalesced with stem R at wing base, but almost immediately diverging towards and fusing with C, finally diverging back and joining R more distally; pterostigma well developed; RS forked distally; M with 4 branches; CU dividing basally; CUA forked distally; CUP well developed. VISHNIAKOVA, 1976. *Perm.–Cret.*

Archipsylla HANDLIRSCH, 1906b, p. 503 [*A. primitiva*; SD ENDERLEIN, 1909, p. 772]. Fore wing with pterostigma relatively short, about as long as wide; crossvein between RS and M1+2 near midwing. ENDERLEIN, 1929; VISHNIAKOVA, 1976. *Jur.*, Europe (Germany), USSR (Kazakh).——FIG. 124,2. *A. turanica* MARTYNOV, USSR; fore wing, ×10 (adapted from Martynov, 1926b and Vishniakova, 1976).

Archipsyllodes VISHNIAKOVA, 1976, p. 83 [*A. speciosis*; OD]. Pterostigma short; crossvein rs+m as in *Eopsylla*; areola postica very slender. *Cret.*, USSR (Asian RSFSR).

Archipsyllopsis VISHNIAKOVA, 1976, p. 83 [*A. baisica*; OD]. Very similar to *Archipsyllodes*, but pterostigma longer. *Cret.*, USSR (Asian RSFSR).

Eopsylla VISHNIAKOVA, 1976, p. 78 [*Dichentomum sojanensis* BECKER-MIGDISOVA, 1962a, p. 102; OD]. Similar to *Archipsylla*, but pterostigma more slender and crossvein from RS joining M before it divides into M1+2 and M3+4. *Perm.*, USSR (European RSFSR).

Family SURIJOKOPSOCIDAE
Becker-Migdisova, 1961

[Surijokopsocidae BECKER-MIGDISOVA, 1961b, p. 284]

Fore wing much wider distally than basally; vein M with 5 branches; basal parts of CUA and R+M thickened, forming cell at wing base; anal area very narrow. Hind wing and body unknown. *Perm.*

Surijokopsocus BECKER-MIGDISOVA, 1961b, p. 284 [*S. radtshenkoi*; OD]. SC close to R; costal area very wide at base; distal branches of CUA recurved. *Perm.*, USSR (Asian RSFSR).——FIG. 124,3. *S. radtshenkoi*; fore wing, ×7 (Becker-Migdisova, 1961b).

Family LOPHIONEURIDAE
Tillyard, 1921

[Lophioneuridae TILLYARD, 1921c, p. 417] [=Cyphoneuridae CARPENTER, 1932b, p. 18; Zoropsocidae TILLYARD, 1935a, p. 273]

Fore wing with vein SC short or absent, extending at most slightly beyond level of origin of RS and terminating on costal margin; RS with 2 branches; stem of M coalesced with stem R; M with 2 branches; CUA with weak posterior branch or unbranched. Hind wing little known, only about two-thirds length of fore wing; CUP and 1A apparently absent. Head broad, without prolongation as in Psocidiidae; antennae reaching only to about midwing. *Perm.*

Lophioneura TILLYARD, 1921c, p. 417 [*L. ustulata*; OD]. RS3+4 directed posteriorly, terminating at wing apex; M forked more deeply than RS. *Perm.*, Australia (New South Wales).——FIG. 124,4. *L. ustulata*; fore wing, ×10 (Tillyard, 1921c).

Austrocypha TILLYARD, 1935a, p. 277 [*A. abrupta*; OD]. Fore wing with SC apparently absent; distal part of R strongly bent anteriorly; stems of RS, M, and CUA arising from stem R and continuing nearly parallel; CUA strongly curved sigmoidally. Hind wing about half length of fore wing; RS and M forked; CU and 1A

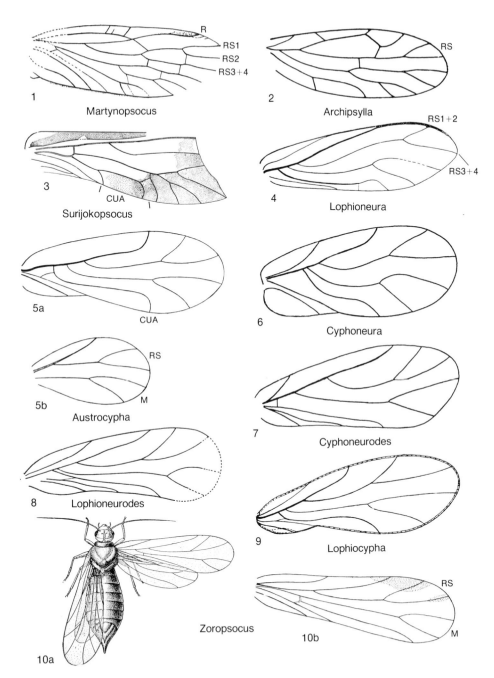

FIG. 124. Martynopsocidae, Archipsyllidae, Surijokopsocidae, and Lophioneuridae (p. 194–197).

absent. *Perm.*, Australia (New South Wales).——FIG. 124,5. **A. abrupta; a,* fore and *b,* hind wings, ×18 (Tillyard, 1935a).

Cyphoneura CARPENTER, 1932b, p. 18 [**C. permiana*; OD]. Fore wing nearly oval; R curved strongly toward anterior margin; branches of M directed posteriorly; CUA unbranched, sigmoidally curved. *Perm.*, USA (Kansas).——FIG. 124,6. **C. permiana*; fore wing, ×26 (Carpenter, 1932b).

196   *Hexapoda*

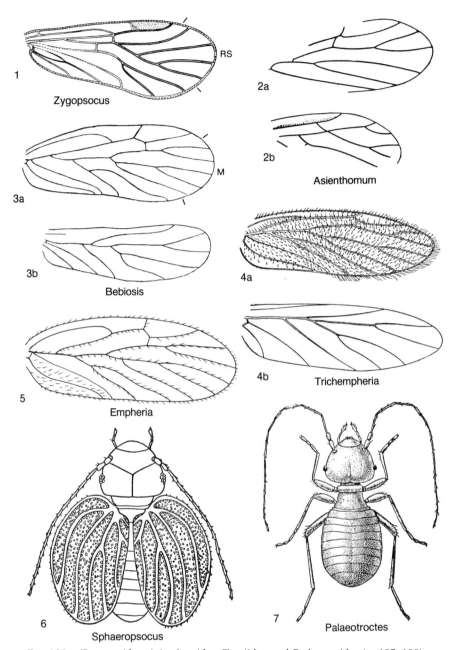

Fig. 125. Zygopsocidae, Asienthomidae, Trogiidae, and Pachytroctidae (p. 197–198).

Cyphoneurodes BECKER-MIGDISOVA, 1953b, p. 281 [**Cyphoneura reducta* CARPENTER, 1932b, p. 19; OD]. Fore wing subtriangular; R not so strongly curved as in *Cyphoneura*; branches of M directed distally; CUA straight. *Perm.*, USA (Kansas).——FIG. 124,7. **C. reducta* (CARPENTER); fore wing, ×22 (Carpenter, 1932b).

Lophiocypha TILLYARD, 1935a, p. 274 [**L. permiana*; OD]. Similar to *Lophioneura* but with RS arising much farther from base of wing, and SC extending about to level of origin of RS. *Perm.*, Australia (New South Wales).——FIG. 124,9. **L. permiana*; fore wing, ×18 (Tillyard, 1935a).

Lophioneurodes BECKER-MIGDISOVA, 1953b, p. 280 [*L. sarbalensis; OD]. CUA arising from stem of R+M; both branches of RS directed anteriorly; RS forked more deeply than M. *Perm., USSR (Asian RSFSR).*——FIG. 124,8. *L. sarbalensis; fore wing, ×18 (Becker-Migdisova, 1962a).

Zoropsocus TILLYARD, 1935a, p. 273 [*A. delicatulus; OD]. CUA independent of R+M, arising from common CU stem and connected to M by short crossvein; branches of RS directed anteriorly. *Perm., USSR (Asian RSFSR), Australia (New South Wales).*——FIG. 124,10a. *Z. tomiensis* BECKER-MIGDISOVA, USSR; whole insect, ×22 (Becker-Migdisova, 1962a). —— FIG. 124,10b. *Z. delicatulus*, Australia; fore wing, ×22 (Tillyard, 1935a).

Family ZYGOPSOCIDAE Tillyard, 1935

[Zygopsocidae TILLYARD, 1935a, p. 271]

Fore wing membranous with heavy veins; vein RS with 4 branches; M with 2 branches; CUA unbranched. Hind wing unknown. *Perm.*

Zygopsocus TILLYARD, 1935a, p. 271 [*Z. permianus; OD]. SC terminating on R near midwing; CUA strongly curved distally; wing margin thick. *Perm., Australia (New South Wales).*——FIG. 125,1. *Z. permianus; fore wing, ×14 (Tillyard, 1935a).

Family ASIENTHOMIDAE Martynov, 1926

[Asienthomidae MARTYNOV, 1926b, p. 1364, footnote, *nom. subst. pro* Lithentomidae MARTYNOV, 1926b, p. 1364]

Fore wing with vein CUA unbranched, M deeply forked, and RS forked. *Jur.*

Asienthomum MARTYNOV, 1926b, p. 1364, footnote, *nom. subst. pro* Lithentomum MARTYNOV, 1926b, p. 1364, *non* SCUDDER, 1867 [*Lithentomum praecox* MARTYNOV, 1926b, p. 1364; OD] [=*Lithopsocus* KARNY, 1930, p. 435, obj.]. Little-known wings. Fore wing with pterostigma about 4 times longer than wide; RS connected to R by an oblique crossvein at base of pterostigma. *Jur., USSR (Kazakh).*——FIG. 125,2. *A. praecox* (MARTYNOV); *a,* fore and *b,* hind wings, ×24 (Rohdendorf, 1962a).

Family TROGIIDAE Enderlein, 1911

[Trogiidae ENDERLEIN, 1911, p. 295]

Antennae with more than 20 segments; tarsi with 3 segments, claws smooth, without teeth; scales absent from body and wings; fore wing commonly broadly rounded distally, rarely absent. *Oligo.–Holo.*

Trogium ILLIGER, 1798, p. 500. *Holo.*

Bebiosis ENDERLEIN, 1911, p. 344 [*B. pertinens; OD]. Similar to *Empheria*, but last segment of maxillary palpus much enlarged and broadened. *Oligo.,* Europe (Baltic).——FIG. 125,3. *B. pertinens; a,* fore and *b,* hind wings, ×3 (Enderlein, 1911).

Empheria HAGEN, 1854, p. 225 [*E. reticulata; OD]. Hairs on membrane of fore wing restricted to area between CUP and anal margin. PICTET & HAGEN, 1856. *Oligo.,* Europe (Baltic).——FIG. 125,5. *E. reticulata; fore wing, ×35 (Enderlein, 1911).

Trichempheria ENDERLEIN, 1911, p. 345 [*Empheria villosa* HAGEN, 1882, p. 221; OD]. Hairs generally distributed on fore wing membrane. *Oligo.,* Europe (Baltic).——FIG. 125,4. *T. villosa* (HAGEN); *a,* fore and *b,* hind wings, ×26 (Becker-Migdisova & Vishniakova, 1962).

Family LEPIDOPSOCIDAE Enderlein, 1903

[Lepidopsocidae ENDERLEIN, 1903, p. 206]

Antennae and tarsi as in Trogiidae, but claws with preapical tooth; scales usually present on wings and body. *Pleist.–Holo.*

Lepidopsocus ENDERLEIN, 1903, p. 328. *Holo.*

Thylacella ENDERLEIN, 1911, p. 349 [*T. eversiana; OD]. Wings and body without scales; hind margin of fore wing evenly curved for its entire length. *Pleist.,* Africa.

Thylax HAGEN, 1866b, p. 172 [*T. fimbriatum; OD]. Similar to *Thylacella*, but hind margin of fore wing angular near middle. ENDERLEIN, 1911. *Pleist.,* Africa.

Family PSYLLIPSOCIDAE Kolbe, 1884

[*nom. transl.* ENDERLEIN, 1903, p. 208, *ex* Psyllipsocini KOLBE, 1884, p. 38]

Antennae and tarsi as in Trogiidae, but veins CUP and 1A meeting at point at wing margin. *Mio.–Holo.*

Psyllipsocus SELYS-LONGCHAMPS, 1872, p. 136. Parts of fore and hind wings. [Generic assignment doubtful.] COCKERELL, 1916a. *Mio.,* Burma–*Holo.*

Family PACHYTROCTIDAE Enderlein, 1905

[*nom. transl.* PEARMAN, 1936, p. 60, *ex* Pachytroctinae ENDERLEIN, 1905a, p. 46]

Antennae usually with 15 segments; tarsi with 3 segments; body normally convex, short; legs long and slender; hind femur not broadened. Wings often reduced or absent. *Oligo.–Holo.*

Pachytroctes ENDERLEIN, 1905, p. 46. *Holo.*
Palaeotroctes ENDERLEIN, 1911, p. 350 [**Atropos succinica* HAGEN, 1882, p. 231; OD]. Eyes very small; meso- and metathorax fused. *Oligo.,* Europe (Baltic).——FIG. 125,7. **P. succinicus* (HAGEN); complete insect, ×60 (Becker-Migdisova, 1962b).
Psylloneura ENDERLEIN, 1903, p. 317. Complete insect. [Generic assignment doubtful.] COCKERELL, 1919e. *Mio.,* Burma–*Holo.*
Sphaeropsocus HAGEN, 1882, p. 226 [**S. kunowi*; OD]. Fore wings forming short, broad elytra extending to end of abdomen; venation much reduced. *Oligo.,* Europe (Baltic).——FIG. 125,6. **S. kunowi*; complete insect, ×54 (Becker-Migdisova & Vishniakova, 1962).

Family LIPOSCELIDAE Enderlein, 1911

[Liposcelidae ENDERLEIN, 1911, p. 350]

Similar to Pachytroctidae, but body flattened and long; legs short; hind femur broad and flat; commonly wingless. *Oligo.–Holo.*

Liposcelis MOTSCHULSKY, 1852, p. 19. ENDERLEIN, 1911. *Oligo.,* Europe (Baltic)–*Holo.*

Family AMPHIENTOMIDAE Enderlein, 1903

[Amphientomidae ENDERLEIN, 1903, p. 332]

Antennae with 12 or 13 segments; tarsi with 3 segments; body and wings usually scaled; fore femora with row of teeth; vein CUA attached to M. *Oligo.*

Amphientomum PICTET, 1854, p. 376 [**A. paradoxum*; OD]. Claws with 2 teeth; abdomen with very small scales. *Oligo.,* Europe (Baltic). ——FIG. 126,*1*. **A. paradoxum; a,* fore and *b,* hind wings; ×35 (Enderlein, 1911).
Electrentomum ENDERLEIN, 1911, p. 337 [**E. klebsianum*; OD]. Wings and body entirely without scales. SMITHERS, 1972. *Oligo.,* Europe (Baltic).——FIG. 126,*2*. **E. klebsianum;* fore wing, ×15 (Enderlein, 1911).
Parelectrentomum ROESLER, 1940a, p. 228 [**P. priscum*; OD]. Similar to *Electrentomum,* but hind wing with closed middle cell; microtrichia on wing membranes very weakly developed. SMITHERS, 1972. *Oligo.,* Europe (Baltic). ——FIG. 126,*4*. **P. priscum; a,* fore and *b,* hind wings, ×15 (Roesler, 1940a).

Family EPIPSOCIDAE Pearman, 1936

[Epipsocidae PEARMAN, 1936, p. 60]

Antennae usually with 13 segments; tarsi with 2 segments; fore wing completely lacking a crossvein from vein R to RS below pterostigma; 1 anal vein. *Oligo.–Holo.*

Epipsocus HAGEN, 1866c, p. 203. ENDERLEIN, 1911. *Oligo.,* Europe (Baltic)–*Holo.*

Family PSOCIDAE Leach, 1815

[*nom. transl.* STEPHENS, 1829a, p. 312, *ex* Psocides LEACH, 1815, p. 139]

Antennae usually with 13 segments; tarsi with 2 segments; vein CUA1 in fore wing united with M by brief coalescence or by short crossvein. *Oligo.–Holo.*

Psocus LATREILLE, 1796, p. 99. ENDERLEIN, 1911; COCKERELL, 1921d. *Oligo.,* Europe (Baltic), England–*Holo.*
Copostigma ENDERLEIN, 1903, p. 229. ENDERLEIN, 1911. *Oligo.,* Europe (Baltic)–*Holo.*
Trichadenotecnum ENDERLEIN, 1909, p. 329. ENDERLEIN, 1911, 1929. *Oligo.,* Europe (Baltic)–*Holo.*

Family MESOPSOCIDAE Enderlein, 1903

[Mesopsocidae ENDERLEIN, 1903, p. 206]

Similar to Psocidae, but tarsi with 3 segments; vein CUA1 not united with M or absent. *Oligo.–Holo.*

Mesopsocus KOLBE, 1880, p. 184. *Holo.*
Elipsocus HAGEN, 1866, p. 207. ENDERLEIN, 1911. *Oligo.,* Europe (Baltic)–*Holo.*
Philotarsus KOLBE, 1880, p. 184. ENDERLEIN, 1911. *Oligo.,* Europe (Baltic)–*Holo.*

Family PSEUDOCAECILIIDAE Pearman, 1936

[Pseudocaeciliidae PEARMAN, 1936, p. 60]

Similar to Mesopsocidae, but tarsi with 2 segments. *Oligo.–Holo.*

Pseudocaecilius ENDERLEIN, 1903, p. 260. *Holo.*
Archipsocus HAGEN, 1882, p. 222. ENDERLEIN, 1911. *Oligo.,* Europe (Baltic)–*Holo.*
Electropsocus ROESLER, 1940b, p. 244 [**E. unguidens*; OD]. Veins and margin of fore wing hairy; antennae shorter than body; 3 pairs of gonapophyses in female; hypandrium of male evenly rounded. SMITHERS, 1972. *Oligo.,* Europe (Baltic).

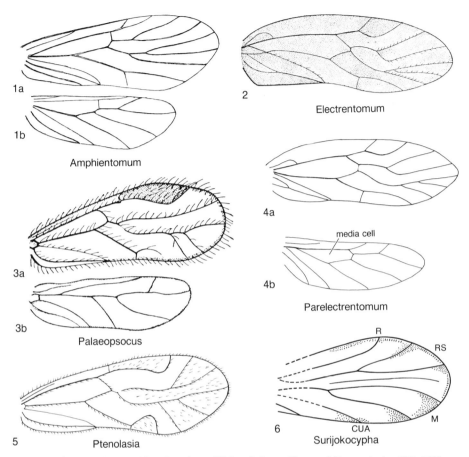

Fig. 126. Amphientomidae, Pseudocaeciliidae, Polypsocidae, and Uncertain (p. 198–200).

Palaeopsocus Kolbe, 1883, p. 190 [*Psocus tener Hagen in Pictet & Hagen, 1856, p. 60; OD]. Veins of fore wing each with single row of hairs; antennae twice as long as fore wing. *Oligo.*, Europe (Baltic).——Fig. 126,3. *P. tener* (Hagen); *a*, fore and *b*, hind wings, ×40 (Becker-Migdisova & Vishniakova, 1962).

Family POLYPSOCIDAE Pearman, 1936

[Polypsocidae Pearman, 1936, p. 60]

Antennae usually with 13 segments; tarsi with 2 or 3 segments; vein CUA2 of fore wing very short. *Oligo.–Holo.*

Polypsocus Hagen, 1866, p. 211. *Holo.*
Caecilius Curtis, 1837, p. 648. Enderlein, 1911; Navás, 1914. *Oligo.*, Europe (Baltic)–*Holo.*
Kolbea Bertkau, 1883, p. 128. Enderlein, 1911. *Oligo.*, Europe (Baltic)–*Holo.*

Ptenolasia Enderlein, 1911, p. 321 [*Caecilius pilosus Hagen, 1882, p. 283; OD]. Distal half of fore wing membrane covered with hairs. *Oligo.*, Europe (Baltic). —— Fig. 126,5. *P. pilosa* (Hagen); fore wing, ×15 (Enderlein, 1911).

Family UNCERTAIN

The following genera, apparently belonging to the order Psocoptera, are too poorly known to permit assignment to families.

Parapsocus Scudder, 1890, p. 117 [*P. disjunctus*; OD]. Little-known insect, probably psocopterous. Smithers, 1972. *Eoc.*, USA (Colorado).
Psococicadellopsis Becker-Migdisova, 1962a, p. 101 [*P. primitiva*; OD]. Little-known wing; RS apparently unbranched. *Trias.*, USSR (Issik-Kul).
Surijokocypha Becker-Migdisova, 1961b, p. 280 [*S. surijokovensis*; OD]. Little-known wings;

M forked more deeply than RS; CUA unbranched. SMITHERS, 1972. *Perm.,* USSR (Asian RSFSR).
——FIG. 126,6. **S. surijokovensis;* fore wing, ×20 (Becker-Migdisova, 1961b).

Vitriala BECKER-MIGDISOVA, 1961b, p. 282 [**V. nigriapex;* OD]. Little-known wing; M apparently unbranched. *Perm.,* USSR (Asian RSFSR).

Order ZORAPTERA
Silvestri, 1913

[Zoraptera SILVESTRI, 1913b, p. 205]

Very small insects, with mandibulate, chewing mouthparts; antennae moniliform, with 9 segments; maxillae and labium normal; prothorax well developed, larger than mesothorax and metathorax; legs well developed; tarsi with 2 segments; most individuals apterous, but winged individuals of both sexes appear in nearly all species; fore wing with greatly reduced venation, consisting of 3 unbranched veins (R, RS, and M) and a forked CUA; hind wing much smaller and with only 2 veins (RS and M); both wings may be shed along basal fracture lines; abdomen with 11 distinct segments and a pair of short cerci. Nymphs similar to adults in general form, some with developing wing buds. Adults and nymphs occur chiefly in decaying wood and rich humus and are apparently fungivorous. *Holo.*

This is a very small order, all known species belonging to one genus, *Zorotypus*. They appear to be highly specialized relicts of a basically primitive group, probably related to the Psocoptera.

Order MALLOPHAGA
Nitzsch, 1818

[Mallophaga NITZSCH, 1818, p. 280]

Small, apterous insects, with body dorsoventrally compressed; head large but diversely shaped; eyes reduced; antennae with 3 to 5 segments, either filiform or capitate; mouthparts with prominent, biting, dentate mandibles; prothorax well developed; mesothorax and metathorax small and frequently fused; legs moderately short, the tarsi with 1 or 2 segments and usually with 2 claws; abdomen with 8 to 10 segments. Nymphs and adults similar in general appearance, both ectoparasites on birds or, more rarely, mammals; they feed on fragments of epidermal products, such as feathers or hairs. *Holo.*

This is a small order of ectoparasites, with somewhat fewer specializations than the Anoplura. They appear to be related to the Anoplura and Psocoptera.

Order ANOPLURA Leach, 1815

[Anoplura LEACH, 1815, p. 64]

Small, apterous insects, the body dorsoventrally compressed; head relatively small; eyes usually much reduced or absent; antennae short, with 3 to 5 segments; mouthparts highly modified for sucking blood, with 3 piercing stylets; thorax small, segmentation indistinct; legs short but well developed, the single tarsal segment bearing a strong claw; abdomen with 9 segments; cerci absent. The nymphs and adults, which are similar in general appearance, are ectoparasites on mammals and feed on blood. *Holo.*

This is a small order of highly specialized ectoparasites, often treated as a suborder of the order Phthiraptera, with the Mallophaga constituting a second suborder. In either case, the ancestral stock of the Anoplura is uncertain, although the Psocoptera are probably nearer to that ancestral line than any other known order (KRISTENSEN, 1981).

Although the Anoplura are not known prior to the Holocene, two well-preserved males of *Neohaematopinus relictus* DUBININ have been found on the frozen body of a gopher (*Citellus*) in Indigirka, USSR (Asian RSFSR). The age of the gopher was determined as about 10,000 years (DUBININ, 1948).

Order CALONEURODEA
Handlirsch, 1937

[*nom. transl.* MARTYNOV, 1938a, p. 75, *ex* Caloneuroidea HANDLIRSCH, 1937, p. 64]

Fore and hind wings commonly similar in venation and texture, hind wings lacking an

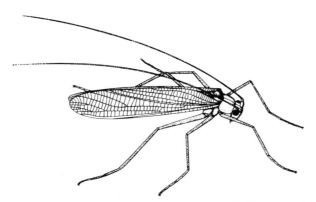

Fig. 127. Caloneurodea; restoration of *Paleuthygramma tenuicornis* Martynov, Paleuthygrammatidae, Permian of USSR, ×1.8 (Sharov, 1966a).

expanded anal area; base of vein CUA coalesced with stem of M for short distance; M commonly with 2 branches; CUA and CUP typically very close together, parallel, and unbranched; 2 or more unbranched anal veins; crossveins numerous, evenly distributed. Body slender; antennae filiform, multisegmented (Fig. 127); head hypognathous or prognathous; mouthparts mandibulate; legs long and slender; tarsi with 5 segments. Females with short, one-segmented cerci. Immature stages unknown. *U. Carb.–Perm.*

The Caloneurodea have generally been regarded as relatives of the Protorthoptera (Handlirsch, 1906b; Martynov, 1938a; Carpenter, 1943a, 1961, 1970; Martynova, 1962b; Burnham, 1984). However, Sharov (1966a) and Rasnitsyn (1980b) consider them to have been endopterygote insects, close to the base of the neuropteroids and the Coleoptera. Evidence for the affinities of the Caloneurodea is admittedly inconclusive and will probably remain so until much more is known about their body structure. At present, the few morphological features of the body mentioned above are known only in two genera (*Paleuthygramma* and *Synomaloptila*). However, since the case for the endopterygote status of the Caloneurodea seems to me to be very weak, I have followed here the more generally accepted view of the order's position.

Although the fore and hind wings do not usually show much difference in texture, in one family, the Caloneuridae, the fore wings are at least slightly tegminous, both membrane and veins being thicker than those of the hind wings. The wing venation throughout the order is considerably more reduced than that of the Protorthoptera and most other generalized orthopteroids. The convexities and concavities of the veins are unusually strong except for the media. The media is neutral, nearly flat, as in most other neopterous insects. The most prominent feature of the venation in most species is the close proximity of the strongly convex CUA and the strongly concave CUP.

This is one of the small extinct orders of insects, including only 16 genera distributed among 9 families. The genus *Genopteryx* Scudder (1885d), placed by Richardson (1956) in the Caloneurodea, has recently been synonymized by Burnham (1983) with *Gerarus* Scudder (1885d) of the order Protorthoptera.

Family CALONEURIDAE
Handlirsch, 1906

[Caloneuridae Handlirsch, 1906b, p. 140]

Wings with vein SC terminating well before apex of wing; CUA and CUP very close together and nearly parallel. M forked dichotomously; 4 anal veins; crossveins numerous. *U. Carb.*

Caloneura Brongniart, 1885a, p. 59 [*C. dawsoni*; OD] [=*Confusio* Handlirsch, 1919b, p. 547 (type, *Homaloneura royeri* Meunier, 1911a, p. 119)]. Fore wing with costal space abruptly narrowed basally; RS with 5 or 6 main branches; CUA and CUP unbranched. *U. Carb.*, Europe (France).——Fig. 128,8. *C. dawsoni*; *a*, fore and *b*, hind wings, ×1.2 (Carpenter, 1961).

Family ANOMALOGRAMMATIDAE
Carpenter, 1943

[*nom. correct.* Rohdendorf, 1957, p. 82, *pro* Anomalogrammidae Carpenter, 1943a, p. 74]

Vein SC terminating at midwing; M deeply forked; 3A absent. *Perm.*

Anomalogramma Carpenter, 1943a, p. 75 [*A. parvum*; OD]. RS forked to about half length of fork of M. *Perm.*, USA (Kansas).——Fig. 128,*10*. *A. parvum*; wing, ×6 (Carpenter, 1943a).

Family APSIDONEURIDAE
Carpenter, 1961

[Apsidoneuridae Carpenter, 1961, p. 151]

Wings with vein SC extending about to wing apex; M forked broadly, the anterior branch (M1+2) arching strongly away from posterior branch; 3 anal veins. *U. Carb.–Perm.*

Apsidoneura Carpenter, 1943a, p. 72 [*A. flexa*; OD]. Fore wing much more slender than in *Caloneura* and narrowed basally; RS with 2 or 3 main branches. Burnham, 1984. *U. Carb.*, Europe (France); *Perm.*, USA (Kansas).——Fig. 128,*9*. *A. flexa*; fore wing, ×2 (Carpenter, 1943a).
Homaloptila Handlirsch, 1919b, p. 546 [*Homaloneura similis* Meunier, 1911a, p. 118; OD]. Fore wing nearly as broad as in *Caloneura*; RS with 4 main branches. *U. Carb.*, Europe (France).——Fig. 128,*7*. *H. similis* (Meunier); *a*, fore and *b*, hind wings, ×2 (Carpenter, 1961).

Family EUTHYGRAMMATIDAE
Martynov, 1928

[*nom. correct.* Rohdendorf, 1957, p. 82, *pro* Euthygrammidae Martynov, 1928b, p. 49]

Similar to Paleuthygrammatidae but with vein CUA remote from CUP and CUP very close to 1A. *Perm.*

Euthygramma Martynov, 1928b, p. 50 [*E. parallelum*; OD]. RS unbranched. *Perm.*, USSR (European RSFSR).——Fig. 128,*6*. *E. parallelum*; wing, ×2.7 (Martynov, 1938a).

Family PALEUTHYGRAMMATIDAE
Carpenter, 1943

[*nom. correct.* Rohdendorf, 1957, p. 82, *pro* Paleuthygrammidae Carpenter, 1943a, p. 70]

Wings long and slender; vein SC terminating not far from wing apex; RS arising near midwing, with 2 or 3 branches; M separating from stem R well before origin of RS; CUA and CUP straight and very close together. *Perm.*

Paleuthygramma Martynov, 1930b, p. 42 [*P. tenuicornis*; OD]. RS branched only at wing apex. *Perm.*, USSR (Asian RSFSR), USA (Kansas).——Fig. 128,*5a*. *P. tenuicornis*, USSR; wing of holotype as preserved, ×2.5 (Martynov, 1930b) (see also Fig. 127).——Fig. 128,*5b,c*. *P. acutum* Carpenter, Kansas; *b*, fore and *c*, hind wings, ×3 (Carpenter, 1943a).
Pseudogramma Carpenter, 1943a, p. 70 [*Euthygramma aberrans* Martynov, 1938a, p. 73; OD]. M unbranched. *Perm.*, USSR (European RSFSR).——Fig. 128,*4*. *P. aberrans* (Martynov); wing, ×3 (Martynov, 1938a).
Vilvia Zalessky, 1933, p. 137 [*V. densinervosa*; OD]. Crossveins numerous and very close together. *Perm.*, USSR (Asian RSFSR).
Vilviopsis Martynov, 1938a, p. 73 [*V. extensa*; OD]. RS with 2 long branches. [Probably synonymous with *Paleuthygramma*.] *Perm.*, USSR (European RSFSR).——Fig. 128,*3*. *V. extensa*; wing, ×2 (Martynov, 1938a).

Family PERMOBIELLIDAE
Tillyard, 1937

[*nom. transl.* Martynov, 1938b, p. 76, *ex* Permobiellinae Tillyard, 1937a, p. 101]

Wings moderately slender; vein SC terminating slightly beyond midwing; RS with 3 branches; CUA and CUP close together proximally but diverging distally. *U. Carb.–Perm.*

Permobiella Tillyard, 1937a, p. 101 [*P. perspicua*; OD]. R extending to wing apex; M forking well beyond origin of RS; crossveins strongly convex. Carpenter, 1943a. *Perm.*, USA (Kansas).——Fig. 129,*3*. *P. perspicua*; wing as preserved, ×5 (Carpenter, 1943a).
Pseudobiella Carpenter, 1970, p. 407 [*P. fasciata*; OD]. Similar to *Permobiella*, but crossveins relatively weak, not strongly convex; front margin of wing concave near end of SC; wings with dark transverse band. *U. Carb.*, USA (New Mexico).——Fig. 129,*1*. *P. fasciata*; fore wing, ×5 (Carpenter, 1970).

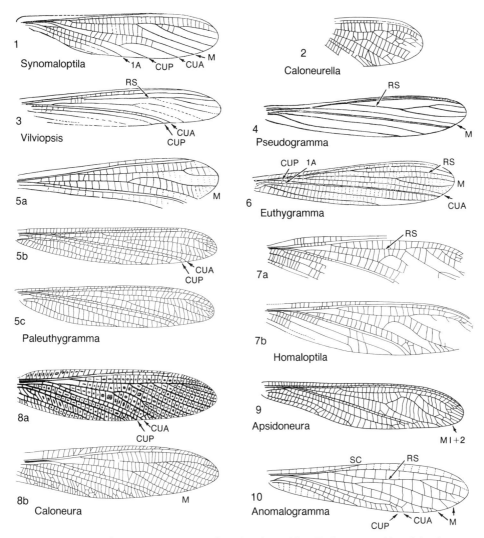

FIG. 128. Caloneuridae, Anomalogrammatidae, Apsidoneuridae, Euthygrammatidae, Paleuthygrammatidae, Synomaloptilidae, and Uncertain (p. 202–204).

Family AMBONEURIDAE
Carpenter, 1980

[Amboneuridae CARPENTER, 1980, p. 111]

Apparently related to the Permobiellidae, but veins CUA and CUP not close together; crossveins forming a coarse network in strong relief over the wing. [Ordinal assignment uncertain.] *U. Carb.*

Amboneura CARPENTER, 1980, p. 112 [*A. klosei*; OD]. Vein RS with 5 terminal branches; M forked to level of midwing. *U. Carb.*, USA (Pennsylvania). ——FIG. 129,4a,b. *A. klosei*; a, photograph of wing showing strong relief of crossveins, ×2.4; b, drawing of venation as preserved in holotype, ×2.4 (both Carpenter, 1980).

Family PLEISIOGRAMMATIDAE
Carpenter, 1943

[*nom. correct.* ROHDENDORF, 1957, p. 82, *pro* Pleisiogrammidae CARPENTER, 1943a, p. 73]

Wings nearly oval, broader than in Paleuthygrammatidae, narrowed basally; vein

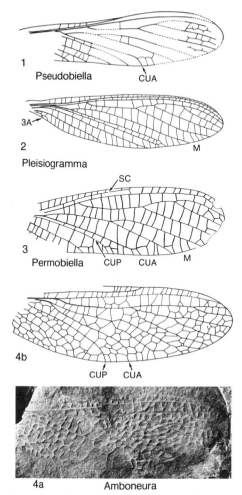

Fig. 129. Permobiellidae, Amboneuridae, and Pleisiogrammatidae (p. 202–204).

SC extending well beyond midwing; 3A vestigial or absent. *Perm.*

Pleisiogramma CARPENTER, 1943a, p. 73 [**P. mediale*; OD]. M unbranched. *Perm.*, USA (Kansas).——FIG. 129,2. **P. mediale*; wing, ×3.2 (Carpenter, 1943a).

Family SYNOMALOPTILIDAE Martynov, 1938

[Synomaloptilidae MARTYNOV, 1938a, p. 76]

Related to Euthygrammatidae, but veins CUA and CUP anastomosed for a considerable distance. *Perm.*

Synomaloptila MARTYNOV, 1938a, p. 76 [**S. longipennis*; OD]. SC terminating well before wing apex; separation of CUA and CUP at about level of midwing. *Perm.*, USSR (Asian RSFSR).——FIG. 128,1. **S. longipennis*; wing, ×2 (Martynov, 1938a).

Family UNCERTAIN

The following genera, apparently belonging to the order Caloneurodea, are too poorly known to permit assignment to families.

Caloneurella CARPENTER, 1934, p. 324 [**C. carbonaria*; OD]. Apical wing fragment, probably related to *Caloneura*. *U. Carb.*, USA (Pennsylvania).——FIG. 128,2. **C. carbonaria*; ×1.8 (Carpenter, 1934).

Pruvostiella HANDLIRSCH, 1922, p. 82 [**Euthyneura lecomtei* PRUVOST, 1919, p. 115; OD]. Small wing fragment. *U. Carb.*, Europe (France).

Order MIOMOPTERA Martynov, 1927

[Miomoptera MARTYNOV, 1927d, p. 101, *emend.* MARTYNOV, 1938b, p. 138]

Small to very small insects, with wings nearly homonomous. Fore wing with vein SC ending before or at midwing; R commonly with a distal twig; RS arising before midwing, with at least 3 terminal branches; M commonly coalesced with CUA basally to varying amounts, but diverging in basal third of wing; M deeply forked; CUA with 2 or 3 terminal branches; CUP unbranched; 2 anal veins typically present. Hind wing similar in form to fore wing, without an anal lobe or fan; M usually arising from CU very near wing base. Body structure little known; head of moderate size; mouthparts apparently mandibulate; antennae conspicuous, relatively thick, with 15 to 20 segments; tarsi with 4 segments (*Palaeomantis*); cerci short. Immature stages unknown. *U. Carb.–Perm.*

The status of this order is uncertain. As originally proposed by MARTYNOV (1927d) it included five Permian families, previously placed in the order Protorthoptera, but it was based mainly on one of them, the Palaeomanteidae (=Delopteridae). The following year TILLYARD (1928b), obviously unaware of MARTYNOV's article, proposed the new order Protoperlaria for the same series of families except the Palaeomanteidae. Subsequent studies of extensive collections of

Palaeomanteidae and Lemmatophoridae (Carpenter, 1933a, 1935a; Martynov, 1938b; Martynova, 1958, 1961b, 1962a) supported the view that the Palaeomanteidae were sufficiently distinctive to justify ordinal separation and that the Lemmatophoridae and related families were in reality part of the order Protorthoptera.

The Miomoptera, as exemplified by the Palaeomanteidae, stand apart from the protorthopterous families, with which they have been associated, by two distinct features. One of these is the absence of the anal lobe or fan on the hind wing, probably a secondary condition, as in the Isoptera and Embioptera. The other is the very small number of crossveins and their virtual absence from the costal area of both wings.

Several families have been added to the Miomoptera since Martynov's original publication on the order. Two of these, Archaemiopteridae (Guthörl, 1939) and Metropatoridae Handlirsch (1906a) almost certainly belong in the order (Martynova, 1958, 1961b; Carpenter, 1965). The evidence for the others, however, is very weak and in my opinion insufficient to justify their inclusion in the Miomoptera. The family Permembiidae Tillyard (1937b), originally described in the Psocoptera, has been transferred to the Miomoptera by Kukalová (1963a) and Riek (1973, 1976a); *Permembia* itself is known from a very few, poorly preserved specimens (Carpenter, 1976), with a venation that has little in common with that of the Palaeomanteidae. Some details of body structure are preserved in two specimens, but since almost nothing is known of the body of the Palaeomanteidae we have no basis for comparing those details. The family Permosialidae Martynov (1928b), originally in the Neuroptera, has been placed in the Miomoptera by Riek (1976a) and Rasnitsyn (1977c), as have the families Permonkidae Rasnitsyn (1977c) and Palaeomantiscidae Rasnitsyn (1977c). These families, however, are characterized by broadened anal areas or anal lobes on the hind wings and numerous crossveins on both wings including the costal areas. Since no revisions of the definition or diagnosis of the Miomoptera have been proposed for the accommodation of these families, the order is treated here essentially as it was defined by Martynov (1938b) and Martynova (1961b, 1962a). The families Permembiidae, Permosialidae, Permonkidae, and Palaeomantiscidae, along with a nymphal form, *Permonympha* Sharov (1957b), are included under Neoptera, Order Uncertain.

The homologies of the wings of the Miomoptera have not been definitely determined. In the fore wing, veins R and CUA are clearly convex, but RS and M show no definite topography. Whether the branches of M represent MA and MP (Kukalová, 1963a) or only one of these veins is uncertain; they are designated here as M1+2 and M3+4. The amount of anastomosis of M with CUA in the fore wing varies from genus to genus; in some species (e.g., *Permodelopterum obscurum* Kukalová; see Fig. 131,4a), M appears to arise independently of CUA, which joins it later; in others (e.g., *Palaeomantis minutum* (Sellards); see Fig. 131,1b) the stem of M seems to be coalesced with that of CUA from the very base of the wing. In all species, however, M diverges from CUA before midwing. In some individual wings, M1+2 seems to arise from RS or RS3+4; this may be a specific or even a generic characteristic, and there is some evidence that it occurs as an individual variation (Carpenter, 1939).

The Miomoptera are among the smallest insects known from the Upper Carboniferous and Permian. However, their affinities are not clear. They are generally considered to have been related to the Psocoptera, although Rasnitsyn (1980b) believes them to have been endopterygote insects, close to the ancestral stock of the Hymenoptera. Martynov concluded (1938b) that they were an early, aberrant branch of protorthopterous or perlarian stock.

Family METROPATORIDAE
Handlirsch, 1906

[Metropatoridae Handlirsch, 1906a, p. 681]

Hind wing nearly oval; vein SC short, weakly developed, and close to R; RS forked

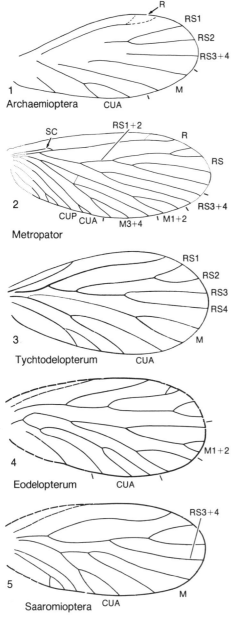

Fig. 130. Metropatoridae and Archaemiopteridae (p. 206).

before midwing; RS1+2 with 4 terminal branches; RS3+4 forked once; M apparently arising from CUA at base of wing and forked almost to level of origin of RS; CUA with short terminal fork. Fore wing and body unknown. *U. Carb.*

Metropator HANDLIRSCH, 1906a, p. 681 [*M. pusillus*; OD]. RS3+4 more deeply forked than RS1+2; M1+2 forked distally; M3+4 forked twice. TILLYARD, 1926c; CARPENTER, 1965. *U. Carb.*, USA (Pennsylvania).——FIG. 130,2. *M. pusillus*; hind wing, ×6.5 (Carpenter, 1965).

Family ARCHAEMIOPTERIDAE Guthörl, 1939

[Archaemiopteridae GUTHÖRL, 1939, p. 320]

Fore and hind wings broadly oval; vein RS1+2 forked, at least distally. Body unknown. *U. Carb.–Perm.*

Archaemioptera GUTHÖRL, 1939, p. 320 [*A. carbonaria*; OD]. RS3+4 unbranched; M forked to more than half its length; CUA with distal fork only. *U. Carb.*, Europe (Germany).——FIG. 130,1. *A. carbonaria*; fore(?) wing, ×10 (Carpenter, new, based on holotype).

Eodelopterum SCHMIDT, 1962, p. 838 [*E. priscum*; OD]. Similar to *Tychtodelopterum*, but CUA with 3 terminal branches; M1+2 terminating at wing apex. GUTHÖRL, 1963. *U. Carb.*, Europe (Germany).——FIG. 130,4. *E. priscum*; hind wing, ×12 (Guthörl, 1963).

Saaromioptera GUTHÖRL, 1963, p. 22 [*S. jordani*; OD]. Similar to *Tychtodelopterum*, but RS3+4 unbranched. *U. Carb.*, Europe (Germany).——FIG. 130,5. *S. jordani*; fore(?) wing, ×11 (Guthörl, 1963).

Tychtodelopterum MARTYNOVA, 1958, p. 70 [*T. relictum*; OD]. Similar to *Archaemioptera*, but RS3+4 deeply forked; CUA forked to at least half its length. *Perm.*, USSR (Asian RSFSR).——FIG. 130,3. *T. relictum*; hind wing, ×16 (Martynova, 1962a).

Family PALAEOMANTEIDAE Handlirsch, 1906

[*nom. correct.* BRUES, MELANDER, & CARPENTER, 1954, p. 811, *pro* Palaeomantidae HANDLIRSCH, 1906b, p. 348] [=Delopteridae SELLARDS, 1909, p. 168]

Fore wing membranous or coriaceous; vein SC usually ending before midwing, less commonly near midwing; RS usually with 3 terminal branches; basal stem of M apparently independent of both R and CU in some genera, but anastomosed with one of these in other genera; M forked deeply, usually to midwing; rarely, M1+2 apparently anastomosed with stem of RS (or connected by crossvein), resembling additional branch of RS; CUA with 2 or 3 terminal branches; distinct marginal indentation at end of CUP in some genera. Hind wing similar to fore wing except for differences characteristic of the order. Antennae with about 15 short segments; tarsi with 4 segments; abdomen short, with wings projecting far beyond abdomen at rest; cerci very short. *U. Carb.–Perm.*

Palaeomantis HANDLIRSCH, 1904b, p. 4 [*P. schmidti*; OD] [=*Delopterum* SELLARDS, 1909, p. 168 (type, *D. minutum*); *Pseudodelopterum* MARTYNOV, 1928b, p. 66 (type, *Delopterum latum* SELLARDS, 1909); *Pseudomantis* MARTYNOV, 1928b, p. 73 (type, *P. minuta*); *Leptoneurula* MARTYNOV, 1928b, p. 77 (type, *L. insignis*); *Delopsocus* TILLYARD, 1928f, p. 474 (type, *D. elongatus*); *Miomantisca* ZALESSKY, 1956a, p. 275 (type, *M. clara*); *Miomatoneurites* ZALESSKY, 1956a, p. 278 (type, *M. sylvaensis*); *Delopteriella* ZALESSKY, 1956a, p. 284 (type, *D. graciosa*); *Stefanomioptera* GUTHÖRL, 1962a, p. 67 (type, *S. hangardi*)]. Fore wing slender, membranous; SC terminating at about midwing; RS diverging from R in basal third of wing and usually with branches RS1, RS2, and RS3+4; M separating from CUA near level of origin of RS; M1+2 and M3+4 diverging near midwing; CUA forked; crossveins very few and weakly developed; posterior margin of wing either smoothly curved or with an indentation at end of CUP. Hind wing slightly shorter than fore wing; SC short, usually terminating about one-third wing length from base. Hind wing similar in form to fore wing but apparently lacking a separate and distinct CUP; CUA not preserved as a definite, convex vein as in fore wing, but forming with CUP a compound vein (CU) and preserved as a strong ridge within a furrow. Head with large compound eyes; first tarsal segment longer than others; cerci short. CARPENTER, 1933a, 1967c; GUTHÖRL, 1962a; MARTYNOVA, 1962a; KUKALOVÁ, 1963a. *U. Carb.,* Europe (Germany); *Perm.,* USA (Kansas, Oklahoma), Europe (Czechoslovakia), USSR (European RSFSR).——FIG. 131,*1a. P. hangardi* (GUTHÖRL), U. Carb., Germany; fore wing, ×9 (Carpenter, 1967c).——FIG. 131,*1b,c. *P. minutum* (SELLARDS), Perm., Kansas; *b,* fore and *c,* hind wings, ×12 (Carpenter, 1933a).

Miomatoneura MARTYNOV, 1927d, p. 106 [*M. frigida*; OD]. Fore wing as in *Palaeomantis,* but M1+2 arising from stem of RS or connected to it by crossvein; CUA with 2 or 3 terminal branches. Hind wing unknown. MARTYNOVA, 1961b, 1962a; KUKALOVÁ, 1963a. *Perm.,* USSR (European RSFSR), Europe (Czechoslovakia).——FIG. 131,*2a. *M. frigida,* Perm., USSR; fore wing, ×8 (Martynova, 1962a).——FIG. 131,*2b. M. candida* KUKALOVÁ, Perm., Czechoslovakia; fore wing, ×14 (Kukalová, 1963a).

Miomatoneurella MARTYNOVA, 1958, p. 71 [*M. reducta*; OD]. Fore wing similar to that of *Miomatoneura,* but RS1+2 unbranched. MARTYNOVA, 1961b. *Perm.,* USSR (Asian RSFSR).——FIG. 131,*3. *M. reducta*; fore wing, ×15 (Martynova, 1962a).

Permodelopterum KUKALOVÁ, 1963a, p. 25 [*P. obscurum*; OD]. Fore wing similar to that of *Perunopterum* but broader and base of M apparently coalesced with R basally. Hind wing and body unknown. *Perm.,* Europe (Czechoslovakia). ——FIG. 131,*4a. *P. obscurum*; fore wing, ×8 (Kukalová, 1963a).——FIG. 131,*4b. P. lumbiforme* KUKALOVÁ; fore wing, ×12 (Kukalová, 1963a).

Perunopterum KUKALOVÁ, 1963a, p. 16 [*P. peruni*; OD]. Fore wing membranous or distinctly coriaceous, densely covered with minute hairs, and more slender than in *Palaeomantis*; indentation of hind margin at end of CUP pronounced; SC terminating near midwing; R with distal twig; RS arising before midwing, typically with branches RS1, RS2, and RS3+4; stem of M free from R basally; M arising from CUA at about level of origin of RS, with 2 long branches; CUA forked. Hind wing similar to fore but with SC shorter and weaker and costal area narrower; M separating from CUA nearer wing base. Both wings tend to show more crossveins than in *Palaeomantis,* even having reticulation in anal area. Body structure apparently much as in *Palaeomantis*; short cerci present. *Perm.,* Europe (Czechoslovakia).——FIG. 131,*5a. *P. peruni*; fore wing, ×12 (Kukalová, 1963a).——FIG. 131,*5b. P.(?) corium* KUKALOVÁ; fore wing, ×8 (Kukalová, 1963a).

Order THYSANOPTERA Haliday, 1836

[Thysanoptera HALIDAY, 1836, p. 439]

Small or minute insects, with slender body (Fig. 132); head usually quadrangular; compound eyes small but prominent, with relatively large, rounded facets; ocelli commonly present; antennae with 6 to 10 segments; labrum and labium forming a short cone, containing as stylets the left mandible (right one absent or vestigial) and extensions of the 2 maxillae; maxillary and labial palpi present; prothorax free from mesothorax and well developed; wings usually present, nearly homonomous, membranous but very narrow, often strap-shaped, with not more than 2 longitudinal veins; wings fringed with long setae, at least along posterior margins; both brachypterous and apterous individuals may occur in some species; legs short, with 1 to 2 tarsal segments; abdomen elongate, seg-

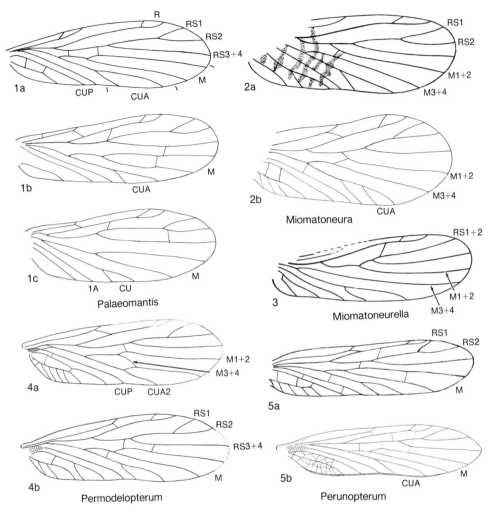

Fig. 131. Palaeomanteidae (p. 206–207).

mentation distinct; ovipositor present (Terebrantia) or absent (Tubulifera); cerci absent. Perm.–Holo.

The Thysanoptera, or thrips, feed by lacerating the surface of plants with their stylets and drawing the plant juices through the mouth cone. Some species are predaceous on small arthropods. The nymphs hatch from eggs laid on or in plant tissue; they resemble the adults in general form and feeding habits. Nymphs of most species pass through two or three quiescent stages (prepupa and pupa) in which wing pads occur.

The thrips, like the Psocoptera, are a distinctive and homogeneous order of insects. Two suborders, Terebrantia and Tubulifera, are generally recognized; the most obvious differences between them are in the presence or absence of the ovipositor in the female, and in the shape of the terminal abdominal segment. However, other differences, of more phylogenetic significance, are found in the detailed structure of the mouthparts and wings. Such morphological evidence indicates that the Terebrantia are more primitive than the Tubulifera and that the family Aeolothripidae of the Terebrantia is the most primitive of the recent families.

Thysanoptera are well represented in the Tertiary deposits, mainly the Baltic amber; these Oligocene species have been studied in detail by two authorities on recent thrips, BAGNALL and PRIESNER. Two pre-Tertiary thrips have been described, one (*Liassothrips*) from the Jurassic (MARTYNOV, 1927b) and the other (*Permothrips*) from the Permian (MARTYNOV, 1935a). That these are thrips seems almost certain, although their subordinal positions are obscure. The fossil record of the Thysanoptera shows little to date about the evolution of the order. The Permian species appear to have had somewhat larger wings than any existing species, but no veins are preserved and the structure of the mouthparts is unknown. The classification used here is that of PRIESNER (1949).

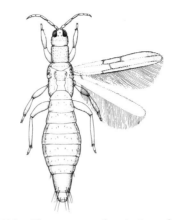

FIG. 132. Thysanoptera; dorsal view of whole insect, *Desmothrips propinquus*, Holocene, ×20 (CSIRO, 1970).

Suborder TEREBRANTIA Haliday, 1836

[Terebrantia HALIDAY, 1836, p. 439]

Terminal abdominal segment conical (rarely tubular) in females, bluntly rounded in male; female with sawlike ovipositor; fore wings nearly always with at least 1 longitudinal vein (in addition to costal vein) extending to apex. *Perm.–Holo.*

Family PERMOTHRIPIDAE Martynov, 1935

[Permothripidae MARTYNOV, 1935a, p. 334]

Head somewhat extended; pronotum transverse; legs short, tibiae more slender than femora; wings broad and long, extending beyond abdomen; ovipositor apparently present. [Subordinal position of the family uncertain; it probably represents an extinct suborder.] *Perm.*

Permothrips MARTYNOV, 1935a, p. 334 [**P. longipennis*; OD]. Abdomen narrowed distally, not tubular. *Perm.,* USSR (Asian RSFSR).——FIG. 133,5. **P. longipennis*; general form of body and wings, ×16 (Martynov, 1935a).

Family AEOLOTHRIPIDAE Uzel, 1895

[Aeolothripidae UZEL, 1895, p. 42]

Wings broad and rounded at apex; ovipositor curved upward; antennae with 9 segments. *Eoc.–Holo.*

Aeolothrips HALIDAY, 1836, p. 451. COCKERELL, 1917b; BAGNALL, 1924a; PRIESNER & QUIÉVREUX, 1935. *Oligo.,* England, Europe (France, Germany)–*Holo.*
Archankothrips PRIESNER, 1924, p. 132 [**A. pugionifer*; OD]. Similar to *Ankothrips* (recent), but hind angles of prothorax with short bristles; ninth antennal segment with 4 pale, transverse sutures. PRIESNER, 1949. *Oligo.,* Europe (Baltic).
Eocranothrips BAGNALL, 1926, p. 17 [**Melanothrips annulicornis* BAGNALL, 1923, p. 36; OD]. Similar to *Cranothrips* (recent), but all antennal segments simple, without projections. PRIESNER, 1949. *Oligo.,* Europe (Baltic).
Lithadothrips SCUDDER, 1875b, p. 221 [**L. vetusta*; OD]. Similar to *Orothrips* (recent), but fore wings widened toward apex. BAGNALL, 1924a; PRIESNER, 1949. *Eoc.,* USA (Utah); *Oligo.,* Europe (Germany).
Melanthrips HALIDAY, 1836, p. 450. SCUDDER, 1890. ?*Eoc.,* USA (Colorado)–*Holo.*
Palaeothrips SCUDDER, 1875b, p. 222 [**P. fossilis*; OD]. Apparently related to *Rhipidothrips* (recent); antennae with 7 segments, apical segments not conical. PRIESNER, 1949. *Eoc.,* USA (Utah).
Promelanthrips PRIESNER, 1930, p. 113 [**P. spiniger*; OD]. Similar to *Ankothrips* (recent), but hind angles of prothorax with one long bristle only. USINGER, 1942. *Oligo.,* Europe (Baltic).
Rhipidothripoides BAGNALL, 1923, p. 36 [**R. abdominalis*; OD]. Similar to *Rhipidothrips* (recent), but ninth segment of abdomen unusually elongate and third, fourth, and fifth antennal segments of about equal length. *Oligo.,* Europe (Baltic).
Stenurothrips BAGNALL, 1914, p. 483 [**S. succineus*; OD]. Ovipositor straight or nearly so; terminal abdominal segment tubular; setae of hind margin of fore wing wavy. [Family position

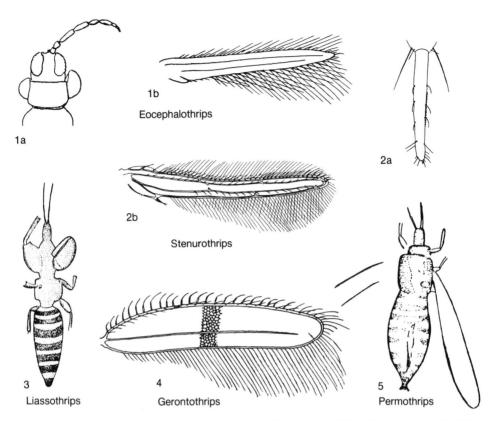

Fig. 133. Permothripidae, Aeolothripidae, Thripidae, Uncertain, and Liassothripidae (p. 209–212).

uncertain; possibly belonging to Heterothripidae.] BAGNALL, 1923; PRIESNER, 1949; STANNARD, 1956. *Oligo.*, Europe (Baltic).——FIG. 133,2. *S. bagnalli* STANNARD; *a,* terminal abdominal segment, ×66; *b,* fore wing, ×55 (Stannard, 1956).

Family THRIPIDAE Stephens, 1829

[Thripidae STEPHENS, 1829b, p. 363]

Similar to Heterothripidae, but antennae with 6 to 9 segments and with slender sense cones; cones simple or forked; tarsal claws, if present, at apex of first or second tarsal segment. *Oligo.–Holo.*

Thrips LINNÉ, 1761, p. 266. *Holo.*
Amorphothrips BAGNALL, 1924c, p. 252 [*A. klebsi; OD]. Similar to *Procerothrips,* but eye occupying whole side of head; pronotum transverse, without setae; hind legs very long and stout. BAGNALL, 1923; PRIESNER, 1949. *Oligo.*, Europe (Baltic).
Anaphothrips UZEL, 1895, p. 142. PRIESNER, 1930; PALMER, 1957. *Oligo.*, Europe (Baltic); *Mio.*, USA (California)–*Holo.*

Frankliniella KARNY, 1910, p. 46. PRIESNER & QUIÉVREUX, 1935. *Oligo.*, Europe (France)–*Holo.*
Gerontothrips PRIESNER, 1949, p. 41, *nom. subst. pro Archaeothrips* PRIESNER, 1924, p. 138, *non* FIELD, 1910 [*Archaeothrips latipennis* PRIESNER, 1924, p. 138; OD]. Wings very broad, entire surface reticulate. PRIESNER, 1930. *Oligo.*, Europe (Baltic).——FIG. 133,4. *G. latipennis* (PRIESNER); fore wing, ×66 (Priesner, 1924).
Heliothrips HALIDAY, 1836, p. 443. BAGNALL, 1924b. *Oligo.*, Europe (Baltic)–*Holo.*
Hercinothrips BAGNALL, 1932, p. 506. STANNARD, 1956. *Oligo.*, Europe (Baltic)–*Holo.*
Homothrips BAGNALL, 1915, p. 588. BAGNALL, 1924b. *Oligo.*, Europe (Baltic)–*Holo.*
Lipsanothrips PRIESNER, 1930, p. 119 [*L. skwarrae; OD]. Antennae with 8 segments, sixth segment much shorter than fifth; wing with 1 or 2 longitudinal veins. *Oligo.*, Europe (Baltic).
Oxythrips UZEL, 1895, p. 141. PRIESNER, 1924, 1930. *Oligo.*, Europe (Baltic)–*Holo.*
Praedendrothrips PRIESNER, 1924, p. 139 [*P. avus; OD]. Antennae with 9 segments, last 4 clearly separate from one another; posterior angles of

pronotum with at least 1 conspicuous bristle. PRIESNER, 1930. *Oligo.,* Europe (Baltic).
Procerothrips BAGNALL, 1924c, p. 252 [**P. cylindricornis*; OD]. Antennae with 8 segments and style with 2 segments; third through sixth antennal segments with parallel sides and of same width. BAGNALL, 1923. *Oligo.,* Europe (Baltic).
Selenothrips KARNY, 1911, p. 180. BAGNALL, 1923. *Oligo.,* Europe (Baltic)–*Holo.*
Taeniothrips SERVILLE, 1843, p. 644. BAGNALL, 1924b; PRIESNER, 1930, 1949. *Oligo.,* Europe (Baltic)–*Holo.*
Telothrips PRIESNER, 1930, p. 116 [**T. klebsi*; OD]. Similar to *Praedendrothrips* but with sixth antennal segment large and stout and seventh, eighth, and ninth segments minute. PRIESNER, 1949. *Oligo.,* Europe (Baltic).

Family HETEROTHRIPIDAE
Bagnall, 1912

[Heterothripidae BAGNALL, 1912, p. 222] [=Hemithripidae BAGNALL, 1923, p. 37; Stenurothripidae BAGNALL, 1923, p. 37; Opadothripidae BAGNALL, 1927, p. 562]

Wings narrow, usually pointed distally; ovipositor curved downward; antennae with 9 or 10 segments; fore tarsi usually with clawlike appendage at base of second segment. *Oligo.–Holo.*

Heterothrips HOOD, 1908, p. 361. *Holo.*
Electrothrips BAGNALL, 1924c, p. 251 [**E. hystrix*; OD]. Cephalic, pronotal, and wing bristles abnormally long and stout; wings and legs long and slender. *Oligo.,* Europe (Baltic).
Hemithrips BAGNALL, 1923, p. 37 [**H. femoralis*; OD]. Similar to *Heterothrips* (recent), but third and fourth antennal segments cylindrical. BAGNALL, 1924a; PRIESNER, 1949. *Oligo.,* Europe (Baltic, Germany).
Opadothrips PRIESNER, 1924, p. 133 [**O. fritschianus*; OD]. Similar to *Oligothrips* (recent), but antennal segments more elongate; terminal segment slender. BAGNALL, 1924a, 1927; PRIESNER, 1949. *Oligo.,* Europe (Baltic).

Family MEROTHRIPIDAE
Hood, 1914

[Merothripidae HOOD, 1914, p. 17]

Wings narrow, pointed distally, surface smooth (not pubescent); ovipositor curved downward; pronotum with dorsal longitudinal sutures; anterior and posterior femora greatly enlarged. *Oligo.–Holo.*

Merothrips ZIMMERMANN, 1900, p. 12. Antennae with 8 segments. PRIESNER, 1924. *Oligo.,* Europe (Baltic)–*Holo.*

Praemerothrips PRIESNER, 1930, p. 130 [**P. hoodi*; OD]. Antennae with 9 segments. PRIESNER, 1949. *Oligo.,* Europe (Baltic).

Family UNCERTAIN

The following genera, apparently belonging to the order Thysanoptera, suborder Terebrantia, are too poorly known to permit assignment to families.

Calothrips OUSTALET, 1873, p. 24 [**C. scudderi*; OD]. Little-known thysanopteron, probably belonging to the Terebrantia. *Oligo.,* Europe (France).
Eocephalothrips BAGNALL, 1924a, p. 161 [**Thrips capito* SCHLECHTENDAL, 1887, p. 579; OD] [=*Protothrips* PRIESNER, 1924, p. 136 (type, *P. speratus*)]. Head quadrate; wings moderately broad, apex pointed. PRIESNER, 1949. *Oligo.,* Europe (Baltic).——FIG. 133,*1*. *E. speratus* (PRIESNER); *a,* head and prothorax; ×66; *b,* fore wing, ×66 (Priesner, 1924).

Suborder TUBULIFERA
Haliday, 1836

[Tubulifera HALIDAY, 1836, p. 459]

Terminal abdominal segments of both sexes almost always tubular; female without ovipositor; fore wing without definite costal vein and with only a vestige of another longitudinal vein, long fringe present. *Oligo.–Holo.*

Family PHLAEOTHRIPIDAE
Uzel, 1895

[Phlaeothripidae UZEL, 1895, p. 42]

Characteristics of suborder. *Oligo.–Holo.*

Phlaeothrips HALIDAY, 1836, p. 441. SCHLECHTENDAL, 1887; BAGNALL, 1924a, 1929. *Oligo.,* Europe (Baltic)–*Holo.*
Cephenothrips PRIESNER, 1930, p. 135 [**C. laticeps*; OD]. Similar to *Pygidiothrips* (recent), but wing bristles short and knobbed. USINGER, 1942; PRIESNER, 1949. *Oligo.,* Europe (Baltic).
Hoplothrips AMYOT & SERVILLE, 1843, p. 640. BAGNALL, 1929; PRIESNER, 1949. *Oligo.,* Europe (Baltic)–*Holo.*
Liotrichothrips BAGNALL, 1929, p. 97 [**L. hystrix*; OD]. Head longer than pronotum, broader than long; cheeks with few prominent setae; antennae long, with third and fourth segments subequal. Similar to *Ethirothrips* (recent), but legs as in *Liothrips* (recent). PRIESNER, 1949. *Oligo.,* Europe (Baltic).
Necrothrips PRIESNER, 1924, p. 147 [**N. nanus*;

OD]. Similar to *Austrothrips* (recent), but eyes very large, protruding, and consisting of many facets. USINGER, 1942; PRIESNER, 1949. *Oligo.,* Europe (Baltic).

Proleeuwenia PRIESNER, 1924, p. 148 [**P. succini*; OD]. Wings reduced (female); similar to *Idiothrips* (recent), but antennae with 8 segments. USINGER, 1942; PRIESNER, 1949. *Oligo.,* Europe (Baltic).

Schlechtendalia BAGNALL, 1929, p. 96 [**S. longitubus*; OD]. Similar to *Phlaeothrips*, but tenth abdominal segment substantially longer than head; fifth antennal segment with a projection; wing bristles blunt. PRIESNER, 1949. *Oligo.,* Europe (Baltic).

Symphyothrips HOOD & WILLIAMS, 1915, p. 131. PRIESNER, 1924, 1949. *Oligo.,* Europe (Baltic)–*Holo.*

Treherniella WATSON, 1923, p. 81. PRIESNER, 1930, 1949. *Oligo.,* Europe (Baltic)–*Holo.*

Suborder UNCERTAIN

The genus described below, apparently belonging to the order Thysanoptera, is too poorly known to permit assignment to suborders.

Family LIASSOTHRIPIDAE Priesner, 1949

[Liassothripidae PRIESNER, 1949, p. 34] [=Mesothripidae MARTYNOV, 1927b, p. 768]

Antennae thin, with at least 7 segments; head narrow; anterior femora very broad; wings unknown. *Jur.*

Liassothrips PRIESNER, 1949, p. 34, *nom. subst. pro Mesothrips* MARTYNOV, 1927b, p. 768, *non* ZIMMERMANN, 1900 [**Mesothrips crassipes* MARTYNOV, 1927b, p. 768; OD]. Little-known thysanopteron; abdomen apparently constricted basally. *Jur.*, USSR (Kazakh).——FIG. 133,3. **L. crassipes* (MARTYNOV); body, ×16 (Martynov, 1927b).

HEMIPTEROID EXOPTERYGOTES

Order HEMIPTERA Linné, 1758

[Hemiptera LINNÉ, 1758, p. 434] [=Hemipsocoptera ZALESSKY, 1937e, p. 51; Palaeohemiptera HANDLIRSCH, 1904b, p. 2]

Exopterygote Neoptera, mostly small to very small, with much morphological diversity. Head opisthognathous or prognathous; compound eyes usually present but diverse in size; two ocelli commonly present, rarely three or none; antennae typically with five segments or less, rarely with as many as ten; mouthparts haustellate, consisting of two pairs of maxillary stylets in a segmented, rostrate labium. Pronotum of moderate size, often diversely modified; meso- and metathorax well developed. Legs usually cursorial, but forelegs of some genera raptorial, vestigial, or absent; tarsi commonly with three segments, rarely with two or one. Wings usually present, but very different in the two suborders. Wing venation quite generalized in primitive forms but much reduced in most families; fore wings of suborder Homoptera usually of uniform texture, those of suborder Heteroptera partly membranous and partly coriaceous. Abdomen well developed; ovipositor usually present. Nymphs resembling adults in basic body structure. *Perm.–Holo.*

This is the largest of the exopterygote orders, and it has apparently been a major order at least since the Triassic. All available evidence suggests that the Hemiptera are most closely related to the Psocoptera, which were well represented in the Permian. The order Hemiptera has traditionally been divided into two suborders, Homoptera and Heteroptera, the members of both groups having the same distinctive, haustellate mouthparts. Both suborders are also represented in the Permian, but the Homoptera have by far the more extensive record in that period.

The wings provide the best means of distinguishing the members of the two suborders. The homologies of the main veins are clear throughout both suborders, even in those in which the venation is much reduced. However, there has been much convergence in the reduction process. In part because of this, the family and generic classifications of the Hemiptera, especially of the Homoptera, have been based mainly on body features, such as the detailed structure of the rostrum, number and size of ocelli, tarsal segmentation, and integumentary details. Since fossils do not usually show such structures, the family position of many of the extinct genera is uncertain.

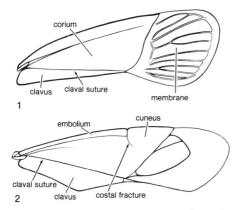

Fig. 134. Hemiptera; wing structure of the suborder Heteroptera.——*1. Nazara* sp., Pentatomidae. (adapted from CSIRO, 1970).——*2. Megacoelum* sp., Miridae (adapted from CSIRO, 1970).

The fore wings of the two suborders, except for a few primitive families in each one, differ mainly in texture. Those of the Homoptera are usually of a uniform or nearly uniform texture. Exceptions are found in a few primitive families, such as the Palaeontinidae and Prosbolidae, in which the fore wing is partly divided into two areas by an irregular nodal line (see Figs. 137,*1* and 139,*4a*). The proximal area is usually more coriaceous than the distal part. A slight break (costal break or indentation) in the costal margin appears to be functionally associated with the nodal line. The anal area, commonly termed the clavus in the Hemiptera, is usually a distinct region of the fore wing. The fore wings of the Heteroptera, usually termed hemelytra, have a more complex structure. The distal portion of the wing is membranous, but the proximal part is coriaceous, consisting of a large, anterior, triangular section (corium) and a relatively small region (clavus), separated from the corium by the claval suture (Fig. 134,*1*). These areas are diverse in form and size in families and genera. The costal area of the corium may be separated from the rest of it by vein M+R, forming the embolium; and a more distal part (cuneus) of the corium may be separated by a costal fracture (Fig. 134,*2*). The hind wings of the Hemiptera are generally more specialized than the fore pair, often with a very different shape. Polymorphism of wings, including aptery, occurs commonly in the order.

The great majority of the Hemiptera are phytophagous, but some Heteroptera are active predators. Immature stages have essentially the same feeding habits as the adults.

The geological record of the Hemiptera is very extensive, including almost a hundred families, two-thirds of which are Homoptera. A surprisingly large number of entomologists, specialists on the systematics of existing Hemiptera, have contributed to our knowledge of this fossil record and of the phylogeny of the order. There is, however, much difference of opinion among them about the systematic position of many of the extinct genera.

Suborder HOMOPTERA
Leach, 1815

[Homoptera Leach, 1815, p. 124]

Fore wing of uniform texture or nearly so, not sharply differentiated into membranous and coriaceous areas; wings typically held sloping over the sides of the body at rest. *Perm.–Holo.*

Family DUNSTANIIDAE
Tillyard, 1916

[Dunstaniidae Tillyard in Tillyard & Dunstan, 1916, p. 31]

Fore wing sharply separated into tegminous basal part and membranous distal area; nodal break prominent; vein SC long, terminating on costal margin; R and RS curved; RS unbranched; clavus broad, triangular; 1A and 2A long, extending to hind margin. Hind wing little known, smaller than fore wing, with rounded anal area. Head, compound eyes, and pronotum relatively large. Relatively large insects. Affinities uncertain, but apparently closely related to the Palaeontinidae. Tillyard, 1918d; Becker-Migdisova, 1949b; Evans, 1956; Becker-Migdisova & Wootton, 1965; Riek, 1976b. *Trias.*

Hexapoda

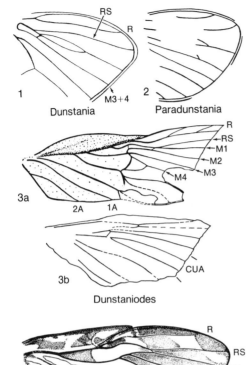

FIG. 135. Dunstaniidae (p. 214).

Dunstania TILLYARD in TILLYARD & DUNSTAN, 1916, p. 31 [*D. pulchra*; OD]. M3+4 not forked, RS apparently joined to M by a short, oblique crossvein. [The genus has been reported from South Africa (RIEK, 1976b), but the generic position of the species described is very uncertain.] *Trias.*, Australia (Queensland).——FIG. 135,*1*. **D. pulchra*; fore wing as preserved, ×1.5 (Evans, 1956).

Dunstaniodes BECKER-MIGDISOVA & WOOTTON, 1965, p. 64 [**D. elongatus*; OD]. Fore wing elongate; costal margin of basal half of wing strongly convex. *Trias.*, USSR (Kirghiz).——FIG. 135,*3*. **D. elongatus*; *a*, fore and *b*, hind wings as preserved, ×3.0 (Becker-Migdisova & Wootton, 1965).

Dunstaniopsis TILLYARD, 1918d, p. 584 [**D. triassica*; OD]. Little-known genus, based on incomplete fore wing; apex apparently more pointed than in *Dunstania*. EVANS, 1956; BECKER-MIGDISOVA & WOOTTON, 1965. *Trias.*, Australia (Queensland).

Paradunstania TILLYARD, 1918d, p. 585 [**P. affinis*; OD]. Little-known genus, based on fragment of fore wing; probably a synonym of *Dunstania*. EVANS, 1956; BECKER-MIGDISOVA & WOOTTON, 1965. *Trias.*, Australia (Queensland).——FIG. 135,*2*. **P. affinis*; fore wing as preserved, ×1.5 (Evans, 1956).

Siksteliana BECKER-MIGDISOVA & WOOTTON, 1965, p. 68 [**S. popovi*; OD]. Little-known genus, based on fore wing. Similar to *Dunstaniodes*, but costal margin of basal half nearly straight. *Trias.*, USSR (Kirghiz).——FIG. 135,*4*. **S. popovi*; fore wing, ×3 (Becker-Migdisova & Wootton, 1965).

Family PALAEONTINIDAE Handlirsch, 1906

[Palaeontinidae HANDLIRSCH, 1906b, p. 618] [=Cicadomorphidae EVANS, 1956, p. 222]

Fore wing as in Dunstaniidae, with membranous, distal part of wing broader and longer than basal, tegminous part; vein SC usually weakly developed, commonly with branches or suggestions of branches; R and M separating before or close to midwing; R and RS nearly straight. Hind wing with a prominent indentation on costal margin; M1 commonly coalesced for short interval with RS; M with 4 branches. Head small, narrow, pronotum wide; body generally with numerous hairs. *Perm.–Jur.*

Palaeontina BUTLER, 1873, p. 126 [**P. oolitica*; OD]. Little-known genus, based on fore wing. M with 4 branches, M1+2 and M3+4 forking at about same level. [The genus was excluded from Homoptera by EVANS (1956) but included here by BECKER-MIGDISOVA (1962b) and POPOV (1980b).] *Jur.*, England.——FIG. 136,*1*. **P. oolitica*; fore wing, ×0.8 (Handlirsch, 1906b).

Asiocossus BECKER-MIGDISOVA, 1962a, p. 89 [**A. subcostalis*; OD]. Little-known genus, based on fragment of fore wing. SC free from R+M except for very base, branched; R+M and stem of R very short. *Trias.*, USSR (Kirghiz).——FIG. 136,*3*. **A. subcostalis*; fore wing base, ×2.5 (Becker-Migdisova, 1962b).

Cicadomorpha MARTYNOV, 1926b, p. 1357 [**C. punctulata*; OD]. SC coalesced with R+M at base; area between M and CUA very broad, without crossveins; CU slightly arched at base. *Jur.*, USSR (Kazakh).——FIG. 136,*7*. **C. punctulata*; fore wing, ×1.0 (Becker-Migdisova, 1962b).

Fletcheriana EVANS, 1956, p. 224 [**F. triassica*; OD]. Fore wing as in *Pseudocossus*, but costal area much broader; SC lying alongside R basally; RS arising from R remote from wing base. [The assignment of a species from the Triassic of South Africa (RIEK, 1976b) to this genus is very uncertain.] *Trias.*, Australia (New South Wales). ——FIG. 136,*2*. **F. triassica*; *a*, fore wing; *b*, hind wing, ×1.0 (Evans, 1956).

Hemiptera—Homoptera

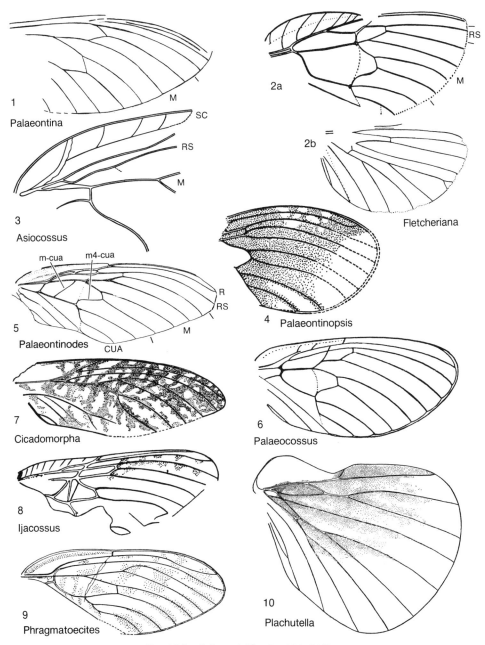

FIG. 136. Palaeontinidae (p. 214–216).

Ijacossus BECKER-MIGDISOVA, 1950, p. 1106 [**I. suchanovae*; OD]. Little-known genus, based on fore wing. Similar to *Palaeontinodes*, but SC with several branches. [Family assignment uncertain.] *Jur.*, USSR (Asian RSFSR).——FIG. 136,8. **I. suchanovae*; fore wing, ×1 (Becker-Migdisova, 1962b).

Palaeocicadopsis T'AN, 1980, p. 161 [**P. chinensis*; OD]. Fore wing similar to that of *Cicadomorpha*, but M branching near wing base. *Perm.*, China (Inner Mongolia).

Palaeocossus OPPENHEIM, 1885, p. 333 [**P. jurassicus*; OD]. Fore wing without nodal indentation; wing broadly oval; distal margin of basal median

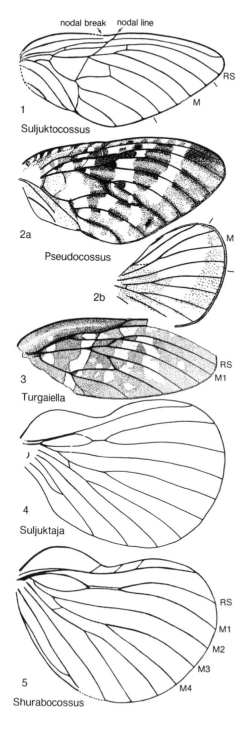

FIG. 137. Palaeontinidae (p. 216–217).

cell (between M and CUA) smoothly curved; hind margin strongly convex. *Jur.*, USSR (Asian RSFSR).——FIG. 136,6. **P. jurassicus*; fore wing, ×1.5 (Evans, 1956).

Palaeontinodes MARTYNOV, 1937a, p. 166 [**P. shabarovi*; OD]. Fore wing triangular; costal indentation weak; SC without branches; crossvein m-cua long; basal median cell divided by crossveins (not shown in figure). BECKER-MIGDISOVA, 1949b; EVANS, 1956. *Jur.*, USSR (Tadzhik, Asian RSFSR).——FIG. 136,5. *P. angarensis* BECKER-MIGDISOVA & WOOTTON; fore wing, ×1 (Becker-Migdisova & Wootton, 1965).

Palaeontinopsis MARTYNOV, 1937a, p. 167 [**P. latipennis*; OD]. Little-known genus. Fore wing apparently oval and with rounded apex. EVANS, 1956; BECKER-MIGDISOVA & WOOTTON, 1965. *Jur.*, USSR (Tadzhik).——FIG. 136,4. **P. latipennis*; fore wing, ×1.5 (Becker-Migdisova, 1962b).

Phragmatoecites OPPENHEIM, 1885, p. 333 [**P. damesi*; OD]. Fore wing with costal margin straight or only slightly curved; nodal indentation weak. EVANS, 1956; BECKER-MIGDISOVA & WOOTTON, 1965. *Jur.*, USSR (Asian RSFSR). ——FIG. 136,9. **P. damesi*; fore wing, ×2.5 (Becker-Migdisova, 1962b).

Plachutella BECKER-MIGDISOVA, 1949b, p. 11 [**P. rotundata*; OD]. Little-known genus, based on hind wing. M2 close to M3+4 at one point but not coalesced with it. BECKER-MIGDISOVA, 1950; BECKER-MIGDISOVA & WOOTTON, 1965. *Jur.*, USSR (Kazakh, Tadzhik).——FIG. 136,10. **P. rotundata*; hind wing, ×2.5 (Becker-Migdisova, 1949b).

Pseudocossus MARTYNOV, 1931d, p. 94 [**P. zemcuznicovi*; OD]. Fore wing triangular, with pronounced indentation of costal margin at nodal break; SC free from R+M at base, branched; RS arising from R near wing base; distinct bands of coloration. Hind wing rounded, much smaller than fore wing. *Jur.*, USSR (Asian RSFSR, Kazakh).——FIG. 137,2. *P. tugaiensis* BECKER-MIGDISOVA & WOOTTON, Kazakh; *a,* fore and *b,* hind wings, ×1.5 (Becker-Migdisova & Wootton, 1965).

Shurabocossus BECKER-MIGDISOVA, 1949b, p. 15 [**S. gigas*; OD]. Hind wing similar to that of *Plachutella,* but M2 coalesced with M3+4 for a considerable interval before separating. *Jur.*, USSR (Tadzhik).——FIG. 137,5. **S. gigas*; hind wing, ×1.5 (Becker-Migdisova, 1962b).

Suljuktaja BECKER-MIGDISOVA, 1949b, p. 17 [**S. turkestanensis*; OD]. Hind wing as in *Shurabocossus* but with the coalesced parts of 1A and 2A at least as long as the free portions. *Jur.*, USSR (Kirghiz).——FIG. 137,4. **S. turkestanensis*; hind wing, ×2 (Becker-Migdisova, 1962b).

Suljuktocossus BECKER-MIGDISOVA, 1949b, p. 8 [**S.

prosboloides; OD]. Fore wing as in *Phragmatoectites* but more nearly triangular and with apex nearly pointed. *Jur.*, USSR (Kirghiz).——FIG. 137,*1*. **S. prosboloides*; fore wing, ×1.5 (Becker-Migdisova, 1962b).

Turgaiella BECKER-MIGDISOVA & WOOTTON, 1965, p. 70 [**T. pomerantsevae*; OD]. Fore wing as in *Palaeontinodes*, but wing oval and basal median cell not divided by crossveins; crossvein m-cua very short. *Jur.*, USSR (Kazakh).——FIG. 137,*3*. **T. pomerantsevae*; fore wing, ×1.5 (Becker-Migdisova & Wootton, 1965).

Family MESOGEREONIDAE Tillyard, 1921b

[Mesogereonidae TILLYARD, 1921b, p. 272]

Fore wing slender, with well-developed submarginal (ambient) vein and coriaceous border; veins SC and R close together and to costal margin; RS arising before fork of M1+2; crossvein m4-cua near wing base and almost longitudinal in position. Hind wing little known, much smaller than fore wing. Body structure unknown. EVANS, 1956; BECKER-MIGDISOVA & WOOTTON, 1965. *Trias.*

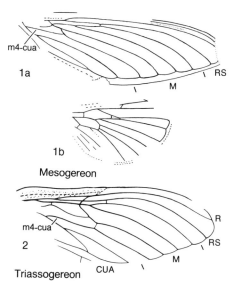

FIG. 138. Mesogereonidae (p. 217).

Mesogereon TILLYARD in TILLYARD & DUNSTAN, 1916, p. 33 [**M. neuropunctatum*; OD]. RS joined to M1 by a short crossvein; M3+4 forking more basally than M1+2. *Trias.*, Australia (New South Wales).——FIG. 138,*1a*. *M. superbum* TILLYARD; fore wing, ×1.2 (Evans, 1956).——FIG. 138,*1b*. *M. shepherdi* TILLYARD; hind wing, ×1.2 (Evans, 1956).

Triassogereon RIEK, 1976b, p. 808 [**T. distinctum*; OD]. Fore wing as in *Mesogereon*, but fork of M3+4 close to fork of M1+2. *Trias.*, South Africa.——FIG. 138,*2*. **T. distinctum*; fore wing, ×1.6 (Riek, 1976b).

Family PROSBOLIDAE Handlirsch, 1906

[Prosbolidae HANDLIRSCH, 1906b, p. 390] [=Sojaneuridae BECKER-MIGDISOVA, 1946, p. 750]

Fore wing: distal part commonly membranous; costal area broad; vein SC usually forming an anterior branch submarginal to costal margin and more rarely an indistinct, short branch that parallels R+M and even part of R; forks of M and CUA usually shallow. Hind wing: costal margin usually deeply excised near middle, convex basally and distally; anal region extended posteriorly. Body structure unknown. BECKER-MIGDISOVA, 1940, 1947, 1962b; EVANS, 1956. *Perm.–Trias.*

Prosbole HANDLIRSCH, 1904b, p. 2 [**P. hirsuta*; OD] [=*Prosbolina* HANDLIRSCH, 1937, p. 132 (type, *Prosbole biexcisa* MARTYNOV, 1928b, p. 7)]. Fore wing: nodal break and nodal line present; R, M, and CUA dividing at about same level. Hind wing: M with at least 4 branches. *Perm.*, USSR (European and Asian RSFSR).——FIG. 139,*4a*. **P. hirsuta*; fore wing, ×1.6 (Evans, 1956).——FIG. 139,*4b*. *P. reducta* MARTYNOV; fore wing, ×3.5 (Becker-Migdisova, 1940).——FIG. 139,*4c*. *P. breviata* BECKER-MIGDISOVA; hind wing, ×2.6 (Becker-Migdisova, 1940).

Austroprosbole EVANS, 1943b, p. 181 [**A. maculata*; OD]. Fore wing with nodal break and nodal line; RS curved posteriorly, touching M1+2 at point of fork; CUA with a shallow, distal fork. EVANS, 1956. *Perm.*, Australia (New South Wales).——FIG. 139,*5*. **A. maculata*; fore wing, ×4 (Evans, 1943b).

Austroprosboloides RIEK, 1973, p. 527 [**A. vandijki*; OD]. Little-known genus; fore wing similar to *Austroprosbole*, but RS touching M1 beyond fork and M3+4 connected to CUA distally. RIEK, 1976a. *Perm.*, South Africa.——FIG. 140,*6*. **A. vandijki*; fore wing, ×4 (Riek, 1973).

Beaufortiscus RIEK, 1976a, p. 779 [**B. dixi*; OD]. Fore wing very similar to that of *Prosbole*; anal

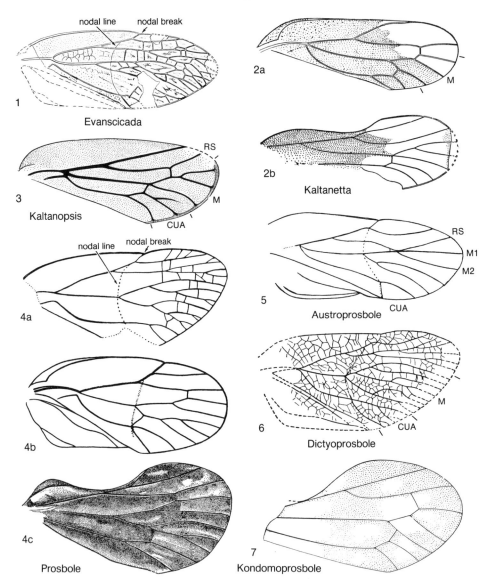

Fig. 139. Prosbolidae (p. 217–220).

area with 3 veins. [Probably a synonym of *Prosbole*.] *Perm.*, South Africa.

Dictyoprosbole MARTYNOV, 1935d, p. 443 [**D. membranosa*; OD]. Fore wing membranous, covered with a network of crossveins; M and CUA dividing at level of origin of RS; RS, M, and CUA with branching as in *Orthoprosbole*. EVANS, 1956. *Perm.*, USSR (Asian RSFSR).——FIG. 139,6. **D. membranosa*; fore wing, ×1.5 (Becker-Migdisova, 1960).

Evanscicada BECKER-MIGDISOVA, 1962b, p. 170, nom. subst. pro *Evansia* BECKER-MIGDISOVA, 1961c, p. 323, non CAMBRIDGE, 1900 [**Evansia speciosa* BECKER-MIGDISOVA, 1961c, p. 323; OD]. Fore wing narrow; basal part tegminous; RS arising at level of forking of M; numerous crossveins distally and indication of network near basal-central part of wing. *Perm.*, USSR (Asian RSFSR).——FIG. 139,1. **E. speciosa*; fore wing, ×2.5 (Becker-Migdisova, 1962b).

Falsia BECKER-MIGDISOVA, 1946, p. 750 [**F. chimaera*; OD]. Similar to *Sojanoneura*, but first

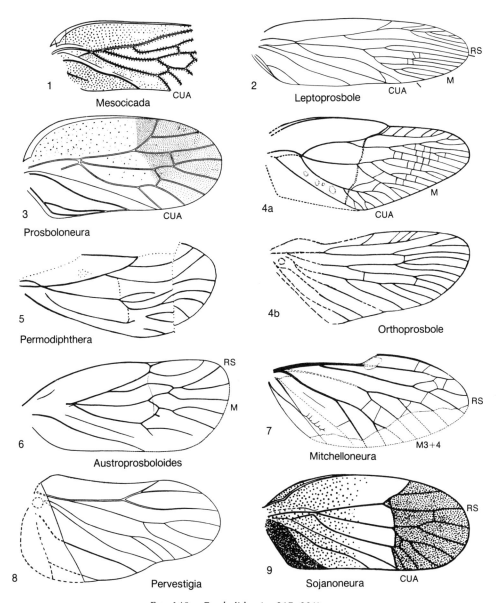

FIG. 140. Prosbolidae (p. 217–221).

and second tarsal segments of same size. BECKER-MIGDISOVA, 1946. *Perm.,* USSR (European RSFSR).

Kaltanetta BECKER-MIGDISOVA, 1961c, p. 303 [**K. nigra*; OD]. Fore wing slender, apex symmetrically curved; RS arising well before level of forking of M and of CUA; M with 3 branches. Hind wing slender distally; marginal indentation deep and wide; M with 3 branches. *Perm.,* USSR (Asian RSFSR).——FIG. 139,*2*. **K. nigra; a,* fore and *b,* hind wings, ×6.5 (Becker-Migdisova, 1961c).

Kaltanopsis BECKER-MIGDISOVA, 1961c, p. 300 [**K. ornata*; OD]. Fore wing similar to that of *Kaltanetta,* but costal margin strongly curved and R continuing in a straight line from its stem; longitudinal veins unusually thick. *Perm.,* USSR (Asian RSFSR).——FIG. 139,*3*. **K. ornata;* fore wing, ×8 (Becker-Migdisova, 1961c).

Kondomoprosbole BECKER-MIGDISOVA, 1961c, p.

Hexapoda

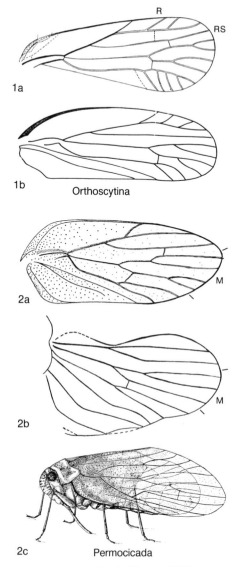

Fig. 141. Prosbolidae (p. 220).

315 [*K. pictata*; OD]. Hind wing: anterior margin with prominent bulge distally; R a straight continuation of stem R; M with 3 short branches. *Perm.*, USSR (Asian RSFSR).——Fig. 139,7. **K. pictata*; hind wing, ×8.5 (Becker-Migdisova, 1962b).

Leptoprosbole RIEK, 1976b, p. 812 [**L. lepida*; OD]. Fore wing elongate; M with 8 terminal branches. [Family assignment doubtful.] *Trias.*, South Africa.——Fig. 140,2. **L. lepida*; fore wing, ×1.5 (Riek, 1976b).

Mesocicada BECKER-MIGDISOVA, 1962a, p. 90 [**M. verrucosa*; OD]. Little-known fore wing; nodal break absent; M with 4 branches; CUA with small fork. [Family assignment doubtful.] *Trias.*, USSR (Kirghiz).——Fig. 140,1. **M. verrucosa*; fore wing, ×14 (Becker-Migdisova, 1962a).

Mitchelloneura TILLYARD, 1921c, p. 414 [**M. permiana*; OD]. Little-known hind wing; RS with irregular distal branches; M with M1, M2, and M3+4; CUA deeply forked. EVANS, 1956. *Perm.*, Australia (New South Wales).——Fig. 140,7. **M. permiana*; hind wing, ×3.2 (Tillyard, 1921c).

Neurobole RIEK, 1976a, p. 779 [**N. ramosa*; OD]. Little-known genus, based on small apical fragment of wing. [Family assignment doubtful.] *Perm.*, South Africa.

Orthoprosbole MARTYNOV, 1935d, p. 445 [**O. congesta*; OD]. Fore wing strongly narrowed in distal half; RS and M with numerous branches; nodal break prominent. Hind wing little known; distal part elongate; M and CUA with numerous branches. BECKER-MIGDISOVA, 1961c. *Perm.*, USSR (Asian RSFSR).——Fig. 140,4a. *O. triangularis* (MARTYNOV); fore wing, ×2.5 (Becker-Migdisova, 1962b).——Fig. 140,4b. **O. congesta*; hind wing, ×3.5 (Becker-Migdisova, 1962b).

Orthoscytina TILLYARD, 1926a, p. 9 [**O. mitchelli*; OD]. Fore wing slender, oval; anal area long; RS arising just before midwing; M and CUA forked at distal third of wing; R with several oblique branches to costal margin. Hind wing little known. EVANS, 1956; RIEK, 1976a. *Perm.*, Australia (New South Wales), Africa (South Africa), USSR (Asian RSFSR).——Fig. 141,1a. **O. mitchelli*, Australia; fore wing, ×6 (Tillyard, 1926a).——Fig. 141,1b. *O. suchovi* BECKER-MIGDISOVA, USSR; fore wing, ×6 (Becker-Migdisova, 1961c).

Permocicada MARTYNOV, 1928b, p. 19 [**P. umbrata*; SD BECKER-MIGDISOVA, 1940, p. 29] [=*Permocicadopsis* BECKER-MIGDISOVA, 1940, p. 54 (type, *Permocicada angusta* MARTYNOV, 1935c, p. 15)]. Fore wing with weak venation; RS arising before forking of M and CUA; M with 3 or 4 branches. Hind wing with deeply indented costal margin having nearly symmetrical slopes. ZALESSKY, 1929, 1932b; EVANS, 1956; BECKER-MIGDISOVA, 1961c, 1962b. *Perm.*, USSR (European and Asian RSFSR).——Fig. 141,2. *P. integra* BECKER-MIGDISOVA; *a*, fore wing, ×4; *b*, hind wing, ×4; *c*, reconstruction, ×3 (Becker-Migdisova, 1940).

Permodiphthera TILLYARD, 1926a, p. 24 [**P. robusta*; OD]. Little-known genus. Fore wing with RS unbranched; branches of M apparently strongly curved. *Perm.*, Australia (New South Wales).——Fig. 140,5. **P. robusta*; fore wing, ×6 (Evans, 1956).

Pervestigia BECKER-MIGDISOVA, 1961c, p. 318 [**P.

veteris; OD]. Hind wing: anterior margin without distal hump; M with 3 branches; CUA with narrow fork distally. *Perm.*, USSR (Asian RSFSR).——FIG. 140,8. **P. veteris*; hind wing, ×3 (Becker-Migdisova, 1961c).

Prosbolomorpha RIEK, 1974c, p. 21 [**P. clara*; OD]. Fore wing as in *Austroprosbole*, but RS not coalesced with M; M3+4 forking at its point of origin. [Probably a synonym of *Austroprosbole*.] *Trias.*, South Africa.

Prosboloneura BECKER-MIGDISOVA, 1961c, p. 305 [**P. colorata*; OD]. Fore wing shaped as in *Sojanoneura*, but CUA more deeply forked and M with 3 branches. *Perm.*, USSR (Asian RSFSR). ——FIG. 140,3. *P. kondonensis* BECKER-MIGDISOVA; fore wing, ×8 (Becker-Migdisova, 1961c).

Sojanoneura MARTYNOV, 1928b, p. 22 [**S. edemskii*; SD BECKER-MIGDISOVA, 1940, p. 44]. Fore wing oval, bluntly rounded; RS arising nearer wing apex than in *Dictyoprosbole*; M with 3 or 4 branches. Hind wing little known, with only a slight bulging of the costal margin basally; M with 2 or 3 branches. MARTYNOV, 1935c; EVANS, 1956. *Perm.*, USSR (European and Asian RSFSR).——FIG. 140,9. *S. stigmata* MARTYNOV; fore wing, ×4 (Becker-Migdisova, 1962b).

Family CICADOPROSBOLIDAE Evans, 1956

[Cicadoprosbolidae EVANS, 1956, p. 222]

Apparently related to Prosbolidae. Fore wing with vein M forking at midwing; RS arising before midwing; short, supplementary veins between branches of R; nodal line distinct, crossing RS remote from origin of RS and crossing M beyond its first fork. *Trias.*

Cicadoprosbole BECKER-MIGDISOVA, 1947, p. 445 [**C. sogutensis*; OD]. Fore wing oval, apex slightly asymmetrical; branches of M and CUA slightly curved and parallel. [Originally placed in the family Prosbolidae but transferred to a new family, Cicadoprosbolidae, by EVANS (1956) and later to the Tettigarctidae by BECKER-MIGDISOVA, 1962b.] *Trias.*, USSR (Kirghiz). ——FIG. 142,4. **C. sogutensis*; fore wing, ×3.5 (Becker-Migdisova, 1947).

Family TETTIGARCTIDAE Distant, 1905

[Tettigarctidae DISTANT, 1905, p. 280]

Fore wing with transparent membranous area; costa broadly sclerotized; apical border narrow; venation much as in *Cicadidae*; vein SC with a short, hook-shaped anterior branch

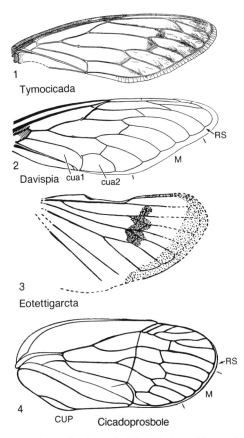

FIG. 142. Cicadoprosbolidae, Tettigarctidae, and Cicadidae (p. 221–222).

basally and a posterior branch coalesced with R+M and R. Hind wing: front margin with shallow indentation and convex area. Body structure: stridulatory organ present in both sexes. [This family is known by only one living genus, *Tettigarcta*, occurring in Australia. There is a reliable Tertiary record of the family, but the several Mesozoic genera that have been placed here are very poorly known and are assigned in this account to the category, family uncertain.] EVANS, 1956; WHALLEY, 1983. *Eoc.–Holo.*

Tettigarcta WHITE, 1845, p. 412. *Holo.*

Eotettigarcta ZEUNER, 1944a, p. 110 [**E. scotica*; OD]. Hind wing similar to that of *Tettigarcta* (recent), but indentation of costal margin much longer; origin of posterior branch of RS more remote from base of wing. Fore wing unknown.

WHALLEY, 1983. *Eoc.*, Scotland.——FIG. 142,*3*. **E. scotica*; hind wing, ×2.5 (Zeuner, 1944a).

Family CICADIDAE Leach, 1815

[Cicadidae LEACH, 1815, p. 124]

Fore wing with costal area reduced to a narrow strip or absent; apical parts of distal forks of veins aligned to form a submarginal vein along the outer and hind margins; anal area narrow and short. Hind wing much smaller than fore wing; anterior margin smooth; submarginal vein formed as in fore wing; anal-jugal area slightly broader than in fore wing. Stridulatory organs (tymbals) present on dorsum of first abdominal segment, at least in males. [A fragmented specimen from the Eocene of France was described as *beauchampi* by Piton (1940a); this species was placed in the existing genus *Chemsitica* STÅL (=*Rihana* DISTANT). However, the fossil does not show enough structural detail for family assignment. See also *Liassocicada* under Homoptera, family Uncertain.] COOPER 1941; WHALLEY, 1983. *Paleoc.–Holo.*

Cicada LINNÉ, 1758, p. 434. COOPER, 1941; WHALLEY, 1983. *Oligo.*, USA (Colorado); *Mio.*, Europe (Yugoslavia, Germany)–*Holo.*

Davispia COOPER, 1941, p. 288 [**D. bearcreekensis*; OD]. Similar to *Tibicen*; cell cua2 broad but slightly more than twice as long as wide; apical margin of cell cua1 evenly and shallowly curving into cell cua2. WHALLEY, 1983. *Paleoc.*, USA (Montana).——FIG. 142,*2*. **D. bearcreekensis*; fore wing, ×1.0 (Cooper, 1941).

Lithocicada COCKERELL, 1906c, p. 457 [**L. perita*; OD]. Similar to *Cicada*, but cubital cell of fore wing with pointed or narrowly truncate apex. COOPER, 1941. *Oligo.*, USA (Colorado).

Platypedia UHLER, 1888, p. 23. COCKERELL, 1908a; COOPER, 1941. *Oligo.*, USA (Colorado)–*Holo.*

Tibicen LATREILLE, 1825, p. 426. SCUDDER, 1892; COOPER, 1941. *Oligo.*, USA (Colorado)–*Holo.*

Tymocicada BECKER-MIGDISOVA, 1954, p. 799 [**T. gorbunovi*; OD]. Fore wing similar to that of *Cosmopsaltivia* (recent), but CUA with longer anterior branch; cell between R and RS slightly broader. *Mio.*, USSR (Asian RSFSR).——FIG. 142,*1*. **T. gorbunovi*; fore wing, ×1.4 (Becker-Migdisova, 1954).

Family SCYTINOPTERIDAE Handlirsch, 1906b

[Scytinopteridae HANDLIRSCH, 1906b, p. 391]

Fore wing tegminous; costal margin commonly thickened basally; veins usually thin; vein SC obsolescent; branches of M and CUA short; crossveins few; only a few closed cells between R, M, and CUA. Hind wing with costal margin with at most a shallow covexity at base of wing; RS unbranched; M and CUA distally branched. Body little known, apparently as in Cicadellidae. *Perm.–Trias.*

Scytinoptera HANDLIRSCH, 1904b, p. 3 [**S. kokeni*; OD] [=*Anomoscyta* MARTYNOV, 1928b, p. 34 (type, *A. reducta*); *Permocixius* MARTYNOV, 1928b, p. 36 (type, *P. kazanensis*); *Scytinopterula* HANDLIRSCH, 1937, p. 115 (type, *Scytinoptera curta* ZALESSKY, 1929, p. 28)]. Fore wing with posterior branch of SC short, forming a sharp curve at level of R+M, or absent; M and CUA with distal forks, forming series of small, marginal cells, usually subequal; anal-jugal region strongly widened. Hind wing with costal margin with conspicuous but gradual convexity near base; no prominent marginal concavity or excision. Pronotum with lateral projections. [RIEK (1976b) has described a late Triassic species (*distorta*) in the genus *Scytinoptera*, but there is really no evidence to justify that placement.] *Perm.*, USSR (European and Asian RSFSR).——FIG. 143,*5a,b*. *S. kaltanica* BECKER-MIGDISOVA; *a*, fore and *b*, hind wings, ×10 (Becker-Migdisova, 1962b). ——FIG. 143,*5c*. *S. picturata* BECKER-MIGDISOVA; fore wing, ×8 (Becker-Migdisova, 1961c).

Anaprosbole BECKER-MIGDISOVA, 1960, p. 28 [**A. ivensis*; OD]. Fore wing with costal margin relatively broad basally; RS arising well beyond midwing; branches of M1+2 much longer than branches of M3+4; CUA with 3 terminal branches. [Family assignment uncertain.] *Perm.*, USSR (European RSFSR).——FIG. 143,*6*. **A. ivensis*; fore wing, ×5.0 (Becker-Migdisova, 1960).

Anomaloscytina DAVIS, 1942, p. 112 [**A. metapteryx*; OD]. Hind wing with costal margin with distinct but gentle concavity; SC short but distinct; anal area extensive. [Family position uncertain.] *Perm.*, Australia (New South Wales). ——FIG. 143,*7*. **A. metapteryx*; hind wing, ×6.5 (Davis, 1942).

Elliptoscarta TILLYARD, 1926a, p. 16 [**E. ovalis*; OD]. Fore wing oval, with apex evenly rounded; costal area (between C and R) broad; R dichotomously forked; M with 5 branches; CUA forked. *Perm.*, Australia (New South Wales).——FIG. 143,*1*. **E. ovalis*; fore wing, ×8.2 (Tillyard, 1926a).

Homaloscytina TILLYARD, 1926a, p. 16 [**H. plana*; OD]. Fore wing as in *Anaprosbole*, but CUA with only 2 terminal branches and connected to M by a crossvein; apex of wing bluntly rounded. EVANS, 1943b. *Trias.*, Australia (New South Wales). ——FIG. 143,*4*. **H. plana*; fore wing, ×8 (Evans, 1943b).

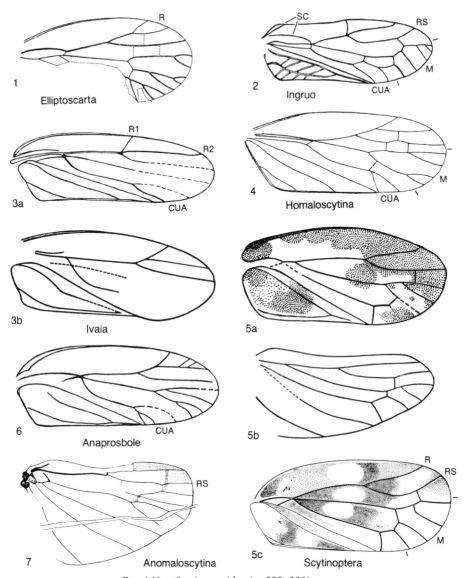

Fig. 143. Scytinopteridae (p. 222–223).

Ingruo BECKER-MIGDISOVA, 1960, p. 19 [*I. lanceolata*; OD]. Fore wing very narrow; posterior branch of SC short, merging with R; CUA dividing at level of origin of RS; fork of CUA large. [Family assignment doubtful]. *Perm.,* USSR (European RSFSR).——FIG. 143,*2*. **I. lanceolata*; fore wing, ×16 (Becker-Migdisova, 1960).

Ivaia BECKER-MIGDISOVA, 1960, p. 25 [*I. indistincta*; OD]. Fore wing moderately broad; costal area (between C and R) broad; R straight; CUA in brief contact with M, then diverging; M apparently unbranched. *Perm.,* USSR (European RSFSR).——FIG. 143,*3a*. *I. procucopoides* BECKER-MIGDISOVA; fore wing, ×6.5 (Becker-Migdisova, 1960).——FIG. 143,*3b*. **I. indistincta*; fore wing, ×8 (Becker-Migdisova, 1962b).

Kaltanospes BECKER-MIGDISOVA, 1961c, p. 344 [**K. kuznetskiensis*; OD]. Fore wing as in *Ingruo,* but CUA dividing much further distally of origin of RS. *Perm.,* USSR (Asian RSFSR).——FIG. 144,*6*. **K. kuznetskiensis*; fore wing and body, ×10 (Becker-Migdisova, 1961c).

Mesonirvana EVANS, 1956, p. 191 [**M. abrupta*; OD]. Fore wing: R with several branches; crossvein m-cu joined to CUA1; RS unbranched. *Trias.,* Australia (Queensland).——FIG. 144,*7*. **M. abrupta*; fore wing, ×5 (Evans, 1956).

Mesothymbris EVANS, 1956, p. 191 [**M. perkinsi*;

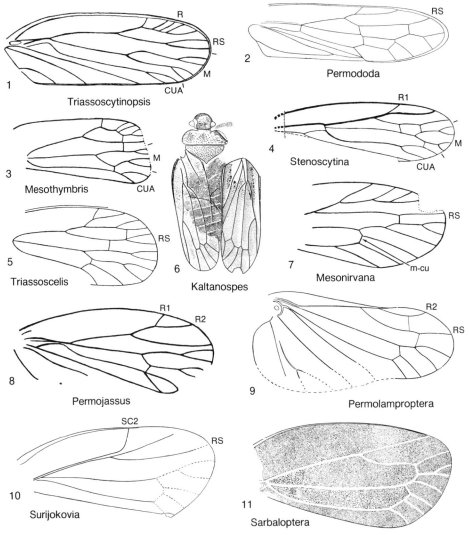

FIG. 144. Scytinopteridae (p. 223–225).

OD]. Fore wing as in *Triassoscytinopsis*, but M1+2 forming almost straight continuation of M; M3+4 bent towards CUA; crossvein m-cu joined to CUA. *Trias.*, Australia (Queensland).——FIG. 144,*3*. **M. perkinsi*; fore wing, ×6 (Evans, 1956).

Permododa BECKER-MIGDISOVA, 1961c, p. 347 [**P. membracoides*; OD]. Fore wing very slender; several closed cells between RS and M and between M and CUA. *Perm.*, USSR (Asian RSFSR).——FIG. 144,*2*. **P. membracoides*; fore wing, ×10 (Becker-Migdisova, 1961c).

Permojassus TILLYARD, 1926a, p. 7 [**P. australis*; OD] [=*Permojassula* HANDLIRSCH, 1937, p. 115, obj.] Fore wing similar to that of *Homaloscytina*, but anal area apparently much narrower. EVANS, 1956. *Perm.*, Australia (New South Wales).——FIG. 144,*8*. **P. australis*; fore wing, ×8 (Evans, 1956).

Permolamproptera BECKER-MIGDISOVA, 1961c, p. 340 [**P. grandis*; OD]. Hind wing similar to that of *Scytinoptera* but with R2 curved distally; anal area extended. *Perm.*, USSR (Asian RSFSR).——FIG. 144,*9*. **P. grandis*; hind wing, ×5.7 (Becker-Migdisova, 1961c).

Sarbaloptera BECKER-MIGDISOVA, 1961c, p. 328 [**S. sarbalensis*; OD]. Fore wing with asymmetrical apex; costal area (between C and R) very broad. *Perm.*, USSR (Asian RSFSR).——FIG. 144,*11*. **S. sarbalensis*; fore wing, ×6.5 (Becker-Migdisova, 1961c).

Stenoscytina TILLYARD, 1926a, p. 15 [**S. aus-

traliensis; OD]. Little-known genus. Fore wing narrow; M with 4 branches; CUA curving abruptly posteriorly after diverging from M. [Family assignment uncertain.] *Perm.,* Australia (New South Wales).——Fig. 144,*4.* **S. australiensis*; fore wing, ×6.5 (Tillyard, 1926a).

Surijokovia BECKER-MIGDISOVA, 1961c, p. 342 [**S. lata*; OD]. Little-known fore wing; posterior branch of SC long, mostly parallel to R, then diverging anteriorly to termination on costal margin; RS apparently unbranched. *Perm.,* USSR (Asian RSFSR).——Fig. 144,*10.* **S. lata*; fore wing, ×16 (Becker-Migdisova, 1961c).

Triassoscelis EVANS, 1956, p. 192 [**T. anomala*; OD]. Fore wing as in *Mesonirvana,* but RS forked. *Trias.,* Australia (Queensland).——Fig. 144,*5.* **T. anomala*; fore wing, ×5 (Evans, 1956).

Triassoscytina EVANS, 1956, p. 179 [**T. incompleta*; OD]. Fore wing as in *Homaloscytina,* but M forking just beyond level of origin of RS. [Family position uncertain.] *Trias.,* Australia (Queensland).

Triassoscytinopsis EVANS, 1956, p. 190 [**T. stenulata*; OD]. Fore wing with apex evenly rounded; R with at least 4 parallel branches distally; RS with from 2 to 4 branches; M with 4 branches. *Trias.,* Australia (Queensland).——Fig. 144,*1.* *T. aberrans* EVANS; fore wing, ×6 (Evans, 1956).

Tychtoscytina BECKER-MIGDISOVA, 1952, p. 179 [**T. kuznetskiensis*; OD]. Fore wing little known, with wide costal area; R1 straight. *Perm.,* USSR (Asian RSFSR).

Family BITURRITIIDAE Metcalf, 1951

[Biturritiidae METCALF, 1951, p. 11]

Fore wing sclerotized; no marginal border; vein M unbranched; radial cell divided by a crossvein. Hind wing nearly of uniform width, with very slight concavity of costal margin. *Trias.–Holo.*

Biturritia GODING, 1930, p. 39. *Holo.*
Absoluta BECKER-MIGDISOVA, 1962a, p. 92 [**A. distincta*; OD]. Hind wing with base of CUA nearly or completely coalesced with stem of M. [Family assignment doubtful.] *Trias.,* USSR (Kirghiz).——Fig. 145,*1.* **A. distincta*; hind wing, ×12 (Becker-Migdisova, 1962b).

Family CICADELLIDAE Latreille, 1802

[Cicadellidae LATREILLE, 1802a, p. 257] [=Jasscopidae HAMILTON, 1971, p. 943]

Fore wing tegminous; several to many closed cells; CUA usually with wide distal fork. Hind wing narrowed distally; submar-

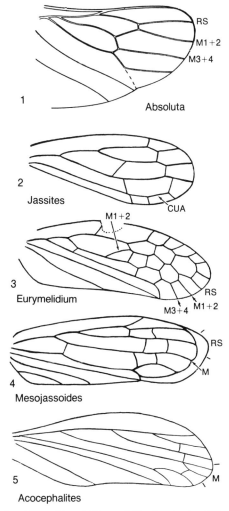

FIG. 145. Biturritiidae and Cicadellidae (p. 225–227).

ginal vein conspicuous, with relatively wide margin; ano-jugal area broad. *Trias.–Holo.*

Cicadella DUMÉRIL, 1806, p. 266. BERVOETS, 1910; COCKERELL, 1920c; BECKER-MIGDISOVA, 1951. *Eoc.,* USA (Colorado); *Oligo.,* Europe (Baltic); *Mio.,* USSR (European RSFSR)–*Holo.*

Acocephalites MEUNIER, 1904e, p. 119 [**A. breddini*; OD]. Little-known genus, based on fore wing with strongly arched costal margin and a venation similar to that of *Mesojassoides*; M with distal fork. *Jur.,* Europe (Spain).——Fig. 145,*5.* **A. breddini*; fore wing, ×14 (Handlirsch, 1907).

Agallia CURTIS, 1833, p. 193. SCUDDER, 1890. *Oligo.,* USA (Colorado)–*Holo.*

Aphrodes CURTIS, 1833, p. 193. SCUDDER, 1890;

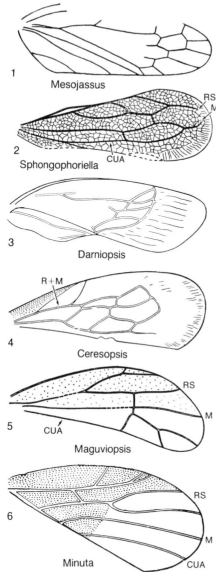

Fig. 146. Eurymelidae and Membracidae (p. 227–228).

STATZ, 1950a. *Eoc.*, USA (Wyoming); *Oligo.–Mio.*, Europe (Germany)–Holo.

Batrachomorphus LEWIS, 1834, p. 51, *nom. correct.* AGASSIZ, 1846, *ex Batracomorphus* LEWIS, 1834. STATZ, 1950a. *Oligo.*, Europe (Germany)–Holo.

Cicadula ZETTERSTEDT, 1838, p. 296. SCUDDER, 1890. *Eoc.*, USA (Wyoming)–*Holo.*

Coelidia GERMAR, 1821, p. 75. SCUDDER, 1890, 1895b. *Eoc.*, USA (Wyoming), Canada (British Columbia); *Oligo.*, USA (Colorado)–*Holo.*

Deltocephalus BURMEISTER, 1838, p. 5. STATZ, 1950a. *Oligo.*, Europe (Germany)–*Holo.*

Durgades DISTANT, 1912, p. 608. BECKER-MIGDISOVA & MARTYNOVA, 1951. *Mio.*, USSR (Kirghiz)–*Holo.*

Eurymelidium TILLYARD, 1919c, p. 884 [*E. australe*; OD]. SC apparently absent; $M1+2$ anastomosed with part of RS. EVANS, 1956. *Trias.*, Australia (Queensland).——FIG. 145,3. *E. australe*; fore wing, ×10 (Evans, 1956).

Euscelis BRULLÉ, 1832, p. 109. STATZ, 1950a; PIERCE, 1963. *Oligo.*, Europe (Germany); *Mio.*, USA (California)–*Holo.*

Gypona GERMAR, 1821, p. 73. [Generic assignment of fossil doubtful.] SCUDDER, 1890. *Oligo.*, USA (Colorado)–*Holo.*

Gyponites STATZ, 1950a, p. 10 [*G. pronota*; OD]. Little-known cicadellid; head short and broad; ocelli large; pronotum long, with parallel sides; scutellum shorter than pronotum. *Oligo.*, Europe (Germany).

Homopterulum HANDLIRSCH, 1907, p. 641 [*Cercopidium signoreti* WESTWOOD, 1854, p. 396; SD CARPENTER, herein]. Little-known genus, based on fore wing. [Family assignment doubtful.] EVANS, 1956. *Jur.*, England.

Idiocerus LEWIS, 1834, p. 47. STATZ, 1950a. *Oligo.*, Europe (Germany)–*Holo.*

Jascopus HAMILTON, 1971, p. 944 [*J. notabilis*; OD]. Little-known genus, based on nymph. [Type of family *Jascopidae* HAMILTON.] EVANS, 1972. *Cret.*, Canada (Manitoba).

Jassites HANDLIRSCH, 1907, p. 642 [*Cicada punctatus* BRODIE, 1845, p. 33; OD]. Little-known genus, based on fore wing. CUA with very short branches. EVANS, 1956. *Jur.*, England.——FIG. 145,2. *J. punctatus*; fore wing, ×5.5 (Evans, 1956).

Jassopsis SCUDDER, 1890, p. 312 [*J. evidens*; OD]. Little-known genus, similar to *Thamnotettix*. Scutellum not more than half the length of thorax. *Oligo.*, USA (Colorado).

Jassus FABRICIUS, 1803, p. 85. BERVOETS, 1910; MEUNIER, 1920c; PITON, 1940a; STATZ, 1950a. *Eoc.*, Europe (France); *Oligo.*, Europe (Baltic, Germany)–*Holo.*

Lavrushinia COCKERELL, 1925g, p. 10 [*L. elegantula*; OD]. Little-known genus, based on long and narrow fore wing; marginal vein very close to wing margin. *Mio.*, USSR (Asian RSFSR).

Macropsis LEWIS, 1834, p. 49. BERVOETS, 1910; STATZ, 1950a. *Oligo.*, Europe (Baltic, Germany)–*Holo.*

Maleojassus ZEUNER, 1941a, p. 90 [*M. primitivus*; OD]. Fore wing as in *Stonasla* (recent), but RS smoothly curved, not bent at junction with M1; M almost straight. *Eoc.*, Scotland.

Megophthalmus CURTIS, 1833, p. 193. STATZ, 1950a. *Oligo.*, Europe (Germany)–*Holo.*

Mesojassoides OMAN, 1937, p. 38 [*M. gigantea*;

OD]. Fore wing as in *Coelidia* but with additional crossveins; M unbranched. *Cret.*, USA (Colorado).——FIG. 145,4. **M. gigantea*; fore wing, ×4.5 (Evans, 1956).

Miochlorotettix CARPENTER, herein [**M. gibroni* PIERCE, 1963, p. 73; OD]. Similar to *Chlorotettix* (recent), but prothorax strongly arched forward and scutellum extending back between wings about as far as forwards. The original generic name, *Miochlorotettix,* was a *nomen nudum* (PIERCE, 1963).] *Mio.*, USA (California).

Miomesamia PIERCE, 1963, p. 81 [**M. juliae*; OD]. Similar to *Ulope* (recent). Face wide, eyes prominent; antennae at sides of front sutures, opposite outer corners of eyes. *Mio.*, USA (California).

Oligogypona STATZ, 1950a, p. 8 [**O. haupti*; OD]. Similar to *Gypona,* but head broad, somewhat narrower than pronotum; costal margin of fore wing strongly arched. *Oligo.*, Europe (Germany).

Oligoidiocerus STATZ, 1950a, p. 15 [**O. pronotumalis*; OD]. Similar to *Idiocerus* but with richer venation and unmarked fore wing. *Oligo.*, Europe (Germany).

Oligopenthimia STATZ, 1950a, p. 9 [**O. ovalis*; OD]. Similar to *Penthimia* (recent). Head short, as wide as pronotum; scutellum long, reaching to the middle of the abdomen. *Oligo.*, Europe (Germany).

Phlepsius FIEBER, 1866, p. 503. PIERCE, 1963. *Mio.*, USA (California)–*Holo.*

Protochlorotettix PIERCE, 1963, p. 78 [**P. calico*; OD]. Similar to *Chlorotettix* (recent), but with last sternum completely divided. *Mio.*, USA (California).

Tetigonia BLANCHARD, 1852, p. 282. STATZ, 1950a. *Oligo.*, Europe (Germany)–*Holo.*

Tettigella CHINA & FENNAH, 1945, p. 711. SCUDDER, 1890; STATZ, 1950a. *Eoc.*, USA (Wyoming); *Oligo.*, USA (Colorado), Europe (Germany)–*Holo.*

Thamnotettix ZETTERSTEDT, 1838, p. 292. COCKERELL, 1920c, 1924a, 1925a; STATZ, 1950a. *Eoc.*, USA (Colorado); *Oligo.*, Europe (Germany); *Mio.*, USA (Colorado)–*Holo.*

Typhlocyba GERMAR, 1833, p. 180. GERMAR & BERENDT, 1856. *Oligo.*, Europe (Baltic)–*Holo.*

Family EURYMELIDAE
Amyot & Serville, 1843

[Eurymelidae AMYOT & SERVILLE, 1843, p. 554]

Fore wing hyaline or opaque and coriaceous; venation often reticulate; vein RS absent; M1+2 retained as separate vein, extending to apex; M3+4 usually unbranched; CUA forked. *Trias.–Holo.*

Eurymela LE PELETIER & SERVILLE, 1828, p. 603. *Holo.*

Mesojassus TILLYARD in TILLYARD & DUNSTAN, 1916, p. 34 [**M. ipsviciensis*; OD]. Little-known genus, based on fore wing. Fork of CUA marginal, very shallow. EVANS, 1956. *Trias.*, Australia (Queensland).——FIG. 146,1. **M. ipsviciensis*; fore wing, ×8.4 (Evans, 1956).

Family MEMBRACIDAE
Rafinesque, 1815

[Membracidae RAFINESQUE, 1815, p. 121]

Fore wing usually membranous, except for basal region; clavus distinct, claval suture along vein 1A; ends of veins usually forming a scalloped submarginal line, the terminal marginal membrane (limbus) extending beyond the veins; veins usually clear and marked by punctures; M either free basally or coalesced in part with stem of R or CUA; cells usually irregular; venation highly diverse. Hind wing well developed, but usually shorter than fore wing; limbus usually present; venation usually similar to that of fore wing. Pronotum extensively developed, often prolonged posteriorly and concealing the scutellum, the wings, and even the entire abdomen; antennae minute, bristlelike; tarsi with 3 segments. *Trias.–Holo.*

This is a very large and diversified family. Fossil forms, which are usually known only from wings, are often difficult to classify because of the variability in the venation, especially that of the fore wings. Much difference of opinion exists among specialists in Homoptera about the generic lines. The taxonomic groups used here are essentially those employed in the *General Catalogue of the Homoptera* (METCALF & WADE, 1966).

Membracis FABRICIUS, 1775, p. 675. *Holo.*

Ceresopsis BECKER-MIGDISOVA, 1958, p. 66 [**C. costalis*; OD]. Fore wing broader than in *Darniopsis,* with conspicuous sclerotized area between costal margin and R+M basally; 3 apical cells. *Trias.*, USSR (Kirghiz).——FIG. 146,4. **C. costalis*; fore wing, ×10 (Becker-Migdisova, 1958).

Darniopsis BECKER-MIGDISOVA, 1958, p. 65. [**D. tragopea*; OD]. Fore wing elongate, with very wide limbus; costal margin only slightly convex; 4 apical cells; M and CUA with common stem; anal area large, triangular. *Trias.*, USSR (Kirghiz).——FIG. 146,3. **D. tragopea*; fore wing, ×10 (Becker-Migdisova, 1958).

Maguviopsis BECKER-MIGDISOVA, 1953c, p. 463 [**M. kotchnevi*; OD]. Fore wing with costal-

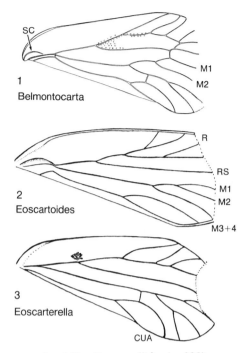

Fig. 147. Eoscarterellidae (p. 228).

distal margin broadly curved; anterior half of wing sclerotized; R and M not coalesced with CUA; M unbranched, straight. *Jur.*, USSR (Asian RSFSR).——Fig. 146,5. **M. kotchnevi*; fore wing, ×16 (Becker-Migdisova, 1953c).

Minuta BECKER-MIGDISOVA, 1958, p. 64 [**M. heteropterata*; OD]. Little-known genus; fore wing short, sclerotized from midwing to front margin; R, M, and CUA merged with CUP basally; CUA long; no apical cells. *Trias.*, USSR (Kirghiz).——Fig. 146,6. **M. heteropterata*; fore wing, ×20 (Becker-Migdisova, 1958).

Sphongophoriella BECKER-MIGDISOVA, 1958, p. 63 [**S. reticulata*; OD]. Fore wing elongate, reticulate; M and CUA not coalesced to form a common stem; venation reduced; cells elongate-oval; anal area narrow, 3 apical cells. *Trias.*, USSR (Kirghiz).——Fig. 146,2. **S. reticulata*; fore wing, ×34 (Becker-Migdisova, 1958).

Family AETALIONIDAE Spinola, 1850

[Aetalionidae SPINOLA, 1850, p. 53]

Similar to Cicadellidae; fore wing with vein RS absent; M1+2 (or M1 and M2) extending to apex of wing; M3+4 usually forked; CUA unbranched. Hind wing with RS absent. *Oligo.–Holo.*

Aetalion LATREILLE, 1810, p. 263. STATZ, 1950a. *Oligo.*, Europe (Germany)–*Holo.*

Family EOSCARTERELLIDAE Evans, 1956

[Eoscarterellidae EVANS, 1956, p. 220]

Fore wing with vein RS arising from R about one-third wing length from base; R with at least 2 branches; CUA separate from M. *Perm.–Trias.*

Eoscarterella EVANS, 1956, p. 220 [**E. media*; OD] [=*Prosbolopsites* BECKER-MIGDISOVA, 1960, p. 90 (type, *P. tillyardi*)]. Fore wing tegminous and rugose, broadest in distal half; RS and M parallel for most of their lengths; M with 4 branches. EVANS, 1961. *Trias.*, Australia (Queensland).——Fig. 147,3. **E. media*; fore wing, ×5.5 (Evans, 1956).

Belmontocarta EVANS, 1958, p. 112 [**B. perfecta*; OD]. Fore wing with SC very short, curving distally towards R+M; M1 and M2 longer than M3 and M4; CUA curved and joined to base of M by short crossvein. *Perm.*, Australia (New South Wales).——Fig. 147,1. **B. perfecta*; fore wing, ×4.5 (Evans, 1958).

Eoscartoides EVANS, 1956, p. 220 [**E. bryani*; OD]. Fore wing with complete marginal border; R and M arched basally; M1+2 forked. EVANS, 1961. *Trias.*, Australia (Queensland).——Fig. 147,2. **E. bryani*; fore wing, ×4.5 (Evans, 1961).

Family PROCERCOPIDAE Handlirsch, 1906

[Procercopidae HANDLIRSCH, 1906b, p. 500]

Fore wing slender, at least three times as long as wide; vein RS arising in basal third of wing; M and CUA branching in distal third of wing, their branches short. Hind wing very little known. EVANS, 1956. *Trias.–Jur.*

Procercopis HANDLIRSCH, 1906b, p. 500 [**P. alutacea*; SD BECKER-MIGDISOVA, 1962b, p. 180]. Fore wing elongate, about 4 times as long as broad; M with at least 3 branches; several crossveins in distal part of wing. *Trias.*, USSR (Kirghiz); *Jur.*, Europe (Germany).——Fig. 148,3. *P. longipennis* BECKER-MIGDISOVA, *Trias.*; fore wing, ×4 (Becker-Migdisova, 1962b).

Procercopina MARTYNOV, 1937a, p. 99 [**P. asiatica*; OD]. Fore wing as in *Procercopis* but relatively broader; only one crossvein between adjacent veins. EVANS, 1956. *Jur.*, USSR (Kirghiz).——Fig. 148,5. **P. asiatica*; fore wing, ×4.6 (Becker-Migdisova, 1962b).

Hemiptera—Homoptera

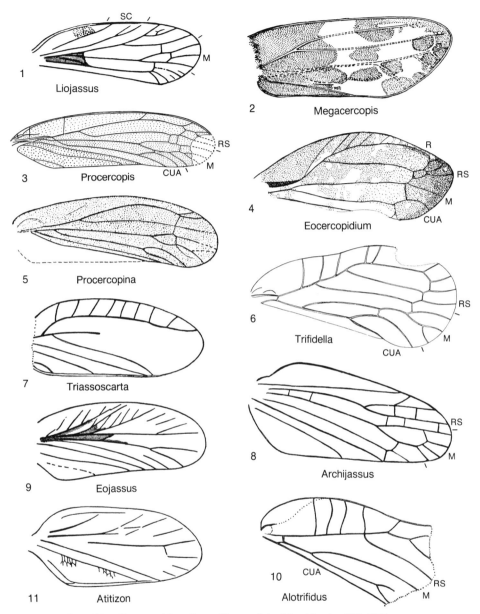

FIG. 148. Procercopidae, Cercopidae, and Archijassidae (p. 228–233).

Family DYSMORPHOPTILIDAE Handlirsch, 1906

[Dysmorphoptilidae HANDLIRSCH, 1906b, p. 492]

Tegmen of irregular form, abruptly narrowed distally, strongly sclerotized; vein SC apparently fused with R; several short branches from R to costal margin; RS arising before midwing. EVANS, 1956. *Trias.–Jur.*

Dysmorphoptila HANDLIRSCH, 1906b, p. 492 [**Belostoma liasina* GIEBEL, 1856, p. 371; OD]. Broad portion of tegmen extending only to about midwing; M with only one distal fork. EVANS, 1956. *Jur.,* Europe (Germany).

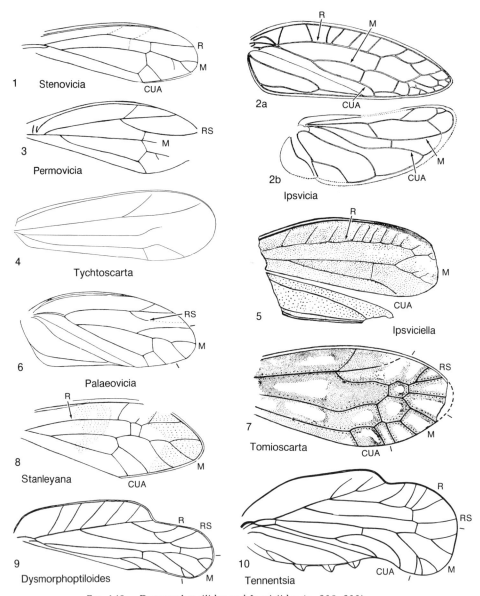

FIG. 149. Dysmorphoptilidae and Ipsviciidae (p. 230–233).

Dysmorphoptiloides EVANS, 1956, p. 218 [*D. elongata*; OD]. Tegmen as in *Dysmorphoptila*, but broad portion extending nearer to apex; M with 2 distal forks. RIEK, 1974b. *Trias.,* Australia (Queensland), South Africa.——FIG. 149,9. *D. elongata*; tegmen, ×3.4 (Evans, 1956).

Mesoatracis BECKER-MIGDISOVA, 1949b, p. 40 [*M. reducta*; OD]. Tegmen as in *Dysmorphoptiloides* but with shorter distal area; M with 3 terminal branches. BECKER-MIGDISOVA, 1962b. *Jur.,* USSR (Tadzhik).

Tennentsia RIEK, 1976b, p. 813 [*Dysmorphoptiloides protuberans* RIEK, 1974c, p. 22; OD]. Fore wing similar to that of *Dysmorphoptiloides,* but SC with several distal branches and RS unbranched; M and CU apparently connected basally by a crossvein. *Trias.,* South Africa.——FIG. 149,10. *T. protuberans*; fore wing, ×2.3 (Riek, 1976b).

Family CERCOPIDAE
Westwood, 1838

[Cercopidae Westwood, 1838, p. 39]

Head narrower than pronotum, usually as wide as anterior margin of scutellum; ocelli on disc of crown, each at posterior end of sulcus; length and width of eyes almost equal; antennae originating in cavities below anterior margin of head; postclypeus commonly protuberant. Fore wings usually coriaceous. [The Aphrophoridae are included here, as a subfamily, because of the difficulty of recognizing the distinguishing features in the fossils.] Evans, 1956. *Trias.–Holo.*

Cercopis Fabricius, 1775, p. 688. [Numerous extinct species from Tertiary deposits and described before 1900 were placed in the genus, but their assignment to *Cercopis* has not been generally accepted (see Handlirsch, 1907, p. 1072-1074). However, a few, well-preserved specimens appear to justify at least tentative placement in the genus.] Scudder, 1890; Cockerell, 1920a, 1927b; Evans, 1956. *Eoc.,* Canada (British Columbia), USA (Colorado, Wyoming), USSR (Asian RSFSR)–*Holo.*

Alotrifidus Evans, 1956, p. 216 [**A. interruptus*; OD]. Fore wing as in *Trifidella,* but costal margin arching basally and RS arising further distally. *Trias.,* Australia (Queensland).——Fig. 148,*10.* **A. interruptus*; fore wing, ×10 (Evans, 1956).

Aphrophora Germar, 1821, p. 48. Cockerell, 1922f, 1925g; Pongrácz, 1928; Piton, 1936c; Théobald, 1937a; Becker-Migdisova, 1964. *Eoc.,* Europe (Baltic, France), Canada (British Columbia); *Oligo.,* England, Europe (France); *Mio.,* USSR (Asian RSFSR)–*Holo.*

Cercopites Scudder, 1890, p. 316 [**C. calliscens* Scudder, 1890, p. 316; SD Carpenter, herein]. Head relatively small; thorax hexagonal; fore wing more than twice as long as broad. *Eoc.,* USA (Wyoming), Canada (British Columbia).

Clastoptera Germar, 1839, p. 187. Scudder, 1890. *Oligo.,* USA (Colorado)–*Holo.*

Dawsonites Scudder, 1895b, p. 18 [**D. veter*; OD]. Similar to *Palecphora,* but RS arising at midwing. *Mio.,* Canada (British Columbia).

Eocercopidium Zeuner, 1944a, p. 116, *nom. subst.* pro *Eocercopis* Zeuner, 1941a, p. 88, *non* Handlirsch, 1939 [**Eocercopis maculata* Zeuner, 1941a, p. 88; OD]. Fore wing similar to that of *Aphrophora,* but R strongly bent anteriorly near base; preradial part of wing very wide, crossed by pectinate branches from R; radial-median area very broad; M separating from CUA very near to base. *Eoc.,* Scotland.——Fig. 148,*4.* **E. maculata* (Zeuner); fore wing, ×6.4 (Zeuner, 1944a).

Megacercopis Cockerell, 1925g, p. 9 [**M. optima*; OD]. Little-known fore wing with venation similar to that of *Stenecphora,* but apex much more pointed. *Mio.,* USSR (Asian RSFSR).——Fig. 148,*2.* **M. optima*; fore wing, ×2.5 (Cockerell, 1925g).

Palaeoptysma Scudder, 1895b, p. 21 [**P. venosa*; OD]. Little-known fore wing, related to *Aphrophora* but very slender. *Eoc.,* Canada (British Columbia).

Palaphrodes Scudder, 1890, p. 333 [**P. irregularis* Scudder, 1890, p. 333; SD Carpenter, herein]. Fore wing as in *Cercopis,* but head very obtuse and rounded in front, narrower distally than thorax. Cockerell, 1908k. *Oligo.,* USA (Colorado).

Palecphora Scudder, 1890, p. 324 [**P. communis* Scudder, 1890, p. 324; SD Carpenter, herein]. Fore wing longer and more slender than that of *Palaphrodes*; costal margin less arched. Cockerell, 1908k. *Oligo.,* USA (Colorado).

Petrolystra Scudder, 1878a, p. 530 [**P. gigantea* Scudder, 1878a, p. 530; SD Carpenter, herein]. Large insects; head large, flat dorsally, twice as broad as long, the front broadly convex; scutellum very small, about half as long as thorax. Scudder, 1890. *Oligo.,* USA (Colorado).

Philagra Stål, 1863, p. 593. Cockerell, 1925g. *Mio.,* USSR (Asian RSFSR)–*Holo.*

Ptyelus Le Peletier & Serville, 1828, p. 608. Théobald, 1937a. *Oligo.,* Europe (France)–*Holo.*

Ptysmaphora Scudder, 1895b, p. 21 [**P. fletcheri*; OD]. Fore wing as in *Palaeoptysma* but with costal margin straighter. *Eoc.,* Canada (British Columbia).

Sinophora Melichar, 1902, p. 113 [**S. maculosa*; OD]. Becker-Migdisova, 1964. *Mio.,* USSR (Asian RSFSR)–*Holo.*

Stenecphora Scudder, 1895b, p. 17 [**S. punctulata*; OD]. Fore wing with very broad apex, slender clavus; RS arising near base. *Eoc.,* Canada (British Columbia).

Stenolocris Scudder, 1895b, p. 19 [**S. venosa*; OD]. Little-known fore wing, with very strong costal vein and RS arising at wing base. [Family assignment doubtful.] *Mio.,* Canada (British Columbia).

Triassoscarta Tillyard, 1919c, p. 874 [**T. subcostalis*; OD]. Little-known genus, based on incomplete tegmen. SC apparently absent; R long, nearly parallel with costal margin and connected to costal margin by about 8 subequal crossveins. [Originally placed in the *Scytinopteridae* but transferred to *Cercopidae* by Evans (1956).] *Trias.,* Australia (Queensland).——Fig. 148,*7.* **T. subcostalis*; fore wing, ×6 (Evans, 1956).

Triecphora Amyot & Serville, 1843, p. 561. Woodward, 1879. *Eoc.,* England–*Holo.*

Trifidella Evans, 1956, p. 215 [*T. perfecta; OD]. Fore wing tegminous, coarsely rugose; several long veinlets between wing margin and R; M and CUA fused basally; CUA forked. Trias., Australia (Queensland).——Fig. 148,6. *T. perfecta; fore wing, ×10 (Evans, 1961).

Family IPSVICIIDAE Tillyard, 1919

[Ipsviciidae Tillyard, 1919c, p. 878] [=Stenoviciidae Evans, 1956, p. 205]

Fore wing uniformly sclerotized; costal margin thick and flattened; vein R consisting usually of R and less commonly of RS; R joined to M by a prominent crossvein; M and CUA usually arising from a common basal stem; M typically branched; CUA and CUP apparently unbranched. Hind wing (known only in Ipsvicia) strongly curved anteriorly in distal area; CUA branched. Body unknown. Perm.–Trias.

The systematic position of this family is obscure. Tillyard (1919c) originally assigned it to the Homoptera, close to the extinct family Syntonopteridae, but later (1926d) transferred it to the Fulgoroidea of the Homoptera. Subsequently, it has been placed in the Heteroptera by Evans (1956), in the Homoptera (Auchenorrhyncha) by Becker-Migdisova (1962b), in the Homoptera (Peloridioidea) by China (1962), and in the Homoptera (Cercopoidea) by Evans (1963). Also, eight of the genera discussed below (Stenovicia, Permocentrus, Permagra, Permonia, Stanleyana, Palaeovicia, Apheloscyta, and Permoscarta) were placed in a new family, Stenoviciidae, by Evans (1956), although most of these were previously assigned to the Ipsviciidae (Evans, 1943b). Becker-Migdisova (1962b) concluded that the new family is unnecessary, and I have followed her treatment in retaining these genera in the Ipsviciidae.

Ipsvicia Tillyard, 1919c, p. 878 [*I. jonesi; OD]. R with several anterior branches to costa near middle of tegmen. Tillyard, 1923b. Trias., Australia (Queensland).——Fig. 149,2. *I. jonesi; a, tegmen; b, hind wing, ×4 (Evans, 1956).

Apheloscyta Tillyard, 1922b, p. 458 [*A. mesocampta; OD]. Branches of all veins of tegmen very short. [Family assignment doubtful.] Evans, 1956; Becker-Migdisova, 1962b. Perm., Australia (New South Wales).

Ipsviciella Becker-Migdisova, 1962a, p. 100 [*I. asiatica; OD]. Tegmen with rounded apex; R nearly straight, with several parallel branches to costal margin; CUA unbranched, merging with M basally. Trias., USSR (Kirghiz).——Fig. 149,5. *I. asiatica; tegmen, ×6.5 (Becker-Migdisova, 1962b).

Ipsviciopsis Tillyard, 1922b, p. 464 [*I. elegans; OD]. RS separating from R near base of tegmen. Evans, 1963. Trias., Australia (Queensland).

Palaeovicia Evans, 1943b, p. 189 [*P. incerta; OD]. Tegmen: RS short; M with 3 branches. Evans, 1956; Becker-Migdisova, 1962b. Perm., Australia (New South Wales).——Fig. 149,6. *P. incerta; tegmen, ×8 (Evans, 1943b).

Permagra Evans, 1943a, p. 7 [*P. distincta; OD]. Tegmen as in Tomioscarta but lacking closed cells. Evans, 1956; Becker-Migdisova, 1962b. Perm., Australia (New South Wales).

Permocentrus Evans, 1956, p. 207 [*Permoscarta trivenulata Tillyard, 1926a, p. 19; OD]. Tegmen with M and CUA independent basally. Becker-Migdisova, 1962b. Perm., Australia (New South Wales).

Permoscarta Tillyard, 1918b, p. 726 [*P. mitchelli; OD]. Little-known genus. Tegmen as in Permocentrus but with 2 crossveins between M and CUA. Evans, 1943a, 1956; Becker-Migdisova, 1962b. Trias., Australia (Queensland).

Permovicia Evans, 1943b, p. 189 [*P. obscura; OD]. Tegmen with RS broadly curved. Evans, 1956. Perm., Australia (New South Wales). ——Fig. 149,3. *P. obscura; ×10 (Evans, 1943b).

Stanleyana Evans, 1943b, p. 188 [*S. pulchra; OD]. Tegmen with RS apparently absent; M and CUA coalesced basally; M with 3 branches. Evans, 1956. Perm., Australia (New South Wales). ——Fig. 149,8. *S. pulchra; tegmen, ×6.5 (Evans, 1943b).

Stenovicia Evans, 1943b, p. 188 [*S. angustata; OD]. Tegmen as in Ipsvicia but much more slender; R long, arising at about midwing; M with 2 very short branches; CUA and M coalesced basally. [Type of family Stenoviciidae Evans, 1956.] Perm., Australia (New South Wales).——Fig. 149,1. *S. angustata; fore wing, ×8 (Evans, 1943b).

Tomioscarta Becker-Migdisova, 1961c, p. 350 [*T. surijokovensis; OD]. Tegmen with R branched at point of origin of RS; several closed cells between M, CUA, and RS. Becker-Migdisova, 1962b. Perm., USSR (Asian RSFSR).——Fig. 149,7. *T. surijokovensis; fore wing, ×6.5 (Becker-Migdisova, 1961c).

Tychtoscarta Becker-Migdisova, 1961c, p. 350 [*T. sokolovensis; OD]. Little-known genus.

Tegmen long and narrow; RS unbranched and continuing in a straight line from stem of R; M unbranched; CUA forked distally. BECKER-MIGDISOVA, 1962b. *Perm.,* USSR (Asian RSFSR).——FIG. 149,4. **T. sokolovensis;* fore wing, ×8 (Becker-Migdisova, 1961c).

Family ARCHIJASSIDAE Becker-Migdisova, 1962

[Archijassidae BECKER-MIGDISOVA, 1962a, p. 95]

Fore wing very wide, in some species with triangular costal area traversed by vein SC; SC usually divided into 2 long branches; RS present; numerous crossveins between branches of R and M; anal area wide, triangular. *Jur.*

Archijassus HANDLIRSCH, 1906b, p. 501 [**Cercopidium heeri* GEINITZ, 1880, p. 529; SD CARPENTER, herein]. Fore wing with costal margin strongly angular; RS arising beyond midwing; M with 4 branches. EVANS, 1956. *Jur.,* Europe (Germany).——FIG. 148,8. **A. heeri* (GEINITZ); fore wing, ×8 (Evans, 1956).

Atitizon HANDLIRSCH, 1939, p. 144 [**A. jassoides;* OD]. Fore wing very broad; costal margin strongly curved but not angular basally; RS arising at midwing. *Jur.,* Europe (Germany). —— FIG. 148,11. **A. jassoides;* fore wing, ×8 (Handlirsch, 1939).

Eojassus HANDLIRSCH, 1939, p. 145 [**E. indistinctus;* OD]. Little-known genus, based on fore wing; costal margin smoothly curved. *Jur.,* Europe (Germany).——FIG. 148,9. **E. indistinctus;* fore wing, ×6.5 (Handlirsch, 1939).

Liojassus HANDLIRSCH, 1939, p. 146 [**L. affinis;* OD]. Fore wing: SC with 2 long branches; RS arising at midwing; costal margin smoothly curved; M with 3 branches. [Family assignment doubtful.] *Jur.,* Europe (Germany). —— FIG. 148,1. **L. affinis;* fore wing, ×6.5 (Handlirsch, 1939).

Family HYLICELLIDAE Evans, 1956

[Hylicellidae EVANS, 1956, p. 195]

Fore wing as in Hylicidae (recent), with M coalesced basally with CUA, but CUA1 present and coalesced with part of M3+4 distally. *Trias.*

Hylicella EVANS, 1956, p. 195 [**H. colorata;* OD] [=*Hylicellites* BECKER-MIGDISOVA, 1962a, p. 95, (type, *Hylicella reducta* EVANS)]. CUA with abrupt basal bend; 2 crossveins between RS and M1+2; 1 crossvein between M1+2 and M3+4. *Trias.,* Australia (Queensland).——FIG. 150,6. **H. colorata;* fore wing, ×5 (Evans, 1956).

Family MUNDIDAE Becker-Migdisova, 1960

[Mundidae BECKER-MIGDISOVA, 1960, p. 31]

Fore wing weakly tegminous, without pits; veins thick; RS, M, and CUA with prominent projections; costal area and anal area broad. *Perm.*

Mundus BECKER-MIGDISOVA, 1960, p. 31 [**M. nodosus;* OD]. Fore wing relatively broad, with asymmetrical, blunt apex; R diverging abruptly at midwing toward costal margin, forking; R2 parallel to RS. *Perm.,* USSR (European RSFSR). —— FIG. 150,5. **M. nodosus;* fore wing, ×8 (Becker-Migdisova, 1960).

Family PEREBORIIDAE Zalessky, 1930

[*nom. correct.* BRUES, MELANDER, & CARPENTER, 1954, p. 813 (*pro* Pereboridae ZALESSKY, 1930, p. 1026)] [=Permoglyphidae EVANS, 1943b, p. 183]

Fore wing membranous; veins R, RS, and CUA with extensive branching. BECKER-MIGDISOVA, 1962b. *Perm.–Trias.*

Pereboria ZALESSKY, 1930, p. 1021 [**P. bella;* OD]. Little-known genus, based on fore wing. R with close pectinate branching; crossveins numerous, irregular; wing large, about 40 mm long. EVANS, 1956; BECKER-MIGDISOVA, 1962b. *Perm.,* USSR (Asian RSFSR).——FIG. 150,9. **P. bella;* fore wing, ×1.5 (Becker-Migdisova, 1962b).

Crosbella EVANS, 1956, p. 192 [**C. elongata;* OD]. Fore wing as in *Permobrachus,* but M more extensively branched. *Trias.,* Australia (Queensland).——FIG. 150,1. **C. elongata;* fore wing, ×4.5 (Evans, 1956).

Kaltanopibrocha BECKER-MIGDISOVA, 1961c, p. 357 [**K. boreoscytinoides;* OD]. Little-known genus, based on hind wing fragment. Costal margin almost straight; R directed posteriorly in apical region, pectinately branched; M forking before RS. [Family assignment doubtful.] *Perm.,* USSR (Asian RSFSR). —— FIG. 150,10. **K. boreoscytinoides;* hind wing, ×4.5 (Becker-Migdisova, 1961c).

Neuropibrocha BECKER-MIGDISOVA, 1961c, p. 356 [**N. ramisubcostalis;* OD]. Fore wing as in *Pereboria,* but R with fewer pectinate branches and less dense reticulation of branches of RS, M, and CUA; area between stems R and M with few crossveins. *Perm.,* USSR (Asian RSFSR). —— FIG. 150,7. **N. ramisubcostalis;* fore wing, ×2.0 (Becker-Migdisova, 1961c).

Permobrachus EVANS, 1943b, p. 183 [**Permodipthera dubia* TILLYARD, 1926a, p. 24; OD]. Fore wing shaped as in *Scytophara,* but R1 curv-

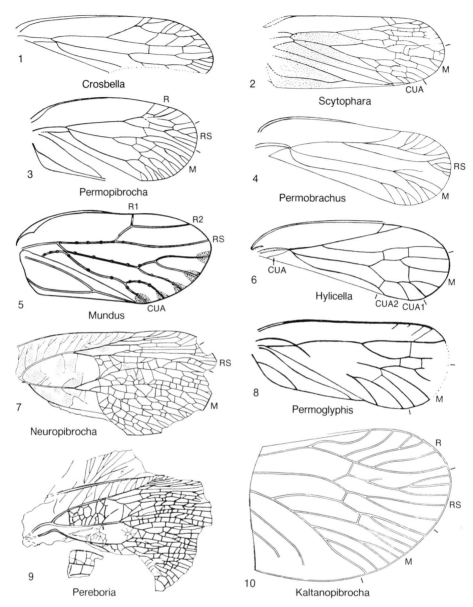

FIG. 150. Hylicellidae, Mundidae, and Pereboriidae (p. 233-235).

ing abruptly to anterior margin; branching of R2 pectinate; M branching well beyond midwing. *Perm.*, Australia (New South Wales).——FIG. 150,4. *P. magnus* EVANS; fore wing, ×3.5 (Evans, 1943b).

Permoglyphis TILLYARD, 1926a, p. 22 [*P. belmontensis*; OD]. Little-known genus, based on fore wing; similar to *Permopibrocha* but apparently with less branching of R, RS, and M; costal margin nearly straight. *Trias.*, Australia (New South Wales).——FIG. 150,8. *P. belmontensis*; fore wing, ×4.5 (Evans, 1956).

Permopibrocha MARTYNOV, 1935c, p. 18 [*P. ramosa*; OD]. Fore wing as in *Pereboria*, but R with fewer branches; M more deeply forked than CUA; fore wing small. *Perm.*, USSR (European RSFSR).——FIG. 150,3. *P. ramosa*; fore wing, ×3.5 (Martynov, 1935c).

Scytophara MARTYNOV, 1937b, p. 36 [*S. extensa*; OD]. Fore wing more slender than in *Permopi-*

brocha; costal margin straight beyond base; M forking at about level of origin of RS. *Perm., USSR (European RSFSR).*——FIG. 150,2. **S. extensa*; fore wing, ×6.5 (Martynov, 1937b).

Family FULGORIDIIDAE Handlirsch, 1939

[*nom. transl.* BECKER-MIGDISOVA, 1962b, p. 184, *ex* Fulgoridiinae HANDLIRSCH, 1939, p. 122]

Fore wing tegminous; costal margin only slightly arched; vein SC long, without branches; RS arising at about midwing; CUA forking well before origin of RS; crossveins few. Hind wing a little shorter than fore wing; anal area very broad; RS simple or with short fork; 1A arched away from CUP. EVANS, 1956; BECKER-MIGDISOVA, 1962b. *Jur.*

Fulgoridium HANDLIRSCH, 1906b, p. 496 [**Phryganidium balticum* GEINITZ, 1880, p. 527; OD] [=*Fulgoridulum* HANDLIRSCH, 1939, p. 140 (type, *F. egens*)]. Fore wing slender; usually with maculations; SC close to margin; R with a series of short branches distally; CUA with several long branches. BODE, 1953; EVANS, 1956. *Jur.,* Europe (Germany).——FIG. 151,*2a*. *F. punctatum* HANDLIRSCH; fore wing, ×10 (Handlirsch, 1939).——FIG. 151,*2b*. *F. reductum* HANDLIRSCH; hind wing, ×10 (Handlirsch, 1939).
Metafulgoridium CARPENTER, herein [**M. spilotum* HANDLIRSCH, 1939, p. 139; OD]. Fore wing as in *Fulgoridium,* but CUA2 unbranched. [The original generic name, *Metafulgoridium,* was a *nomen nudum* (HANDLIRSCH, 1939).] *Jur.,* Europe (Germany).——FIG. 151,*1*. **M. spilotum*; fore wing, ×6.5 (Handlirsch, 1939).

Family LOPHOPIDAE Stål, 1866

[Lophopidae STÅL, 1866, p. 130]

Head markedly narrower than pronotum; vertex usually narrow; pronotum short and broad, tricarinate. Fore wing coriaceous, with conspicuous venation and supernumerary longitudinal veins and crossveins; wing usually elongate; apical margin broadly rounded; claval veins united before apex. Fore and middle tibiae usually compressed. *Jur.–Holo.*

Lophops SPINOLA, 1838, p. 205. *Holo.*
Eofulgoridium MARTYNOV, 1937a, p. 164 [**E. kisylkiense*; OD]. Fore wing with SC about midway between C and R; M dividing at midwing; M with 3 branches. Hind wing little known; costal margin concave; RS arising beyond midwing; M and CUA dividing beyond midwing. EVANS, 1956; BECKER-MIGDISOVA, 1962b. *Jur.,* USSR

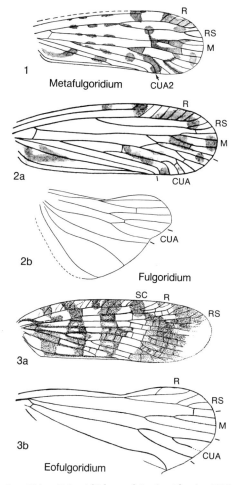

FIG. 151. Fulgoridiidae and Lophopidae (p. 235).

(Kirghiz).——FIG. 151,*3a*. **E. kisylkiense*; fore wing, ×4.2.——FIG. 151,*3b*. *E. proximum* MARTYNOV; hind wing, ×5 (Martynov, 1937a).
Scoparidea COCKERELL, 1920c, p. 243 [**S. nebulosa*; OD]. Fore wing with RS parallel to R; apical region with numerous, parallel veins; no regular gradate series of veins. [Family assignment doubtful.] *Eoc.,* USA (Colorado).

Family CIXIIDAE Spinola, 1838

[Cixiidae SPINOLA, 1838, p. 204]

Head not elongate; antennae with 2 segments, bearing flagella; wings unusually well developed. In fore wing, veins SC, R, and M with common stem; claval suture distinct; claval veins united into a claval stem. *Perm.–Holo.*

Cixius LATREILLE, 1804, p. 168. [The assignment of a Jurassic species from England to this genus is very doubtful (FENNAH, 1961), as is that of the several species in Baltic amber (GERMAR & BERENDT, 1856).] SCUDDER, 1890. *Eoc.*, USA (Wyoming); *Oligo.*, USA (Colorado)–*Holo.*

Asiocixius BECKER-MIGDISOVA, 1962a, p. 97 [**A. fulgoroides*; OD]. Fore wing membranous, except at base; costal margin smoothly rounded; R2 curved toward RS and giving rise to several veinlets; RS forked distally; M forking beyond midwing and with extensive pectinate branching; CUA with a long fork. *Trias.*, USSR (Kirghiz). ——FIG. 152,6. **A. fulgoroides*; fore wing, ×5 (Becker-Migdisova, 1962b).

Boreocixius BECKER-MIGDISOVA, 1955, p. 1100 [**B. sibiricus*; OD]. Fore wing with costal margin strongly thickened; RS arising very near wing base; R and RS with very short branches; fork of CUA long and curved. *Trias.*, USSR (Asian RSFSR). —— FIG. 152,12. **B. sibiricus*; fore wing, ×10 (Becker-Migdisova, 1962b).

Cixiella BECKER-MIGDISOVA, 1962a, p. 98 [**C. reducta*; OD]. Fore wing weakly tegminous, distal portion membranous; RS arising near midwing; M forking beyond level of origin of RS, with 3 terminal branches, and forming a large, closed cell; CUA curved basally. *Trias.*, USSR (Kirghiz). —— FIG. 152,8. **C. reducta*; fore wing, ×10 (Becker-Migdisova, 1962b).

Cycloscytina MARTYNOV, 1926b, p. 1349 [**C. delutinervis*; OD]. Fore wing tegminous, elongate; costal margin only slightly curved; R with a series of branches as in *Mesocixiella* but shorter; M joined to RS distally by a recurved branch. Hind wing little known; M with 2 long branches, arising before midwing. EVANS, 1956. *Trias.*, USSR (Kirghiz); *Jur.*, USSR (Kazakh, Tadzhik).——FIG. 152,3. **C. delutinervis*, Jur., Kazakh; fore wing, ×6 (Becker-Migdisova, 1962b).

Diaplegma SCUDDER, 1890, p. 288 [**D. abductum* SCUDDER, 1890, p. 290; SD CARPENTER, herein]. Similar to *Cixius*, but RS arising near midwing, each of its forks dividing into 2 or 3 distal, curved branches. *Oligo.*, USA (Colorado).

Eofulgorella COCKERELL, 1909j, p. 172 [**E. bradburyi*; OD]. Fore wing resembling that in *Oliarus* but elongate and with costal margin concave; crossveins forming a very regular series. [Family assignment doubtful.] *Eoc.*, USA (Colorado).

Eoliarus COCKERELL, 1925a, p. 10 [**E. quadristictus*; OD]. Similar to *Oliarus*, but RS arising well before the pterostigmal area and giving rise to 4 very oblique branches anteriorly. *Eoc.*, USA (Colorado).

Hyalesthes SIGNORET, 1865, p. 128 [**H. obsoletus*; OD]. STATZ, 1950a. *Oligo.*, Europe (Germany)–*Holo.*

Mesocixiella BECKER-MIGDISOVA, 1949b, p. 38 [**M. asiatica*; OD]. Fore wing with costal margin only slightly curved; R with a series of parallel branches leading to margin; RS arising before midwing with 3 or 4 terminal branches; M forked beyond midwing. EVANS, 1956. *Trias.*, USSR (Kirghiz); *Jur.*, USSR (Kazakh).——FIG. 152,7. **M. asiatica*; fore wing, ×6.5 (Becker-Migdisova, 1962b).

Mesocixius TILLYARD, 1919c, p. 876 [**M. triassicus*; OD]. Fore wing with RS forking about halfway between origin of RS and wing apex; fork of M less distal. EVANS, 1956. *Trias.*, Australia (Queensland).——FIG. 152,10. **M. triassicus*; fore wing, ×5.4 (Tillyard, 1919c).

Mundopoides COCKERELL, 1925g, p. 11 [**M. cisthenaria*; OD]. Similar to *Mundopa* (recent), having nearly straight costal and outer margins, the apex being obliquely truncate; SC terminating at midwing. *Mio.*, USSR (Asian RSFSR).

Myndus STÅL, 1862, p. 307. COCKERELL, 1926b. *Oligo.*, England–*Holo.*

Oeclixius FENNAH, 1963, p. 43 [**O. amphion*; OD]. Similar to *Oecleus* (recent) but with long, slender tibiae; pterostigma only moderately developed; tegminal veins distinctly granulate. *Mio.*, Mexico (Chiapas). ——FIG. 152,5. **O. amphion*; fore wing, ×13 (Fennah, 1963).

Oliarites SCUDDER, 1890, p. 293 [**Mnemosyne terrentula* SCUDDER, 1878b, p. 773; OD]. Little-known genus, with head less than half as broad as thorax; veins forming a weak reticulation distally. [Family assignment doubtful.] *Eoc.*, USA (Wyoming).

Oliarus STÅL, 1862, p. 306. COCKERELL, 1910b. *Oligo.*, Europe (Baltic)–*Holo.*

Oligonila CARPENTER, herein [**O. defectuosa* THÉOBALD, 1937a, p. 258; OD]. Fore wing as in *Anila* (recent) but lacking the oblique vein in the costal area. [The original generic name, *Oligonila*, was a *nomen nudum* (THÉOBALD, 1937a).] *Oligo.*, Europe (France).

Permocixiella BECKER-MIGDISOVA, 1961c, p. 361 [**P. venosa*; OD]. Fore wing elongate, costal margin nearly straight; R2 straight; branches of CUA nearly straight. *Perm.*, USSR (Asian RSFSR).——FIG. 152,4. **P. venosa*; fore wing, ×5.4 (Becker-Migdisova, 1961c).

Protoliarus COCKERELL, 1920c, p. 243 [**P. hamatus*; OD]. Similar to *Oliarus* but without a stigmatic spot on wings. COCKERELL, 1924a; COCKERELL & LEVEQUE, 1931. *Eoc.*, USA (Colorado).

Scytocixius MARTYNOV, 1937b, p. 34 [**S. mendax*; OD]. Fore wing broader distally than basally; costal margin smoothly curved; R2 strongly arched away from margin; RS similarly arched but less strongly; M with 3 distal branches; CUA forking at the level of origin of RS. *Perm.*, USSR (Asian RSFSR).——FIG. 152,1. **S. mendax*; fore wing, ×10 (Becker-Migdisova, 1962b).

Surijokocixius BECKER-MIGDISOVA, 1961c, p. 359

Hemiptera—Homoptera

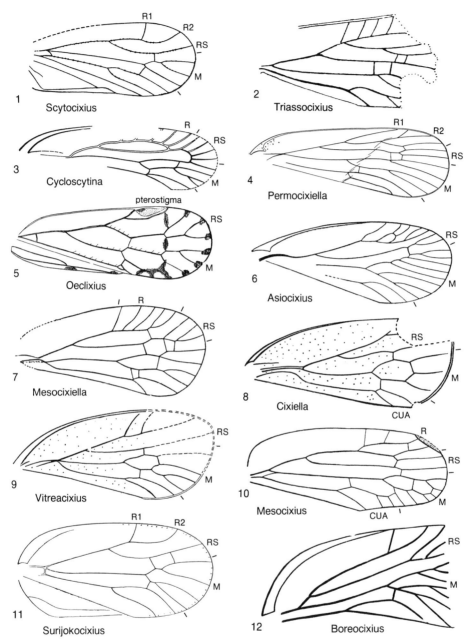

Fig. 152. Cixiidae (p. 236–238).

[*S. tomiensis; OD]. Fore wing broad, with broadly rounded apex; costal margin thickened; R2 strongly curved; RS unbranched; branches of CUA long and curved. Evans, 1956. *Perm.*, USSR (Asian RSFSR).——Fig. 152,*11*. *S. tomiensis*; fore wing, ×15 (Becker-Migdisova, 1961c).

Triassocixius Tillyard, 1919c, p. 878 [*T. australicus*; OD]. Little-known genus, based on fragment of fore wing; R forked close to the origin of RS; oblique crossveins from R to costal margin. [Family position uncertain.] *Trias.*, Australia (Queensland). —— Fig. 152,*2*. *T. australicus*; fore wing, ×5.5 (Evans, 1956).

Vitreacixius Becker-Migdisova, 1962a, p. 99 [*V.

ellipticus; OD]. Fore wing weakly tegminous; similar to *Cixiella*, but RS arising slightly more basally; M with 4 branches, closed cell smaller than in *Cixiella*. *Trias.*, USSR (Kirghiz).—— FIG. 152,9. **V. ellipticus*; fore wing, ×6 (Becker-Migdisova, 1962b).

Family ACHILIDAE Stål, 1866

[Achilidae STÅL, 1866, p. 130]

Head usually small; frons and clypeus large. Hind tibiae elongate; second segment of hind tarsus large. Fore wing well developed, basal two-thirds thickened; veins SC and R united for a short interval basally; SC with 2 or more short branches leading to costal margin, forming stigmatic area; R branched only apically, connected to M by 2 or more crossveins; M with at least 3 branches; clavus short, claval veins united to form claval stem. Hind wing moderately large. *Oligo.–Holo.*

Achilus KIRBY, 1819, p. 474. *Holo.*
Elidiptera SPINOLA, 1839, p. 304. SCUDDER, 1890. *Oligo.*, USA (Colorado)–*Holo.*
Protepiptera USINGER, 1939, p. 66 [**P. kaweckii*; OD]. Similar to *Epiptera* (recent) but with vertex distinctly in front of eyes; posterior margin of vertex concavely arcuate. *Oligo.*, Europe (Baltic).

Family RICANIIDAE Amyot & Serville, 1843

[Ricaniidae AMYOT & SERVILLE, 1843, p. 527]

Head usually as wide as the pronotum; vertex short and broad; clypeus much narrower than frons. Fore wing large, broadly triangular; costal margin usually nearly straight; costal area broad with numerous crossveins; basal area of clavus without pustules; venation diverse; veins R, M, and CU typically with numerous branches, with 1 or 2 subapical lines of gradate crossveins. Hind wing smaller than fore wing and with reduced venation. Basal segment of hind tarsus very small, without lateral spines. *Trias.–Holo.*

Ricania GERMAR, 1818, p. 221. DALMAN, 1826; GIEBEL, 1862; SCUDDER, 1890. *Eoc.*, Canada (British Columbia); *Oligo.*, Europe (Baltic)–*Holo.*
Cotradechites FENNAH, 1968, p. 144 [**C. lithinus*; OD]. Similar to *Cotrades* (recent), but tegmen twice as long as broad; costal area broad, with dense venation. *Paleoc.*, USA (North Dakota).
Dilaropsis COCKERELL, 1920c, p. 244 [**D. ornatus*; OD]. Fore wing broad, triangular; costal margin slightly convex; SC ending about two-thirds wing length from base; M diverging abruptly from R near origin of RS. *Eoc.*, USA (Colorado).
Eobladina HAUPT, 1956, p. 13 [**E. antiqua*; OD]. Little-known genus, based on fore wing; costal area wide distally; SC joined to R at base by curved crossvein, forming a very short basal cell; RS arising well before midwing. *Eoc.*, Europe (Germany).——FIG. 153,2. **E. antiqua*; fore wing, ×6 (Haupt, 1956).
Eoricania HENRIKSEN, 1922b, p. 24 [**E. danica*; OD]. Fore wing as in *Ricania* (recent), but 1A and 2A joined proximally beyond wing base. *Eoc.*, Europe (Denmark).——FIG. 153,4. **E. danica*; fore wing, ×2.5 (Henriksen, 1922b).
Hammapteryx SCUDDER, 1890, p. 298 [**H. reticulata*; OD]. Fore wing subtriangular; costal margin arched at base; numerous crossveins from SC to margin; R with at least 2 arcuate branches distally; RS arising well before midwing. COCKERELL, 1920a, 1920b; COCKERELL & SANDHOUSE, 1921; HENRIKSEN, 1922b; PITON, 1940a. *Eoc.*, USA (Colorado, Wyoming), Europe (Denmark, France), England.——FIG. 153,3. *H. paucistriata* HENRIKSEN, Denmark; fore wing, ×4 (Henriksen, 1922b).
Ludibrium BECKER-MIGDISOVA, 1962a, p. 100 [**L. ludus*; OD]. Hind wing little known; RS apparently arising distally as a continuation of stem R; M forked to about midwing. *Trias.*, USSR (Kirghiz).——FIG. 153,5. **L. ludus*; hind wing, ×6 (Becker-Migdisova, 1962a).
Neoricania CARPENTER, 1990, p. 131, *nom. subst. pro Eoricania* HAUPT, 1956, p. 12, *non* HENRIKSEN, 1922b [**Eoricania reticulata* HAUPT; OD]. Fore wing with costal space much narrower than in *Eoricania*; SC much closer to C. *Eoc.*, Europe (Germany).
Scolypopites TILLYARD, 1923a, p. 17 [**S. bryani*; OD]. Fore wing as in *Scolypopa* (recent), but SC shorter, reaching only to a little beyond midwing; only one series of gradate veins. *Mio.*, Australia (Queensland).——FIG. 153,6. **S. bryani*; fore wing, ×3.5 (Tillyard, 1923a).

Family NOGODINIDAE Melichar, 1898

[Nogodinidae MELICHAR, 1898, p. 204]

Head about as wide as pronotum; frons longer than wide. Fore wing large, usually broadest towards apex, coriaceous or hyaline, with numerous veins and crossveins; costal area with several crossveins; basal cell usually large; clavus not punctulate; claval stem reaching apex of fore wing. Hind tibiae with lateral spines; second segment of hind tarsus small, with a pair of spines distally. *Eoc.–Holo.*

Nogodina STÅL, 1859, p. 326. *Holo.*
Detyopsis COCKERELL, 1920c, p. 242 [*D. scudderi*; OD]. Fore wing much as in *Detya* (recent); veinlets from SC to costal margin numerous; RS forking well before midwing. *Eoc.*, USA (Colorado).
Tritophania JACOBI, 1937, p. 188 [*T. patruelis*; OD]. Similar to *Gaetulia* (recent), but frons without a keel; pterostigma absent. *Oligo.*, Europe (Baltic).——FIG. 153,*1*. *T. patruelis*; whole insect, ×3.4 (Jacobi, 1937).

Family FULGORIDAE Latreille, 1807

[Fulgoridae LATREILLE, 1807, p. 163]

Head usually large and simple, but often with prominent, cephalic process; postclypeus large, triangular; compound eyes large. Fore wing well developed, with numerous supernumerary veins and crossveins; hind wing with the anal and jugal areas reticulate. *Eoc.–Holo.*

Fulgora LINNÉ, 1767, p. 703. *Holo.*
Callospilopteron COCKERELL, 1920c, p. 245 [*C. ocellatum*; OD]. Fore wing broad, with obtuse apex; costal area much reduced; SC short; anterior veinlets from SC and R very oblique; ocelliform spots near outer margin. [Family assignment doubtful.] *Eoc.*, USA (Wyoming).
Eucophora SPINOLA, 1839, p. 200. SCUDDER, 1895b. *Eoc.*, Canada (British Columbia)–*Holo.*
Lystra FABRICIUS, 1803, p. 56. SCUDDER, 1890. [Generic assignment of fossil doubtful.] *Eoc.*, USA (Wyoming)–*Holo.*
Nyktalos METCALF, 1952, p. 230, *nom. subst. pro Nyctophylax* SCUDDER, 1890, p. 279, *non* FITZINGER, 1860 [*Nyctophylax uhleri* SCUDDER; OD]. Large species of uncertain affinities; head with a stout, recurved process; legs stout; femora and tibiae carinate. *Oligo.*, USA (Colorado).
Poiocera LAPORTE, 1832, p. 221. GERMAR & BERENDT, 1856. *Oligo.*, Europe (Baltic)–*Holo.*

Family FLATIDAE Spinola, 1838

[Flatidae SPINOLA, 1838, p. 205]

Head narrower than thorax; lateral edges of face not angular. Fore wing with costal area having crossveins; basal area of clavus granulate; clavus often open, claval veins separate or joined apically. Hind tibiae without a movable spur; first hind tarsomere short,

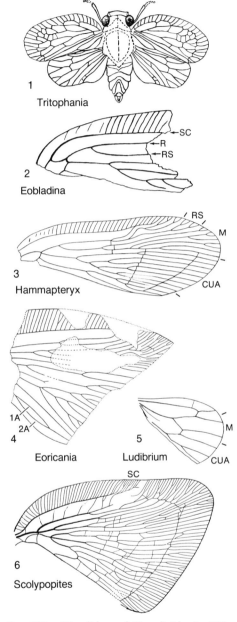

FIG. 153. Ricaniidae and Nogodinidae (p. 238–239).

second one small with a spine on each side. *Eoc.–Holo.*

Flata FABRICIUS, 1798, p. 511. *Holo.*
Aphaena GUÉRIN & MÉNEVILLE, 1833, p. 452. SCUDDER, 1890; COCKERELL, 1920c. *Eoc.*, USA (Wyoming); *Oligo.*, USA (Colorado)–*Holo.*

Hexapoda

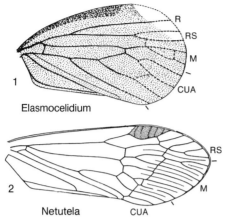

FIG. 154. Issidae and Dictyopharidae (p. 240–241).

Ficarasites SCUDDER, 1890, p. 301 [*F. stigmaticum*; OD]. Little-known genus; costal area narrow, with oblique veinlets; few crossveins. *Eoc.*, USA (Wyoming).

Giselia HAUPT, 1956, p. 14 [*G. multifurcata*; OD]. Fore wing as in *Uxantis* (recent); SC curved away from margin as it approaches midwing; R and M fused basally, separating early, with RS arising about one-sixth wing length from base; CUA apparently with a deep fork. *Eoc.*, Europe (Germany).

Lechaea STÅL, 1866, p. 236. HENRIKSEN, 1922b. *Eoc.*, Europe (Denmark)–*Holo.*

Ormenis STÅL, 1862, p. 68. COCKERELL, 1926a. *Tert. (epoch unknown)*, Argentina. *Holo.*

Poekilloptera LATREILLE, 1796, p. 90. COCKERELL, 1921d. *Oligo.*, England–*Holo.*

Thaumastocladius COCKERELL & SANDHOUSE, 1921, p. 456 [*T. simplex*; OD]. Fore wing as in *Gaga* (recent); costal area broad, with numerous oblique veinlets; R branching apically; M and CUA coalesced to about midwing; CUP distinctly forked. [Family position doubtful.] COCKERELL, 1924a. *Eoc.*, USA (Wyoming).

Family ARAEOPIDAE Metcalf, 1938

[Araeopidae METCALF, 1938, p. 281]

Head usually small; antennae short, usually not longer than head and thorax combined. Fore wing diverse, ranging from brachypterous, with reduced venation, to fully developed, with normal venation; vein SC typically with 2 branches; R coalesced with SC for about half its length, then coalesced with part of M; M usually with 3 branches; CU with 3 branches. Hind wing usually present, sometimes reduced; SC and R coalesced for more than half their lengths; M unbranched. Hind femora and tibiae elongate; spur well developed at apex of tibia, either spinelike or much enlarged and complex. *Eoc.–Holo.*

Araeopus SPINOLA, 1839, p. 336. COCKERELL, 1924a; STATZ, 1950a. *Eoc.*, USA (Colorado); *Oligo.*, Europe (Germany)–*Holo.*

Amagua COCKERELL, 1924a, p. 3 [*A. fortis*; OD]. Fore wing as in *Stenocranus* (recent); wing of uniform width, narrow; crossveins m-cu long. *Mio.*, USSR (Asian RSFSR).

Chloriona FIEBER, 1866, p. 519. BECKER-MIGDISOVA, 1964. *Mio.*, USSR (Asian RSFSR)–*Holo.*

Liburnia STÅL, 1866, p. 179. COCKERELL, 1917h. *Mio.*, Burma–*Holo.*

Family ISSIDAE Spinola, 1838

[Issidae SPINOLA, 1838, p. 158]

Head usually at least as wide as thorax; lateral margins of thorax not keeled; anterior margin of pronotum rounded and extended. Fore wing usually with reduced venation and often small; costal area small, without crossveins, or absent; base of costal margin not strongly curved; clavus not granulate. Hind tibiae with 2 to 4 spines; second hind tarsomere with a spine on each side. *Jur.–Holo.*

Issus FABRICIUS, 1803, p. 99. BERVOETS, 1910. *Oligo.*, Europe (Baltic)–*Holo.*

Elasmocelidium MARTYNOV, 1926b, p. 1355 [*E. rotundatum*; OD]. Fore wing short, much broadened distally; SC nearly parallel to costal margin; costal margin thickened; RS arising well before midwing; RS and M forked distally; anal area extending only to about midwing. BODE, 1953; EVANS, 1956; BECKER-MIGDISOVA, 1962b. *Jur.*, USSR (Kazakh); Europe (Germany).—— FIG. 154,*1*. *E. rotundatum*, Kazakh; fore wing, ×6.3 (Becker-Migdisova, 1962b).

Issites HAUPT, 1956, p. 16 [*I. glaber*; OD]. Fore wing as in *Issus* (recent) but without the dense reticulation. *Eoc.*, Europe (Germany).

Mesotubilustrium BECKER-MIGDISOVA, 1949b, p. 35 [*M. asiaticum*; OD]. Similar to *Elasmocelidium*, but RS arising near midwing. *Jur.*, USSR (Kazakh).

Tetragonidium BODE, 1953, p. 194 [*T. parallelogramma*; OD]. Fore wing as in *Elasmocelidium*, but M with more branches. *Jur.*, Europe (Germany).

Family DICTYOPHARIDAE Spinola, 1838

[Dictyopharidae SPINOLA, 1838, p. 202]

Head relatively large; structural details of vertex and frons diverse. Legs usually slender and elongate; hind tibiae commonly with 3 to 5 stout spines; second hind tarsal segment large, with a row of small spines at apex. Fore wing either normal or reduced; vein SC and R coalesced beyond basal area of wing; R branching irregularly distally; an irregular transverse line commonly formed by series of crossveins in apical third of wing. Hind wing usually large, with irregular venation. EMELJANOV, 1983. *Cret.–Holo.*

Dictyophara GERMAR, 1833, p. 175. [The family assignment of "Dictyophara" *scudderi* PITON (1940a), from the Eocene of France, is uncertain.] SCUDDER, 1890; BECKER-MIGDISOVA, 1964; EMELJANOV, 1983. *Mio.*, USSR (Asian RSFSR)–*Holo.*

Chanithus AMYOT, 1847, p. 160. BECKER-MIGDISOVA, 1964; EMELJANOV, 1983. *Mio.*, USSR (Asian RSFSR)–*Holo.*

Florissantia SCUDDER, 1890, p. 293 [**F. elegans*; OD]. Little-known genus, apparently related to *Dictyophara*. [Originally placed in Cixiidae by SCUDDER (1890); transferred to Dictyopharidae by EMELJANOV (1983).] COCKERELL, 1909a. *Oligo.*, USA (Colorado).

Netutela EMELJANOV, 1983, p. 84 [**N. annunciator*; OD]. Similar to *Cladodiptera* (recent), but clavus of fore wing without crossveins; M forking distally of origin of RS. *Cret.*, USSR (Asian RSFSR).——FIG. 154,2. **N. annunciator*; fore wing, ×6.5 (Emeljanov, 1983).

Family ARCHESCYTINIDAE Tillyard, 1926

[Archescytinidae TILLYARD, 1926g, p. 385] [=Permopsyllidae TILLYARD, 1926g, p. 390; Lithoscytinidae CARPENTER, 1933a, p. 436; Maueriidae ZALESSKY, 1937e, p. 54; Permoscytinopsidae ZALESSKY, 1939, p. 36; Uraloscytinidae ZALESSKY, 1939, p. 40; Maripsocidae ZALESSKY, 1939, p. 44; Kaltanaphididae SZELEGIEWICZ, 1971, p. 63]

Fore and hind wings membranous, similar in size and almost alike in venation. Fore wing with vein SC very close and parallel to R+M, R, and R1; R forming a pterostigma; RS originating at about midwing; M usually with at least 3 branches; CUA arising from stem CU, then directed towards R+M, which it touches at the point of separation of M; CUA forked; anal area small. Hind wing similar to fore wing except that CUA arises as an independent vein from the wing base and is not directed towards R+M. Head hypognathous; beak long; antennae long, multisegmented; ovipositor prominent in some genera at least. SZELEGIEWICZ & POPOV, 1978. *Perm.*

Archescytina TILLYARD, 1926g, p. 385 [**A. permiana*; OD] [=*Maueria* ZALESSKY, 1937e, p. 54 (type, *M. sylvensis*); *Permoscytinopsis* ZALESSKY, 1939, p. 36 (type, *P. maueriaeformis*)]. Fore wing with costal margin nearly straight except near base; SC close and parallel to R; R+M arched anteriorly; R2 parallel to RS; M usually with 3 branches. Antennae long and slender, with about 25 segments; beak long; forelegs with thickened femora; female with long, retractible ovipositor. CARPENTER, 1931b, 1939; ZALESSKY, 1937e, 1939; BECKER-MIGDISOVA, 1961c, 1961d, 1962b. *Perm.*, USA (Kansas), USSR (European and Asian RSFSR).——FIG. 155,1a. *Archescytina* sp., USSR; lateral view of body, ×6 (Becker-Migdisova, 1961d).——FIG. 155,1b,c. **A. permiana*, Kansas; b, fore wing; c, hind wing, ×6.5 (Carpenter, 1939).

Bekkerscytina EVANS, 1958, p. 111 [**B. primitiva*; OD]. Similar to *Eoscytina*, but RS arising nearer to origin of M. *Perm.*, Australia (New South Wales).——FIG. 155,10. **B. primitiva*; fore wing, ×6.3 (Evans, 1958).

Eoscytina EVANS, 1958, p. 109 [**E. migdisovae*; OD]. Similar to *Archescytina*, but fork of CUA very deep and broad and stem of CUA, as it leaves CUP, sigmoidally curved. *Perm.*, Australia (New South Wales). —— FIG. 155,9. **E. migdisovae*; fore wing, ×6 (Evans, 1958).

Kaltanaphis BECKER-MIGDISOVA, 1959a, p. 107 [**K. permiensis*; OD]. Little-known genus, based on fragment of hind wing. [Originally assigned to Permaphidopseidae; placed in new family, Kaltanaphididae, by SZELEGIEWICZ, 1971; transferred to Archescytinidae by SZELEGIEWICZ & POPOV, 1978.] *Perm.*, USSR (Asian RSFSR).

Kaltanoscytina BECKER-MIGDISOVA, 1959a, p. 105 [**K. nigra*; OD]. Wings as in *Archescytina*, but R longer and straighter in both pairs. BECKER-MIGDISOVA, 1961c; SZELEGIEWICZ & POPOV, 1978. *Perm.*, USSR (Asian RSFSR).——FIG. 155,8. **K. nigra*; fore wing, ×7 (Becker-Migdisova, 1961c).

Maripsocus ZALESSKY, 1939, p. 44 [**M. ambiguus*; OD]. Little-known fore wing; venation as in *Archescytina*, but M apparently with 2 branches. EVANS, 1956. *Perm.*, USSR (European RSFSR).

Paleoscytina CARPENTER, 1931b, p. 118 [**P. brevistigma*; OD]. Similar to *Archescytina*, but CUA of fore wing unbranched. BECKER-MIGDISOVA, 1961c. *Perm.*, USA (Kansas), USSR (Asian RSFSR).——FIG. 155,3. **P. brevistigma*; fore wing, ×18 (Carpenter, 1933a).

Permopsylla TILLYARD, 1926g, p. 390 [**P. americana*; OD] [=*Lithoscytina* CARPENTER, 1933a, p. 436 (type, *L. cubitalis*)]. Fore wing as in *Archescytina* but relatively broader; costal margin slightly concave at level of origin of M. BECKER-MIGDISOVA, 1960, 1961c, 1962b. *Perm.*, USA (Kansas), USSR (European and Asian RSFSR). —— FIG. 155,7. **P. americana*; fore wing, ×16 (Carpenter, 1931b).

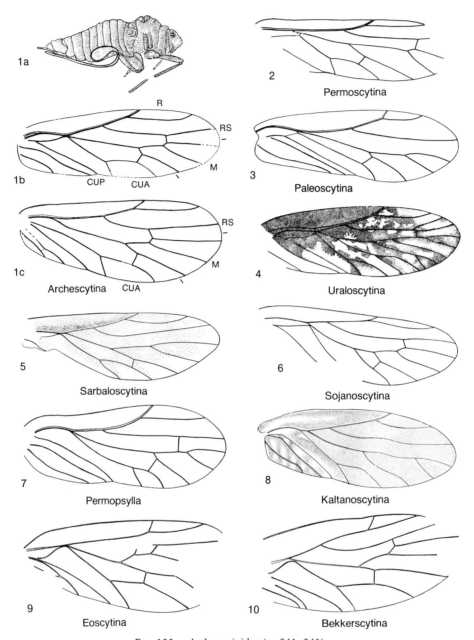

FIG. 155. Archescytinidae (p. 241–243).

Permopsyllopsis ZALESSKY, 1939, p. 38 [*P. rossica*; OD]. Little-known fore wing; venation as in *Archescytina*, but RS straight. BECKER-MIGDISOVA, 1960. *Perm.,* USSR (Asian RSFSR).

Permoscytina TILLYARD, 1926g, p. 387 [*P. kansasensis*; OD]. Similar to *Archescytina*, but SC and R nearly straight basally; proximal branch of M arising at about level of origin of RS. CARPENTER, 1939. *Perm.,* USA (Kansas).——FIG. 155,2. *P. kansasensis*; fore wing, ×4.2 (Carpenter, 1939).

Sarbaloscytina BECKER-MIGDISOVA, 1959a, p. 104 [*S. angustipennis*; OD]. Similar to *Archescytina*, but stem R+M short and nearly straight.

BECKER-MIGDISOVA, 1961c. *Perm.*, USSR (Asian RSFSR).——FIG. 155,5. **S. angustipennis*; fore wing, ×4.5 (Becker-Migdisova, 1961c).

Sojanoscytina MARTYNOV, 1933c, p. 885 [**S. grandis*; OD] [=*Ivascytina* MARTYNOV, 1933c, p. 888 (type, *I. difficilis*)]. Fore wing similar to that of *Archescytina*, but M with 4 or more branches. *Perm.*, USSR (European RSFSR).——FIG. 155,6. **S. grandis*; fore wing, ×3.4 (Becker-Migdisova, 1961c).

Tshekardaella BECKER-MIGDISOVA, 1960, p. 59 [**T. tshekardaensis*; OD; =*Tchecardaella tchecardaensis* BECKER-MIGDISOVA, 1948a, p. 130, *nom. nud.*]. Little-known genus, based on wing and body fragments. Fore wing as in *Archescytina* but shorter and more nearly oval. BECKER-MIGDISOVA, 1962b; SZELEGIEWICZ & POPOV, 1978. *Perm.*, USSR (Asian RSFSR).

Uraloscytina ZALESSKY, 1939, p. 40 [**U. prosbolioides*; OD]. Fore wing as in *Archescytina*, but M more extensively branched and with proximal branch arising about the level of origin of RS. [Type of family Uraloscytinidae ZALESSKY, 1939.] *Perm.*, USSR (Asian RSFSR).——FIG. 155,4. *U. multinervosa* BECKER-MIGDISOVA; fore wing, ×4 (Becker-Migdisova, 1962b).

Family BOREOSCYTIDAE Becker-Migdisova, 1949

[Boreoscytidae BECKER-MIGDISOVA, 1949a, p. 171]

Little-known family. Fore wing much broader distally than basally; vein M with at least 3 branches. Hind wing and body unknown. *Perm.*

Boreoscyta BECKER-MIGDISOVA, 1949a, p. 172 [**B. nefasta*; OD]. Fore wing triangular; RS with pectinate branches directed to costal margin. ROHDENDORF, 1957. *Perm.*, USSR (European RSFSR).——FIG. 156,4. *B. mirabilis* BECKER-MIGDISOVA; fore wing, ×6.5 (Becker-Migdisova, 1949a).

Archescytinopsis BECKER-MIGDISOVA, 1949a, p. 175 [**Sojanoscytina latipennis* MARTYNOV, 1933c, p. 887; OD]. Fore wing not so markedly triangular as in *Boreoscyta*; RS without pectinate branches. *Perm.*, USSR (European RSFSR).——FIG. 156,3. **A. latipennis* (MARTYNOV); fore wing, ×6.5 (Becker-Migdisova, 1949a).

Family PINCOMBEIDAE Tillyard, 1922

[Pincombeidae TILLYARD, 1922a, p. 282]

Little-known family of uncertain affinities. Fore(?) wing triangular; veins M and CUA originating at same point on R; anal area apparently very narrow. Hind wing appar-

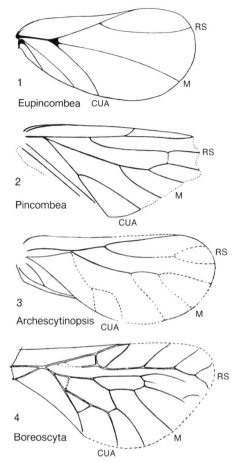

FIG. 156. Boreoscytidae and Pincombeidae (p. 243).

ently smaller than fore; R, M, and CUA diverging from same place. Body unknown. *Perm.*

Pincombea TILLYARD, 1922a, p. 282 [**P. mirabilis*; OD]. Fore(?) wing: M with 3 branches; CUA forked to half its length; one crossvein between M and RS, none between CUA and M. EVANS, 1956. *Perm.*, Australia (New South Wales). —— FIG. 156,2. **P. mirabilis*; fore(?) wing, ×16 (Tillyard, 1922a).

Eupincombea DAVIS, 1942, p. 114 [**E. postica*; OD]. Hind wing: RS, M, and CUA unbranched; costal area triangular. EVANS, 1956. *Perm.*, Australia (New South Wales).——FIG. 156,1. **E. postica*; hind wing, ×20 (Davis, 1942).

Protopincombea EVANS, 1943b, p. 193 [**P. obscura*; OD]. Fore wing as in *Pincombea*, but 2 crossveins between RS and M and one between

M and CUA. Evans, 1956. *Perm.,* Australia (New South Wales).

Family PROTOPSYLLIDIIDAE Carpenter, 1931

[Protopsyllidiidae Carpenter, 1931b, p. 119] [=Permaphidiopseidae Becker-Migdisova, 1960, p. 57]

Fore wing variable in shape; vein SC not a distinct vein; RS typically unbranched; stem of M fused with CUA; anal area small but distinct and coriaceous; CUP straight. Hind wing smaller than fore wing. Body structure little known; legs slender. *Perm.–Jur.*

Protopsyllidium Tillyard, 1926a, p. 26 [**P. australe*; OD]. Fore wing with RS arising well before midwing; M with 2 branches. Evans, 1956. *Perm.,* Australia (New South Wales).——Fig. 157,9. **P. australe*; fore wing, ×16 (Tillyard, 1926a).

Asiopsyllidium Becker-Migdisova, 1959a, p. 113 [**A. unicum*; OD]. Fore wing much wider distally than basally; RS arising well before midwing; M with 2 branches; CUA with a narrow fork. *Trias.,* USSR (Kirghiz).——Fig. 157,6. **A. unicum*; fore wing, ×10 (Becker-Migdisova, 1959a).

Belpsylla Evans, 1943b, p. 192 [**B. reticulata*; OD]. Fore wing broad distally; M with 3 straight branches; one crossvein between RS and M1+2 and another between RS and M1; CUA with small fork; anal area with Y-shaped vein. *Perm.,* Australia (New South Wales). —— Fig. 157,10.**B. reticulata*; fore wing, ×12 (Evans, 1943b).

Cicadellopsis Martynov, 1937a, p. 107 [**C. incerta*; OD]. Fore wing with costal margin strongly convex; RS arising near wing base; M forked; CUA with small distal fork. Evans, 1956; Becker-Migdisova, 1962b. *Trias.–Jur.,* USSR (Kirghiz).——Fig. 157,8. **C. incerta,* Jur.; fore wing, ×13 (Martynov, 1937a).

Cicadopsyllidium Becker-Migdisova, 1959a, p. 112 [**C. elongatum*; OD]. Little-known genus. Fore wing narrow; pterostigma apparently absent; RS arising well before midwing; M and CUA apparently fused basally. [Family assignment doubtful.] *Trias.,* USSR (Kirghiz).

Clavopsyllidium Davis, 1942, p. 117 [**C. minutum*; OD]. Fore wing as in *Protopsyllidium,* but M with 3 branches; CUA1 arched. Evans, 1943b, 1956. *Perm.,* Australia (New South Wales).——Fig. 157,7. **C. minutum*; fore wing, ×18 (Davis, 1942).

Permaphidopsis Becker-Migdisova, 1960, p. 58 [**P. sojanensis*; OD]. Little-known genus, based on hind wing. Wing broad distally; M coalesced basally with CUA; CUA forked distally with strongly curved CUA1. Szelegiewicz & Popov, 1978. *Perm.,* USSR (European RSFSR).

Permopsyllidium Tillyard, 1926a, p. 27 [**P. mitchelli*; OD]. RS arising near midwing; M with 3 branches. Carpenter, 1931b. *Perm.,* Australia (New South Wales).——Fig. 157,5. **P. mitchelli*; fore wing, ×14 (Tillyard, 1926a).

Permopsyllidops Davis, 1942, p. 116 [**P. stanleyi*; OD]. Fore wing similar to *Protopsyllidium,* but CUP absent or poorly developed; M with 3 branches. Evans, 1956. *Perm.,* Australia (New South Wales).——Fig. 157,1. **P. stanleyi*; fore wing, ×15 (Davis, 1942).

Permopsylloides Evans, 1943b, p. 193 [**P. insolita*; OD]. Fore wing of uniform width; costal area wide; RS arising before midwing, curved; M apparently with 2 branches; CUA sinuate; anal area with Y-shaped vein. Evans, 1956. *Perm.,* Australia (New South Wales).——Fig. 157,4. **P. insolita*; fore wing, ×12 (Evans, 1943b).

Permothea Tillyard, 1926a, p. 28 [**P. latipennis*; OD]. Fore wing much as in *Protopsyllidium,* but M with 3 branches. Carpenter, 1931b; Evans, 1956. *Perm.,* Australia (New South Wales).

Permotheella Davis, 1942, p. 116 [**P. scytinopteroides*; OD]. RS strongly curved; M with 3 branches; anal veins forming Y-shaped vein. Evans, 1943b, 1956. *Perm.,* Australia (New South Wales).——Fig. 157,3. **P. scytinopteroides*; fore wing, ×14 (Davis, 1942).

Propatrix Becker-Migdisova, 1960, p. 55 [**P. psylloides*; OD; =*P. psylloides* Becker-Migdisova, 1948a, p. 130, *nom. nud.*]. Fore wing with long pterostigmal area. RS arising at midwing; M with 3 branches; CUA with wide fork. Becker-Migdisova, 1962b; Szelegiewicz & Popov, 1978. *Perm.,* USSR (European RSFSR). ——Fig. 157,2. **P. psylloides*; fore wing and body, ×8 (Becker-Migdisova, 1960).

Psocopsyllidium Davis, 1942, p. 115 [**P. media*; OD]. Fore wing as in *Protopsyllidium* but more slender. Evans, 1943b, 1956. *Perm.,* Australia (New South Wales).

Psocoscytina Davis, 1942, p. 112 [**P. bifida*; OD]. Similar to *Protopsyllidium,* but M with 3 branches; RS arising at midwing with distal fork. Evans, 1956. *Perm.,* Australia (New South Wales).——Fig. 158,2. **P. bifida*; fore wing, ×12 (Davis, 1942).

Psyllidella Evans, 1943b, p. 192 [**P. magna*; OD]. Fore wing with RS arising beyond midwing; M with 3 long branches; costal margin sinuate. *Perm.,* Australia (New South Wales).——Fig. 158,5. **P. magna*; fore wing, ×10 (Evans, 1943b).

Psyllidiana Evans, 1943b, p. 192 [**P. davisia*; OD] [=*Protopsyllops* Evans, 1943b, p. 192 (type, *P. minuta*)]. Fore wing as in *Protopsyl-*

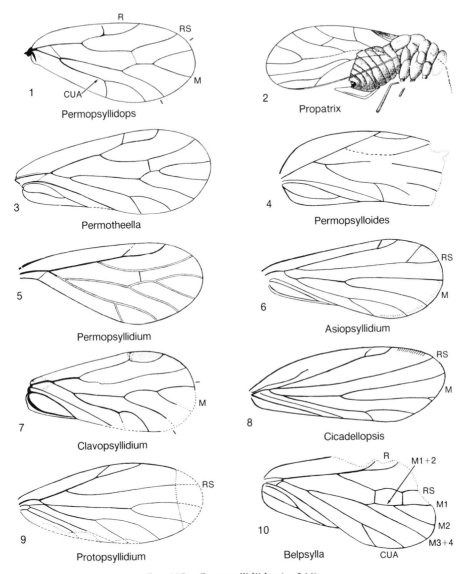

Fig. 157. Protopsyllidiidae (p. 244).

lidium, but RS arising near midwing and very straight; CUA deeply forked. EVANS, 1956. *Perm.*, Australia (New South Wales).——FIG. 158,*1*. **P. davisia*; fore wing, ×22 (Evans, 1943b).

Tomiopsyllidium BECKER-MIGDISOVA, 1959a, p. 112 [**T. iljinskiense*; OD]. Fore wing slender, triangular; RS arising just before midwing, curving away from R distally. BECKER-MIGDISOVA, 1961c. *Perm.*, USSR (Asian RSFSR).——FIG. 158,*4*. **T. iljinskiense*; fore wing, ×22 (Becker-Migdisova, 1960).

Triassopsylla TILLYARD, 1918b, p. 753 [**T. plecioides*; OD]. Little-known genus, based on wing fragment; RS curved; M with 3 branches. EVANS, 1956. *Trias.*, Australia (New South Wales).

Triassothea EVANS, 1956, p. 236 [**T. analis*; OD]. Fore wing as in *Protopsyllidium*, but RS arising near wing base; M+CUA very short; M with distal fork. *Trias.*, Australia (Queensland).

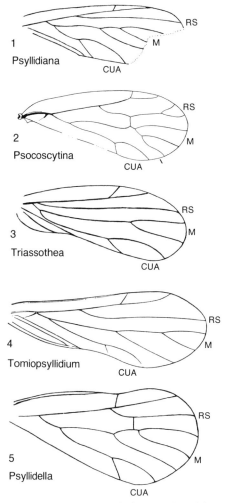

FIG. 158. Protopsyllidiidae (p. 244–246).

——FIG. 158,3. *T. analis*; fore wing, ×14 (Evans, 1956).

Family GENAPHIDIDAE
Handlirsch, 1907

[Genaphididae HANDLIRSCH, 1907, p. 643]

Little-known family. Antennae with 7 segments, bearing annular, secondary sense organs (rhinaria). Fore wing with all veins of nearly same thickness; M arising at level of base of pterostigma, with 3 terminal branches; CUA with short base, arising from stem R+M. HEIE, 1967, 1985; SHAPOSHNIKOV, 1979b, 1980. *Jur.*

Genaphis HANDLIRSCH, 1907, p. 643 [*Aphis valdensis*; OD]. Little-known genus. RS arising near middle of pterostigma. HEIE, 1967. *Jur.*, England. —— FIG. 159,*1*. *G. valdensis*; fore wing, ×18 (Heie, 1967).

Juraphis SHAPOSHNIKOV, 1979b, p. 66 [*J. crassipes*; OD]. Fore wing with RS arising slightly distally of middle of pterostigma. Antennae and legs stout. HEIE, 1985. *Jur.*, USSR (Kazakh). —— FIG. 159,*4*. *J. crassipes*; fore and hind wings, ×18 (Shaposhnikov, 1979b).

Family CANADAPHIDIDAE
Richards, 1966

[*nom. transl.* KONONOVA, 1976, p. 119, *ex* Canadaphidinae RICHARDS, 1966, p. 757]

Head dorsoventrally flattened, prolonged anteriorly; antennal bases ventral, in front of compound eyes; antennae with 5 to 6 segments; rostrum apparently very short; tarsi long; ovipositor well developed; siphuncles and cauda apparently not present. Fore wing with vein M with two forks. Hind wing relatively large. *Cret.*

Canadaphis ESSIG in CARPENTER & others, 1937, p. 19 [*C. carpenteri*; OD]. M of fore wing arising near origin of CUA1; CUA1 slightly sinuate; tarsi with 2 segments. HEIE, 1967, 1981; KONONOVA, 1976. [A record of this genus (*C. mordvilkoi* KONONOVA, 1976, p. 120) from the Cretaceous of USSR (Asian RSFSR) is very questionable. See KONONOVA, 1976, and HEIE, 1985.] *Cret.*, Canada (Manitoba).——FIG. 159,*2*. *C. carpenteri*; dorsal view, ×35 (Essig in Carpenter & others, 1937).

Alloambria RICHARDS, 1966, p. 756 [*A. caudata*; OD]. Antennae with at least 5 segments. Fore wing with CUA1 and CUA2 arising independently from stem SC+R+M; CUA1 sinuate. Tarsi with 2 segments. *Cret.*, Canada (Manitoba).——FIG. 159,*3*. *A. caudata*; dorsal view, ×50 (Richards, 1966).

Pseudambria RICHARDS, 1966, p. 758 [*P. longirostris*; OD]. Antennae with 6 segments. Fore wing with CUA1 sinuate; CUA2 very weakly developed. HEIE, 1981, 1985. *Cret.*, Canada (Manitoba).

Family PALAEOAPHIDIDAE
Richards, 1966

[*nom. transl.* KONONOVA, 1976, p. 121, *ex* Palaeoaphidinae RICHARDS, 1966, p. 750]

Similar to Canadaphididae, but antennae with 7 segments; ovipositor well developed. Fore wing with vein RS arising from proxi-

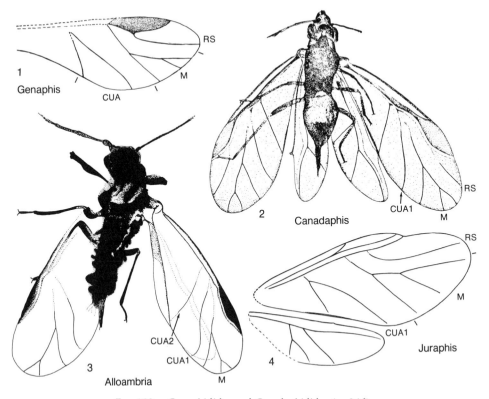

FIG. 159. Genaphididae and Canadaphididae (p. 246).

mal third of pterostigma; hind wing relatively shorter than in Canadaphididae. HEIE, 1985. *Cret.*

Palaeoaphis RICHARDS, 1966, p. 750 [*P. archimedia*; OD]. Little-known genus. Media of fore wing incomplete basally; legs with short hairs. [The assignment of *P. incognata* KONONOVA, 1976, p. 121 (Cretaceous of USSR) to the family Palaeoaphididae is very uncertain.] HEIE, 1985. *Cret.*, Canada (Manitoba).——FIG. 160,*1*. *P. archimedia*; fore wing, ×45 (Richards, 1966).

Ambaraphis RICHARDS, 1966, p. 752 [*A. costalis*; OD]. Similar to *Palaeoaphis,* but apical tarsal segments with long, conspicuous preapical setae. HEIE, 1985. *Cret.*, Canada (Manitoba).

Family SHAPOSHNIKOVIIDAE Kononova, 1976

[Shaposhnikoviidae KONONOVA, 1976, p. 122]

Little-known family. Antennae with 7 segments, its total length only half that of fore wing. Fore wing: vein M with 3 terminal branches; CUA1 and CUA2 widely separated basally. HEIE, 1981, 1985. *Cret.*

Shaposhnikovia KONONOVA, 1976, p. 122 [*S. electri*; OD]. Fore wing with M arising from base of pterostigma. Second segment of fore tarsus about one-fourth as long as tibia. HEIE, 1981. *Cret.*, USSR (Asian RSFSR).

Family OVIPARASIPHIDAE Shaposhnikov, 1979

[Oviparasiphidae SHAPOSHNIKOV, 1979b, p. 75]

Antennae with annular, secondary sense organs (rhinaria). Fore wing with vein RS arising from middle of pterostigma; M with 3 branches; CUA1 and CUA2 originating separately from a common stem (SC+R+M). Ovipositor large. *Cret.*

Oviparasiphum SHAPOSHNIKOV, 1979b, p. 75 [*O. jakovlevi*; OD]. Rhinaria on antennae forming convex rings. Femora stout. *Cret.*, Mongolia.

Family TAJMYRAPHIDIDAE
Kononova, 1975

[Tajmyraphididae Kononova, 1975, p. 795]

Antennae with 4 to 6 segments. Fore wing broadly rounded distally; pterostigma short, vein RS not connected to it; M with one fork; CUA1 about three times as long as CUA2. Heie, 1985. *Cret.*

Tajmyraphis Kononova, 1975, p. 796 [*T. zherichini*; OD]. Antennae with 5 or 6 segments. *Cret.,* USSR (Asian RSFSR).
Jantardakhia Kononova, 1975, p. 804 [*J. electri*; OD]. Antennae with 5 segments. Fore wing with bases of CUA1 and CUA2 widely separated. *Cret.,* USSR (Asian RSFSR).
Khatangaphis Kononova, 1975, p. 803 [*K. sibirica*; OD]. Similar to *Tajmyraphis,* but antennae with 4 or 5 segments; pterostigma of fore wing very short. *Cret.,* USSR (Asian RSFSR).
Retinaphis Kononova, 1975, p. 801 [*R. glandulosa*; OD]. Similar to *Tajmyraphis,* but antennae longer, with 6 segments. *Cret.,* USSR (Asian RSFSR).

Family MINDARIDAE Tullgren, 1909

[Mindaridae Tullgren, 1909, p. 58]

Cauda subtriangular. Fore wing with pterostigma narrow, pointed, extending to apex of wing; vein RS arising from the proximal part of pterostigma. *Cret.–Holo.*

Mindarus Koch, 1857, p. 277 [=*Pterostigma* Buckton, 1883, p. 178 (type, *P. recurvus*); *Schizoneuroides* Buckton, 1883, p. 178 (type, *S. scudderi*); *Sychnobrochus* Scudder, 1890, p. 268 (type, *S. reviviscens*)]. Baker, 1922; Heie, 1967, 1969b, 1985. *Oligo.,* Europe (Baltic), USA (Colorado)–*Holo.*
Nordaphis Kononova, 1977, p. 593 [*N. sukatchevae*; OD]. Little-known genus. Antennae with 6 segments. Fore wing with pterostigma very elongate; RS straight; M with one fork. Legs long. [Placed in Drepanosiphidae by Kononova but transferred to Mindaridae by Heie (1985).] *Cret.,* USSR (Asian RSFSR).

Family HORMAPHIDIDAE
Mordvilko, 1908

[Hormaphididae Mordvilko, 1908, p. 364]

Antennae with 3 to 5 segments, much shorter than body; antennae of alate form with narrow, ringlike, secondary rhinaria. Fore wing with veins CUA1 and CUA2 arising from same point on SC+R+M. *Oligo.–Holo.*

Hormaphis Osten-Sacken, 1861, p. 422. *Holo.*
Electrocornia Heie, 1972, p. 249 [*E. antiqua*; OD]. Little-known genus, based on nymph. Antennae with 5 segments; head and pronotum fused; frons with 2 hornlike processes. [Originally placed in Thelaxidae but later transferred to Hormaphididae (Heie, 1985).] *Oligo.,* Europe (Baltic).

Family ELEKTRAPHIDIDAE
Steffan, 1968

[Elektraphididae Steffan, 1968, p. 11]

Antennae with 5 segments. Fore wing with vein RS greatly reduced; M typically without branches; CUA1 and CUA2 arising from stem CUA or originating independently from stem SC+R+M. Kononova, 1976. *Cret.–Oligo.*

Schizoneurites Cockerell, 1915, p. 487 [*S. brevirostris*; OD] [=*Antiquaphis* Heie, 1967, p. 88 (type, *A. robustus*); *Elektraphis* Steffan, 1968, p. 11 (type, *E. polykrypta*)]. Fore wing with CUA1 and CUA2 arising from common stem CUA. Antennae with transverse folds. Heie, 1967, 1976, 1980, 1985; Steffan, 1968. *Oligo.,* Europe (Baltic), England.——Fig. 160,5. *S. robustus* (Heie), Baltic; fore wing, ×34 (Heie, 1967).
Antonaphis Kononova, 1977, p. 589 [*A. brachycera*; OD]. Antennae short, with 5 segments. Fore wing with RS long, slightly curved; M branched once. [Originally placed in Pemphigidae but transferred to Elektraphididae by Heie (1985).] *Cret.,* USSR (Asian RSFSR).
Tajmyrella Kononova, 1976, p. 118 [*T. cretacea*; OD]. Similar to *Schizoneurites,* but CUA1 and CUA2 arising independently from stem SC+R+M. Heie, 1981. *Cret.,* USSR (Asian RSFSR).

Family THELAXIDAE Baker, 1920

[Thelaxidae Baker, 1920, p. 21]

Antennae with 5 segments. Media of fore wing with 2 terminal branches. Hind wing with two oblique veins. *Oligo.–Holo.*

Thelaxes Westwood, 1840, p. 118. *Holo.*
Palaeothelaxes Heie, 1967, p. 42 [*P. setosa*; OD]. Little-known genus. All body segments of apterous form with very thick, large setae; frons of alate form with similar large setae. *Oligo.,* Europe (Baltic).

Family ANOECIIDAE Tullgren, 1909

[Anoeciidae Tullgren, 1909, p. 186]

Antennae commonly with 6 segments, and in alate forms with oval or subcircular sec-

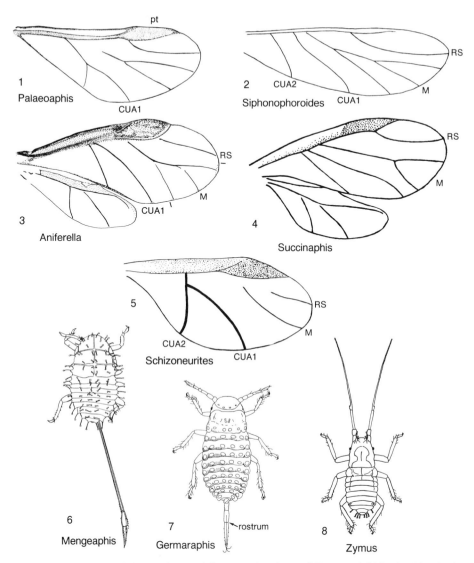

FIG. 160. Palaeoaphididae, Elektraphididae, Pemphigidae, and Drepanosiphidae (p. 247–251).

ondary rhinaria; marginal tubercles present on prothorax and some abdominal segments. Pterostigma of fore wing not more than four times longer than its width. *Oligo.–Holo.*

Anoecia KOCH, 1857, p. 275. *Holo.*
Berendtaphis HEIE, 1971, p. 262 [*Lachnus cimicoides* GERMAR & BERENDT, 1856, p. 5; OD]. Little-known genus, based on apterous form. Antennae with 6 segments, distal segments conspicuously thickened; head and pronotum not fused. *Oligo.*, Europe (Baltic).

Family PEMPHIGIDAE Koch, 1857

[Pemphigidae KOCH, 1857 p. viii]

Antennae short, usually with 6 segments and with one very short terminal process. Fore wing with vein M unbranched or with one fork. Hind wing with 1 or 2 oblique veins. *Oligo.–Holo.*

Pemphigus HARTIG, 1839, p. 645. *Holo.*
Eriosoma LEACH, 1818, p. 60. HEIE, 1968a, 1969a,

1969c, 1985. *Mio./Plioc.,* Europe (Germany)– Holo.

Germaraphis HEIE, 1967, p. 47 [*Lachnus dryoides* GERMAR & BERENDT, 1856, p. 29; OD]. Little-known genus, based mainly on apterous specimens. Antennae with 5 or 6 segments, the second one at least as long as the fourth. [Originally placed in Phloeomyzidae by HEIE (1967) and BECKER-MIGDISOVA (1973) but transferred to Pemphigidae by HILLE RIS LAMBERS (1980) and HEIE (1985).] HEIE, 1969b, 1972, 1985; BECKER-MIGDISOVA, 1973. *Oligo.,* Europe (Baltic). —— FIG. 160,7. **G. dryoides*; apterous specimen, reconstruction, ×30 (Heie, 1967).

Succinaphis HEIE, 1967, p. 173 [**S. flauensgaardi*; OD]. Apparently similar to *Pemphigus* (recent). Media of fore wing branched; wing membrane with fine reticulation. HEIE, 1985. *Oligo.,* Europe (Baltic). —— FIG. 160,4. **S. flauensgaardi*; fore and hind wings, ×45 (Heie, 1967).

Family DREPANOSIPHIDAE
Koch, 1857

[Drepanosiphidae KOCH, 1857, p. vii]

Closely allied to the Aphididae. Secondary transverse or circular rhinaria usually present on third antennal segment of alate females. Fore wing: vein M with 2 or 3 terminal branches. Hind wing with 2 or 3 oblique veins. HEIE, 1980, 1982. *Cret.–Holo.*

Drepanosiphum KOCH, 1855, p. 201. *Holo.*

Aixaphis HEIE, 1970b, p. 115 [**Tetraneura oligocenica* THÉOBALD, 1937a, p. 16; OD]. Antennae about half body length, with 6 segments. Fore wing: M with 3 terminal branches; CUA1 and CUA2 arising independently from stem SC+R+M, their bases relatively remote. [Family assignment doubtful.] HEIE, 1985. *Oligo.,* Europe (France).

Aniferella RICHARDS, 1966, p. 759 [**A. bostoni*; OD]. Antennae with 5 segments. Fore wing with well developed pterostigma; RS nearly straight; M with 2 forks; CUA1 and CUA2 arising separately from stem SC+R+M. HEIE, 1981, 1985. *Cret.,* Canada (Alberta). —— FIG. 160,3. **A. bostoni*; fore and hind wings, ×35 (Richards, 1966).

Balticaphis HEIE, 1967, p. 160 [**B. exsiccata*; OD]. Little-known genus, based on apterous form. Antennae with 5 or 6 segments. Fore femora thickened. HEIE, 1985. *Oligo.,* Europe (Baltic).

Balticomaraphis HEIE, 1967, p. 167 [**B. latens*; OD]. Little-known genus, based on cast cuticle of nymph. Antennae with 6 segments; ocular tubercles well developed. *Oligo.,* Europe (Baltic).

Conicaudus HEIE, 1972, p. 255 [**C. longipes*; OD]. Little-known genus, based on alate form. Antennae about as long as body. M of fore wing with 3 or 4 terminal branches. Tarsi very long. *Oligo.,* Europe (Baltic).

Cretacallis SHAPOSHNIKOV, 1979a, p. 730, footnote [**C. polysensoria*; OD]. Antennae with 6 segments. Fore wing: M with 3 terminal branches; CUA1 and CUA2 originating independently from stem SC+R+M. [Family position doubtful.] *Cret.,* Mongolia.

Electrocallis HEIE, 1967, p. 147 [**E. bakeri*; OD] [=*Dimeraphis* BECKER-MIGDISOVA, 1973, p. 87 (type, *D. arnoldii*)]. Antennae of alate form much longer than body and composed of 6 segments. Fore wing with pterostigma short; M with 3 terminal branches; CUA1 and CUA2 arising separately from stem SC+R+M. Fore femora thicker than the others. *Oligo.,* Europe (Baltic).

Megantennaphis HEIE, 1967, p. 142 [**M. hauniensis*; OD]. Antennae with 6 segments and much longer than body. Fore wing with pterostigma long, pointed; RS almost straight; M with 3 terminal branches. Fore and hind femora large and strong. BECKER-MIGDISOVA, 1973. *Oligo.,* Europe (Baltic).

Megapodaphis HEIE, 1967, p. 155 [**M. monstrabilis*; OD]. Antennae with 6 segments and at least as long as body. Fore wing: M with 2 terminal branches. Fore femora strongly thickened. HEIE, 1972, 1985. *Oligo.,* Europe (Baltic).

Mengeaphis HEIE, 1967, p. 113 [**Lachnus glandulosus*; OD]. Little-known genus, based on immature nymphs. Antennae with 4 segments; rostrum at least twice the length of body. BECKER-MIGDISOVA, 1973. *Oligo.,* Europe (Baltic). —— FIG. 160,6. **M. glandulosus*; dorsal view, ×85 (Becker-Migdisova, 1973).

Oligocallis HEIE, 1967, p. 133 [**O. larssoni*; OD]. Little-known genus, based on alate form. Similar to *Pterasthenica* (recent), but venation of fore wing less reduced in *Oligocallis*. HEIE, 1972. *Oligo.,* Europe (Baltic).

Oryctaphis SCUDDER, 1890, p. 266 [**O. lesueuri*; OD]. Little-known genus, possibly a synonym of *Siphonophoroides*. HEIE, 1985. *Oligo.,* USA (Colorado).

Palaeophyllaphis HEIE, 1967, p. 97 [**P. longirostris*; OD]. Antennae with 6 segments. Fore wing: M with 2 or 3 terminal branches; pterostigma slightly pointed but short. GERMAR & BERENDT, 1856; HEIE, 1972, 1985. *Oligo.,* Europe (Baltic).

Palaeosiphon HEIE, 1967, p. 119 [**Aphis hirsuta* GERMAR & BERENDT, 1856, p. 6; OD]. Little-known genus. Antennae of apterous form with 5 segments. Fore wing: M with 3 terminal branches. Hind wing with only one oblique vein.

Head and first two thoracic segments of alate form with long, curved, hornlike projections. HEIE, 1971. *Oligo.,* Europe (Baltic).

Siphonophoroides BUCKTON, 1883, p. 176 [*S. antiqua*; OD][=*Archilachus* BUCKTON, 1883, p. 177 (type, *A. pennata*); *Aphantaphis* SCUDDER, 1890, p. 253 (type, *S. exsuca*); *Cataneura* SCUDDER, 1890, p. 245 (type, *C. absens*); *Amalancon* SCUDDER, 1890, p. 270 (type, *A. lutosus*)]. Antennae slender, longer than body. Fore wing with RS very long, relatively straight, arising from proximal half of pterostigma; M with 3 terminal branches. COCKERELL, 1908u, 1909b; HEIE 1967, 1985. *Eoc.,* Europe (Denmark); *Oligo.,* USA (Colorado).——FIG. 160,2. **S. antiqua*; fore wing, ×14 (Heie, 1967).

Sternaphis HEIE, 1972, p. 257 [**S. electricola*; OD]. Fore wing with RS short and straight; M with 2 terminal branches. *Oligo.,* Europe (Baltic).

Succaphis HEIE, 1967, p. 110 [**S. holgeri*; OD]. Little-known genus, based on apterous form. Head and pronotum separated; antennae with 4 segments; rostrum longer than body. [Family assignment doubtful.] HEIE, 1985. *Oligo.,* Europe (Baltic).

Tertiaphis HEIE, 1969b, p. 144 [**T. haentzscheli*; OD]. Antennae with 6 segments and shorter than body. Fore wing: M with 2 terminal branches; CUA1 and CUA2 arising separately from stem SC+R+M. HEIE, 1985. *Oligo.,* Europe (Baltic).

Zymus HEIE, 1972, p. 254 [**Z. succinicola*; OD]. Little-known genus, based on nymph. Antennae with 4 segments and with long, filamentous terminal segment; head and pronotum fused; strong bristles on head and posterior part of abdomen. *Oligo.,* Europe (Baltic).——FIG. 160,8. **Z. succinicola*; dorsal view of nymph, ×24 (Heie, 1972).

Family APHIDIDAE Latreille, 1802

[Aphididae LATREILLE, 1802a, p. 263]

Compound eyes large in all instars; antennae commonly with 6 segments (rarely with 5), at least half length of body. Fore wing: vein RS with 2 or 3 terminal branches; CUA and CUP arising independently from stem R+M+CU. Hind wing commonly with 2 oblique veins, rarely only one. Wings slanted at rest. *Cret.–Holo.*

Aphis LINNÉ, 1758, p. 451. *Holo.*
Aphidocallis KONONOVA, 1977, p. 595 [**A. caudatus*; OD]. Antennae with 5 segments. Fore wing with pterostigma short, extending only to about level of midwing; M with 3 terminal branches. *Cret.,* USSR (Asian RSFSR).

Baltichaitophorus HEIE, 1967, p. 180 [**B. jutlandicus*; OD]. Little-known genus, based on apterous forms. Antennae with 6 segments, about as long as body. HEIE, 1980. *Oligo.,* Europe (Baltic).

Diatomyzus HEIE, 1970a, p. 163 [**D. eocaenicus*; OD]. Little-known genus, based on alate specimens. Similar to several existing genera, but RS of fore wing unusually long. *Eoc.,* Europe (Denmark).

Pseudamphorophora HEIE, 1967, p. 175 [**P. succini*; OD]. Little-known genus, based on apterous forms. [Family assignment doubtful.] HEIE, 1971, 1980. *Oligo.,* Europe (Baltic).

Family LACHNIDAE Koch, 1857

[Lachnidae KOCH, 1857, p. vii]

Similar to Anoeciidae, but prothorax and abdominal segments lacking marginal tubercles. Pterostigma of fore wing commonly much longer than 4 times its width. *Mio./ Plio.–Holo.*

Lachnus BURMEISTER, 1835, p. 92. *Holo.*
Longistigma WILSON, 1909, p. 385. HEIE & FRIEDRICH, 1971; HEIE, 1985. *Mio./Plio.,* Iceland–*Holo.*

Family ALEYRODIDAE Westwood, 1840

[Aleyrodidae WESTWOOD, 1840, p. 442]

Wings slightly thickened, commonly covered with a powdery wax. Fore wing venation weakly formed, only veins R and M extending to distal part of wing. Antennae with 7 segments; terminal abdominal segment with a large, dorsal opening, associated with storage of honey dew. *Oligo.–Holo.*

Aleyrodes LATREILLE, 1796, p. 93. [Generic assignment of fossil doubtful.] MENGE, 1856; SCHLEE, 1970. *Oligo.,* Europe (Baltic)–*Holo.*
Aleurodicus DOUGLAS, 1892, p. 32. [Generic assignment of fossil doubtful.] COCKERELL, 1919e; SCHLEE, 1970. *Mio.,* Burma–*Holo.*

Family COLEOSCYTIDAE Martynov, 1935

[Coleoscytidae MARTYNOV, 1935c, p. 24]

Fore wing oval, weakly coriaceous, membranous distally; subcostal area abruptly widened at base; costal margin at right angles to wing axis at this point; vein SC marginal;

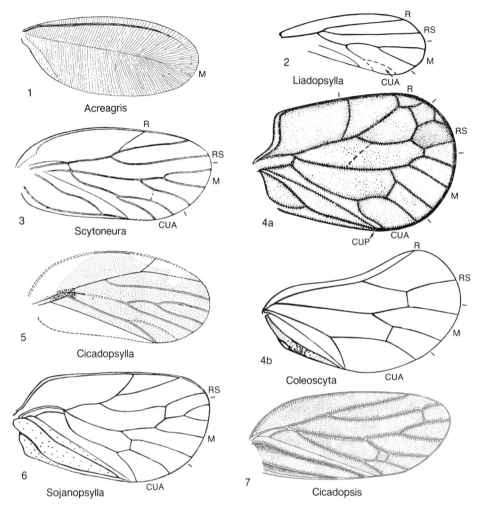

FIG. 161. Coleoscytidae, Cicadopsyllidae, Psyllidae, and Margarodidae (p. 252-254).

R long, with a branch to costal margin near midwing; M and CUA distally branched; CUP straight, unbranched; 1A and 2A with a common stem. Hind wing membranous, widened distally, more slender than fore wing, with concave anterior margin; subcostal area very narrow; M forked, CUA with a very wide fork; anal area narrow. Head hypognathous; eyes not divided. Hind coxae large. *Perm.*

Coleoscyta CARPENTER, herein [*C. rotundata* MARTYNOV, 1935c, p. 24; OD] [=*Coleoscytodes* CARPENTER, herein (type, *C. venosa* MARTYNOV, 1935c, p. 24; OD)]. Fore wing very broad, costal margin thickened; RS with distal fork. [The original generic names, *Coleoscyta* and *Coleoscytodes*, were *nomina nuda* (MARTYNOV, 1935c).] BECKER-MIGDISOVA, 1962b. *Perm.*, USSR (European RSFSR).——FIG. 161,4a. *C. rotundata*; fore wing, ×8. —— FIG. 161,4b. *C. venosa* (MARTYNOV); hind wing, ×8 (Becker-Migdisova, 1960).

Sojanopsylla BECKER-MIGDISOVA, 1960, p. 45 [*S. brevipennis*; OD]. Fore wing as in *Coleoscyta*, but subcostal area gradually widened basally and R and RS longer; M with 3 or 4 branches. *Perm.*, USSR (European and Asian RSFSR).——FIG. 161,6. *S. brevipennis*; fore wing, ×4.5 (Becker-Migdisova, 1960).

Family CICADOPSYLLIDAE
Martynov, 1931

[nom. transl. MARTYNOV, 1935c, p. 16, ex Cicadopsyllinae MARTYNOV 1931c, p. 172]

Fore wing elongate oval, commonly membranous; subcosta apparently close to costal margin; RS long, ending near wing apex. Hind wing with M apparently arising from stem of R; CUA originating independently of R+M. Head hypognathous, with protuberances on vertex. Hind coxae conical, elongate. BECKER-MIGDISOVA, 1962b. *Perm.*

Cicadopsylla MARTYNOV, 1931c, p. 173 [*C. permiana*; OD]. Fore wing with M forking near level of midwing. *Perm.,* USSR (European RSFSR).——FIG. 161,5. *C. permiana*; fore wing, ×4 (Becker-Migdisova, 1962b).
Cicadopsis BECKER-MIGDISOVA, 1959a, p. 110 [*C. rugosipenna*; OD]. Similar to *Cicadopsylla,* but R without distal, anterior branch. *Perm.,* USSR (Asian RSFSR).——FIG. 161,7. *C. rugosipenna*; fore wing, ×8 (Becker-Migdisova, 1962b).
Scytoneura MARTYNOV, 1935c, p. 16 [*S. elliptica*; OD]. Fore wing similar to *Cicadopsylla,* but M dividing more distally. BECKER-MIGDISOVA, 1962b. *Perm.,* USSR (Asian RSFSR).——FIG. 161,3. *S. elliptica*; fore wing, ×3 (Becker-Migdisova, 1962b).
Scytoneurella ZALESSKY, 1939, p. 39 [*S. major*; OD]. Fore wing membranous, costal margin slightly convex; M dividing distally of fork of CUA, with 3 short branches. BECKER-MIGDISOVA, 1962b. *Perm.,* USSR (Asian RSFSR).

Family PSYLLIDAE Latreille, 1807

[Psyllidae LATREILLE, 1807, p. 168]

Fore wing usually coriaceous; costal area broad; veins M and CUA united to form a basal stem; RS arising from R independently; M and CU usually arising as a common stem; RS unbranched; M and CUA forked. Hind wing smaller and more slender, with R and M unbranched. Antennae with 9 to 10 segments. *Jur.–Holo.*

Psylla GEOFFREY, 1762, p. 482. BECKER-MIGDISOVA, 1964. *Oligo.,* England; *Mio.,* USSR (European RSFSR)–*Holo.*
Agonoscena ENDERLEIN, 1914, p. 234. BECKER-MIGDISOVA, 1964. *Mio.,* USSR (European RSFSR)–*Holo.*
Catopsylla SCUDDER, 1890, p. 277 [*C. prima*; OD]. Little-known genus. Fore wing as in *Psylla,* but cell of CU much longer. *Oligo.,* USA (Colorado).
Liadopsylla HANDLIRSCH, 1920, p. 213 [*L. geinitzi*; OD]. Fore wing oval, membranous; R and RS long, parallel; stem of R short; fork of M long. MARTYNOV, 1926b; BECKER-MIGDISOVA, 1949b. *Jur.,* Europe (Germany), USSR (Asian RSFSR).——FIG. 161,2. *L. tenuicornis* MARTYNOV, USSR; fore wing, ×20 (Martynov, 1926b).
Livilla CURTIS, 1836, p. 625. COCKERELL, 1921d. *Oligo.,* England–*Holo.*
Necropsylla SCUDDER, 1890, p. 276 [*N. rigida*; OD]. Little-known genus; fore wing as in *Psyllopsis* (recent) but subtriangular. COCKERELL, 1911b, 1915. *Oligo.,* USA (Colorado), England.
Psyllites COCKERELL, 1914f, p. 636 [*P. crawfordi*; OD]. Little-known genus, probably a synonym of *Catopsylla.* *Oligo.,* USA (Colorado).
Retroacizzia HESLOP-HARRISON, 1961, p. 504. BECKER-MIGDISOVA, 1964. *Mio.,* USSR (European RSFSR)–*Holo.*
Strophingia ENDERLEIN, 1914, p. 233. *Oligo.,* Europe (Baltic)–*Holo.*
Trioza FÖRSTER, 1848, p. 67. BECKER-MIGDISOVA, 1964. *Mio.,* USSR (European RSFSR)–*Holo.*

Family COCCIDAE Fallén, 1814

[Coccidae FALLÉN, 1814, p. 23]

Adults with marked sexual dimorphism. Males with fore wings normally developed; hind wings reduced or halterlike. Females apterous; antennae diverse, commonly much reduced; abdominal spiracles absent. *Oligo.–Holo.*

Coccus LINNÉ, 1758, p. 455. MENGE, 1856; COCKERELL, 1906b; BECKER-MIGDISOVA, 1962b. *Oligo.,* Europe (Baltic)–*Holo.*

Family ORTHEZIIDAE
Amyot & Serville, 1843

[Ortheziidae AMYOT & SERVILLE, 1843, p. 619]

Similar to Coccidae. Females with body clearly segmented; antennae with distinct segmentation; abdominal spiracles present. *Oligo.–Holo.*

Orthezia BOSC, 1784, p. 173. *Holo.*
Ochyrocoris MENGE, 1856, p. 17 [*O. electrina*; OD]. Little-known genus, probably a synonym of *Orthezia* (recent). COCKERELL, 1906a; BECKER-MIGDISOVA, 1962b. *Oligo.,* Europe (Baltic).

Family MARGARODIDAE
Cockerell, 1899

[Margarodidae Cockerell, 1899, p. 390]

Males commonly winged, with few unbranched veins. Females with convex body, strongly sclerotized, with clear segmentation; abdomen with an anal tube or a sclerotized ring, lacking setae. *Cret.–Holo.*

Margarodes Guilding, 1829, p. 118. *Holo.*
Acreagris Koch in Koch & Berendt, 1854, p. 123 [**A. crenata*; OD]. Female adult: antennae with 9 segments; body entirely or nearly devoid of setae; tarsi two-segmented. Male adult: compound eyes; wings with a single vein parallelling the costal margin to wing apex; M delicate, bisecting the wing diagonally; hind wing reduced to slender halteres; antennae with at least 8 segments; tarsi one-segmented; abdomen with long threads of wax arising from clusters of dorsal ducts. Ferris, 1941. *Oligo.*, Europe (Baltic). ——Fig. 161,*1*. **A. crenata*; fore wing of male, ×6 (Ferris, 1941).
Electrococcus Beardsley, 1969, p. 271 [**E. canadensis*; OD]. Male small; antennae with 10 segments, pedicel conspicuously enlarged; legs long and slender; compound eye reduced to a single row of ommatidia. Fore wing well developed, with R and M distinct. *Cret.*, Canada (Manitoba).

Family PSEUDOCOCCIDAE
Cockerell, 1905

[Pseudococcidae Cockerell, 1905, p. 193]

Similar to the Coccidae. Females typically covered with a mealy or filamentous, waxy secretion, commonly protruding as short lateral and long anal filaments; legs well developed. Males apterous or winged, typically with two long caudal wax filaments. *Oligo.–Holo.*

Pseudococcus Westwood, 1840, p. 118. *Holo.*
Puto Signoret, 1875, p. 394. Cockerell, 1908g. *Oligo.*, Europe (Baltic)–*Holo.*

Family UNCERTAIN

The following genera, apparently belonging to the suborder Homoptera, are too poorly known to permit assignment to families.

Anconatus Buckton, 1883, p. 177 [**A. dorsuosus*; OD]. Little-known aphidoid of uncertain affinities. Heie, 1967, 1985. *Oligo.*, USA (Colorado).
Annulaphis Shaposhnikov, 1979b, p. 73 [**A. rasnitsyni*; OD]. Little-known genus, based on incomplete specimens; apparently related to *Ellinaphis*. [Originally placed in Palaeoaphididae, but transferred by Heie (1985) to family uncertain.] *Cret.*, USSR (Asian RSFSR).
Aphidioides Motschulsky, 1856, p. 29 [**A. succifera*; OD]. Little-known aphidoid genus, based on apterous form. Heie, 1967, 1985. *Oligo.*, Europe (Baltic).
Aphidulum Handlirsch, 1939, p. 163 [**A. pusillum*; OD]. Little-known genus. Heie, 1967. *Jur.*, England.
Archeglyphis Martynov, 1931a, p. 89 [**A. crassinervis*; OD]. Little-known wing fragment. Becker-Migdisova, 1961c; Rohdendorf & Rasnitsyn, 1980. *Perm.*, USSR (Asian RSFSR).
Archipsyche Handlirsch, 1906b, p. 624 [**A. eichstattensis*; OD]. Little-known genus, apparently similar to *Limacodites*. *Jur.*, Europe (Germany).
Austroscytina Evans, 1943b, p. 181 [**A. imperfecta*; OD]. Little-known wing, possibly related to *Archescytinidae*. *Perm.*, Australia (New South Wales).
Beaconiella Evans, 1963, p. 21 [**B. fennahi*; OD]. All principal veins of fore and hind wings multibranched; possibly a fulgoroid. Riek, 1973. *Trias.*, Australia (New South Wales).
Beloptesis Handlirsch, 1906b, p. 625 [**B. oppenheimi*; OD]. Fore wing markedly triangular, nearly as broad as long; venation apparently as in *Limacodites*. Hind wing small, oval. Evans, 1956. *Jur.*, Europe (Germany).
Bernaea Schlee, 1970, p. 18 [**B. neocomica*; OD]. Female with head wider than pronotum; median ocellus present; antennae with 7 segments, the third segment much longer than distal segments. Veins absent on hind wing, represented by lines of pigment. [Placed by Schlee in "Aleyrodina *sensu lato*," without family assignment.] *Cret.*, Lebanon.
Borisrohdendorfia Becker-Migdisova, 1959b, p. 138 [**B. picturata*; OD]. Based on distal fragment of wing. Becker-Migdisova, 1961c. *Perm.*, USSR (Asian RSFSR).
Cercopidium Westwood, 1854, p. 394 [**C. hahni* Westwood, 1854, p. 394; SD Carpenter, herein]. Little-known genus, based on wing fragment. Heer, 1870a; Henricksen, 1922b. *Jur.*, England; *Eoc.*, Greenland.
Chiliocycla Tillyard, 1919c, p. 868 [**C. scolopoides*; OD]. Fore wing with strongly thickened costal border; RS present, arising before midwing; closed cell between M1+2 and M3+4; CUA connected to base of M by crossvein. [Type of family Chiliocyclidae Evans, 1956, p. 209.] Evans, 1956, 1961. *Trias.*, Australia (Queensland).——Fig. 162,*1*. **C. scolopoides*; fore wing, ×4.5 (Evans, 1956).
Cicadellites Heer, 1853a, p. 119 [**C. pallidus* Heer, 1853a, p. 119; SD Carpenter, herein]. Little-

Hemiptera—Homoptera

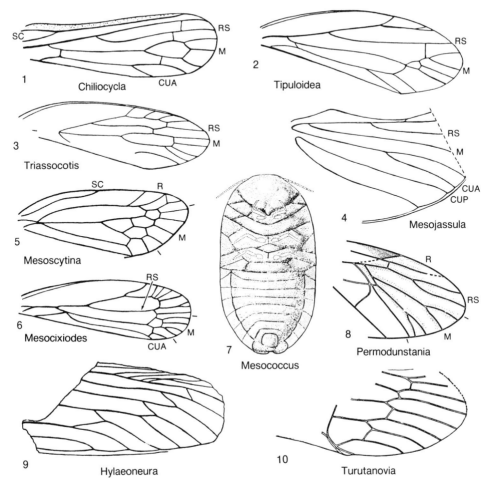

FIG. 162. Uncertain (p. 254–259).

known homopteron, possibly belonging to the Cercopidae. PITON & THÉOBALD, 1935. *Oligo.,* Europe (France); *Mio.,* Europe (Croatia).

Cicadellium WESTWOOD, 1854, p. 394 [**C. dipsas*; SD HANDLIRSCH, 1907, p. 641] [=*Pseudodelphax* HANDLIRSCH, 1907, p. 641 (type, *Delphax pulcher* BRODIE, 1845, p. 33)]. Little-known genus, based on fore wing. EVANS, 1956. *Jur.,* England.

Cixiites HANDLIRSCH, 1906b, p. 498 [*ced*C. liasinus*; OD]. Little-known wing, possibly related to Fulgoridiidae. BECKER-MIGDISOVA, 1962b. *Jur.,* Europe (Germany).

Cixioides HANDLIRSCH, 1906b, p. 640 [*Cixius maculatus* BRODIE, 1845, p. 33; OD]. Little-known fore wing, possibly related to Cixiidae. *Jur.,* England.

Diphtheropsis MARTYNOV, 1937a, p. 110 [*D. incerta*; OD]. Little-known genus, based on incomplete fore wing with nearly straight costal margin and long R + M. EVANS, 1956. *Jur.,* USSR (Kirghiz).

Dysmorphoscartella RIEK, 1973, p. 527 [*D. lobata*; OD]. Little-known genus, based on distal fragment of wing. RIEK, 1976a. [Originally placed in Eoscartarellidae.] *Perm.,* South Africa.

Echinaphis COCKERELL, 1913f, p. 229 [*E. rohweri*; OD]. Little-known genus, based on hind wing and body fragments; apparently related to Greenideidae (recent) and Drepanosiphidae. HEIE, 1967, 1970b, 1985. *Oligo.,* USA (Colorado).

Electromyzus HEIE, 1972, p. 250 [*E. acutirostris*; OD]. Fore wing with RS very slightly curved; M with 2 terminal branches, arising close to point of origin of CUA1 and CUA2. HEIE, 1985. *Oligo.,* Europe (Baltic).

Ellinaphis SHAPOSHNIKOV, 1979b, p. 71 [*E. incognita*; OD]. Little-known genus, originally placed in Palaeoaphididae but transferred by Heie (1985) to category of family uncertain. *Cret.,* USSR (Asian RSFSR).

Eochiliocycla Davis, 1942, p. 114 [*E. angusta*; OD]. Little-known genus, based on fore wing; possibly fulgoroid. Evans, 1956. *Perm.,* Australia (New South Wales).

Eocicada Oppenheim, 1888, p. 229 [*E. microcephala*; OD]. Little-known genus, based on wing fragment. Evans, 1956. *Jur.,* Europe (Germany).

Eopsyllidium Davis, 1942, p. 114 [*E. delicatulum*; OD]. Little-known hind wing, with CUA free from M basally. Possibly related to the Protopsyllidiidae. Evans, 1956. *Perm.,* Australia (New South Wales).

Fulgoridiella Becker-Migdisova, 1962a, p. 96 [*F. raetica*; OD]. Little-known wing fragment, possibly related to Fulgoridiidae. *Trias.,* USSR (Kirghiz).

Fulgoropsis Martynov, 1937a, p. 165 [*F. dubiosa*; OD]. Little-known genus, based on wing fragment, possibly related to Fulgoridiidae. Becker-Migdisova, 1962b. *Jur.,* USSR (Kirghiz).

Geranchon Scudder, 1890, p. 248 [*Lachnus petrorum* Scudder, 1877b, p. 279; OD]. Little-known genus, possibly belonging to Aphidoidea. Heie, 1967, 1985. *Eoc.,* Canada (British Columbia).

Gryllites Germar, 1842, p. 82 [*G. dubius*; OD]. Little-known genus, originally placed in Orthoptera. Hagen, 1862; Assmann, 1877; Popov, 1971. *Jur.,* Europe (Germany).

Hastites Cockerell, 1922f, p. 161 [*H. muiri*; OD]. Little-known genus. Fore wing elongate; R apparently with a short distal branch; M dividing distally, with 3 terminal branches; CUA with 3 terminal branches. *Oligo.,* England.

Heidea Schlee, 1970, p. 9 [*H. cretacica*; OD]. Male with head about same width as pronotum; median ocellus present; third antennal segment about as long as distal segments. Vein present in hind wing. [Considered by Schlee to be related to the existing and Tertiary Aleurodidae but differing markedly in several traits.] *Cret.,* Lebanon.

Hooleya Cockerell, 1922f, p. 160 [*H. indecisa*; OD]. Little-known fore wing; costal margin broad; SC apparently separating from R before midwing, and giving rise to a series of short, oblique veinlets to costal margin. *Oligo.,* England.

Homopterites Handlirsch, 1906b, p. 499 [*H. anglicus*; OD]. Little-known fore wing. *Jur.,* England.

Hylaeoneura Lameere & Severin, 1897, p. 37 [*H. lignei*; OD]. Little-known genus, based on distal fragment of fore wing. R with several long, pectinate branches to costal margin; M with 3 branches. *Cret.,* Europe (Belgium). —— Fig. 162,9. **H. lignei*; fore wing, ×2.5 (Handlirsch, 1907).

Hypocixius Cockerell, 1926a, p. 501 [*H. oblitescens*; OD]. Little-known genus, based on incomplete fore wing. Possibly related to Cixiidae. *Tert. (epoch unknown),* Argentina (Jujuy).

Jurocallis Shaposhnikov, 1979b, p. 68 [*J. longipes*; OD]. Antennae tapering from base to apex; RS arising from distal part of pterostigmal area; M arising from base of pterostigma and with 3 terminal branches. [Originally placed in Drepanosiphidae.] Heie, 1985. *Cret.,* USSR (Asian RSFSR).

Kaltanocicada Becker-Migdisova, 1961c, p. 291 [**K. dunstanioides*; OD]. Little-known hind wing, with broadly rounded apex and wide concavity of front wing margin; CUA with long fork. Becker-Migdisova, 1962b. *Perm.,* USSR (Asian RSFSR).

Kaltanoscyta Becker-Migdisova, 1959a, p. 110 [**K. reticulata*; OD]. Little-known fragment of fore wing, strongly coriaceous and with dense reticulation over wing. Possibly related to Coleoscytidae. *Perm.,* USSR (Asian RSFSR).

Karabasia Martynov, 1926b, p. 1356 [*K. paucinervis*; OD]. Little-known insect, possibly related to Jassidae. *Jur.,* USSR (Kazakh).

Karajassus Martynov, 1926b, p. 1352 [*K. crassinervis*; OD]. Little-known insect, possibly close to Cicadellidae. Becker-Migdisova, 1962b. *Jur.,* USSR (Kazakh).

Kisylia Martynov, 1937a, p. 109 [*K. psylloides*; OD]. Little-known genus, based on fore wing. Nodus and nodal line absent; stem of R slightly shorter than R+M; CUA not coalesced with M. *Jur.,* USSR (Kirghiz).—— Fig. 163,4. **K. psylloides*; fore wing, ×3 (Becker-Migdisova, 1962b).

Larssonaphis Heie, 1967, p. 168 [*L. obnubila*; OD]. Little-known aphidoid genus. Heie, 1985. *Oligo.,* Europe (Baltic).

Liassocicada Bode, 1953, p. 201 [*L. antecedens*; OD]. Little-known genus, based mainly on body structure. Rostrum elongate, extending at least to middle of abdomen. [Liassocicada was redefined by Whalley (1983) and provisionally placed in the Cicadidae. However, I doubt that our very slight knowledge of the body structures of these Jurassic and Triassic specimens justifies the extension of the range of the Cicadidae to another 150 million years before the Paleocene. Accordingly, the genus *Liassocicada* is herein provisionally placed in the Homoptera, family uncertain.] Whalley, 1983. *Trias.,* England; *Jur.,* Europe (Germany).

Limacodites Handlirsch, 1906b, p. 622 [*L. mesozoicus*; OD]. Little-known genus, based on wing fragments. Probably related to *Eocicada*. *Jur.,* Europe (Germany).

Lithecphora Scudder, 1890, p. 329 [*L. unicolor* Scudder, 1890, p. 329; SD Carpenter, herein]. Little-known insect, with slender fore wing. *Oligo.,* USA (Colorado).

Lithopsis SCUDDER, 1878b, p. 773 [*L. fimbriata*; OD]. Body stout; head not produced between the eyes. Tegmina extending well beyond abdomen. SCUDDER, 1890; COCKERELL, 1921b; PONGRÁCZ, 1935; PITON, 1940a. *Eoc.*, USA (Wyoming), Europe (France, Germany).

Locrites SCUDDER, 1890, p. 323 [*L. copei* SCUDDER, 1890, p. 323; SD CARPENTER, herein]. Little-known homopteron; head large, protuberant; scutellum equiangular. HEER, 1853a. *Oligo.*, USA (Colorado); *Mio.*, Europe (Croatia).

Margaroptilon HANDLIRSCH, 1906b, p. 499 [*M. woodwardi* HANDLIRSCH, 1906b, p. 499; SD CARPENTER, herein]. Little-known wings, with numerous small maculations; possibly a fulgoroid. BODE, 1953; EVANS, 1956. *Jur.*, England, Europe (Germany).

Mesaleuropsis MARTYNOV, 1937a, p. 108 [*M. venosa*; OD]. Little-known wings. Fore wing rounded distally; pterostigma absent; M with 2 branches; CUA apparently unbranched. Hind wing about half as long as fore wing, with unbranched RS and M. *Jur.*, USSR (Tadzhik).

Meshemipteron COCKERELL, 1915, p. 476 [*M. incertum*; OD]. Little-known genus, based on small fragment of wing. *Jur.*, England.

Mesocicadella EVANS, 1956, p. 193 [*M. venosa*; OD]. Little-known genus, based on fragment of fore wing. Several parallel, oblique veins between R and wing margin; M with numerous branches. [Originally placed in the Scytinopteridae but moved to family uncertain by EVANS in 1961.] *Trias.*, Australia (Queensland).——FIG. 163,2. *M. venosa*; fore wing, ×3.5 (Evans, 1956).

Mesocixiodes TILLYARD, 1922b, p. 462 [*M. termioneura*; OD]. Fore wing with SC very close to costal margin; RS present; M forking in distal part of wing, with a small, closed cell between forks. EVANS, 1956. *Trias.*, Australia (Queensland).——FIG. 162,6.*M. termioneura*; fore wing, ×5.2 (Evans, 1956).

Mesococcus BECKER-MIGDISOVA, 1959a, p. 110 [*M. asiaticus*; OD]. Based on wingless form (female?); body oval; legs greatly reduced; abdomen with 9 visible segments. BECKER-MIGDISOVA, 1962b. *Trias.*, USSR (Kirghiz).—— FIG. 162,7. *M. asiaticus*; whole insect, ×24 (Becker-Migdisova, 1959a).

Mesodiphthera TILLYARD, 1919c, p. 873 [*M. grandis*; OD]. Little-known genus, based on small fragment of fore wing. CUA anastomosed with M basally. [Placed in Tropiduchidae by TILLYARD (1922b) and in Homoptera, family uncertain, by EVANS (1956).] *Trias.*, Australia (Queensland).——FIG. 163,3.*M. grandis*; fore wing, ×3.5 (Tillyard, 1919c).

Mesojassula EVANS, 1956, p. 203 [*M. marginata*; OD]. Hind wing with costal margin with marked medial depression; M unbranched; CUA with 2 equal branches; marginal vein present. *Trias.*,

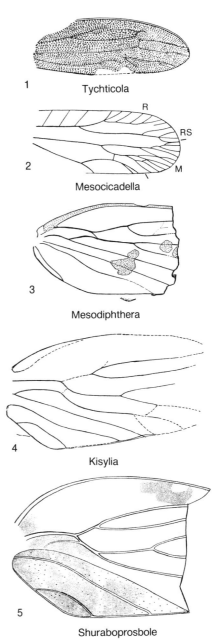

1 Tychticola

2 Mesocicadella

3 Mesodiphthera

4 Kisylia

5 Shuraboprosbole

FIG. 163. Uncertain (p. 256–259).

Australia (Queensland). —— FIG. 162,4. *M. marginata*; hind wing, ×4.5 (Evans, 1956).

Mesoledra EVANS, 1956, p. 211, *nom. subst. pro Mesojassus* HANDLIRSCH, 1939, p. 145, *non* TILLYARD, 1916 [*Mesojassus pachyneurus* HANDLIRSCH, 1939, p. 145; OD]. Little-known genus, based on incomplete wing; possibly related to Cicadellidae. *Jur.,* Europe (Germany).

Mesoscytina TILLYARD, 1919c, p. 871 [**M. australis*; OD]. Fore wing with SC distinct, long; RS apparently arising in very distal part of wing; M dividing at midwing and forming a closed cell at fork. Possibly related to Scytinopteridae. EVANS, 1956. *Trias.,* Australia (Queensland).——FIG. 162,5.**M. australis*; fore wing, ×5.2 (Evans, 1956).

Meuniera PITON, 1936c, p. 1 [**M. haupti*; OD]. Little-known genus, based on fragment of fore wing. RS arising well before midwing; basal stem of M free from R. COOPER, 1941. *Eoc.,* Europe (France).

Pachypsyche HANDLIRSCH, 1906b, p. 623 [**Palaeontina vidale* MEUNIER, 1902e, p. 9; OD]. Little-known genus. Fore wing rectangular; anterior margin straight, without nodal break; venation as in *Limacodites. Jur.,* Europe (Spain).

Palaeoforda KONONOVA, 1977, p. 588 [**P. tajmyrensis*; OD]. Little-known genus. Antennae with 6 segments. Fore wing with RS arising from distal part of pterostigma; M unbranched. Legs short. [Placed in Pemphigidae by KONONOVA but transferred to family uncertain by HEIE (1985).] *Cret.,* USSR (Asian RSFSR).

Parafulgoridium HANDLIRSCH, 1939, p. 138 [**Fulgoridium simplex* GEINITZ, 1880, p. 528; OD]. Little-known genus, based on poorly preserved fore wing. *Jur.,* Europe (Germany).

Parajassus BODE, 1953, p. 200 [**P. hattorfensis*; OD]. Little-known wing. BECKER-MIGDISOVA, 1962b. *Jur.,* Europe (Germany).

Perissovena RIEK, 1976a, p. 775 [**P. heidiae*; OD]. Little-known genus, based on hind wing. *Perm.,* South Africa.

Permocapitus EVANS, 1943b, p. 195 [**P. globulus*; OD]. Little-known genus, based on head. Head oval, eyes globular; transverse ridge between eyes. *Perm.,* Australia (New South Wales).

Permocephalus EVANS, 1943a, p. 8 [**P. knighti*; OD]. Little-known insects, known only by fragments of head. *Perm.,* Australia (New South Wales).

Permodunstania BECKER-MIGDISOVA, 1961c, p. 290 [**P. prosboloides*; OD]. Distal fragment of fore wing; RS forked; M4 free from M3 distally. *Perm.,* USSR (Asian RSFSR).——FIG. 162,8. **P. prosboloides*; fore wing, ×2.7 (Becker-Migdisova, 1961c).

Petropteron COCKERELL, 1912b, p. 94 [**P. mirandum*; OD]. Little-known genus, based on wing fragment; possibly a fulgoroid. *Cret.,* USA (Colorado).

Phragmatoecicossus BECKER-MIGDISOVA, 1949b, p. 11 [**P. shurabensis*; OD]. Little-known genus, based on fragment of fore wing. Probably related to Paleontinidae. *Jur.,* USSR (Asian RSFSR).

Plecophlebus COCKERELL, 1917h, p. 327 [**P. nebulosus*; OD]. Little-known genus, based on wing and fragments of body. [Originally placed in Trichoptera, but transferred to Homoptera, family uncertain, by BOTOSANEANU, 1981.] *Mio.,* Burma.

Prolystra OPPENHEIM, 1888, p. 228 [**P. lithographica*; OD]. Little-known genus, probably close to *Limacodites.* EVANS, 1956. *Jur.,* Europe (Germany).

Prosbolopsis MARTYNOV, 1935c, p. 19 [**P. ovalis*; OD]. Little-known insect, with reduced venation in tegmen. [Type of family Prosbolopseidae BECKER-MIGDISOVA, 1946.] EVANS, 1956; BECKER-MIGDISOVA, 1962b. *Perm.,* USSR (European RSFSR).

Protopsyche HANDLIRSCH, 1906b, p. 623 [**P. braueri*; OD]. Little-known genus, similar to *Limacodites. Jur.,* Europe (Germany).

Reticulocicada BECKER-MIGDISOVA, 1961c, p. 362 [**R. brachyptera*; OD]. Little-known tegmen, with coarse reticulation; possibly a fulgoroid. *Perm.,* USSR (Asian RSFSR).

Sbenaphis SCUDDER, 1890, p. 250 [**S. quesneli*; OD]. Little-known aphidoid genus. HEIE, 1967, 1985. *Eoc.,* Canada (British Columbia).

Shuraboprosbole BECKER-MIGDISOVA, 1949b, p. 23 [**S. plachutai*; OD]. Little-known genus, based on wing fragment. Basal stem of R only about half as long as R+M; RS arising well before midwing; CUA anastomosed with M for a short distance. *Jur.,* USSR (Tadzhik).——FIG. 163,5. **S. plachutai*; fore wing as preserved, ×2.5 (Becker-Migdisova, 1949b).

Stenoglyphis EVANS, 1947b, p. 432 [**S. kimblensis*; OD]. Little-known genus, possibly related to Scytinopteridae. EVANS, 1956. *Perm.,* Australia (New South Wales).

Tingiopsis BECKER-MIGDISOVA, 1953c, p. 461 [**T. reticulata*; OD]. Little-known genus, based on incomplete fore wing with fine reticulation. [Originally placed in Tingidae (Heteroptera) but transferred to Homoptera, probably Cercopidae, by EVANS (1957).] *Trias.,* USSR (Tadzhik).

Tipuloidea WIELAND, 1925, p. 23 [**T. rhaetica*; OD]. Little-known genus, based on fore wing. Costal margin arched; SC apparently absent; RS arising before midwing; closed median cell very small. [Originally placed in order Diptera.] EVANS, 1956. *Trias.,* Argentina.——FIG. 162,2. **T. rhaetica*; fore wing, ×2 (Evans, 1956).

Triassoaphis EVANS, 1956, p. 238 [**T. cubitus*; OD]. Little-known genus, based on wing fragment. [Originally placed in Aphididae but transferred to Aphidoidea, family uncertain, by BECKER-MIGDISOVA & AIZENBERG (1962).] RICHARDS, 1966; HEIE, 1967, 1981; SHAPOSHNIKOV, 1979a. *Trias.,* Australia (Queensland).

Triassocotis EVANS, 1956, p. 194 [**T. australis*; OD]. Little-known genus, based on distal half of tegmen. Tegmen narrow; R with 4 branches;

RS unbranched; M with 4 branches and a cell included between M1+2 and M3+4. [Originally placed in Scytinopteridae but transferred to family uncertain by EVANS (1961).] *Trias.,* Australia (Queensland).——FIG. 162,*3*. **T. australis*; fore wing, ×4.5 (Evans, 1956).

Triassojassus TILLYARD, 1919c, p. 887 [**T. proavittus*; OD]. Little-known genus, based on incomplete tegmen. Costal margin unusually convex; RS unbranched; M with 5 branches. [Originally placed in the Jassidae, but EVANS transferred first (1956) to the Chilocyclidae and later (1961) to family uncertain.] *Trias.,* Australia (New South Wales).

Turutanovia BECKER-MIGDISOVA, 1949b, p. 21 [**T. karatavia*; OD]. Little-known genus, based on distal fragment of fore wing. BECKER-MIGDISOVA, 1962b. *Jur.,* USSR (Kazakh).——FIG. 162,*10*. **T. karatavia*; fore wing as preserved, ×2 (Becker-Migdisova, 1962b).

Tychticola BECKER-MIGDISOVA, 1952, p. 181 [**T. longipenna*; OD]. Little-known genus, based on incomplete fore wing. Wing apparently long and narrow; RS long and parallel to R2. *Perm.,* USSR (Asian RSFSR).——FIG. 163,*1*. **T. longipenna*; fore wing, ×5 (Becker-Migdisova, 1962b).

Suborder HETEROPTERA
Latreille, 1810

[Heteroptera LATREILLE, 1810, p. 433]

Fore wing typically with the proximal part strongly coriaceous and the distal part membranous, forming a hemelytron; wings usually held flat over abdomen at rest. *Perm.–Holo.*

Family PROGONOCIMICIDAE
Handlirsch, 1906

[Progonocimicidae HANDLIRSCH, 1906b, p. 493] [=Eocimicidae HANDLIRSCH, 1906b, p. 494; Actinocytinidae EVANS, 1956, p. 244; Cicadocoridae BECKER-MIGDISOVA, 1958, p. 60]

Small species, dorsoventrally flattened; pronotum distinctly broader than long; fore wing apparently of uniform texture; veins RS and M coalesced basally; SC apparently coalesced with stem of R basally, diverging toward costal margin near midwing; M with 2 to 4 branches; CUA with 2 to 3 branches. [Placed by POPOV (1980a) in suborder Peloridiina, along with the Peloridiidae (recent).] *Perm.–Jur.*

Progonocimex HANDLIRSCH, 1906b, p. 494 [**P. jurassicus*; OD] [=*Eocimex* HANDLIRSCH, 1906b, p. 494 (type, *E. liasinus*)]. Fore wing with rounded apex; clavus broad, nearly triangular; M with 3 branches. BECKER-MIGDISOVA, 1962b; POPOV & WOOTTON, 1977. *Jur.,* Europe (Germany).——FIG. 164,*8*. *Progonocimex; a, *P. jurassicus,* dorsal view; *b, P. liasinus* (HANDLIRSCH), fore wing, both ×9 (Popov & Wootton, 1977).

Actinoscytina TILLYARD, 1926a, p. 18 [**A. belmontensis*; OD] [=*Pseudipsvicia* HANDLIRSCH, 1939, p. 17 (type, *P. ala*)]. Little-known genus. Tegmen similar to that of *Progonocimex,* but more slender, anterior margin less curved; SC curving directly toward anterior margin of wing. EVANS, 1956; POPOV & WOOTTON, 1977. *Perm.,* Australia (New South Wales).——FIG. 164,*5*. **A. belmontensis*; tegmen, ×8 (Evans, 1956).

Archicercopis HANDLIRSCH, 1939, p. 142 [**A. falcata*; OD]. Anterior margin of fore wing strongly convex basally; precostal area broad; wing apex pointed and directed anteriorly. EVANS, 1956; BECKER-MIGDISOVA, 1962b; POPOV & WOOTTON, 1977. *Jur.,* Europe (Germany).——FIG. 164,*6*. **A. falcata*; fore wing, ×13 (Popov & Wootton, 1977).

Cicadocoris BECKER-MIGDISOVA, 1958, p. 62 [**C. kuliki*; OD]. Tegmen with smoothly curved anterior margin; M with 3 branches; M3+4 unbranched. EVANS, 1961; POPOV, 1982. *Trias.,* USSR (Kirghiz).——FIG. 164,*9*. **C. kuliki*; restoration, ×10 (Becker-Migdisova, 1958).

Eocercopis HANDLIRSCH, 1939, p. 142 [**E. ancyloptera*; OD] [=*Cercoprisca* HANDLIRSCH, 1939, p. 143 (type, *C. similis*); *Cercopinus* HANDLIRSCH, 1939, p. 143 (type, *C. ovalis*)]. Fore wing with very convex and thickened costal margin; apex pointed; clavus broad and nearly triangular. EVANS, 1956; BECKER-MIGDISOVA, 1958; POPOV & WOOTTON, 1977. *Jur.,* Europe (Germany).——FIG. 164,*7*. **E. ancyloptera*; fore wing, ×13 (Popov & Wootton, 1977).

Heterojassus EVANS, 1961, p. 23 [**H. membranaceus*; OD]. Tegmen oval; SC and R terminating on costal margin near level of midwing. *Trias.,* Australia (Queensland).——FIG. 164,*4*. **H. membranaceus*; fore wing, ×19 (Evans, 1961).

Heteroscytina EVANS, 1956, p. 245 [**H. tillyardi*; OD]. Fore wing narrowed apically, much as in *Actinoscytina,* but costal area narrower and crossveins forming a more nearly complete transverse series. WOOTTON, 1963. *Trias.,* Australia (Queensland).

Hexascytina WOOTTON, 1963, p. 250 [**H. transecta*; OD]. Little-known genus, apparently similar to *Progonocimex,* based on incomplete tegmen. SC diverging from stem R near midwing at almost a 90° angle; anterior margin of tegmen distinctly convex. *Trias.,* Australia (Queensland).

Microscytinella WOOTTON, 1963, p. 251 [**M. radians*; OD]. Little-known genus, based on small

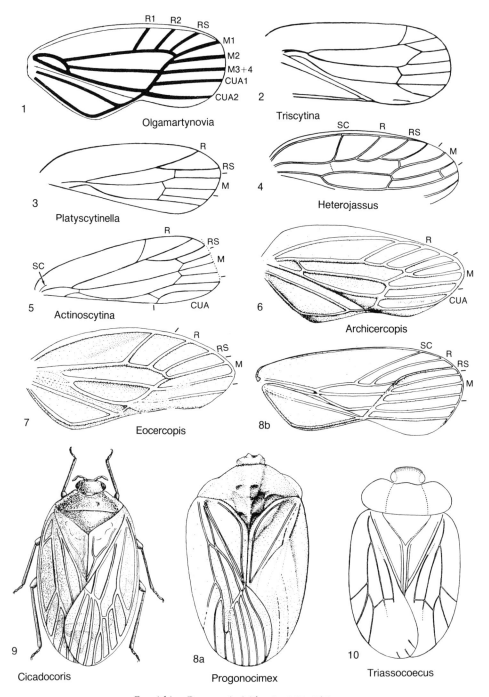

FIG. 164. Progonocimicidae (p. 259–260).

Hemiptera—Heteroptera

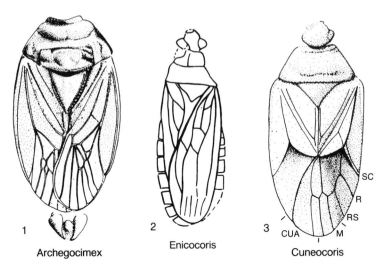

Fig. 165. Archegocimicidae, Enicocoridae, and Cuneocoridae (p. 261–262).

fragment of tegmen. *Trias.*, Australia (Queensland).

Olgamartynovia BECKER-MIGDISOVA, 1958, p. 63 [*O. turanica*; OD]. Tegmen as in *Cicadocoris*, but fork of M1+2 longer. POPOV, 1982. *Trias.*, USSR (Kirghiz).——FIG. 164,*1*. *O. complexa* POPOV; tegmen, ×14 (Popov, 1982).

Platyscytinella EVANS, 1956, p. 245 [*P. paradoxa*; OD]. Tegmen shaped as in *Actinoscytina*; SC absent; M1 continuing the nearly straight line of stem M; clavus unknown. WOOTTON, 1963. *Trias.*, Australia (New South Wales).——FIG. 164,*3*. *P. paradoxa*; fore wing, ×15 (Wootton, 1963).

Triassocoecus EVANS, 1963, p. 22 [*T. chinai*; OD]. Little-known genus. Tegmen broader than in *Actinoscytina*; pronotum with large lateral lobes. *Trias.*, Australia (New South Wales).——FIG. 164,*10*. *T. chinai*; dorsal view, ×10 (Evans, 1963).

Triscytina EVANS, 1956, p. 246 [*T. rotundata*; OD]. Similar to *Actinoscytina* but tegmen much broader; costal margin nearly straight; apex evenly rounded. *Trias.*, Australia (New South Wales).——FIG. 164,*2*. *T. rotundata*; tegmen, ×16 (Evans, 1956).

Family ARCHEGOCIMICIDAE Handlirsch, 1906

[Archegocimicidae HANDLIRSCH, 1906b, p. 493] [=Eonabidae HANDLIRSCH, 1920, p. 207; Diatillidae HANDLIRSCH, 1920, p. 210]

Small Heteroptera of uncertain relationships. Body dorsoventrally flattened; pronotum coarsely warty. Fore wing with apex rounded; clavus narrow; costal margin strongly sclerotized; veins SC, R, and M coalesced for about two-fifths length of wing. *Jur.*

Archegocimex HANDLIRSCH, 1906b, p. 493 [*A. geinitzi*; OD] [=*Eonabis* HANDLIRSCH, 1920, p. 207 (type, *E. primitiva*); *Archegocoris* HANDLIRSCH, 1939, p. 114 (type, *A. liadis*)]. Radial complex of fore wing (R and RS) with 3 branches terminating on anterior margin of wing. POPOV & WOOTTON, 1977. *Jur.*, Europe (Germany).——FIG. 165,*1*. *A. geinitzi*; dorsal view, ×14 (Popov & Wootton, 1977).

Anosmus HANDLIRSCH, 1939, p. 115 [*A. spilopterus*; OD]. Costal area of fore wing narrow; radial complex branched. POPOV & WOOTTON, 1977. *Jur.*, Europe (Germany).

Corynecoris BODE, 1953, p. 132 [*C. semigranulatus*; OD]. Little-known genus, based on poorly preserved specimen. [Family assignment doubtful.] POPOV & WOOTTON, 1977. *Jur.*, Europe (Germany).

Diatillus HANDLIRSCH, 1920, p. 210 [*D. debilis*; OD]. Little-known genus with archegocimicid venation. POPOV & WOOTTON, 1977. *Jur.*, Europe (Germany).

Ensphingocoris BODE, 1953, p. 139 [*E. praerotundatus*; OD]. Little-known genus, based on poorly preserved specimen lacking wings; body apparently that of a large archegocimicid. POPOV & WOOTTON, 1977. *Jur.*, Europe (Germany).

Entomecoris BODE, 1953, p. 134 [*A. minor*; OD]. Fore wing with common stem of SC, R, and M curving away from costal margin; wing differentiated into corium and membrane. POPOV & WOOTTON, 1977. *Jur.*, Europe (Germany).

Eurynotis BODE, 1953, p. 134 [*E. incisus*; OD].

Similar to *Somatocoris,* but radial complex with 2 branches. BECKER-MIGDISOVA, 1962b; POPOV & WOOTTON, 1977. *Jur.,* Europe (Germany).

Macropterocoris BODE, 1953, p. 138 [*M. obtusus;* OD]. Little-known genus; head and thorax resembling those of the Archegocimicidae. POPOV & WOOTTON, 1977. *Jur.,* Europe (Germany).

Progonocoris HANDLIRSCH, 1939, p. 115 [*P. pictus;* OD]. Similar to *Anosmus,* but radial complex of fore wing apparently unbranched; costal area long. BECKER-MIGDISOVA, 1962b; POPOV & WOOTTON, 1977. *Jur.,* Europe (Germany).

Somatocoris BODE, 1953, p. 141 [*S. conservatus;* OD]. Similar to *Archegocimex* but smaller; radial complex (R and RS) with 3 branches. BECKER-MIGDISOVA, 1962b; POPOV & WOOTTON, 1977. *Jur.,* Europe (Germany).

Family ENICOCEPHALIDAE Stål, 1858

[Enicocephalidae STÅL, 1858, p. 81]

Similar to the Reduviidae, but head constricted behind eyes; rostrum with 4 segments; fore wings entirely membranous, with longitudinal veins but few crossveins. *Mio.–Holo.*

Enicocephalus WESTWOOD, 1838, p. 22. *Holo.*
Disphaerocephalus COCKERELL, 1917g, p. 361 [*D. constrictus;* OD]. Little-known genus of small, elongate species, with long, thin legs and antennae; body with long, fine pubescence; hind legs long and narrow; tarsi 1-2-2; wings unknown. [Genus based on nymph and adult male.] ŠTYS, 1969. *Mio.,* Burma.

Paenicotechys ŠTYS, 1969, p. 353 [*Enicocephalus fossilis* COCKERELL, 1916a, p. 135; OD]. Similar to *Aenictopechys* (recent), but posterior margin of pronotum excised; eyes contiguous dorsally; middle tarsi with 2 segments. ŠTYS, 1969. *Mio.,* Burma.

Family ENICOCORIDAE Popov, 1980

[Enicocoridae POPOV, 1980a, p. 50]

Apparently related to the Enicocephalidae. Head short, rostrum thick, curved. Tegmen entirely membranous, clavus and corium not differentiated; radial-medial and cubital-anal sectors of veins widely separated at base; veins nearly parallel distally. Legs thin, cursorial. *Cret.*

Enicocoris POPOV, 1980a, p. 50 [*E. manlaicus;* OD]. Head prognathous; pronotum transverse; scutellum much narrower than pronotum; subcostal area of tegmen wide. *Cret.,* Mongolia.
——FIG. 165,*2* *E. manlaicus;* dorsal view, ×9 (Popov, 1980a).

Family DIPSOCORIDAE Dohrn, 1859

[Dipsocoridae DOHRN, 1859, p. 36]

Similar to the Saldidae, but third antennal segment not thickened at base. *Mio.–Holo.*

Dipsocoris HALIDAY, 1855, fig. 61. *Holo.*
Ceratocombus SIGNORET, 1852, p. 542. WYGODZINSKY, 1959. *Mio.,* Mexico (Chiapas)–*Holo.*

Family CUNEOCORIDAE Handlirsch, 1920

[Cuneocoridae HANDLIRSCH, 1920, p. 208]

Small insects. Fore wings reaching end of abdomen and overlapped distally; pronotum wider than long; scutellum triangular. Fore wing not clearly differentiated into corium and membrane; veins M and CU branched. *Jur.*

Cuneocoris HANDLIRSCH, 1920, p. 208 [*C. geinitzi;* OD]. M and CU each with 2 branches. POPOV & WOOTTON, 1977. *Jur.,* Europe (Germany). —— FIG. 165,*3.* *C. geinitzi;* dorsal view, ×22 (Popov & Wootton, 1977).

Family GERRIDAE Leach, 1815

[Gerridae LEACH, 1815, p. 123]

Body slender; rostrum with 4 segments; fore wings without differentiation of corium, membrane, or clavus; posterior femora extending well beyond end of abdomen; claws ante-apical. Semiaquatic. ANDERSEN, 1982b. *Eoc.–Holo.*

Gerris FABRICIUS, 1794, p. 187. COCKERELL, 1909j; HANDLIRSCH, 1910b; THÉOBALD, 1937a. *Oligo.,* USA (Colorado), Canada (British Columbia), Europe (France)–*Holo.*
Metrobates UHLER, 1871, p. 108. SCUDDER, 1890. *Oligo.,* USA (Colorado)–*Holo.*
Telmatrechus SCUDDER, 1890, p. 351 [*Hygrotrechus stali* SCUDDER, 1879a, p. 183B; SD CARPENTER, herein]. Eyes not prominent; first antennal segment only a little longer than second; thorax relatively short; legs very long, with the tibiae equal in length to femora of same leg. *Eoc.,* USA (Wyoming); *Mio.,* Canada (British Columbia).

Family HYDROMETRIDAE Stephens, 1829

[Hydrometridae STEPHENS, 1829, p. 352]

Very slender species; head long and narrow but widened distally; antennae with 4 (rarely 5) segments; legs very long and slender, claws apical; rostrum with 3 segments; tegmen with corium and membrane. *Eoc.—Holo.*

Hydrometra LATREILLE, 1796, p. 86. *Holo.*
Eocenometra ANDERSEN, 1982a, p. 91 [*E. danica*; OD]. Similar to *Bacillometra* (recent) and *Hydrometra* (recent), but first antennal segment much longer than second; thorax relatively short and robust. ANDERSEN, 1982b. *Eoc.*, Europe (Denmark).

Family VELIIDAE
Amyot & Serville, 1843

[Veliidae AMYOT & SERVILLE, 1843, p. 418]

Similar to the Gerridae, but rostrum with 3 segments; posterior femora shorter, extending very little beyond end of abdomen at most. *Oligo.—Holo.*

Velia LATREILLE, 1804, p. 270. MEUNIER, 1914a. *Oligo.*, Europe (France)—*Holo.*
Palaeovelia SCUDDER, 1890, p. 349 [*P. spinosa*; OD]. Similar to *Microvelia* (recent). Head small, recessed to level of eyes in emarginate prothorax; hind legs very short, reaching only tip of abdomen; femora and tibiae of equal lengths; hind tibiae with long spines distally. *Oligo.*, USA (Colorado).
Stenovelia SCUDDER, 1890, p. 349 [*S. nigra*; OD]. Similar to *Palaeovelia*, but hind tibiae without long spines distally. *Oligo.*, USA (Colorado).

Family NOTONECTIDAE
Latreille, 1802

[Notonectidae LATREILLE, 1802a, p. 253]

Aquatic species, similar to the Naucoridae, but forelegs raptorial, and hind tarsi without claws. *Jur.—Holo.*

Notonecta LINNÉ, 1758, p. 439. PITON, 1942; LAUCK, 1960; POPOV, 1964; MARTINI, 1971. *Oligo.*, USA (Colorado), Europe (Germany); *Mio.*, Europe (France)—*Holo.*
Anisops SPINAR, 1837, p. 58. DEICHMÜLLER, 1881; ŠTYS & ŘIHA, 1975a. *Oligo.*, Europe (Czechoslovakia)—*Holo.*
Asionecta POPOV in BECKER-MIGDISOVA & POPOV, 1963, p. 78 [*A. curtipes*; OD]. Similar to *Notonecta* but with first segment of front and middle legs very short. *Jur.*, USSR (Kazakh).——FIG. 166,2. *A. curtipes*; ventral view, ×5 (Popov in Becker-Migdisova & Popov, 1963).
Clematina POPOV, 1964, p. 66 [*Notonecta primaeva* HEYDEN, 1859a, p. 11; OD]. Little-known genus, apparently related to *Clypostemma*. POPOV, 1971; ŠTYS, 1973. *Oligo.*, Europe (Germany).
Clypostemma POPOV, 1964, p. 64 [*C. xyphiale*; OD]. Species of moderate size and of uncertain relationship within the family. Rostrum with 4 segments; tarsi of all legs with 2 segments. POPOV, 1971; ŠTYS, 1973. *Cret.*, USSR (Asian RSFSR).
Enithares SPINOLA, 1837, p. 60. A nymph is only fossil record. [Generic assignment uncertain.] ŠTYS & ŘIHA, 1975a. *Oligo./Mio.*, Europe (Czechoslovakia)—*Holo.*
Liadonecta POPOV, 1971, p. 172 [*L. tomiensis*; OD]. Little-known genus, based on nymph. Body elongate-oval, head transverse; hind tibiae and tarsi of uniform width. *Jur.*, USSR (Asian RSFSR).——FIG. 166,3. *L. tomiensis*; dorsoventral view, ×14 (Popov, 1971).
Nepidium WESTWOOD, 1854, p. 396 [*N. stolones*; OD]. Little-known genus, based on poorly preserved specimen. [Put in Naucoridae by HANDLIRSCH (1906b) and in Notonectidae by POPOV (1971).] *Jur.*, England.
Notonectites HANDLIRSCH, 1906b, p. 639 [*Notonecta elterleini* DEICHMÜLLER, 1886, p. 64; OD]. Little-known genus, apparently close to *Notonecta* and *Anisopus* (recent). POPOV, 1964, 1971; ŠTYS & ŘIHA, 1975a. *Jur.*, Europe (Germany).
Pelonecta POPOV, 1971, p. 170 [*P. solnhofeni*; OD]. Body elongate-oval, widest near base of abdomen; hind tibiae shorter than femora or tarsi; femora thickened, strongly developed. *Jur.*, Europe (Germany).——FIG. 166,1. *P. solnhofeni*; ventral view, ×2.2 (Popov, 1971).
Soevenia STATZ, 1950b, p. 63 [*Notonecta heydeni* DEICHMÜLLER, 1881, p. 328; OD]. Similar to *Anisops*; clypeus fused to frons. Body structure little known. *Oligo.*, Europe (Germany, Czechoslovakia).

Family SCAPHOCORIDAE
Popov, 1968

[Scaphocoridae POPOV, 1968, p. 106]

Body oval; head hypognathous; pronotum large, covering scutellum; tegmen with membrane; hind legs relatively short; tarsi with a single segment and dense hairs. Probably related to the Naucoridae. *Jur.*

Scaphocoris POPOV, 1968, p. 106 [*S. notatus*; OD]. Head strongly transverse from above; scutellum very small, triangular; clavus with distinct anal veins; membrane present; hind tarsi shorter than tibiae. *Jur.*, USSR (Kazakh).

Family NAUCORIDAE Leach, 1815

[Naucoridae LEACH, 1815, p. 123] [=Aphlebocoridae HANDLIRSCH, 1906b, p. 494; Apopnidae HANDLIRSCH, 1920, p. 209]

Antennae four-segmented, shorter than head; fore wing membrane without veins; forelegs raptorial; tarsi with more than one segment; hind tarsi with claws. *Jur.—Holo.*

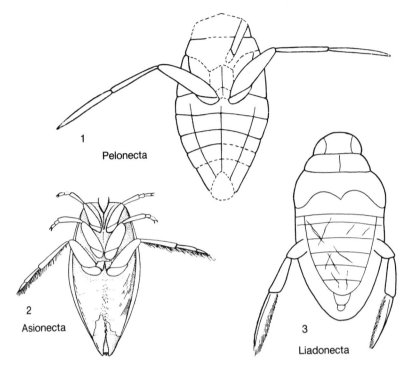

Fig. 166. Notonectidae (p. 263).

Naucoris GEOFFROY, 1762, p. 473. HEER, 1853a; PITON & THÉOBALD, 1937; STATZ, 1950b. *Oligo.*, Europe (Germany); *Mio.*, Europe (France, Croatia)–*Holo.*

Aidium POPOV, 1968, p. 103 [*A. pleurale*; OD]. Anterior margin of pronotum concave; scutellum very large; tegmen, except for clavus, without venation; embolium absent; clavus narrow, hind legs relatively short. POPOV, 1971. *Jur.*, USSR (Kazakh).

Angaronecta POPOV, 1971, p. 146 [*A. longirostris*; OD]. Rostrum very long, reaching hind margin of prothorax; forelegs short; femora of all legs thickened; hind tarsi with single segment. *Jur.*, USSR (Asian RSFSR).——FIG. 167,3. *A. longirostris*; ventral view, ×3.7 (Popov, 1971).

Aphlebocoris HANDLIRSCH, 1906b, p. 495 [*A. nana*; OD]. Fore wing not differentiated into corium and membrane; clavus narrow, nearly quadrilateral. POPOV & WOOTTON, 1977. *Jur.*, Europe (Germany).——FIG. 167,6. *A. punctata* HANDLIRSCH; fore wing, ×11 (Popov & Wootton, 1977).

Apopnus HANDLIRSCH, 1920, p. 209 [*A. magniclavus*; OD]. Little-known genus. Fore wing differentiated into corium and membrane; costal margin convex; clavus broad and triangular. POPOV & WOOTTON, 1977. *Jur.*, Europe (Germany).

Diplonychus LAPORTE, 1832, p. 18. HEER, 1853a. *Mio.*, Europe (Croatia)–*Holo.*

Heleonaucoris POPOV, 1971, p. 149 [*H. maculipennis*; OD]. Clavus of moderate size; embolium narrow and developed only at base of tegmen; border between corium and membrane indistinct; corium spotted. *Jur.*, USSR (Kirghiz). ——FIG. 167,1. *H. maculipennis*; tegmen, ×4.2 (Popov, 1971).

Liadonaucoris POPOV, 1971, p. 144 [*L. rohdendorfi*; OD]. Tegmen longer than abdomen; clavus longer than scutellum; vein R present on tegmen. *Jur.*, USSR (Kirghiz).——FIG. 167,2. *L. rohdendorfi*; ×5.5 (Popov, 1971).

Nectodes POPOV, 1968, p. 105 [*N. maculatus*; OD]. Little-known genus, based on tegmen. Clavus large and broad; embolium distinct, extending for half length of corium; membrane large. *Jur.*, USSR (Kazakh).

Nectonaucoris POPOV, 1968, p. 104 [*N. lariversi*; OD]. Anterior margin of pronotum straight; tegmen without veins; embolium absent; clavus narrow; hind legs relatively short. *Jur.*, USSR (Kazakh).

Sphaerodemopsis HANDLIRSCH, 1906b, p. 543 [*Sphaerodema jurassicum* OPPENHEIM, 1888, p. 235; OD]. Tegmen strongly sclerotized; clavus usually long and heavily sclerotized. POPOV, 1971. *Jur.*, Europe (Germany).——FIG. 167,5. *S. jurassica*; dorsal view, ×2.5 (Popov, 1971).

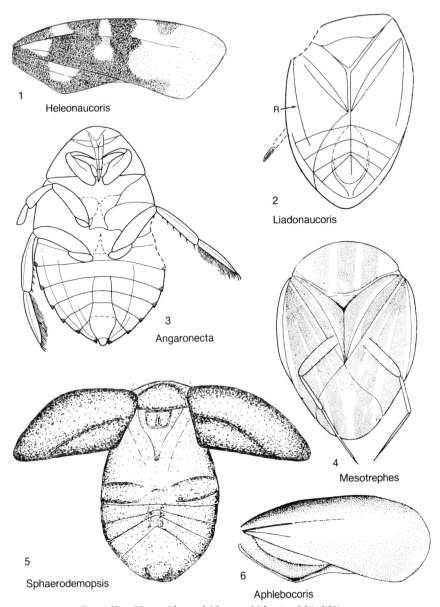

Fig. 167. Naucoridae and Mesotrephidae (p. 264–265).

Family MESOTREPHIDAE
Popov, 1971

[Mesotrephidae Popov, 1971, p. 160]

Small species, related to the Scaphocoridae. Pronotum transverse, convex, elliptical, completely covering head from above. *Cret.*

Mesotrephes Popov, 1971, p. 160 [**M. striata;* OD]. Scutellum very small; tegmen with only one vein, extending along most of costal margin. Hind legs relatively short; tibiae and tarsi thin. *Cret.*, USSR (Kazakh). —— Fig. 167,4. **M. striata;* dorsoventral view, ×20 (Popov, 1971).

Family BELOSTOMATIDAE
Leach, 1815

[Belostomatidae Leach, 1815, p. 123]

Similar to the Nepidae, but antennae with 4 segments; posterior legs adapted for swimming, the tibiae flattened; aquatic. *Jur.–Holo.*

266 *Hexapoda*

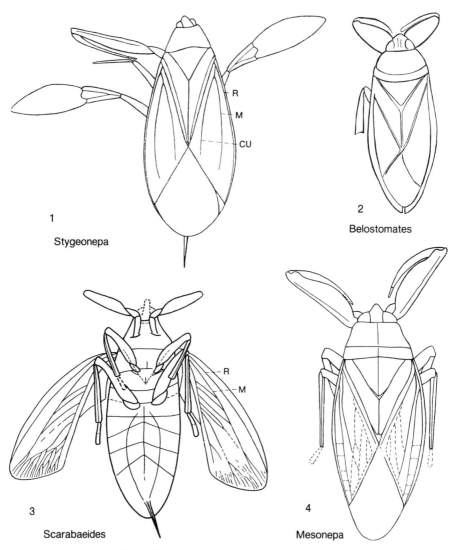

Fig. 168. Belostomatidae (p. 266–267).

Belostoma Latreille, 1807, p. 144. Germar, 1837. *Oligo.*, Europe (Germany)–*Holo.*

Belostomates Schöberlin, 1888, p. 61 [*Belostomum speciosa* Heer, 1865, p. 303; OD]. Little-known genus, with very broad front femora. Heer, 1865. *Mio.*, Europe (Germany).——Fig. 168,2. *B. speciosa*; dorsal view, ×0.6 (Heer, 1865).

Lethocerus Mayr, 1853, p. 17. Řiha & Kukalová, 1967; Popov, 1971. *Oligo.*, USSR (Asian RSFSR); *Mio.*, Europe (Czechoslovakia)–*Holo.*

Mesonepa Handlirsch, 1906b, p. 637 [*Nepa primordialis* Germar, 1839, p. 206; SD Popov, 1971, p. 116]. Similar to *Belostoma*, but fore wing with much larger membranous area; fore tarsi with one segment. Popov, 1971. *Jur.*, Europe (Germany).——Fig. 168,4. *M. primordialis*; dorsal view, ×2 (Popov, 1971).

Scarabaeides Germar, 1839, p. 218 [*S. deperditus*; OD] [=*Mesobelostomum* Haase, 1890a, p. 21, obj.]. Similar to *Lethocerus* (recent), but fore wing with M and R widely separated and remote from costal margin; membranous area of wing without venation. Oppenheim, 1888; Popov, 1971. *Jur.*, Europe (Germany). —— Fig. 168,3. *S. deperditus*; ventral view, ×1.2 (Popov, 1971).

Stygeonepa Popov, 1971, p. 119 [*S. foersteri*; OD].

Related to *Belostoma,* but pronotum more transverse; fore wing with vestiges of R, M, and CU; hind tibiae and one-segmented tarsi forming broad lobes. *Jur.,* Europe (Germany).——FIG. 168,*1.* **S. foersteri*; dorsal view, ×2 (Popov, 1971).

Family NEPIDAE Latreille, 1802

[Nepidae LATREILLE, 1802a, p. 252]

Antennae with 3 segments, shorter than head; membrane of fore wings reticulate; hind legs adapted for walking; tibiae not flattened; aquatic. *Oligo.–Holo.*

Nepa LINNÉ, 1758, p. 440. HEER, 1853a; HUNGERFORD, 1932. *Oligo.,* USA (Colorado); *Mio.,* Europe (Germany)*–Holo.*

Family SHURABELLIDAE Popov, 1971

[Shurabellidae POPOV, 1971, p. 121]

Small species, related to the Corixidae. Pronotum transverse, sculptured; tegmen broad; hind legs relatively slender. *Jur.*

Shurabella BECKER-MIGDISOVA, 1949b, p. 28 [**S. lepyroniopsis*; OD] [=*Coleopteropsis* BECKER-MIGDISOVA, 1949b, p. 31 (type, *C. dolichoptera*)]. Pronotum about three times as wide as long; tegmen strongly sclerotized; vestiges of SC, R, and M present on tegmen. *Jur.,* USSR (Kirghiz).——FIG. 169,*4.* **S. lepyroniopsis*; dorsoventral view, ×10 (Popov, 1971).

Family CORIXIDAE Leach, 1815

[Corixidae LEACH, 1815, p. 124]

Head not inserted into prothorax; antennae shorter than head; fore tarsi consisting of only one spatulate segment. *Jur.–Holo.*

Corixa GEOFFROY, 1762, p. 477. HEER, 1853a; SCUDDER, 1890; SCHLECHTENDAL, 1894. *Oligo.,* USA (Colorado), Europe (Germany); *Mio.,* Europe (Germany)*–Holo.*

Archaecorixa POPOV, 1968, p. 101 [**A. lata*; OD]. Pronotum transverse; corium of tegmen with distinct venation; SC coalesced with R, M, and CU for varying lengths, finally terminating on costal margin; embolium absent. POPOV, 1971. *Jur.,* USSR (Kazakh).

Baissocorixa POPOV, 1966, p. 99 [**B. jaczewskii*; OD]. Similar to *Corixa,* but veins R, M, and CU more strongly developed; head narrow; eyes small; eighth abdominal tergite well developed. *Jur./Cret.,* USSR (Asian RSFSR).

Diacorixa POPOV, 1971, p. 137 [**D. miocaenica*; OD]. Similar to *Sigara* (recent) but with deep furrow along entire length of pronotum; vein CU weakly formed. *Mio.,* USSR (Kirghiz).

Diapherinus POPOV, 1966, p. 97 [**D. ornatipennis*; OD]. Little-known genus, based on tegmen. SC, R, M, and CU visible on corium; anal veins clear on clavus; embolium weakly developed. *Jur./Cret.,* USSR (Asian RSFSR).——FIG. 169,*2.* **D. ornatipennis*; tegmen, ×5 (Popov, 1966).

Gazimuria POPOV, 1971, p. 130 [**G. scutellata*; OD]. Elongate species. Antennae with 4 segments; pronotum not more than three times wider than its length; tegmen with veins R, M, and CU; hind legs densely covered with hairs. *Jur.,* USSR (Asian RSFSR).——FIG. 169,*5.* **G. scutellata*; dorsoventral view, ×6 (Popov, 1971).

Ijanecta POPOV, 1971, p. 132 [**I. angarica*; OD]. Pronotum well developed; scutellum small; fore margin of wing with wide embolium; all veins apparently absent. *Jur.,* USSR (Asian RSFSR). ——FIG. 169,*1.* **I. angarica*; dorsal view, ×1 (Popov, 1971).

Karataviella BECKER-MIGDISOVA, 1949b, p. 25 [**K. brachyptera*; OD]. Pronotum twice as wide as long; only vein 1A on clavus. *Jur.,* USSR (Kazakh).——FIG. 169,*3.* **K. brachyptera*; dorsal view, ×7 (Popov, 1971).

Mesosigara POPOV, 1971, p. 129 [**M. kryshtofovichi*; OD]. Similar to *Baissocorixa,* but fore wing with R coalesced with SC for its entire length; M fused at base with CU. *Cret.,* USSR (Asian RSFSR).——FIG. 169,*7.* **M. kryshtofovichi*; lateral view, ×12 (Popov, 1971).

Sigaretta POPOV, 1971, p. 136 [**Corixa florissantiella* COCKERELL, 1906e, p. 209; OD]. Pronotum large but covering only part of the scutellum; tegmen with well-developed embolium rim; anal vein present on clavus. *Oligo.,* USA (Colorado). ——FIG. 169,*6.* **S. florissantiella* (COCKERELL); dorsal view, ×9 (Popov, 1971).

Family ARADIDAE Brullé, 1835

[Aradidae BRULLÉ, 1835, p. 326]

Body strongly flattened; head porrect; antennae and rostrum with 4 segments; clavus narrowed apically; wing membrane with few or no veins; abdomen broader than wings; tarsi with 2 segments. *Oligo.–Holo.*

Aradus FABRICIUS, 1803, p. 116. GERMAR & BERENDT, 1856; USINGER, 1941; POPOV, 1978. *Oligo.,* Europe (Baltic); *Mio.,* Europe (Croatia)*–Holo.*

Calisius STÅL, 1858, p. 67. USINGER, 1941. *Oligo.,* Europe (Baltic)*–Holo.*

Mezira AMYOT & SERVILLE, 1843, p. 305. USINGER, 1941. *Oligo.,* Europe (Baltic); *Mio.,* Europe (Croatia)*–Holo.*

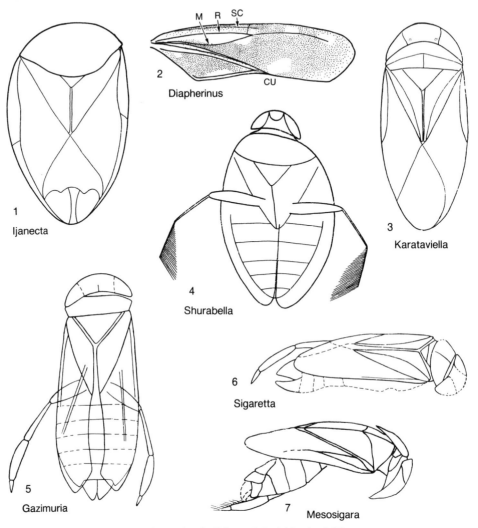

Fig. 169. Shurabellidae and Corixidae (p. 267).

Family SALDIDAE Amyot & Serville, 1843

[Saldidae Amyot & Serville, 1843, p. xlix]

Head shorter than thorax and scutellum; antennae four-segmented, longer than head; third antennal segment thickened at base; rostrum three-segmented; fore wings without reticulate cells, but with 4 or 5 long, closed cells; corium with an embolium; forelegs not raptorial. *Jur.–Holo.*

Salda Fabricius, 1803, p. 113. Germar & Berendt, 1856. *Oligo.*, Europe (Baltic)–*Holo.*

Oligosaldina Carpenter, herein [*O. rottensis* Statz & Wagner, 1950, p. 101; OD]. Fore wing similar to *Chiloxanthus* (recent) but with cells of membrane nearly the same length. [The original generic name, *Oligosaldina*, was a *nomen nudum* (Statz & Wagner, 1950).] *Oligo.*, Europe (Germany).

Saldonia Popov, 1973, p. 704 [*S. rasnitsyni*; OD]. Pronotum transverse; RS close to front margin of tegmen; membrane not present on tegmen; scutellum small, shorter than claval suture. *Jur.*, USSR (Asian RSFSR).

Family COREIDAE Leach, 1815

[Coreidae Leach, 1815, p. 121]

Head much narrower and shorter than prothorax; antennae longer than head, with

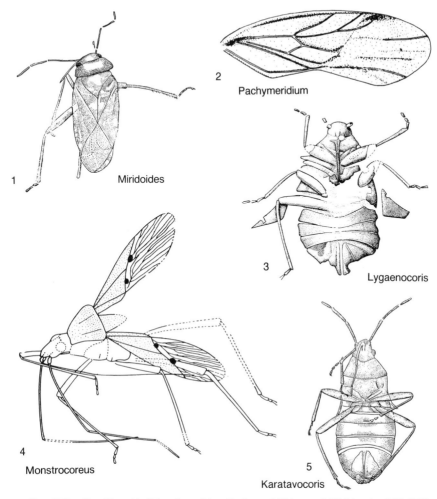

FIG. 170. Coreidae, Alydidae, Lygaeidae, Pachymeridiidae, and Miridae (p. 269–275).

4 segments; fore wing not reticulate, its membrane with many longitudinal veins, often uniting; body stout, legs thick. *Jur.–Holo.*

Coreus FABRICIUS, 1794, p. 120. THÉOBALD, 1937a. *Oligo.,* Europe (France)–*Holo.*
Achrestocoris SCUDDER, 1890, p. 413 [*A. cinerarius*; OD]. Tegmen with large rhomboidal cell at apex of corium. *Oligo.,* USA (Colorado).
Anasa AMYOT & SERVILLE, 1843, p. 209. SCUDDER, 1890. *Oligo.,* USA (Colorado)–*Holo.*
Berytopsis HEER, 1853a, p. 54 [*B. femoralis*; OD]. Little-known coreid, apparently related to *Berytus* (recent). *Mio.,* Europe (Germany).
Corizus FALLÉN, 1814, p. 8. SCUDDER, 1890; COCKERELL, 1926a. *Oligo.,* USA (Colorado); *Tert. (epoch unknown),* Argentina–*Holo.*
Harmostites HEER, 1853a, p. 49 [*H. oeningensis*; OD]. Little-known genus, based on poorly preserved specimen. *Mio.,* Europe (Germany).
Heeria SCUDDER, 1890, p. 430 [*H. gulosa*; SD HANDLIRSCH, 1907, p. 1049]. Similar to *Arencoris* (recent) but with second and third antennal segments unequal. *Oligo.,* USA (Colorado).
Hypselonotus HAHN, 1833, p. 186. HEER, 1853a. *Mio.,* Europe (Germany)–*Holo.*
Jadera STÅL, 1860, p. 59. COCKERELL, 1909j. *Eoc.,* USA (Colorado)–*Holo.*
Karatavocoris BECKER-MIGDISOVA, 1962b, p. 222 [*K. asiatica*; OD]. Head much narrower than pronotum; first antennal segment short, not extending beyond apex of head; femora without spines. *Jur.,* USSR (Kazakh). —— FIG. 170,5. *K. asiatica*; ventral view, ×6 (Becker-Migdisova, 1962b).
Leptoscelis LAPORTE, 1832, p. 31. HEYDEN, 1858. *Oligo.,* Europe (Bavaria)–*Holo.*

Palaeocoris HEER, 1853a, p. 46 [*P. spectabilis*; OD]. Little-known genus, apparently related to *Acanthosoma* (recent). *Mio.*, Europe (Croatia).

Phthinocoris SCUDDER, 1890, p. 414 [*P. colligatus*; SD HANDLIRSCH, 1907, p. 1049]. Similar to *Achrestocoris*, but thorax much longer. *Oligo.*, USA (Colorado).

Piezocoris SCUDDER, 1890, p. 416 [*P. peritus* SCUDDER, 1890, p. 416; SD CARPENTER, herein]. Similar to *Phthinocoris*, but head large, one-half to two-thirds width of thorax. *Oligo.*, USA (Colorado).

Spartocera LAPORTE, 1832, p. 42. HEER, 1853a. *Mio.*, Europe (Croatia)–*Holo*.

Syromastes LATREILLE, 1829, p. 196. HEER, 1853a; STATZ & WAGNER, 1950. *Oligo.*, Europe (Germany); *Mio.*, Europe (Germany)–*Holo*.

Family ALYDIDAE Stål, 1872

[Alydidae STÅL, 1872, p. 53]

Similar to Coreidae, but head nearly as broad and as long as the prothorax, broader than anterior margin of pronotum; body and legs elongate. ŠTYS & ŘIHA, 1977. *Jur.–Holo*.

Alydus FABRICIUS, 1803, p. 248. *Holo*.

Cydamus STÅL, 1858, p. 33. [Generic assignment of fossil doubtful.] SCUDDER, 1890; ŠTYS & ŘIHA, 1975b, 1977. *Oligo.*, USA (Colorado)–*Holo*.

Daclera SIGNORET, 1862, p. 27. [Generic assignment of fossil doubtful.] THÉOBALD, 1937a; ŠTYS & ŘIHA, 1977. *Oligo.*, Europe (France)–*Holo*.

Heeralydus ŠTYS & ŘIHA, 1975b, p. 190 [*H. bucculatus*; OD]. Similar to *Alydus* (recent), but head relatively short and having long bucculae reaching proximally between insertion of antennae and anterior margins of eyes. ŠTYS & ŘIHA, 1977. *Oligo.*, Europe (Germany).

Monstrocoreus POPOV, 1968, p. 109 [*M. quadrimaculatus*; OD]. Antennae thin and long, nearly as long as body; rostrum also very long; tegmen long, with only one distinct vein; legs long and thin, femora about as long as tibiae and much broader; tarsi with 3 segments. ŠTYS & ŘIHA, 1977. *Jur.*, USSR (Kazakh).——FIG. 170,4. *M. quadrimaculatus*; lateral view, ×3 (Popov, 1968).

Orthriocorisa SCUDDER, 1890, p. 429 [*O. longipes*; OD]. Little-known genus, similar to *Leptocoris* (recent). ŠTYS & ŘIHA, 1977. *Oligo.*, USA (Colorado).

Protenor STÅL, 1867, p. 543. [Generic assignment of fossil doubtful.] SCUDDER, 1890; ŠTYS & ŘIHA, 1977. *Oligo.*, USA (Colorado)–*Holo*.

Sulcalydus ŠTYS & ŘIHA, 1975b, p. 186 [*S. kalabisi*; OD]. Similar to *Alydus* (recent), but membrane of tegmen with more veins and apical part of corium longer. *Oligo.*, Europe (Czechoslovakia).

Family MESOPENTACORIDAE Popov, 1968

[Mesopentacoridae POPOV, 1968, p. 112]

Pronotum transverse, anterior corners projecting; anterior margins of tegmen thickened, forming a ridge along entire length of corium; venation vestigial. Tegmina and pronotum coarsely punctate. [Apparently related to the Urostylidae (recent).] *Jur.*

Mesopentacoris POPOV, 1968, p. 112 [*M. costalis*; OD]. Head narrower than pronotum; second antennal segment longest; corium with one vein; tibiae very slender. *Jur.*, USSR (Kazakh).

Family LYGAEIDAE Schilling, 1829

[Lygaeidae SCHILLING, 1829, p. 85]

Head shorter than thorax and scutellum; antennae straight, not elbowed; four to five veins in membrane of fore wing, not forming ante-apical cells. SLATER, 1964. *Jur.–Holo*.

Lygaeus FABRICIUS, 1794, p. 133. HEER, 1853a; SCUDDER, 1890; PITON & THÉOBALD, 1935; THÉOBALD, 1937b; SLATER, 1964. *Oligo.*, USA (Colorado), Europe (France, Germany); *Mio.*, Europe (France, Germany, Croatia)–*Holo*.

Aphanus LAPORTE, 1832, p. 35. THÉOBALD, 1937a; SLATER, 1964. *Oligo.*, Europe (Germany, France)–*Holo*.

Catopamera SCUDDER, 1890, p. 387 [*C. augheyi*; SD SLATER, 1964, p. 1519]. Related to *Myodochina* (recent); head subtriangular, slightly broader than long; antennae slender, no longer than the head and thorax together. SLATER, 1964. *Oligo.*, USA (Colorado).

Cephalocoris HEER, 1853a, p. 61 [*C. pilosus*; OD]. Similar to *Cymus* (recent). SLATER, 1964. *Mio.*, Europe (Germany).

Chilacis FIEBER, 1864, p. 72. STATZ & WAGNER, 1950. *Oligo.*, Europe (Germany)–*Holo*.

Cholula DISTANT, 1882, p. 210. SLATER, 1964. *Eoc.*, USA (Wyoming)–*Holo*.

Cophocoris SCUDDER, 1890, p. 391 [*C. tenebricosus*; OD]. Little-known genus; probably close to *Catopamera*, but head rounded; antennae only half as long as body. SLATER, 1964. *Oligo.*, USA (Colorado).

Coptochromus SCUDDER, 1890, p. 405 [*C. manium*; OD]. Little-known genus; head fully as long as broad and as broad as apex of thorax. SLATER, 1964. *Oligo.*, USA (Colorado).

Cryptochromus SCUDDER, 1890, p. 409 [*C. letatus*; OD]. Related to *Coptochromus*; head large, much broader than long. SLATER, 1964. *Oligo.*, USA (Colorado).

Ctereacoris SCUDDER, 1890, p. 394 [*C. primigenus*; OD]. Little-known genus; probably related to *Catopamera* but with much shorter middle femora. SLATER, 1964. *Oligo.*, USA (Colorado).

Diniella BERGOTH, 1893, p. 202. SLATER, 1964. *Oligo.*, Europe (France, Germany)–*Holo.*

Drymus FIEBER, 1860, p. 178. STATZ & WAGNER, 1950. *Oligo.*, Europe (Germany)–*Holo.*

Eucorites SCUDDER, 1890, p. 392 [*E. serescens*; OD]. Little-known genus; probably related to *Catopamera* but with more rounded head; antennae longer than head and thorax. SLATER, 1964. *Oligo.*, USA (Colorado).

Exitelus SCUDDER, 1890, p. 408 [*E. exsanguis*; OD]. Similar to *Cryptochromus*, but head only a little broader than long. SLATER, 1964. *Oligo.*, USA (Colorado).

Geocoris FALLÉN, 1814, p. 10. SCUDDER, 1890. *Oligo.*, USA (Colorado)–*Holo.*

Heterogaster SCHILLING, 1829, p. 37. HEER, 1853a; SLATER, 1964. *Oligo.*, Europe (France); *Mio.*, Europe (Germany, Croatia)–*Holo.*

Ischnodemus FIEBER, 1837, p. 337. STATZ & WAGNER, 1950. *Oligo.*, Europe (Germany)–*Holo.*

Ligyrocoris STÅL, 1872, p. 51. SCUDDER, 1890. *Oligo.*, USA (Colorado)–*Holo.*

Linnaea SCUDDER, 1890, p. 396 [*L. carcerata*; SD SLATER, 1964, p. 1523]. Little-known lygaeid; head small; antennae as long as combined head and thorax; thorax very broad. *Oligo.*, USA (Colorado).

Lithochromus SCUDDER, 1890, p. 402 [*L. gardneri*; SD SLATER, 1964, p. 1524]. Little-known lygaeid; head as broad as long; antennae about half as long as body. *Oligo.*, USA (Colorado).

Lithocoris SCUDDER, 1890, p. 390 [*L. evulsus*; OD]. Little-known genus, apparently related to *Myodochina* (recent); head large, subrotund. SLATER, 1964. *Oligo.*, USA (Colorado).

Lygaenocoris POPOV, 1961, p. 1211 [*L. prynadai*; OD]. Eighth abdominal segment strongly developed, covering the ninth. *Jur.*, USSR (Kazakh).
——FIG. 170,3. *L. prynadai*; whole insect, ×6.5 (Popov, 1961).

Lygaeosoma SPINOLA, 1837, p. 254. SLATER, 1964. *Oligo.*, Europe (Germany)–*Holo.*

Mesolygaeus PING, 1928, p. 43 [*M. laiyangenis*; OD]. Similar to *Lygaeus*, but veins of fore wing membrane much more prominent. SLATER, 1964. *Cret.*, China (Shantung).

Miogonates SAILER & CARVALHO in A. R. PALMER, 1957, p. 256 [*M. subimpunctatus*; OD]. Similar to *Lethaeus* and other recent lethaeini but with a smoother integument than is characteristic of the related genera. *Mio.*, USA (California).

Necrochromus SCUDDER, 1890, p. 406 [*N. cockerelli*; SD SLATER, 1964, p. 1525]. Body regularly oval; head as broad as apex of thorax. *Oligo.*, USA (Colorado).

Phrudopamera SCUDDER, 1890, p. 388 [*P. wilsoni*; SD SLATER, 1964, p. 1521]. Similar to *Catopamera*, but antennae much longer than combined head and thorax. *Oligo.*, USA (Colorado).

Pionosomus FIEBER, 1860, p. 48. SLATER, 1964. *Oligo.*, Europe (Germany)–*Holo.*

Praenotochilus THÉOBALD, 1937a, p. 289 [*P. parallelus*; OD]. Similar to *Aphanus*, but body more cylindrical and first antennal segment much longer. *Oligo.*, Europe (France).

Procoris SCUDDER, 1890, p. 392 [*P. bechleri*; SD SLATER, 1964, p. 1521]. Little-known genus; probably similar to *Eucorites*, but posterior margin of thorax more truncate. *Oligo.*, USA (Colorado).

Procrophius SCUDDER, 1890, p. 382 [*P. communis*; SD SLATER, 1964, p. 1512]. Similar to *Crophius* (recent) but with shorter antennae. *Oligo.*, USA (Colorado).

Procymophyes SAILER & CARVALHO in A. R. PALMER, 1957, p. 255 [*P. lithax*; OD]. Similar to *Cymophyes* (recent), but eyes well removed from anterior margin of pronotum. *Mio.*, USA (California).

Procymus USINGER, 1940, p. 79 [*P. cockerelli*; OD]. Similar to Cymus (recent), but body short, broad, and covered with cymine punctures. SLATER, 1964. *Oligo.*, USA (Colorado).

Prolygaeus SCUDDER, 1890, p. 405 [*P. inundatus*; OD]. Body very regularly oval. Antennae as long as head and thorax, the first segment not extending beyond frons, the last two segments longer than first two. *Oligo.*, USA (Colorado).

Raglius STÅL, 1872, p. 57. STATZ & WAGNER, 1950. *Oligo.*, Europe (Germany)–*Holo.*

Rhyparochromus HAHN, 1826, p. 17. [Most extinct species included here have uncertain generic positions; they were originally put in *Pachymerus* LEPELETIER & SERVILLE, 1825, which is now placed on the Official Index of Rejected and Invalid Names in Zoology (Op. 676, 1963, ICZN).] SLATER, 1964. *Eoc.*, USA (Wyoming); *Oligo.*, Europe (Germany, Baltic, France), USA (Colorado); *Mio.*, USA (Colorado), Europe (Germany, Croatia)–*Holo.*

Scolopostethus FIEBER, 1860, p. 188. STATZ & WAGNER, 1950. *Oligo.*, Europe (Germany)–*Holo.*

Stenopamera SCUDDER, 1890, p. 385 [*S. tenebrosa*; SD SLATER, 1964, p. 1521]. Little-known genus, related to *Catopamera*. *Oligo.*, USA (Colorado).

Tiromerus SCUDDER, 1890, p. 401 [*T. torpefactus*; OD]. Little-known genus, similar to *Rhyparochromus*; second segment of antennae much longer than third or fourth. *Oligo.*, USA (Colorado).

Trapezonotus FIEBER, 1860, p. 50. SCUDDER, 1890; STATZ & WAGNER, 1950. *Oligo.*, Europe (Germany), USA (Colorado)–*Holo.*

Family BERYTIDAE Fieber, 1851

[Berytidae FIEBER, 1851, p. 9]

Body very slender; head conical, porrect; antennae and rostrum four-segmented; pronotum much longer than wide; scutellum armed; legs very slender; tarsi three-segmented. *Oligo.–Holo.*

Berytus FABRICIUS, 1803, p. 264. *Holo.*
Megalomerium FIEBER, 1859, p. 208. THÉOBALD, 1937a. *Oligo.,* Europe (France)–*Holo.*

Family PYRRHOCORIDAE Fieber, 1860

[Pyrrhocoridae FIEBER, 1860, p. 43]

Body elongate-oval; antennae and beak four-segmented; ocelli absent; membrane with 2 large basal cells, giving rise to several (about 8) branching veins; tarsi three-segmented. *Oligo.–Holo.*

Pyrrhocoris FALLÉN, 1814, p. 9. STATZ & WAGNER, 1950. *Oligo.,* Europe (Germany)–*Holo.*
Dysdercus AMYOT & SERVILLE, 1843, p. 272. SCUDDER, 1890. *Oligo.,* USA (Colorado)–*Holo.*

Family CYDNIDAE Billberg, 1820

[Cydnidae BILLBERG, 1820, p. 7]

Similar to the Pentatomidae, but forelegs fossorial; tibiae strongly spinose, veins of fore wing membrane radiating from base. *Eoc.–Holo.*

Cydnus FABRICIUS, 1803, p. 184. HEER, 1853a; FÖRSTER, 1891; THÉOBALD, 1937a; STATZ, 1950a. *Oligo.,* Europe (Germany, France)–*Holo.*
Crocistethus FIEBER, 1860, p. 84. STATZ & WAGNER, 1950. *Oligo.,* Europe (Germany)–*Holo.*
Cyrtomenus AMYOT & SERVILLE, 1843, p. 90. SCUDDER, 1890. *Eoc.,* USA (Wyoming)–*Holo.*
Necrocydnus SCUDDER, 1890, p. 443 [*N. amyzonus* SCUDDER, 1890, p. 443; SD CARPENTER, herein]. Head only slightly sunk into prothorax; anterior-lateral angles of thorax rounded. *Eoc.,* USA (Wyoming); *Oligo.,* USA (Colorado).
Procydnus SCUDDER, 1890, p. 438 [*P. quietus* SCUDDER, 1890, p. 438; SD CARPENTER, herein]. Very similar to *Stenopelta,* but body less than twice as long as broad. *Eoc.,* USA (Wyoming); *Oligo.,* USA (Colorado).
Sehirus AMYOT & SERVILLE, 1843, p. 96. STATZ & WAGNER, 1950. *Oligo.,* Europe (Germany)–*Holo.*
Stenopelta SCUDDER, 1890, p. 437 [*Aethus punctulatus* SCUDDER, 1878b, p. 769; OD]. Scutellum triangular, as broad as long; head sunk deeply into prothorax, the depth of the thoracic emargination being about half its width; body more than twice as long as broad. *Eoc.,* USA (Wyoming).
Teleocydnus HENRIKSEN, 1922b, p. 32 [*T. transitorius*; OD]. Similar to *Cydnus* (recent) but with a long, slender scutellum, reaching about to abdominal apex. *Eoc.,* Europe (Denmark).
Thlibomenus SCUDDER, 1890, p. 448 [*T. petreus*; OD]. Similar to *Necrocydnus,* but head even more prominent; anterior emargination of prothorax slight or absent. *Oligo.,* USA (Colorado).

Family SCUTELLERIDAE Leach, 1815

[Scutelleridae LEACH, 1815, p. 121]

Body oval, usually strongly convex; head triangular; 2 ocelli; rostrum four-segmented; scutellum very large, U-shaped; tarsi three-segmented. *Eoc.–Holo.*

Scutellera LAMARCK, 1801, p. 293. *Holo.*
Coptosoma LAPORTE, 1832, p. 73. PITON, 1940a. *Eoc.,* Europe (France)–*Holo.*
Poecilocoris DALLAS, 1848, p. 100. STATZ & WAGNER, 1950. *Oligo.,* Europe (Germany)–*Holo.*
Tectocoris HAHN, 1834, p. 33. HENRIKSEN, 1922b. *Eoc.,* Europe (Denmark)–*Holo.*

Family PACHYMERIDIIDAE Handlirsch, 1906

[Pachymeridiidae HANDLIRSCH, 1906b, p. 495] [=Sisyrochoridae HANDLIRSCH, 1920, p. 210; Psychrochoridae HANDLIRSCH, 1920, p. 298; Hypocimicidae HANDLIRSCH, 1939, p. 119]

Fore wing differentiated into corium and membrane; venation of membrane indistinct; clavus broad and nearly half as long as wing; vein SC separating from R+M near division of R and M. POPOV & WOOTTON, 1977. *Jur.*

Pachymeridium GEINITZ, 1880, p. 529 [*P. dubium*; OD]. Fore wing with SC, R, and M separating at a single point; R branched. POPOV & WOOTTON, 1977. *Jur.,* Europe (Germany).
—— FIG. 170,2. *P. dubium*; fore wing, ×9 (Popov & Wootton, 1977).
Apsicoria HANDLIRSCH, 1939, p. 121 [*A. semideleta*; OD]. Similar to *Sisyrocoris,* but R branched; corium relatively smooth. POPOV & WOOTTON, 1977. *Jur.,* Europe (Germany).
Cathalus HANDLIRSCH, 1939, p. 121 [*C. alutaceus*; OD]. Little-known genus, based on wing fragment; apparently similar to *Sisyrocoris,* but R and M very close together in fore wing; corium less punctate. POPOV & WOOTTON, 1977. *Jur.,* Europe (Germany).
Hypocimex HANDLIRSCH, 1939, p. 119 [*H. membranaceus*; OD]. Little-known genus, based on

poorly preserved specimen; family position doubtful. POPOV & WOOTTON, 1977. *Jur.,* Europe (Germany).

Psychrocoris HANDLIRSCH, 1920, p. 208 [*P. cuneifera*; OD]. Little-known genus, apparently similar to *Sisyrocoris*. Fore wing slender; corium including more than three-fourths of wing surface. POPOV & WOOTTON, 1977. *Jur.,* Europe (Germany).

Sisyrocoris HANDLIRSCH, 1920, p. 210 [*S. rudis*; OD]. Little-known genus. Fore wing coarsely punctate; R unbranched. POPOV & WOOTTON, 1977. *Jur.,* Europe (Germany).

Family PENTATOMIDAE Leach, 1815

[Pentatomidae LEACH, 1815, p. 121]

Body oval; head triangular, porrect, much narrower than thorax; antennae five-segmented; rostrum four-segmented; ocelli present; scutellum extending beyond middle of abdomen, narrowed posteriorly to form triangle; membrane with numerous veins; tarsi two- or three-segmented. *Eoc.–Holo.*

Pentatoma OLIVIER, 1789, p. 25. HEER, 1853a; HEYDEN & HEYDEN, 1865; FÖRSTER, 1891; HANDLIRSCH, 1906b. *Eoc.,* Europe (Greenland); *Oligo.,* Europe (Germany, Baltic); *Mio.,* Europe (Croatia, Germany)–*Holo.*

Acanthosoma CURTIS, 1824, p. 28. HEER, 1853a; FÖRSTER, 1891; PITON & THÉOBALD, 1935. *Oligo.,* Europe (Germany); *Mio.,* Europe (Croatia, France)–*Holo.*

Arma HAHN, 1832, p. 91. FÖRSTER, 1891. *Oligo.,* Europe (Germany)–*Holo.*

Asopus BURMEISTER, 1834, p. 19. PITON, 1940b. *Eoc.,* Europe (France)–*Holo.*

Brachypelta AMYOT & SERVILLE, 1843, p. 89. NOVÁK, 1877. *Oligo.,* Europe (Germany, Czechoslovakia)–*Holo.*

Cacoschistus SCUDDER, 1890, p. 459 [*C. maceratus*; OD]. Similar to *Mataeoschistus* (recent) but with broader head and less prominent frontal area. *Oligo.,* USA (Colorado).

Carpocoris KOLENATI, 1846, p. 45. KUKALOVÁ & RIHA, 1957. *Mio.,* Europe (Czechoslovakia)–*Holo.*

Deryeuma PITON, 1940a, p. 159 [*D. primordialis*; OD]. Pronotum narrowed in front, notched in region of head; antennae five-segmented, the first segment very short, the second very long; tarsi three-segmented. *Eoc.,* Europe (France).

Dinidorites COCKERELL, 1921e, p. 34 [*D. margiformis*; OD]. Body narrow; pronotum and scutellum with numerous, dark punctures. *Eoc.,* USA (Colorado).

Doryderes AMYOT & SERVILLE, 1843, p. 121. PITON, 1940a. *Eoc.,* Europe (France)–*Holo.*

Eurydema LAPORTE, 1832, p. 61. HEER, 1853a; PITON & THÉOBALD, 1935; THÉOBALD, 1937a. *Mio.,* Europe (Germany); *Mio./Plio.,* Europe (France)–*Holo.*

Eurygaster LAPORTE, 1832, p. 68. THÉOBALD, 1937a. *Oligo.,* Europe (France)–*Holo.*

Eysarcoris HAHN, 1834, p. 66. HEER, 1853a; NAORA, 1933b; THÉOBALD, 1937a. *Oligo.,* Europe (France); *Mio.,* Europe (Germany); *Tert. (epoch unknown),* Japan–*Holo.*

Halys FABRICIUS, 1803, p. 180. HEER, 1853a. *Mio.,* Europe (Germany)–*Holo.*

Latahcoris COCKERELL, 1931b, p. 312 [*L. spectatus*; OD]. Head less than one-third width of pronotum; pronotum coarsely punctate, more than twice as long as wide; scutellum with straight sides. *Mio.,* USA (Washington).

Manevalia PITON, 1940a, p. 159 [*M. pachyliformis*; OD]. Little-known genus, apparently related to *Pachylis* (recent). *Oligo.,* Europe (France).

Mesohalys BEIER, 1952, p. 134 [*M. muezenbergiana*; OD]. Pronotum and mesonotum very coarsely punctate; abdominal tergites finely punctate; front margin of pronotum notched. *Mio.,* Europe (Germany).

Neurocoris HEER, 1853a, p. 23 [*N. rotundatus* HEER, 1853a, p. 23; SD CARPENTER, herein]. Little-known genus; pronotum very broad; tegmen very short and broad. SCUDDER, 1885b. *Mio.,* Europe (Croatia).

Nezara AMYOT & SERVILLE, 1843, p. 143. THÉOBALD, 1937a. *Oligo.,* Europe (France); *Mio./Plio.,* Europe (France).

Pachycoris BURMEISTER, 1835, p. 391. HEER, 1853a. *Mio.,* Europe (Germany)–*Holo.*

Palomena MULSANT & REY, 1866, p. 277. MEUNIER, 1915a. *Mio.,* Europe (France)–*Holo.*

Pentatomites SCUDDER, 1890, p. 461 [*P. foliarum*; OD]. Similar to *Polioschistus* but with sides of thorax convex in front of lateral prominences. VERHOEFF, 1917; COCKERELL, 1927d. *Oligo.,* England, USA (Colorado); *Tert. (epoch unknown),* USSR (Asian RSFSR).

Phloeocoris BURMEISTER, 1835, p. 371. HEER, 1853a. *Mio.,* Europe (Croatia)–*Holo.*

Poliocoris KIRKALDY, 1910, p. 130 [*P. amnesis*; OD]. Allied to *Teleoschistus*. Body oval; head longer than wide between eyes; scutellum extending halfway to apex of abdomen and rounded posteriorly. *Oligo.,* USA (Colorado).

Polioschistus SCUDDER, 1890, p. 460 [*P. ligatus* SCUDDER, 1890, p. 460; SD CARPENTER, herein]. General form as in *Euschistus* (recent); head in front of eyes subquadrate; thorax very short, about 4 times as broad as long. *Oligo.,* USA (Colorado).

Poteschistus SCUDDER, 1890, p. 458 [*P. obnubilus*; OD]. Little-known genus, with body regularly ovate. *Oligo.,* USA (Colorado).

Pycanum AMYOT & SERVILLE, 1843, p. 171. PITON, 1940a. *Eoc.,* Europe (France)–*Holo.*

Teleocoris KIRKALDY, 1910, p. 129 [*T. pothetias*; OD]. Head prominent, longer than its width between the eyes; pronotum more than 3 times as wide as base of head; scutellum regularly triangular, half the length of abdomen. *Oligo.*, USA (Colorado).

Teleoschistus SCUDDER, 1890, p. 454 [*T. antiquus*; SD COCKERELL, 1909b, p. 74]. Head nearly half as broad as thorax and broader than long; apical border of prothorax emarginate; scutellum about as long as wide, reaching less than halfway to end of abdomen. COCKERELL, 1909b; HENRIKSEN, 1922b. *Eoc.*, Europe (Denmark); *Oligo.*, USA (Colorado).

Tetyra FABRICIUS, 1803, p. 128. HEER, 1853a. *Mio.*, Europe (Germany)–*Holo*.

Thnetoschistus SCUDDER, 1890, p. 457 [*T. revulsus*; OD] [=*Mataeoschistus* SCUDDER, 1890, p. 459 (type, *M. limigenus* SCUDDER), obj.; the two species are based on counterparts of the same fossil]. Similar to *Euschistus* (recent) but more elongate. *Oligo.*, USA (Colorado).

Tiroschistus SCUDDER, 1890, p. 462 [*T. indurescens*; OD]. Head rounded, with very little extension in front of eyes; antennae 2 times as long as head and thorax together. *Oligo.*, USA (Colorado).

Family ANTHOCORIDAE Amyot & Serville, 1843

[Anthocoridae AMYOT & SERVILLE, 1843, p. xxxvii]

Similar to the Saldidae, but fore wing membrane without long, closed cells; corium with an embolium. *Oligo.–Holo.*

Anthocoris FALLÉN, 1814, p. 9. *Holo.*
Temnostethus FIEBER, 1860, p. 263. STATZ & WAGNER, 1950. *Oligo.*, Europe (Germany)–*Holo*.

Family NABIDAE Costa, 1852

[Nabidae COSTA, 1852, p. 66]

Similar to the Reduviidae but more slender; rostrum with 4 segments; membrane of fore wings with distinctly branched veins or with a few longitudinal veins emitting radiating veins. *Jur.–Holo.*

Nabis LATREILLE, 1802, p. 248. HEER, 1853a, 1865; THÉOBALD, 1937a; JORDAN, 1952. *Oligo.*, Europe (Baltic, France); *Mio.*, Europe (Croatia)–*Holo*.
Karanabis BECKER-MIGDISOVA, 1962b, p. 219 [*K. kiritshenkoi*; OD]. Antennae with 4 segments; pronotum conical, strongly narrowed anteriorly; legs long. *Jur.*, USSR (Kazakh).

Family REDUVIIDAE Latreille, 1807

[Reduviidae LATREILLE, 1807, p. 126]

Head shorter than thorax and scutellum, not constricted behind eyes; antennae four-segmented; rostrum three-segmented; fore wings not reticulate; forelegs raptorial. *Oligo.–Holo.*

Reduvius FABRICIUS, 1775, p. 729. *Holo.*
Eothes SCUDDER, 1890, p. 355 [*E. elegans*; OD]. Related to *Opsicoetus* (recent), but body more slender and terminal antennal segments stout. *Oligo.*, USA (Colorado).
Evagoras BURMEISTER, 1843, p. 368. HEER, 1853a. *Mio.*, Europe (Germany)–*Holo*.
Harpactor LAPORTE, 1832, p. 8. HEER, 1853a. *Mio.*, Europe (Germany, Croatia)–*Holo*.
Limnacis GERMAR, 1856, p. 19. HEER, 1853a. *Oligo.*, Europe (Baltic)–*Holo*.
Miocoris COCKERELL, 1927e, p. 591 [*M. fagi*; OD]. Anterior femora stout; first antennal segment not as long as head. *Oligo.*, USA (Colorado).
Pirates BURMEISTER, 1835, p. 222. HEER, 1853a. *Mio.*, Europe (Germany)–*Holo*.
Poliosphageus KIRKALDY, 1910, p. 130 [*P. psychrus*; OD]. Similar to *Repipta* (recent) but with first antennal segment scarcely longer than head; second segment much longer than first. *Oligo.*, USA (Colorado).
Proptilocerus WASMANN, 1933, p. 1 [*P. dolosus*; OD]. Similar to *Ptilocerus* (recent) but with second antennal segment and the 2 terminal segments longer and thicker. *Oligo.*, Europe (Baltic).
Prostemma LAPORTE, 1832, p. 12. HEER, 1853a. *Mio.*, Europe (Germany)–*Holo*.
Rhinocoris HAHN, 1833, p. 20. STATZ & WAGNER, 1950. *Oligo.*, Europe (Germany)–*Holo*.
Stenopoda LAPORTE, 1832, p. 26. HEER, 1853a. *Mio.*, Europe (Germany)–*Holo*.
Tagalodes SCUDDER, 1890, p. 356 [*T. inermis*; OD]. Similar to *Taglis* (recent) but with shorter thorax and without spines on fore femora. *Oligo.*, USA (Colorado).

Family TINGIDAE Laporte, 1833

[Tingidae LAPORTE, 1833, p. 47]

Head shorter than thorax; antennae shorter than head, with 4 segments; fore wings lace-like, entirely reticulate. *Oligo.–Holo.*

Tingis FABRICIUS, 1803, p. 124. DRAKE & RUHOFF, 1960. *Oligo.*, USA (Colorado); *Mio.*, Europe (Croatia)–*Holo*.
Cantacader AMYOT & SERVILLE, 1843, p. 299. DRAKE, 1950; DRAKE & RUHOFF, 1960. *Oligo.*, Europe (Baltic)–*Holo*.

Celantia DISTANT, 1903, p. 137. COCKERELL, 1921f; DRAKE & RUHOFF, 1960. *Oligo.*, Europe (England)–*Holo.*

Dictyla STÅL, 1874, p. 57. SCUDDER, 1890; DRAKE & RUHOFF, 1960. *Oligo.*, USA (Colorado), Europe (Czechoslovakia); *Mio.*, Europe (Germany)–*Holo.*

Eotingis SCUDDER, 1890, p. 359 [*E. antennata*; OD]. Similar to *Tingis*; pronotum smooth; costal area of fore wing enlarged apically. DRAKE & RUHOFF, 1960. *Oligo.*, USA (Colorado).

Phatnoma FIEBER, 1844, p. 57. DRAKE, 1950; DRAKE & RUHOFF, 1960. *Oligo.*, Europe (Baltic)–*Holo.*

Family MIRIDAE Hahn, 1831

[Miridae HAHN, 1831, p. 234]

Head porrect; eyes large; ocelli absent; antennae and beak with 4 segments, beak not held in a groove; scutellum distinct; membrane of tegmen usually with 2 basal cells, veins otherwise absent from membrane; tarsi two-segmented. *Jur.–Holo.*

Miris FABRICIUS, 1794, p. 183. *Holo.*

Aporema SCUDDER, 1890, p. 369 [*A. praestrictum*; OD]. Little-known genus, probably close to *Phytocoris*; scutellum large, equiangular, with straight sides. *Oligo.*, USA (Colorado).

Calocoris FIEBER, 1858, p. 305. STATZ & WAGNER, 1950. *Oligo.*, Europe (Germany)–*Holo.*

Capsus FABRICIUS, 1803, p. 241. [Generic assignment of species doubtful.] SCUDDER, 1890. *Oligo.*, USA (Colorado)–*Holo.*

Carmelus DISTANT, 1884, p. 297. [Generic assignment of species doubtful.] SCUDDER, 1890. *Oligo.*, USA (Colorado)–*Holo.*

Closterocoris UHLER, 1890, p. 76. [Generic assignment of species doubtful.] SCUDDER, 1890. *Oligo.*, USA (Colorado)–*Holo.*

Fulvius STÅL, 1862, p. 322 [=*Oligocoris* JORDAN, 1944a, p. 8 (type, *O. bidentata*)]. CARVALHO, 1954. *Oligo.*, Europe (Baltic)–*Holo.*

Fuscus DISTANT, 1884, p. 299. [Generic assignment of species doubtful.] SCUDDER, 1890. *Oligo.*, USA (Colorado)–*Holo.*

Hadronema UHLER, 1872, p. 412. SCUDDER, 1890. *Oligo.*, USA (Colorado)–*Holo.*

Jordanofulvius CARVALHO, 1954, p. 188, *nom. subst.* pro *Electrocoris* JORDAN, 1944b, p. 133, *non* USINGER, 1942 [*Electrocoris fuscus* JORDAN, 1944b, p. 133; OD]. Little-known genus, apparently belonging to recent tribe Cylapinae. *Oligo.*, Europe (Baltic).

Lygus HAHN, 1831, p. 28. STATZ & WAGNER, 1950. *Oligo.*, Europe (Germany)–*Holo.*

Miomonalonion SAILER & CARVALHO in PALMER, 1957, p. 257 [*M. conoidifrons*; OD]. Related to *Monalonion* (recent), but frons conately produced between antennae; first antennal segment very thick. PALMER, 1957. *Mio.*, USA (California).

Miridoides BECKER-MIGDISOVA, 1962b, p. 217 [*M. mesozoicus*; OD]. Antennae shorter than body; tegmen reaching to end of abdomen with front margin convex and only 2 veins in corium. *Jur.*, USSR (Kazakh).——FIG. 170,*1*. *M. mesozoicus*; ×10 (Becker-Migdisova, 1962b).

Phytocoris FALLÉN, 1814, p. 10. GERMAR & BERENDT, 1856; THÉOBALD, 1937a. *Oligo.*, Europe (Baltic, France)–*Holo.*

Poecilocapsus REUTER, 1875, p. 73. SCUDDER, 1890. *Oligo.*, USA (Colorado)–*Holo.*

Scutellifer POPOV, 1968, p. 108 [*S. karatauicus*; OD]. Antennae longer than body, its first segment longer than pronotum; scutellum very large; membrane of tegmen without spots; fore femora long, slightly flattened; hind legs very long. *Jur.*, USSR (Kazakh).

Family UNCERTAIN

The following genera, apparently belonging to the suborder Heteroptera, are too poorly known to permit assignment to families.

Cacalydus SCUDDER, 1890, p. 419 [*C. exsterpatus*; SD ŠTYS & ŘÍHA, 1977, p. 180]. Little-known genus; probably a coreid. *Oligo.*, USA (Colorado).

Copidopus HANDLIRSCH, 1906b, p. 635 [*C. jurassicus*; OD]. Little-known genus. Large species; antennae with 5 segments; hind legs with thickened femora. *Jur.*, Europe (Germany).

Coreites HEER, 1853a, p. 56 [*C. crassus* HEER, 1853a, p. 56; SD CARPENTER, herein]. Little-known heteropteron, possibly belonging to Coreidae. PITON & THÉOBALD, 1935. *Oligo.*, Europe (France); *Mio.*, Europe (Croatia).

Cydnopsis HEER, 1853a, p. 13 [*C. haidingeri* HEER, 1853a, p. 13; SD CARPENTER, herein]. Little-known genus; legs without spines. COCKERELL, 1909j; HANDSCHIN, 1937; PITON & RUDEL, 1936. *Eoc.*, USA (Colorado); *Oligo.*, Europe (France)–*Mio.*, Europe (Croatia, Germany).

Deraiocoris BODE, 1953, p. 128 [*D. insculptus*; OD]. Little-known heteropteron; head and thorax punctate. *Jur.*, Europe (Germany).

Dichaspis BODE, 1953, p. 137 [*D. laesa*; OD]. Little-known heteropteron, with small head; wings and venation virtually unknown. *Jur.*, Europe (Germany).

Electrocoris USINGER, 1942, p. 43 [*E. brunneus*; OD]. Cimicoid genus, with ocelli present; 4 free, longitudinal veins in membrane of tegmen; abdominal trichobothria absent. *Oligo.*, Europe (Baltic).

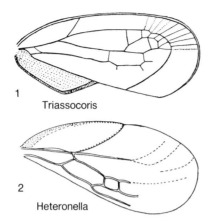

Fig. 171. Uncertain (p. 276–277).

Engerrophorus BODE, 1953, p. 144 [*E. nitidus; OD]. Little-known heteropteron, with small head. *Jur.,* Europe (Germany).

Engynabis BODE, 1953, p. 130 [*E. tenuis; OD]. Little-known genus, based on incomplete wings and body; probably related to Gerridae. POPOV & WOOTTON, 1977. *Jur.,* Europe (Germany).

Eogerridium BODE, 1953, p. 131 [*E. gracile; OD]. Little-known genus, based on fragment of body; legs long and slender. *Jur.,* Europe (Germany).

Etirocoris SCUDDER, 1890, p. 425 [*E. infernalis; OD]. Little-known genus; head elongate, slender, prolonged between antennae. ŠTYS & ŘIHA, 1977. *Oligo.,* USA (Colorado).

Euraspidium BODE, 1953, p. 137 [*E. granulosum; OD]. Little-known heteropteron, with punctations on thorax. *Jur.,* Europe (Germany).

Hadrocoris HANDLIRSCH, 1939, p. 117 [*H. ocutellaris; OD]. Little-known heteropteron, with a large scutellum and punctate head and thorax. [Type of family Hadrocoridae HANDLIRSCH, 1939, p. 116.] *Jur.,* Europe (Germany).

Heteronella EVANS, 1961, p. 22 [*H. marksei; OD]. Little-known genus. Tegmen with suggestion of costal fracture; venation distinct in basal half of tegmen only. *Trias.,* Australia (Queensland).——FIG. 171,2. *H. marksei;* fore wing, ×14 (Evans, 1961).

Ischnocoris BODE, 1953, p. 136 [*I. bitoratus; OD]. Little-known genus, with broad head. *Jur.,* Europe (Germany).

Leptoserinetha THÉOBALD, 1937a, p. 362 [*L. navicularis; OD]. Little-known genus, based on poorly preserved specimen. ŠTYS & ŘIHA, 1977. *Oligo.,* Europe (France).

Liasocoris WENDT, 1940, p. 19 [*L. hainmulleri; OD]. Little-known genus, with prominent scutellum. *Jur.,* Europe (Germany).

Megalocoris BODE, 1953, p. 127 [*M. laticlavus;

OD]. Little-known heteropteron; body large, oval in form; venation unknown. *Jur.,* Europe (Germany).

Ophthalmocoris BODE, 1953, p. 126 [*O. liasscus; OD]. Little-known insect; fore wings apparently membranous. Ordinal assignment doubtful. *Jur.,* Europe (Germany).

Palaeonepidoideus MEUNIER, 1900, p. 13 [*P. carinata; OD]. Little-known heteropteron, possibly belonging to the Nepidae. *Jur.,* Europe (Germany).

Pricecoris PINTO & ORNELLAS, 1974b, p. 296 [*P. beckeras; OD]. Little-known genus, based on poorly preserved specimen; venation not preserved. [Type of family Pricecoridae PINTO & ORNELLAS.] *Cret.,* Brazil (Maranhão).

Probascanion HANDLIRSCH, 1939, p. 118 [*P. megacephalum; OD]. Little-known heteropteron, with relatively large head; venation unknown. [Type of family Probascanionidae HANDLIRSCH, 1939.] *Jur.,* Europe (Germany).

Pronabis BODE, 1953, p. 129 [*P. utroquelaesus; OD]. Little-known genus; fore wing without distinct membranous area. *Jur.,* Europe (Germany).

Protocoris HEER, 1852, p. 15 [*P. planus; OD]. Little-known genus; fore wing with distinct membranous area. [Type of family Protocoridae HANDLIRSCH, 1906b, p. 495.] *Jur.,* Europe (Germany).

Rhepocoris SCUDDER, 1890, p. 426 [*R. praetectus; SD ŠTYS & ŘIHA, 1977, p. 182] [=*Parodarmistus* SCUDDER, 1890, p. 421 (type, *P. collisus* SCUDDER; SD ŠTYS & ŘIHA, 1977)]. Little-known genus, possibly related to family Pyrrhocoridae. *Oligo.,* USA (Colorado).

Stiphroschema BODE, 1953, p. 143 [*S. longealatum; OD]. Little-known heteropteron, with small head and broad thorax; fore wing apparently very thin. *Jur.,* Europe (Germany).

Strobilocoris BODE, 1953, p. 138 [*S. mediocordatus; OD]. Little-known heteropteron; thorax quadrate, with coarse sculpturing; venation unknown. *Jur.,* Europe (Germany).

Tenor SCUDDER, 1890, p. 425 [*C. speluncae; OD]. Little-known genus, based on poorly preserved specimen. ŠTYS & ŘIHA, 1977. *Oligo.,* USA (Colorado).

Trachycoris BODE, 1953, p. 142 [*T. abbreviatus; OD]. Little-known heteropteron; similar to *Strobilocoris* but with broader wings; venation unknown. *Jur.,* Europe (Germany).

Triassocoris TILLYARD, 1922b, p. 466 [*T. myersi; OD]. Tegmen with corium present in central part of wing; membrane submarginal, separated from corium by impressed line; M and CU arising independently from stem R; radiating veins extending from M to distal margin; clavus sharply defined, short. [Type of family Triassocoridae

TILLYARD.] EVANS, 1956. *Trias.*, Australia (Queensland).——FIG. 171,*1*. **T. myersi*; tegmen, ×8 (Tillyard, 1922b).

Suborder UNCERTAIN

The following genera, apparently belonging to the Hemiptera, are too poorly known to permit assignment to suborders.

Family PARAKNIGHTIIDAE Evans, 1950

[Paraknightiidae EVANS, 1950, p. 250]

Paranotal lobes well developed. Tegmen with costal fracture in basal third of wing. Female with well-developed ovipositor. *Perm.*

Paraknightia EVANS, 1943b, p. 185 [**P. magnifica*; OD]. Tegmen with costal margin thickened; R+M dividing about one-fifth wing length from base. [Originally placed in Homoptera but transferred to Heteroptera by EVANS (1950); moved to a different suborder, Peloridiina, by POPOV (1980b) along with the existing family Peloridiidae.] EVANS, 1950; BECKER-MIGDISOVA & POPOV, 1962. *Perm.*, Australia (New South Wales).——FIG. 172. **P. magnifica*; *a,* dorsal view, ×3 (Becker-Migdisova & Popov, 1962); *b,* tegmen, ×4.6 (Evans, 1950).

Family UNCERTAIN

Docimus SCUDDER, 1890, p. 314 [**D. psylloides*; OD]. Little-known genus, based on fragment. HANDLIRSCH, 1907. *Oligo.*, USA (Colorado).

Prosigara SCUDDER, 1890, p. 343 [**P. flabellum*; OD]. Little-known genus, based on poorly preserved specimen. HANDLIRSCH, 1907. *Oligo.*, USA (Colorado).

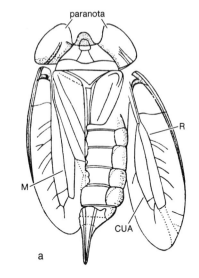
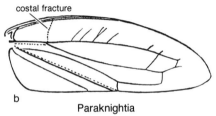

FIG. 172. Paraknightiidae (p. 277).